Lecture Notes
in Computational Science
and Engineering

70

Editors

Timothy J. Barth
Michael Griebel
David E. Keyes
Risto M. Nieminen
Dirk Roose
Tamar Schlick

Michel Bercovier • Martin J. Gander
Ralf Kornhuber • Olof Widlund
Editors

Domain Decomposition Methods in Science and Engineering XVIII

Springer

Michel Bercovier
The Hebrew University of Jerusalem
School of Computer Science and Engineering
Ross Building - Givat Ram Campus
91904 Jerusalem
Israel
berco@cs.huji.ac.il

Martin J. Gander
Université de Genève
Section de mathématiques
2-4 rue du Lièvre
1211 Genève 24
Switzerland
martin.gander@unige.ch

Ralf Kornhuber
Freie Universität Berlin
Institut für Mathematik
Arnimallee 6
14195 Berlin-Dahlem
Germany
kornhube@math.fu-berlin.de

Olof Widlund
Courant Institute
251 Mercer Street
New York, NY 10012
USA
widlund@cims.nyu.edu

ISSN 1439-7358
ISBN 978-3-642-02676-8 e-ISBN 978-3-642-02677-5
DOI: 10.1007/978-3-642-02677-5
Springer Heidelberg Dordrecht London New York

Library of Congress Control Number: 2009930358

Mathematics Subject Classification Numbers (2000): 65F10, 65N30, 65N55

Cover design: deblik, Berlin

Printed on acid-free paper

Springer is part of Springer Science + Business Media (www.springer.com)

Preface

This volume contains a selection of 41 refereed papers presented at the 18th International Conference of Domain Decomposition Methods hosted by the School of Computer Science and Engineering (CSE) of the Hebrew University of Jerusalem, Israel, January 12–17, 2008.

1 Background of the Conference Series

The International Conference on Domain Decomposition Methods has been held in twelve countries throughout Asia, Europe, the Middle East, and North America, beginning in Paris in 1987. Originally held annually, it is now spaced at roughly 18-month intervals. A complete list of past meetings appears below.

The principal technical content of the conference has always been mathematical, but the principal motivation has been to make efficient use of distributed memory computers for complex applications arising in science and engineering. The leading such computers, at the "petascale" characterized by 10^{15} floating point operations per second of processing power and as many Bytes of application-addressable memory, now marshal more than 200,000 independent processor cores, and systems with many millions of cores are expected soon. There is essentially no alternative to domain decomposition as a stratagem for parallelization at such scales. Contributions from mathematicians, computer scientists, engineers, and scientists are together necessary in addressing the challenge of scale, and all are important to this conference.

Though the conference has grown up in the wake of commercial massively parallel processors, it must be remarked that some important applications of domain decomposition are not massively parallel at all. "Gluing together" just two subproblems to effectively exploit a different solver on each is also part of the technical fabric of the conference. Even as multiprocessing becomes commonplace, multiphysics modeling is in ascendancy, so the International Conference on Domain Decomposition Methods remains as relevant and as fundamentally interdisciplinary as ever.

The conference typically draws between 100 and 200 researchers concerned with the large-scale computational solution of PDEs in areas such as fluid dynamics, structural mechanics, biomechanics, geophysics, plasma physics, radiation transport, electricity and magnetism, flows in porous media, and the like. The conference is led by the International Scientific Committee of DDM.ORG under a set of by-laws that appear at the website `www.ddm.org`.

While research in domain decomposition methods is presented at numerous venues, the International Conference on Domain Decomposition Methods is the only regularly occurring international forum dedicated to interdisciplinary technical interactions between theoreticians and practitioners working in the creation, analysis, software implementation, and application of domain decomposition methods.

International Conferences on Domain Decomposition Methods:

- Paris, France, 1987
- Los Angeles, USA, 1988
- Houston, USA, 1989
- Moscow, USSR, 1990
- Norfolk, USA, 1991
- Como, Italy, 1992
- University Park (Pennsylvania), USA, 1993
- Beijing, China, 1995
- Ullensvang, Norway, 1996
- Boulder, USA, 1997
- Greenwich, UK, 1998
- Chiba, Japan, 1999
- Lyon, France, 2000
- Cocoyoc, Mexico, 2002
- Berlin, Germany, 2003
- New York City, USA, 2005
- St. Wolfgang/Strobl, Austria, 2006
- Jerusalem, Israel, 2008

International Scientific Committee on Domain Decomposition Methods:

- Petter Bjørstad, Bergen
- Martin Gander, Geneva
- Roland Glowinski, Houston
- Laurence Halpern, Paris
- Ronald Hoppe, Augsburg and Houston
- Hideo Kawarada, Chiba
- David Keyes, New York
- Ralf Kornhuber, Berlin
- Yuri Kuznetsov, Houston and Moscow
- Ulrich Langer, Linz
- Jacques Périaux, Paris
- Alfio Quarteroni, Lausanne

- Zhong-ci Shi, Beijing
- Olof Widlund, New York
- Jinchao Xu, University Park

2 About the Eighteenth Conference

The eighteenth conference was chaired by Michel Bercovier, Bertold Badler Chair of Scientific Computation at the School of Computer Science and Engineering, and held on the Edmond J. Safra Campus of the Hebrew University, at Givat Ram, Jerusalem. 107 scientists from 15 countries attended. The conference included 12 invited plenary lectures, 22 talks given in five Minisymposia, 30 contributed talks, and two special sessions: one dedicated to the memory of Moshe Israeli, who should have been on the organizing committee, and a special collection of ten talks, organized as Minisymposium 5, given as an "Historical Perspective to Milestones in the Development of Domain Decomposition." Conference details remain on the conference web site `http://www.cs.huji.ac.il/dd18`.

The twelve invited talks were:

- Achi Brandt, Weizmann Institute of Science and University of California at Los Angeles, *Principles of Systematic Upscaling*
- Michael J. Holst, University of California, San Diego: *Analysis and Convergent Adaptive Solution of the Einstein Constraint Equations*
- Ronald W. Hoppe, University of Houston and University of Augsburg: *Adaptive Multilevel Primal-Dual Interior-Point Methods in PDE Constrained Optimization*
- Claude Le Bris, Ecole Nationale des Ponts et Chaussées: *Domain Decomposition and Electronic Structure Calculations: a New Approach*
- Patrick Le Tallec, Ecole Polytechnique: *From Domain Decomposition to Homogenization in the Numerical Modelling of Materials*
- Jan Martin Nordbotten, University of Bergen and Princeton University: *Variational Scale Separation Methods*
- Ilaria Perugia, University of Pavia: *Plane Wave Discontinuous Galerkin Methods*
- Olivier Pironneau, University of Paris-VI: *Numerical Zoom for Multi-Scale Problems*
- Francois-Xavier Roux, ONERA and University of Paris-VI: *Domain Decomposition methods: Industrial Experience at Hutchinson*
- Xuemin Tu, University of California at Berkeley: *Balancing Domain Decomposition Methods by Constraints (BDDC)*
- Olof B. Widlund, Courant Institute, New York University: *Accommodating Irregular Subdomains in Domain Decomposition Theory*
- Jinchao Xu, Pennsylvania State University: *Robust Iterative Methods for Singular and Nearly Singular System of Equations*

The papers in Part I of these proceedings are ordered alphabetically according to the names of the plenary speakers.

The "Milestone" lectures were:

- Olof Widlund, Courant Institute, New York University: *Coarse Space Components of Domain Decomposition Algorithms*
- Petter Bjørstad, University of Bergen: *To Overlap or not to Overlap*
- Roland Glowinski, University of Houston: *On Fictitious Domain Methods*
- Jinchao Xu, Pennsylvania State University: *On the Method of Subspace Corrections*
- Alfio Quarteroni, Ecole Polytechnique Fédérale de Lausanne: *Heterogeneous Domain Decomposition*
- David Keyes, Columbia University: *Domain Decomposition and High Performance Computing*
- Francois-Xavier Roux: *The FETI Method*
- Frédéric Nataf, Ecole Polytechnique: *Optimized Schwarz Methods*
- Xiao-Chuan Cai, University of Colorado and Boulder: *Domain Decomposition Methods for Nonlinear Problems*
- Laurence Halpern, University of Paris 13: *Space-Time Parallel Methods*

These lectures were taped and will remain available at `http://www.cs.huji.ac.il/dd18/video`.

The papers in Part II of these proceedings are ordered according to the order of the five minisymposia, and inside each such group according to the names of the speakers. Part III is organized similarly.

The session dedicated to the memory of Moshe Israeli (1940-2007) included three lectures:

- Amir Averbuch, Tel Aviv University: *Contributions of Prof. Moshe Israeli to Scientific Computing*
- Irad Yavneh, Technion: *Automated Transformations of PDE Systems*
- Roland Glowinski, University of Houston: *Clustering Phenomena for Particulate Flow in Spinning Cylinders*

The Local Organizing Committee Members were:

- Michel Bercovier (Chairman), Hebrew University of Jerusalem
- Amir Averbuch, Tel Aviv University
- Pinhas Z. Bar-Yoseph (IACMM representative), Technion
- Matania Ben-Artzi, Hebrew University of Jerusalem
- Michael S. Engelman, Corporate VP, ANSYS
- Dan Givoli, Technion
- Raz Kupferman, Hebrew University of Jerusalem
- Zohar Yosibash, Ben Gurion University

The Organizers are grateful to the following companies and organizations for their material support:

- Hutchinson Rubber, France
- Bercom, Israel

- Hebrew University, including the Leibniz Research Center for Computer Science and the Edmund Landau Center for Research in Mathematical Analysis
- Cray Ltd., Israel
- SGI, Israel

Thanks are also due to Uri Heinemann, webmaster, Neva Treistman and Naama Yitzhak, administrative assistants who managed all the logistical details and produced the book of abstracts. Mohad Shini and Yehuda Arav oversaw the technical material at the conference and Ouri Bercovier taped the "Milestone" talks. Finally, the organizers would like to thank the Municipality of Jerusalem, and Yigal Amedi, Deputy Mayor of Jerusalem, for the reception at the Town Hall.

3 About Domain Decomposition Methods

Domain decomposition, a form of divide-and-conquer for mathematical problems posed over a physical domain, as in partial differential equations, is the most common paradigm for large-scale simulation on massively parallel distributed, hierarchical memory computers. In domain decomposition, a large problem is reduced to a collection of smaller problems, each of which is easier to solve computationally than the undecomposed problem, and most or all of which can be solved independently and concurrently. Typically, it is necessary to iterate over the collection of smaller problems, and much of the theoretical interest in domain decomposition algorithms lies in ensuring that the number of iterations required is very small. Indeed, the best domain decomposition methods share with their cousins, multigrid methods, the property that the total computational work is linearly proportional to the size of the input data, or that the number of iterations required is at most logarithmic in the number of degrees of freedom of individual subdomains.

Algorithms whose work requirements are linear or log-linear in the size of the input data in this context are said to be "optimal." Near optimal domain decomposition algorithms are now known for many, but certainly not all, important classes of problems that arise science and engineering. Much of the contemporary interest in domain decomposition algorithms lies in extending the classes of problems for which optimal algorithms are known.

Domain decomposition algorithms can be tailored to the properties of the physical system as reflected in the mathematical operators, to the number of processors available, and even to specific architectural parameters, such as cache size and the ratio of memory bandwidth to floating point processing rate.

Domain decomposition has proved to be an ideal paradigm not only for execution on advanced architecture computers, but also for the development of reusable, portable software. The most complex operation in a typical domain decomposition method — the application of the preconditioner — carries out in each subdomain steps nearly identical to those required to apply a conventional preconditioner to the global domain. Hence software developed for the global problem can readily be adapted to the local problem, instantly presenting lots of "legacy" scientific code for to be harvested for parallel implementations. Furthermore, since the majority of data sharing between subdomains in domain decomposition codes occurs in two archetypal communication operations — ghost point updates in overlapping zones between neighboring subdomains, and global reduction operations, as in forming an inner product — domain decomposition methods map readily onto optimized, standardized message-passing environments, such as MPI.

Finally, it should be noted that domain decomposition is often a natural paradigm for the modeling community. Physical systems are often decomposed into two or more contiguous subdomains based on phenomenological considerations, such as the importance or negligibility of viscosity or reactivity, or any other feature, and the subdomains are discretized accordingly, as independent tasks. This physically-based domain decomposition may be mirrored in the software engineering of the corresponding code, and leads to threads of execution that operate on contiguous subdomain blocks. These can be either further subdivided or aggregated to fit the granularity of an available parallel computer.

4 Selected Bibliography of Books and Survey Articles

1. Bjørstad, P., Espedal, M., Keyes, D.E. eds.: *Proc. Ninth Int. Symp. on Domain Decomposition Methods for Partial Differential Equations* (Ullensvang, 1997), Wiley, New York, 1999.
2. Chan, T.F., Glowinski, R., Périaux, J., Widlund, O.B., eds.: *Proc. Second Int. Symp. on Domain Decomposition Methods for Partial Differential Equations* (Los Angeles, 1988), SIAM, Philadelphia, 1989.
3. Chan, T.F., Glowinski, R., Périaux, J., Widlund, O.B., eds.: *Proc. Third Int. Symp. on Domain Decomposition Methods for Partial Differential Equations* (Houston, 1989), SIAM, Philadelphia, 1990.
4. Chan, T., Kako, T., Kawarada, H., Pironneau, O., eds.: *Proc. Twelfth Int. Conf. on Domain Decomposition Methods in Science and Engineering* (Chiba, 1999), DDM.org, Bergen, 2001.
5. Chan, T.F., Mathew, T.P.: *Domain Decomposition Algorithms*, Acta Numerica, 1994, pp. 61-143.
6. Débit, N., Garbey, M., Hoppe, R., Keyes, D., Kuznetsov, Y.A., Périaux, J., eds.: *Proc. Thirteenth Int. Conf. on Domain Decomposition Methods in Science and Engineering* (Lyon, 2000), CINME, Barcelona, 2002.
7. Farhat, C., Roux, F.-X.: *Implicit Parallel Processing in Structural Mechanics*, Computational Mechanics Advances **2**, 1994, pp. 1–124.
8. Glowinski, R., Golub, G.H., Meurant, G.A., Périaux, J., eds.: *Proc. First Int. Symp. on Domain Decomposition Methods for Partial Differential Equations* (Paris, 1987), SIAM, Philadelphia, 1988.
9. Glowinski, R., Kuznetsov, Y.A., Meurant, G.A., Périaux, J., Widlund, O.B., eds.: *Proc. Fourth Int. Symp. on Domain Decomposition Methods for Partial Differential Equations* (Moscow, 1990), SIAM, Philadelphia, 1991.

10. Glowinski, R., Périaux, J., Shi, Z.-C., Widlund, O.B., eds.: *Eighth International Conference of Domain Decomposition Methods* (Beijing, 1995), Wiley, Strasbourg, 1997.
11. Hackbusch, W.: *Iterative Methods for Large Sparse Linear Systems*, Springer, Heidelberg, 1993.
12. Herrera, I., Keyes, D., Widlund, O., Yates, R., eds.: *Proc. Fourteenth Int. Conf. on Domain Decomposition Methods in Science and Engineering* (Cocoyoc, Mexico, 2003), National Autonomous University of Mexico (UNAM), Mexico City, 2003.
13. Keyes, D.E., Chan, T.F., Meurant, G.A., Scroggs, J.S., Voigt, R.G., eds.: *Proc. Fifth Int. Conf. on Domain Decomposition Methods for Partial Differential Equations* (Norfolk, 1991), SIAM, Philadelphia, 1992.
14. Keyes, D.E., Saad, Y., Truhlar, D.G., eds.: *Domain-based Parallelism and Problem Decomposition Methods in Science and Engineering*, SIAM, Philadelphia, 1995.
15. Keyes, D.E., Xu, J., eds.: *Proc. Seventh Int. Conf. on Domain Decomposition Methods for Partial Differential Equations* (State College, 1993), AMS, Providence, 1995.
16. Korneev, V.G., Langer, U.: *Domain Decomposition and Preconditioning*, Chapter 22 in Vol. 1 (Fundamentals) of the "Encyclopedia of Computational Mechanics", E. Stein, R. de Borst and T.J.R. Hughes, eds., Wiley, 2004.
17. Kornhuber, R., Hoppe, R., Périaux, J., Pironneau, O., Widlund, O., Xu, J., eds.: *Proc. Fifteenth Int. Conf. on Domain Decomposition Methods* (Berlin, 2003), Springer, Heidelberg, 2004.
18. Kruis, J.: *Domain Decomposition Methods for Distributed Computing*, Saxe-Coburg, Dun Eaglais, 2005.
19. Lai, C.-H., Bjørstad, P., Cross, M., Widlund, O., eds.: *Proc. Eleventh Int. Conf. on Domain Decomposition Methods* (Greenwich, 1999), DDM.org, Bergen, 2000.
20. Langer, U., Steinbach, O.: *Coupled Finite and Boundary Element Domain Decomposition Methods*, in "Boundary Element Analysis: Mathematical Aspects and Application", M. Schanz and O. Steinbach, eds., Lecture Notes in Applied and Computational Mechanics, Vol. 29, Springer, Berlin, pp. 29–59, 2007.
21. Lebedev, V.I., Agoshkov, V.I.: *Poincaré-Steklov operators and their applications in analysis*, Academy of Sciences USSR, Dept. of Numerical Mathematics, Moskow, 1983 (in Russian).
22. Le Tallec, P.: *Domain Decomposition Methods in Computational Mechanics*, Computational Mechanics Advances **2**, 1994, pp. 121–220.
23. Mandel, J., Farhat, C., Cai, X.-C., eds.: *Proc. Tenth Int. Conf. on Domain Decomposition Methods in Science and Engineering* (Boulder, 1998), AMS, Providence, 1999.
24. Mathew, T.P.A.: *Domain Decomposition Methods for the Numerical Solution of Partial Differential Equations*. Vol. 61 of *Lecture Notes in Computational Science & Engineering*, Springer, Heidelberg, 2008.

25. Nepomnyaschikh, S.: *Domain Decomposition Methods*, In "Lectures on Advanced Computational Methods in Mechanics", J. Kraus and U. Langer, eds., Radon Series on Computational and Applied Mathematics, de Gruyter, Berlin, 2007.
26. Oswald, P.: *Multilevel Finite Element Approximation: Theory and Applications*, Teubner Skripten zur Numerik, Teubner, Stuttgart, 1994.
27. Pavarino, L., Toselli, A.: *Recent Developments in Domain Decomposition Methods*, Vol. 23 of *Lecture Notes in Computational Science & Engineering*, Springer, Heidelberg, 2002.
28. Quarteroni, A., Périaux, J., Kuznetsov, Y.A., Widlund, O.B., eds.: *Proc. Sixth Int. Conf. on Domain Decomposition Methods in Science and Engineering* (Como, 1992), AMS, Providence, 1994.
29. Quarteroni, A., Valli, A.: *Domain Decomposition Methods for Partial Differential Equations*, Oxford, 1999.
30. Saad, Y.: *Iterative Methods for Sparse Linear Systems*, PWS, Boston, 1996.
31. Smith, B.F., Bjørstad, P.E., Gropp, W.D.: *Domain Decomposition: Parallel Multilevel Algorithms for Elliptic Partial Differential Equations*, Cambridge Univ. Press, Cambridge, 1996.
32. Steinbach, O.: *Stability Estimates for Hybrid Coupled Domain Decomposition Methods*, Lecture Notes in Mathematics, Vol. 1809, Springer, Berlin, 2003.
33. Toselli, A., Widlund, O.: *Domain Decomposition Methods—Algorithms and Theory*, Springer, New York, 2005.
34. Widlund, O., Keyes, D.E., eds.: *Proc. Sixteenth Int. Conf. on Domain Decomposition Methods in Sciences and Engineering* (New York City, 2005), Springer, Heidelberg, 2007.
35. Wolhmuth, B.I.: *Discretization Methods and Iterative Solvers Based on Domain Decomposition*, Vol. 17 of *Lecture Notes in Computational Science & Engineering*, Springer, Heidelberg, 2001.
36. Xu, J.: *Iterative Methods by Space Decomposition and Subspace Correction*, SIAM Review **34**, 1991, pp. 581–613.
37. Xu, J., Zou, J.: *Some Nonoverlapping Domain Decomposition Methods*, SIAM Review **40**, 1998, pp. 857–914.

Acknowledgments

The editors would like to thank all authors for their cooperation, the anonymous referees for their valuable labors, and Dr. Martin Peters and Ms. Thanh-Ha Le Thi from Springer Verlag for their support and friendly collaboration in preparing these proceedings. Finally, the beautiful appearance of this manuscript is due primarily to the editorial craftsmanship of Reeva Goldsmith of the Courant Institute of New York University; the Courant Institute donated her services.

Michel Bercovier, Jerusalem
Martin Gander, Geneva
Ralf Kornhuber, Berlin
Olof Widlund, New York

Contents

Part I Plenary Presentations

Part II Minisymposia

Part I

Plenary Presentations

A Domain Decomposition Approach for Calculating the Graph Corresponding to a Fibrous Geometry

Achi Brandt[1], Oleg Iliev[2], and Joerg Willems[3]

[1] Department of Applied Mathematics & Computer Science, The Weizmann Institute of Science, 76100 Rehovot, Israel achi.brand@tweizmann.ac.il
[2] Fraunhofer Institut für Techno- und Wirtschaftsmathematik, Fraunhofer-Platz 1, 67663 Kaiserslautern, Germany oleg.iliev@itwm.fraunhofer.de
[3] Fraunhofer Institut für Techno- und Wirtschaftsmathematik, Fraunhofer-Platz 1, 67663 Kaiserslautern, Germany joerg.willems@itwm.fraunhofer.de

1 Introduction

The effective properties of composite materials/media are in strong demand in engineering, geoscience, and environmental studies to name just a few examples. In [2], we presented an efficient algorithm for computing an approximation of the effective thermal conductivity tensor for high contrast fibrous geometries. The essential idea of the approach is to take into consideration the network-like structure of a given fibrous geometry and to perform all calculations on the induced unstructured grid. More precisely, the intersections of fibers are considered as nodes and the connecting fibers between nodes are considered as edges of an undirected graph. The weight of each edge depends on the diameter and the conductivity of the respective fiber and the distance of the connected nodes. A comparison between the results produced by our algorithm and classical methods, which resolve the fibrous geometry using volumetric elements, yields evidence of its efficiency and reliability for a large class of problems from engineering and science.

In the article at hand the primary focus is on increasing the computational efficiency of the essential preprocessing step, i.e., of setting up the graph. In [2] the computation of the fiber intersections is carried out straightforwardly, i.e., each fiber is compared against any other fiber for intersection. This preprocessing stage, if carried out like this, has a complexity which is quadratic in the number of fibers and can therefore, for samples with very many fibers, become prohibitively expensive.

The idea to reduce the complexity of the straightforward strategy discussed above is to partition the domain into a grid of coarse cells. Then by going along each fiber, we determine the coarse grid cells through which this fiber passes. Once this has been completed we go through each coarse cell and check for intersections only among those fibers passing through one and the same cell. This is done in such a way that two fibers are compared only once, no matter if they mutually lie in several coarse cells. The resulting graph is - except for the ordering of the nodes - identical

M. Bercovier et al. (eds.), *Domain Decomposition Methods in Science and Engineering XVIII*, Lecture Notes in Computational Science and Engineering,
DOI: 10.1007/978-3-642-02677-5, © Springer-Verlag Berlin Heidelberg 2009

to the one computed by the standard approach. The computational cost, however, is significantly reduced.

The remainder of the article at hand is organized as follows: In section 2 we introduce some notations and definitions needed for presenting our argument. After that we give a short description of the problem which we are ultimately interested in solving. In section 4 the algorithm that we use to construct the graphs corresponding to fibrous geometries is discussed. In a subsection we also provide a short analysis of the computational cost of this algorithm. The final section of this paper is devoted to numerical results and conclusions.

2 Preliminaries

For the arguments to follow we would like to introduce some notations and definitions. In order to make the presentation somewhat simpler we restrict our exposition to three spatial dimensions, which is anyway the most interesting case from a practical point of view. In [2] random fibrous geometries and the computation of their effective thermal conductivities are the main targets of consideration. Let us now briefly discuss what exactly we mean by a fiber and a random fibrous geometry.

By a fiber φ, we mean a cylindrical object of finite or infinite length. In particular, a fiber is supposed to have a straight line at its center. To generate a fibrous geometry these objects are randomly "thrown into" our domain Ω and cut-off at the boundary $\partial\Omega$. For simplicity we assume Ω to be brick shaped. The collection of all fibers in Ω is denoted by Φ. Let the set of all intersections of the straight lines at the centers of fibers with $\partial\Omega$ be denoted by $\partial\omega$. The actual numerical generation of our fibrous geometries is done by the GeoDict2008 software (For more information see http://www.geodict.com.) With this random construction different fibers may and in general will intersect.

Now, let ω be the set of points, where two or more fibers cross. For a simpler presentation and to avoid unnecessary technicalities, we assume, that whenever two fibers (i.e., the cylindrical objects) have a nonempty intersection the same holds true for their center lines. For a randomly generated fibrous geometry this assumption will in general not be satisfied. In practice, however, this doesn't pose any serious difficulties. In order to determine whether two fibers cross, we calculate the distance between their center lines. If this distance is smaller than the sum of the fiber radii, we say that the fibers cross and for each of the involved center lines we store the point at which they are closest, i.e., the distance of these points is equal to the distance of the center lines of the involved fibers. The crossing node is then set to be in the middle of these two points. We also define $\overline{\omega} := \omega \cup \partial\omega$ to be the set of all internal and boundary intersections.

Let h be the characteristic distance between adjacent (i.e., adjacent on a fiber) nodes in $\overline{\omega}$ and let d be the characteristic diameter of all fibers in $\overline{\Omega}$. We require $d \ll h$ in order to have a meaningful notion of a graph induced by the fibers (which correspond to the edges of the graph) and their intersections (which correspond to the nodes of the graph).

3 Statement of the Problem

It is a well-known result from homogenization theory (cf. e.g. [3] and the references therein) that the effective conductivity tensor $\widetilde{K}$ for a periodic or statistically homogeneous medium can be calculated by

$$\widetilde{K}\mathbf{e}_i = \langle K \nabla u_i \rangle_{\Omega}, \quad i = 1,2,3,$$

where $\mathbf{e}_i$ is the i-th unit vector, K denotes the fine scale conductivity, $\langle \cdot \rangle_{\Omega}$ is the volumetric average over Ω, and u_i solves

$$\begin{aligned} \nabla \cdot (K \nabla u_i) &= 0 \quad \text{in } \Omega \\ u_i(\mathbf{x}) &= x_i \quad \text{on } \partial\Omega, \end{aligned} \tag{1}$$

with x_i being the i-th component of $\mathbf{x}$.

In [1] it was shown that for a composite medium having a (very) high contrast in conductivities, the effective conductivity tensor of the entire medium can be approximated by solving three (one for each spatial dimension) constant coefficient elliptic problems. These constant coefficient problems are posed only on the highly conductive parts of Ω. Several numerical examples in [1] show that this approach yields very good results for a class of problems interesting from a scientific and engineering point of view.

Departing from the framework in [1] an algorithm was developed in [2] specifically designed for approximating the effective conductivity tensors for high contrast fibrous geometries. The essential idea is to perform all calculations on the graph induced by the underlying fibrous structure. The discrete problems corresponding to (1) read as follows:

$$\begin{aligned} \mathcal{D}(K \mathcal{G} y_i) &= 0 \quad \text{in } \omega \\ y_i &= x_i \quad \text{on } \partial\omega, \end{aligned} \tag{2}$$

where $\mathcal{D}$ and $\mathcal{G}$ are discrete versions of the divergence and gradient operator, respectively, having values on the nodes (●) and faces (|) between adjacent nodes, respectively (see Fig. 1). For a precise definition of $\mathcal{D}$ and $\mathcal{G}$ as well as for an error analysis we would like to refer the reader to [2].

4 A Divide and Conquer Algorithm

The computational bottleneck of the algorithm discussed in [2] is the preprocessing step of setting up the graph, i.e., the computation of the set of intersections ω. If this is done in a straightforward way, meaning by comparing each fiber with every other, the computational cost is $\mathcal{O}(n_{\Phi}^2)$, where n_{Φ} is the number of fibers in Ω. For large geometries with very many fibers this will of course soon become prohibitively expensive.

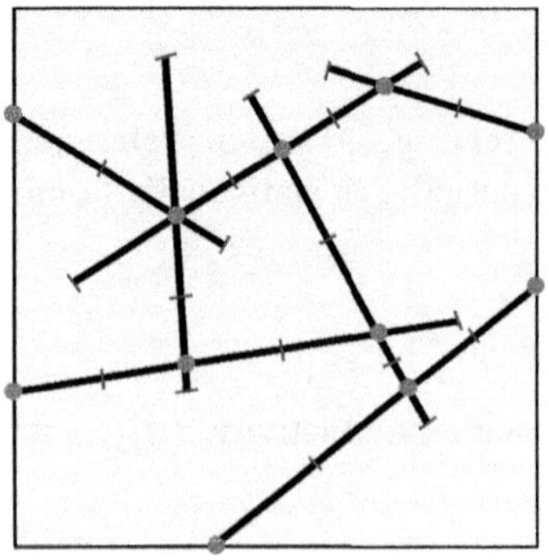

Fig. 1. Fibrous structure with induced nodes and faces.

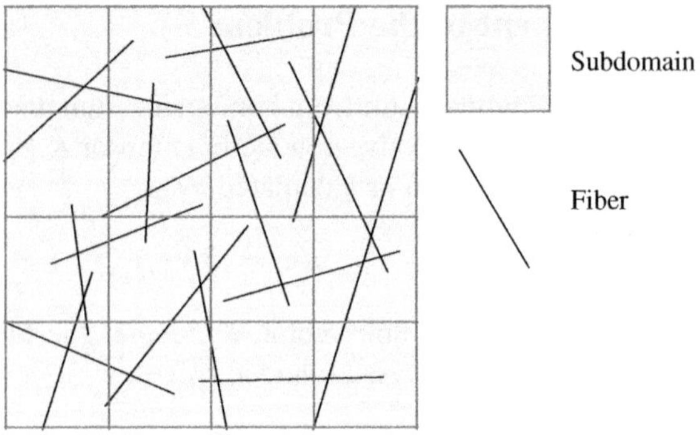

Fig. 2. Subdomains $\Omega_{\mathbf{j}}$ and fibers φ_i.

The idea to cure this problem is to divide our domain Ω into subdomains $\Omega_{\mathbf{j}}$, $\mathbf{j} \in \{1,\dots n_{\Omega,x_1}\} \times \{1,\dots,n_{\Omega,x_2}\} \times \{1,\dots,n_{\Omega,x_3}\} =: \mathscr{J}$, where n_{Ω,x_1}, n_{Ω,x_2}, and n_{Ω,x_3} are the number of subdomains in each spatial direction, $\mathbf{j}$ is a multi-index, and $\bigcup_{\mathbf{j}\in\mathscr{J}} \Omega_{\mathbf{j}} = \Omega$ (cf. Fig. 4). For simplicity, we again suppose that $\Omega_{\mathbf{j}}$ is brick shaped. Then for each fiber we check through which subdomains it passes and construct the sets $\Phi_{\mathbf{j}}$, where $\Phi_{\mathbf{j}}$ denotes the set of fibers passing through $\Omega_{\mathbf{j}}$. Then for each $\mathbf{j} \in \mathscr{J}$ we check for intersections among all $\varphi \in \Phi_{\mathbf{j}}$. In Algorithm 1 we make these considerations more formal.

Remark 1. The condition $\lambda_2(\Omega_{\mathbf{j}} \cap \Omega_{\hat{\mathbf{j}}}) \neq 0$ in step 8 of Algorithm 1 means that we only check adjacent subdomains which have a common face with the previous subdomain. We don't need to take into consideration those adjacent subdomains which only have a common edge or point. This is because fibers are volumetric objects. In particular they have a strictly positive diameter.

Remark 2. It should be noted that the standard straightforward approach of testing each fiber with any other for intersection is a special case of Algorithm 1 – consider the case $\#\mathscr{J} = 1$.

4.1 Numerical Complexity of Algorithm 1

Now, we would like to obtain an estimate of the numerical cost of Algorithm 1 in order to be able to compare it with the complexity of the straightforward approach of checking each fiber with respect to any other one for intersection. It is evident, that this straightforward approach requires $\mathcal{O}(n_{\Phi}^2)$ operations.

Since for general randomly generated fiber geometries the computation of the numerical complexity of Algorithm 1 would go into too much detail concerning the generation of such geometries, we perform our analysis only for one particular structure with regularly arranged infinitely long fibers (cf. Fig. 5). More

$\Phi_{\mathbf{j}} = \emptyset \ \forall \mathbf{j} \in \mathscr{J}$
for $i = 1, \ldots, n_\Phi$ **do**
  Compute an end point $\mathbf{x}_i$ of φ_i and determine $\mathbf{j} \in \mathscr{J}$ such that $\mathbf{x}_i \in \Omega_{\mathbf{j}}$.
  Set $\Phi_{\mathbf{j}} = \Phi_{\mathbf{j}} \cup \{\varphi_i\}$, i.e., add φ_i to the set of fibers passing through $\Omega_{\mathbf{j}}$.
  Set $\tilde{\mathscr{J}} = \{\mathbf{j}\}$. The subdomains corresponding to $\tilde{\mathscr{J}}$ are those intersected by φ_i and having at least one neighbor which hasn't been checked for intersection with φ_i, yet.
  while $\#\tilde{\mathscr{J}} \neq 0$ **do**
    for $\mathbf{j} \in \tilde{\mathscr{J}}$ **do**
      Let $\hat{\mathscr{J}}$ be the set of all $\hat{\mathbf{j}}$ such that $\lambda_2(\Omega_{\mathbf{j}} \cap \Omega_{\hat{\mathbf{j}}}) \neq 0$ and $\varphi_i \notin \Phi_{\hat{\mathbf{j}}}$, where λ_2 is the two-dimensional Lebesgue measure. The subdomains corresponding to $\hat{\mathscr{J}}$ are those neighbors of $\Omega_{\mathbf{j}}$ for which intersection with φ_i hasn't been verified yet.
      for $\hat{\mathbf{j}} \in \hat{\mathscr{J}}$ **do**
        if φ_i crosses $\Omega_{\hat{\mathbf{j}}}$ **then**
          Set $\Phi_{\hat{\mathbf{j}}} = \Phi_{\hat{\mathbf{j}}} \cup \{\varphi_i\}$, i.e., φ_i is added to the set of fibers passing through $\Omega_{\hat{\mathbf{j}}}$.
          Set $\tilde{\mathscr{J}} = \tilde{\mathscr{J}} \cup \{\hat{\mathbf{j}}\}$. Since $\Omega_{\hat{\mathbf{j}}}$ is intersected by φ_i we now in turn need to check the neighbors of $\Omega_{\hat{\mathbf{j}}}$ for intersection with φ_i, too.
        end if
      end for
      Set $\tilde{\mathscr{J}} = \tilde{\mathscr{J}} \backslash \{\mathbf{j}\}$. Since all neighbors of $\Omega_{\mathbf{j}}$ have been checked for intersection with φ_i, $\mathbf{j}$ is removed from $\tilde{\mathscr{J}}$.
    end for
  end while
end for
for $\mathbf{j} \in \mathscr{J}$ **do**
  for $\varphi_i \in \Phi_{\mathbf{j}}$ **do**
    for $\varphi_k \in \Phi_{\mathbf{j}}$ and $k > i$ **do**
      if φ_k and φ_i haven't been tested for intersecting yet **then**
        Test φ_k and φ_i for intersection and add a corresponding node to the graph if the fibers cross.
      end if
    end for
  end for
end for

Algorithm 1: Compute a graph corresponding to a fiber geometry.

precisely, we assume that our domain is the unit cube, i.e., $\Omega = [0,1]^3$. The fibers are defined by connecting the following pairs of points $\{(0, h/2 + i_2 h, h/2 + i_3 h), (1, h/2 + i_2 h, h/2 + i_3 h)\}$, $\{(h/2 + i_1 h, 0, h/2 + i_3 h), (h/2 + i_1 h, 1, h/2 + i_3 h)\}$, and $\{(h/2 + i_1 h, h/2 + i_2 h, 0), (h/2 + i_1 h, h/2 + i_2 h, 1)\}$, for $i_1, i_2, i_3 = 0, 1, \ldots, 1/h - 1$. Here we tacitly assume that $1/h \in \mathbb{N}$. Additionally, we require the diameters of all fibers to be smaller than the side lengths of the subdomains, each of which is assumed to be of equal cubic size and shape. It is evident that the example geometry just described is quite particular. In fact, it can be easily seen that the number of intersections is rather large compared to a random geometry with an equal number of fibers. Despite being artificial we will however see below that this geometry is quite representative in terms of the computational costs of Algorithm 1. Table 1 gives an

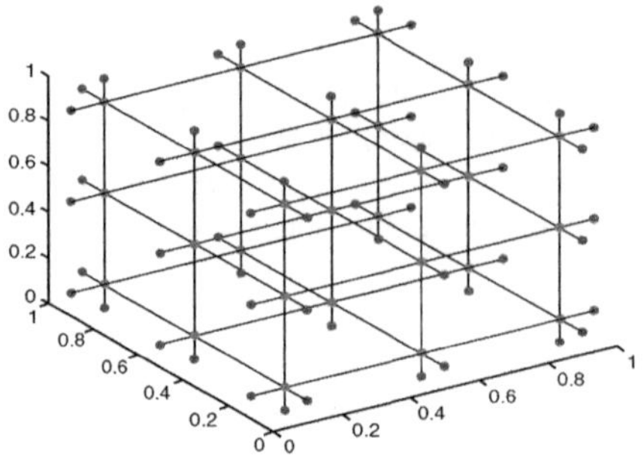

Fig. 3. Interior and boundary nodes for a regular fiber structure.

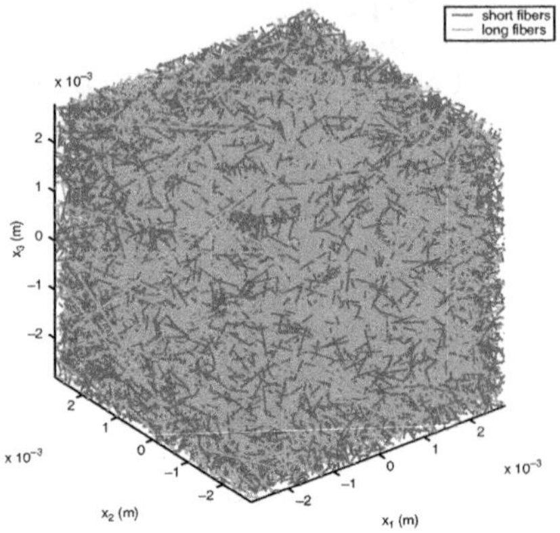

Fig. 4. Geometry with 1% svf and equal parts of long and short fibers.

overview of the computational costs of the different steps of Algorithm 1 when applied to this example geometry.

Based on the information in Table 1 we can see that the total numerical complexity of Algorithm 1 (i.e., steps 1-27) is given by

$$\mathcal{O}(n_\Phi n_{\Omega,x_1}) + \mathcal{O}\left(\frac{n_\Phi^2}{n_{\Omega,x_1}}\right). \tag{3}$$

Thus, we easily deduce that choosing

$$n_{\Omega,x_1} = \mathcal{O}(\sqrt{n_\Phi}) \tag{4}$$

leads to a total numerical complexity of

$$\mathcal{O}(n_\Phi^{3/2}) \tag{5}$$

when applied to our regular example geometry sketched in Fig. 5. This is of course a major improvement compared to the complexity $\mathcal{O}(n_\Phi^2)$ of the standard approach.

Remark 3. It should be noted here that the reasoning above is somewhat specific for our example geometry. For general randomly generated fibrous geometries with multiple fiber lengths and diameters we cannot obtain such a nice and compact formula as in (5). Nevertheless, our considerations above are surprisingly representative for more general cases as a collection of examples in section 5 shows.

5 Numerical Results and Conclusions

Now, let us take a look at the actual numerical performance of Algorithm 1 when applied to large randomly generated fibrous geometries. In order to do this, we first

Steps	Order of Complexity	
10-13	$\mathcal{O}(1)$	
9-15	$\mathcal{O}(1)$	Since $\#\mathcal{J} \leq 6$.
6-17	$\mathcal{O}(n_{\Omega,x_1})$	Since the number of subdomains that each fiber passes through is $\mathcal{O}(n_{\Omega,x_1})$ and each subdomain is checked at most once. Note that $n_{\Omega,x_1} = n_{\Omega,x_2} = n_{\Omega,x_3}$ and that we require the fiber diameters to be smaller than the side lengths of the subdomains.
2-18	$\mathcal{O}(n_\Phi n_{\Omega,x_1})$	
20-26	$\mathcal{O}((\#\Phi_\mathbf{j})^2) = \mathcal{O}\left(\frac{n_\Phi^2}{n_{\Omega,x_1}^4}\right)$	Since in each subdomain of our regular fiber structure (see Fig. 5) there are $\frac{3}{(hn_{\Omega,x_1})^2}$ fibers and in the entire domain Ω there are $\frac{3}{h^2}$ fibers, i.e., $n_\Phi = \frac{3}{h^2}$.
19-27	$\mathcal{O}\left(\#\mathcal{J}\frac{n_\Phi^2}{n_{\Omega,x_1}^4}\right) = \mathcal{O}\left(\frac{n_\Phi^2}{n_{\Omega,x_1}}\right)$	Since $\#\mathcal{J} = n_{\Omega,x_1}^3$ for our cubic domain.

Table 1. Computational cost of Algorithm 1

specify the parameters used in the generation of our structures. All geometries are generated by the GeoDict2008 software using a grid of 2000^3 voxels on Ω, which is chosen to be a cube with side-length 5.6e-3m. Thus, the side-length of a voxel is 2.8e-6m. We consider structures having a solid volume fraction (svf) of 1%, 3%, and 5%, i.e., 1%, 3%, and 5% of Ω are occupied by fibers, respectively. For each of these svf we consider a geometry with equal parts of infinitely long and short fibers ("short" meaning 100 voxels long), one with infinitely long fibers only, and one with short fibers only. We then consider a series of choices for $n_{\Omega,x_1} = n_{\Omega,x_2} = n_{\Omega,x_3}$ and compare the cpu-times needed for setting up the graphs. To get an impression how these fibrous geometries look we refer to Fig. 4, which shows a plot of the structure with 1% svf and with equal parts of short and long fibers.

The tables in Figs. 5–7 show the data specific of the problems under consideration (number of fibers, number of nodes, etc.) and the computational costs for the cases $n_{\Omega,x_1} = n_{\Omega,x_2} = n_{\Omega,x_3} = 1$ and $n_{\Omega,x_1} = n_{\Omega,x_1}^{opt}$, where n_{Ω,x_1}^{opt} is the optimal choice for n_{Ω,x_1} in terms of the time needed for setting up the graph corresponding to the fibrous geometry. (In order to determine n_{Ω,x_1}^{opt} we consider a series of n_{Ω,x_1}, see the top plots in Figs. 5–7.)

As we can see, the reduction of cpu-time when choosing $n_{\Omega,x_1} = n_{\Omega,x_1}^{opt}$ instead of $n_{\Omega,x_1} = 1$ is substantial. For the geometries involving only long fibers the time needed for setting up the graph is roughly cut in half (cf. table in Fig. 5). For the fibrous structure with a solid volume fraction (svf) of 5% and only short fibers the cpu-time for constructing the graph is reduced to less than 0.3% when choosing $n_{\Omega,x_1} = n_{\Omega,x_1}^{opt}$ (table in Fig. 6). Looking at the table in 7 we see that also for geometries consisting of short and long fibers the cpu-time for setting up the graph is reduced by more than one order of magnitude when choosing the optimal n_{Ω,x_1}.

For the instances that we consider we see that by choosing $n_{\Omega,x_1} = n_{\Omega,x_1}^{opt}$ the computational cost of constructing the graph corresponding to our geometry can be

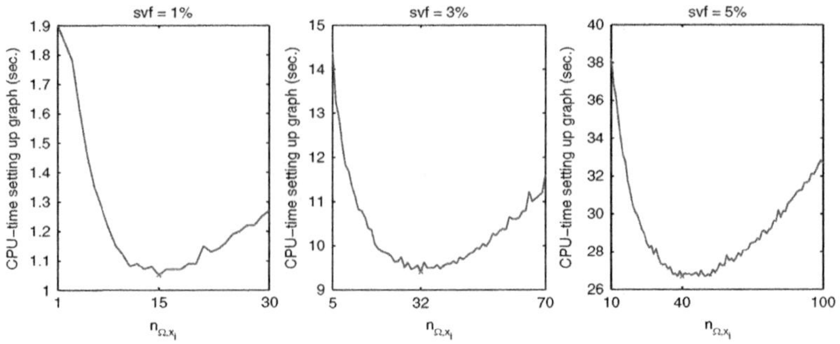

CPU-times needed for setting up the graph for different choices of $n_{\Omega,x_1} = n_{\Omega,x_2} = n_{\Omega,x_3}$.

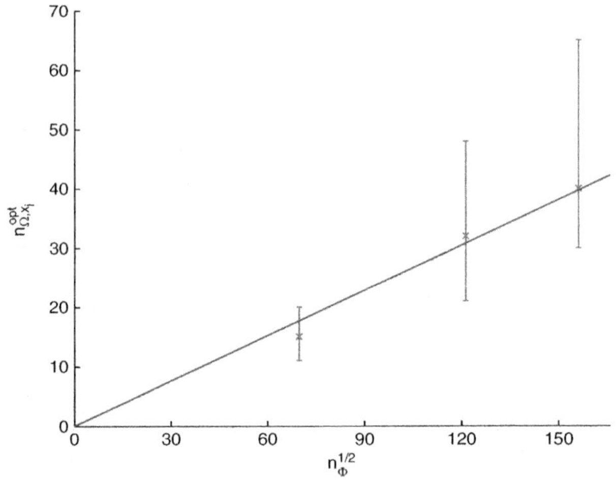

Optimal choice of n_{Ω,x_1}, i.e. n^{opt}_{Ω,x_1} (x), vs. $\sqrt{n_\Phi}$ with 5% deviation margins and a linear leasts-squares fit.

svf short fibers	0%		0%		0%	
svf long fiber	1%		3%		5%	
# fibers	4851		14686		24366	
# interior nodes	15462		132121		351366	
effective conductivity tensor	2.64e-2 - - - 2.65e-2 - - - 2.64e-2		3.13e-2 - - - 3.14e-2 - - - 3.15e-2		3.66e-2 - - - 3.64e-2 - - - 3.64e-2	
# coarse grid cells	1	15^3	1	32^3	1	40^3
total CPU-time (sec.)	2.3e0	1.4e0	2.9e1	1.7e1	1.1e2	6.3e1
CPU-time constructing the graph	1.9e0	1.1e0	1.8e1	9.4e0	5.7e1	2.7e1
CPU-time solving the system	< 1	< 1	1.1e1	7.6e0	5.4e1	3.5e1

Computational results and costs.

Fig. 5. CPU-time analysis and numerical results for geometries with **only long** fibers and solid volume fractions of 1%, 3%, and 5%, respectively.

reduced to the same order of magnitude as the cost needed for solving the arising linear system. (Here we would like to remark that for solving the linear system we employ the ILU preconditioned Conjugate Gradient (CG) solver implemented in the LASPack package (see http://www.mgnet.org/mgnet/codes/laspack/html/laspack.html) using a relative residual reduction of $1e-6$ as stopping criterion.) Before, i.e., when choosing $n_{\Omega,x_1} = 1$, almost the entire computational cost for determining an approximation to the effective thermal conductivity tensor was devoted to setting up the computational graph. Therefore, it was not feasible to spend much effort on speeding up the solution of the arising linear system. Now, with this new approach of dividing Ω into subdomains, we see that in some cases the cpu-time for solving the arising linear system can actually exceed the cpu-time for constructing the graph (cf. tables in Figs. 5-7). With this observation it seems reasonable to also optimize the process of solving the arising linear system - e.g. by employing algebraic multi-grid methods and the like - which is a topic of our further research.

As an interesting side note we would like to remark that in all investigated cases (cf. tables in Figs. 5-7) the cpu-time for solving the linear system also reduces (by around 30%) when choosing the optimal n_{Ω,x_1}. This observation seems surprising, since the graph constructed by Algorithm 1 and the number of CG-iterations required to satisfy the convergence criterion are independent of the choice for n_{Ω,x_1}. The only plausible explanation that we have for this certainly desirable side effect is that for $n_{\Omega,x_1} = n^{opt}_{\Omega,x_1}$ the nodes of the graph are not in the same order as when choosing $n_{\Omega,x_1} = 1$. Apparently, this re-odering of the unknowns speeds up the matrix-vector multiplication of the system matrix, which could be due to a better cache-optimization. Providing a detailed analysis of this issue is, however, beyond the scope of this article.

Looking at the graphs in Figs. 5-7, where the cpu-time for constructing the graph is plotted vs. the choice for n_{Ω,x_1}, we see that there is in fact an optimal choice n^{opt}_{Ω,x_1}. This observation can be explained via (3). When choosing n_{Ω,x_1} larger (smaller) than n^{opt}_{Ω,x_1} the first (second) term of (3) dominates.

Now, we would like to investigate the question, whether relation (4), which we derived for the very regular fiber structure shown in Fig. 5, also holds - at least approximately - for our randomly generated geometries. For this we plot n^{opt}_{Ω,x_1} against $\sqrt{n_\Phi}$ for different fibrous geometries (see lower left plots in Figs. 5-7). Of course, we can only hope for (4) to hold for structures with different solid volume fractions but with the same kind of fibers. Therefore, we only try to verify (4) for these cases. Looking at the least squares linear fit (blue line) in Figs. 5-7, where the fitted line is forced through the origin and thus the only free parameter is its slope, we can see that (4) is indeed quite well satisfied. Nevertheless, the constant involved in (4) is different for different choices of fibers. For the sequence of geometries with svf 1%, 3%, and 5% and only long fibers it is approximated to 2.54e-1, while for the cases of only short fibers it is approximately 9.89e-2. The constant for the geometries involving equal parts of long and short fibers is estimated to 1.11e-1 and thus in-between the two former ones.

In addition to n^{opt}_{Ω,x_1} the lower left plots in Figs. 5-7 also show margins which correspond to those choices for n_{Ω,x_1} for which the cpu-time for setting up the graph

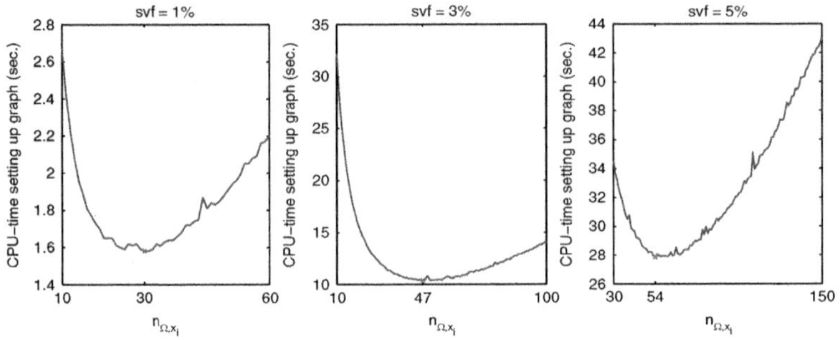

CPU-times needed for setting up the graph for different choices of $n_{\Omega,x_1} = n_{\Omega,x_2} = n_{\Omega,x_3}$.

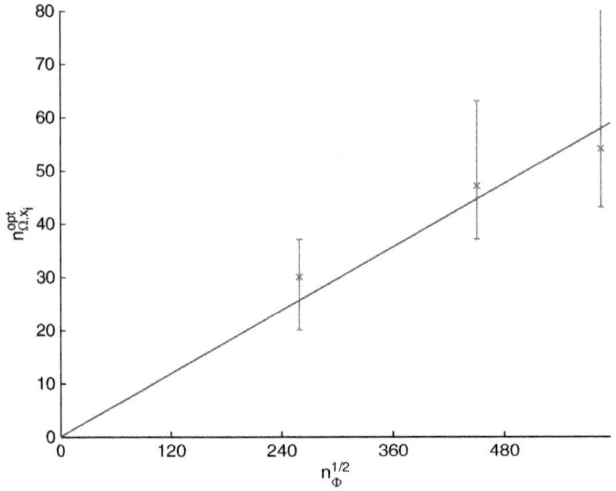

Optimal choice of n_{Ω,x_1}, i.e. n^{opt}_{Ω,x_1} (x), vs. $\sqrt{n_\Phi}$ with 5% deviation margins and a linear leasts-squares fit.

svf short fibers	1%		3%		5%	
svf long fiber	0%		0%		0%	
# fibers	66953		202845		341543	
# interior nodes	2127		113936		378634	
effective conductivity tensor	2.40e-2 - - - 2.40e-2 - - - 2.40e-2		2.40e-2 - - - 2.40e-2 - - - 2.40e-2		2.57e-2 - - - 2.57e-2 - - - 2.57e-2	
# coarse grid cells	1	15^3	1	32^3	1	40^3
total CPU-time (sec.)	3.6e2	2.2e0	3.5e3	5.9e1	9.9e3	1.1e2
CPU-time constructing the graph	3.6e2	1.6e0	3.4e3	1.0e1	9.8e3	2.8e1
CPU-time solving the system	< 1	< 1	6.5e1	4.7e1	1.1e2	7.7e1

Computational results and costs.

Fig. 6. CPU-time analysis and numerical results for geometries with **only short** fibers and solid volume fractions of 1%, 3%, and 5%, respectively.

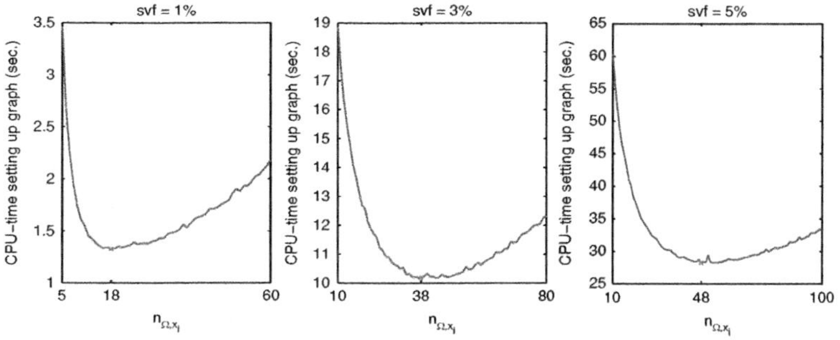

CPU-times needed for setting up the graph for different choices of $n_{\Omega,x_1} = n_{\Omega,x_2} = n_{\Omega,x_3}$.

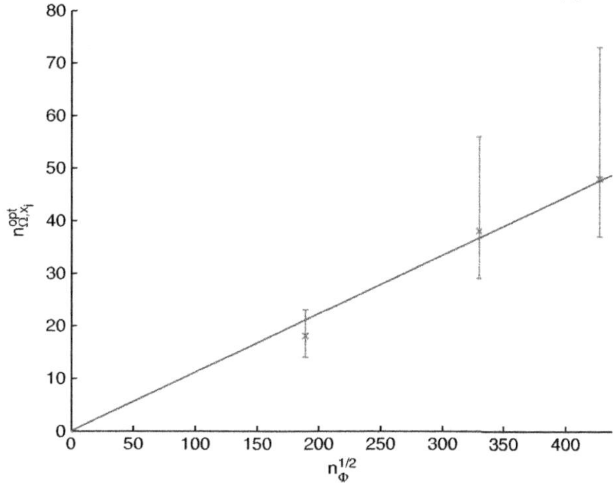

Optimal choice of n_{Ω,x_1}, i.e. n^{opt}_{Ω,x_1} (x), vs. $\sqrt{n_\Phi}$ with 5% deviation margins and a linear leasts-squares fit.

svf short fibers	0.5%		1.5%		2.5%	
svf long fiber	0.5%		1.5%		2.5%	
# fibers	35830		108668		182974	
# interior nodes	14383		136086		366286	
effective conductivity tensor	2.52e-2 - - - 2.52e-2 - - - 2.52e-2		2.80e-2 - - - 2.79e-2 - - - 2.79e-2		3.14e-2 - - - 3.14e-2 - - - 3.15e-2	
# coarse grid cells	1	18^3	1	38^3	1	48^3
total CPU-time (sec.)	9.5e1	2.0e0	1.0e3	2.1e1	2.9e3	7.1e1
CPU-time constructing the graph	9.4e1	1.3e0	9.9e2	1.0e1	2.8e3	2.8e1
CPU-time solving the system	< 1	< 1	1.4e1	1.0e1	6.4e1	4.1e1

Computational results and costs.

Fig. 7. CPU-time analysis and numerical results for geometries with **equal parts of short and long** fibers and solid volume fractions of 1%, 3%, and 5%, respectively.

is at most 5% higher than for n_{Ω,x_1}^{opt}. For practical problems it of course doesn't make sense to apply Algorithm 1 for several choices of n_{Ω,x_1} to determine the optimal one. Instead one is interested in approximating n_{Ω,x_1}^{opt} beforehand, and then use this approximation in the calculations. It is quite obvious that (4) can be used to predict an approximation to n_{Ω,x_1}^{opt}. Furthermore, it should be noted that the margins shown in 5-7 indicate that - especially for large and thus costly geometries - one doesn't really have to approximate n_{Ω,x_1}^{opt} very accurately in order to obtain almost optimal performance. Thus, it seems promising that an automatic way of approximating n_{Ω,x_1}^{opt}, which could then be used in Algorithm 1, can be implemented. This is also an objective of our further research.

On the whole, we would like to conclude that Algorithm 1 constitutes a very powerful enhancement of the approach presented in [2]. The computational costs are significantly reduced, which makes our graph-laplacian approach applicable to even larger geometries containing even more fibers.

References

[1] Ewing, R., Iliev, O., Lazarov, R., Rybak, I., Willems, J.: An efficient approach for upscaling properties of composite materials with high contrast of coefficients. Technical Report 132, Fraunhofer ITWM, 2007.

[2] Iliev, O., Lazarov, R., Willems, J.: A graph-laplacian approach for calculating the effective thermal conductivity of complicated fiber geometries. Technical Report 142, Fraunhofer ITWM, 2008.

[3] Jikov, V.V., Kozlov, S.M., Oleinik, O.A.: *Homogenization of Differential Operators and Integral Functionals*. Springer, 1st ed., 1994.

Adaptive Multilevel Interior-Point Methods in PDE Constrained Optimization

Harbir Antil[1], Ronald H.W. Hoppe[1,2], and Christopher Linsenmann[1,2]

[1] University of Houston, Department of Mathematics
(http://www.math.uh.edu/~rohop/)
[2] University of Augsburg, Institute for Mathematics
(http://scicomp.math.uni-augsburg.de)

Summary. We are concerned with structural optimization problems where the state variables are supposed to satisfy a PDE or a system of PDEs and the design variables are subject to inequality constraints. Within a primal-dual setting, we suggest an all-at-once approach based on interior-point methods. Coupling the inequality constraints by logarithmic barrier functions involving a barrier parameter and the PDE by Lagrange multipliers, the KKT conditions for the resulting saddle point problem represent a parameter dependent nonlinear system. The efficient numerical solution relies on multilevel path-following predictor-corrector techniques with an adaptive choice of the continuation parameter where the discretization is taken care of by finite elements with respect to nested hierarchies of simplicial triangulations of the computational domain. In particular, the predictor is a nested iteration type tangent continuation, whereas the corrector is a multilevel inexact Newton method featuring transforming null space iterations. As an application in life sciences, we consider the optimal shape design of capillary barriers in microfluidic biochips.

1 Introduction

The optimization of structures and systems has a long history that can be traced back to the work of Bernoulli, Euler, Lagrange, and Saint-Venant. It became its own discipline during the second half of the last century when the rapid progress in electronic data processing required the development and implementation of highly efficient and robust algorithmic optimization tools. Nowadays, shape optimization is an indispensable tool for many design issues in aero- and fluid dynamics, electromagnetics, and structural mechanics. The spectrum of analytical and numerical methods is well documented by numerous monographs on the subject that have been published during the past twenty-five years (cf., e.g., Allaire [1], Bendsøe [4], Delfour and Zolesio [7], Haslinger and Mäkinen [17], Mohammadi and Pironneau [23], Sokolowski and Zolesio [26]).

In this paper, we will focus on an all-at-once approach by means of primal-dual interior-point methods. Using classical barrier functions, this results in a parameter

M. Bercovier et al. (eds.), *Domain Decomposition Methods in Science and Engineering XVIII*, Lecture Notes in Computational Science and Engineering,
DOI: 10.1007/978-3-642-02677-5,

dependent nonlinear system which is solved by a multilevel predictor-corrector continuation strategy with an adaptive choice of the continuation steplength along the central path. The predictor relies on a nested iteration type continuation, whereas the corrector features an inexact Newton method involving transforming null space iterations as inner iterations. As a multiscale multiphysics application, we consider the optimal design of capillary barriers in surface acoustic wave driven microfluidic biochips used for hybridization and sequencing in genomics.

2 Optimal Design of Processes and Systems

A typical shape optimization problem associated with a time-independent PDE or a system thereof as the underlying state equation amounts to the minimization of a shape functional J over bounded domains Ω in Euclidean space $\mathbb{R}^d$. The state function u is assumed to satisfy a boundary value problem as described by means of a partial differential operator L, and there may be further equality and/or inequality constraints on the domain described by some function h.

$$\inf_{\Omega} J(u,\Omega), \quad J(u,\Omega) := \int_{\Omega} j(x,u(x))\,dx, \tag{1a}$$

$$\text{subject to } Lu = f \ \text{ in } \Omega, \quad u = g \ \text{ on } \Gamma, \quad h(\Omega) \geq 0. \tag{1b}$$

The inherent difficulty that the minimization is over a certain class of domains instead of a set of functions in an appropriate function space can be circumvented by the so-called shape calculus as developed by Céa, Delfour, Zolésio and others (cf., e.g., Delfour and Zolesio [7]). Denoting by $J_r(\Omega) := J(u(\Omega),\Omega)$ the reduced functional, the necessary optimality conditions can be stated by means of the shape gradient

$$\nabla J_r(\Omega)[V] = \lim_{t \to 0+} \frac{J_r(\Omega_t(V)) - J_r(\Omega)}{t} = \langle \nabla J_r(\Omega), V \rangle ,$$

defined by means of smooth velocity fields V and a family of transformations of Ω under V such that $\Omega_t(V) = T_t(\Omega), T_t(x) = x(t), x'(t) = V(x(t))$. The shape gradient is a distributional derivative admitting the boundary integral representation

$$\langle \nabla J_r(\Omega), V \rangle = \int_{\Gamma} \langle V, \nu \rangle \left\{ j(x,g) + \frac{\partial p}{\partial \nu} \frac{\partial (g-u)}{\partial \nu} \right\} ds,$$

where p stands for the adjoint state satisfying the adjoint state equation $L^* p = \partial j / \partial u(\cdot, u)$. Sufficient optimality conditions invoke the shape Hessian which can also be given a boundary integral representation admitting an interpretation as a pseudo differential operator of order 1 (cf., e.g., Eppler and Harbrecht [10]). The analytical investigation of shape Hessians and the development and implementation of numerical tools based thereon is subject to intensive ongoing research. The numerical methods developed so far require some smoothness of the domain and suffer from a lack of stability otherwise.

Since interior-point methods essentially rely on second order information, in the sequel we will use a more classical approach based on a parametrization of the domain by a finite number of design variables. The boundary Γ is represented by a composite Bézier curve using a certain number of Bézier control points $\alpha \in \mathbb{R}^m, m \in \mathbb{N}$, which serve as design variables. The equality and/or inequality constraints are expressed by means of the design variables. For the finite element approximation of (1a)–(1b) we choose $\hat{\alpha}$ as a reference design and refer to $\hat{\Omega} := \Omega(\hat{\alpha})$ as the associated reference domain. Then, the actual domain $\Omega(\alpha)$ can be obtained from the reference domain $\hat{\Omega}$ by means of a mapping $\Omega(\alpha) = \Phi(\hat{\Omega};\alpha)$. The advantage of using the reference domain $\hat{\Omega}$ is that finite element approximations can be performed with respect to that fixed domain without being forced to remesh for every new set of the design variables. The finite element discretization of (1a)–(1b) with respect to a simplicial triangulation $\mathcal{T}_h(\Omega)$ of the computational domain Ω leads to a finite dimensional optimization problem

$$\inf_{u_h,\alpha} J_h(u_h,\alpha), \tag{2a}$$

$$\text{subject to } L_h u_h = b_h,\ h(\alpha) \geq 0, \tag{2b}$$

where $u_h \in \mathbb{R}^n$ is the finite element approximation of the state u, $J_h(u_h,\alpha)$ the discretized objective functional and $L_h u_h = b_h$ the algebraic system arising from the finite element discretization of the PDE.

The inequality constrained nonlinear programming problem (2a)–(2b) will be numerically solved by adaptive multilevel path-following primal-dual interior-point methods as described in the following subsections. For ease of notation, in the sequel we will drop the subindex h.

3 Adaptive Multilevel Primal-Dual Interior Point Methods

We couple the inequality constraints in (1b) by logarithmic barrier functions with a barrier parameter $\mu = \frac{1}{\tau} > 0$, $\tau \to \infty$, and the equality constraint by a Lagrange multiplier $\lambda \in \mathbb{R}^n$. This leads to the saddle point problem

$$\inf_{u,\alpha} \sup_{\lambda} \mathcal{L}^{(\tau)}(u,\lambda,\alpha) , \tag{3}$$

where $\mathcal{L}^{(\tau)}$ stands for the Lagrangian

$$\mathcal{L}^{(\tau)}(u,\lambda,\alpha) = B^{(\tau)}(u,\alpha) + \langle \lambda, Lu - b \rangle. \tag{4}$$

Here, $B^{(\tau)}(u,\alpha)$ is the so-called barrier function as given by

$$B^{(\tau)}(u,\alpha) := J(u,\alpha) - \frac{1}{\tau} \ln(h(\alpha)) . \tag{5}$$

and $\langle \cdot,\cdot \rangle$ stands for the Euclidean inner product on $\mathbb{R}^n$ (for details cf., e.g., Wright [34]). The central path $\tau \longmapsto x(\tau) := (u(\tau), \lambda(\tau), \alpha(\tau))^T$ is given as the solution of the nonlinear system

$$F(x(\tau),\tau) = \begin{pmatrix} \mathcal{L}_u^{(\tau)}(u,\lambda,\alpha) \\ \mathcal{L}_\lambda^{(\tau)}(u,\lambda,\alpha) \\ \mathcal{L}_\alpha^{(\tau)}(u,\lambda,\alpha) \end{pmatrix} = 0\,, \tag{6}$$

where the subindices refer to the derivatives of the Lagrangian with respect to the primal, the dual, and the design variables. The choice of the barrier parameter strongly influences the performance of the interior-point method. There are static strategies with the Fiacco-McCormick approach as the most prominent one (cf. Fiacco and McCormick [11]), where the barrier parameter is fixed until an approximate solution has been obtained, and there is a variety of dynamic update strategies (cf. Armand et al. [3], El-Bakry et al. [9], Gay et al. [14], Nocedal et al. [24], Tits et al. [27], Ulbrich et al. [28], Vanderbei and Shanno [29]). Convergence properties of the Fiacco-McCormick approach have been studied in Byrd et al. [5] and Wächter and Biegler [30], whereas a convergence analysis of dynamic update strategies has been addressed in Armand et al. [3], El-Bakry et al. [9], Nocedal et al. [24], Ulbrich et al. [28].

We consider the solution of (6) by an adaptive continuation method based on the affine invariant convergence theory of Newton-type methods (see, e.g., Deuflhard [8], Weiser and Deuflhard [31]).

The adaptive continuation method is a predictor-corrector method with an adaptively determined continuation step size in the predictor and Newton's method as a corrector. It relies on the affine invariant convergence theory of Newton and Newton-type methods and ensures that the iterates stay within a neighborhood (contraction tube) of the central path so that convergence to a local minimum of the original minimization problem can be achieved (cf. Fig. 1).

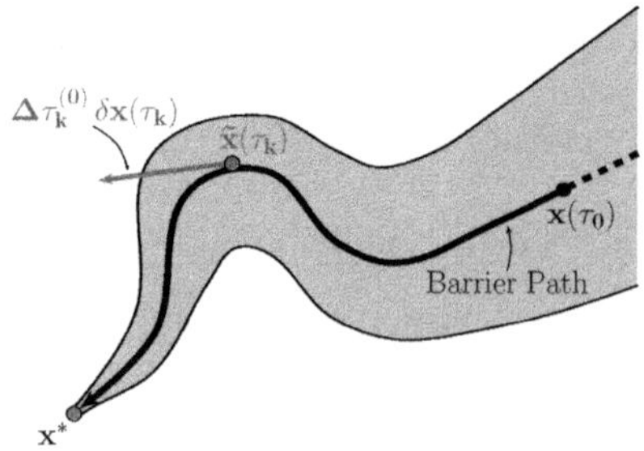

Fig. 1. Predictor step of the adaptive continuation method.

Predictor Step

The predictor step relies on tangent continuation along the trajectory of the Davidenko equation

$$F_x(x(\tau),\tau)x'(\tau) = -F_\tau(x(\tau),\tau) \tag{7}$$

and amounts to the implementation of an explicit Euler step: Given some approximation $\tilde{x}(\tau_k)$ at $\tau_k > 0$, compute $\tilde{x}^{(j_0)}(\tau_{k+1})$, where $\tau_{k+1} = \tau_k + \Delta\tau_k^{(j)}$, according to

$$F_x(\tilde{x}(\tau_k),\tau_k)\delta x(\tau_k) = -F_\tau(\tilde{x}(\tau_k),\tau_k)\ , \tag{8a}$$

$$\tilde{x}^{(j_0)}(\tau_{k+1}) = \tilde{x}(\tau_k) + \Delta\tau_k^{(j)}\delta x(\tau_k)\ , \tag{8b}$$

starting with $j = 0$ ($j \geq 1$ only if required by the correction step (see below)). We use $\Delta\tau_0^{(0)} = \Delta\tau_0$ for some given initial step size $\Delta\tau_0$, whereas for $k \geq 1$ the predicted step size $\Delta\tau_k^{(0)}$ is chosen by

$$\Delta\tau_k^{(0)} := \left(\frac{\|\Delta x^{(j_0)}(\tau_k)\|}{\|\tilde{x}(\tau_k) - \tilde{x}^{(j_0)}(\tau_k)\|} \, \frac{\sqrt{2}-1}{2\Theta(\tau_k)} \right)^{1/2} \Delta\tau_{k-1}\ , \tag{9}$$

where $\Delta\tau_{k-1}$ is the computed continuation step size, $\Delta x^{(j_0)}(\tau_k)$ is the first Newton correction (see below), and $\Theta(\tau_k) < 1$ is the contraction factor associated with a successful previous continuation step.

Corrector Step

As a corrector, we use Newton's method applied to

$$F(x(\tau_{k+1}),\tau_{k+1}) = 0$$

with $\tilde{x}^{(j_0)}(\tau_{k+1})$ from (8b) as a start vector. In particular, for $\ell \geq 0$ (Newton iteration index) and $j_\ell \geq 0$ (j being the steplength correction index) we compute $\Delta\mathbf{x}^{(j_\ell)}(\tau_{k+1})$ according to

$$F_x(\tilde{x}^{(j_\ell)}(\tau_{k+1}),\tau_{k+1})\ \Delta x^{(j_\ell)}(\tau_{k+1}) = -F(\tilde{x}^{(j_\ell)}(\tau_{k+1}),\tau_{k+1}), \tag{10}$$

update $\tilde{x}^{(j_{\ell+1})}(\tau_{k+1}) := \tilde{x}^{(j_\ell)}(\tau_{k+1}) + \Delta x^{(j_\ell)}(\tau_{k+1})$ and compute $\overline{\Delta x}^{(j_\ell)}(\tau_{k+1})$ as the associated simplified Newton correction

$$F_x(\tilde{x}^{(j_\ell)}(\tau_{k+1}),\tau_{k+1})\overline{\Delta x}^{(j_\ell)}(\tau_{k+1}) = -F(\tilde{x}^{(j_\ell)}(\tau_{k+1}) + \Delta x^{(j_\ell)}(\tau_{k+1}),\tau_{k+1}).$$

We monitor convergence of Newton's method by means of

$$\Theta^{(j_\ell)}(\tau_{k+1}) := \|\overline{\Delta x}^{(j_\ell)}(\tau_{k+1})\| / \|\Delta x^{(j_\ell)}(\tau_{k+1})\|\ .$$

In case of successful convergence, we set $\tilde{x}(\tau_{k+1}) := \tilde{x}^{(j_\ell)}(\tau_{k+1})$ with ℓ being the current Newton iteration index, accept the current step size $\Delta\tau_k := \Delta\tau_k^{(j)}$ with current steplength correction index j and proceed with the next continuation step. However, if the monotonicity test

$$\Theta^{(j_\ell)}(\tau_{k+1}) < 1 \tag{11}$$

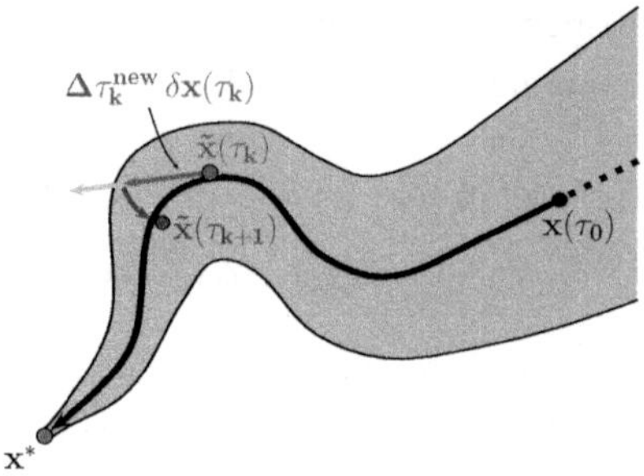

Fig. 2. Correction step of the adaptive continuation method.

fails for some $j_\ell \geq 0$, the predicted steplength $\Delta\tau_k^{(j)}$ has been chosen too large so that the predicted solution $\tilde{\mathbf{x}}^{(j_0)}(\tau_{k+1})$ is not situated within the Kantorovich neighborhood of $x(\tau_{k+1})$, i.e., it is outside the contraction tube around the central path (cf. Fig. 2). The corrector step provides a correction of the steplength for the tangent direction $\delta x(\tau_k)$ such that the new iterate stays within the contraction tube. To do so, the continuation step from (8b) has to be repeated with the reduced step size

$$\begin{aligned} \Delta\tau_k^{(j+1)} &:= \left(\frac{\sqrt{2}-1}{g(\Theta^{(j_\ell)})}\right)^{1/2} \Delta\tau_k^{(j)}\,, \\ g(\Theta) &:= \sqrt{\Theta+1}-1 \end{aligned} \tag{12}$$

until we either achieve convergence or for some prespecified lower bound $\Delta\tau_{min}$ observe

$$\Delta\tau_k^{(j+1)} < \Delta\tau_{\min}\,.$$

In the latter case, we stop the algorithm and report convergence failure.

The Newton steps are realized by an inexact Newton method featuring right-transforming iterations (cf., e.g., Hoppe et al. [18], Hoppe and Petrova [20]). For a discussion of the impact of the inexactness on the pathfollowing we refer to Weiser and Deuflhard [31], sec. 3.2. The derivatives occurring in the KKT conditions and the Hessians are computed by automatic differentiation (cf., e.g., Griewank [15]).

We perform the predictor-corrector scheme in a multilevel framework with respect to a hierarchy of discretizations. We describe the multilevel approach in case of a two-level scheme with the levels $\ell-1$ and ℓ (cf. Fig. 3). Since in multigrid continuation methods it is advantageous to use smaller continuation steps on the coarser grids (cf., e.g., Hackbusch [16], Hoppe and Mittelmann [19]), the prediction is done by nested iteration in such a way that some adaptive continuation steps are performed on the coarser level $\ell-1$ before a predicted value is computed on the finer level ℓ. The corrector is a Newton multigrid method incorporating a two-level PDE solver featuring appropriate smoothers. The iterates are checked for acceptance by the level ℓ monotonicity test. In some more detail, we illustrate the two-level scheme in case of two continuation steps on level $\ell-1$. We assume that approximations $x^{\ell-1}(\tau_k)$ and $x^\ell(\tau_k)$ are available for some continuation parameter τ_k. Firstly, we perform 2 continuation steps with an adaptive choice of the continuation steplengths. Secondly, we use the the level $\ell-1$ approximations $x^{\ell-1}(\tau_k)$ and $x^{\ell-1}(\tau_{k+2})$ as well as the level 1 approximation $x^{\ell}(\tau_k)$ to obtain a level 1 prediction at τ_{k+2}. This approximation is then corrected by the two-level Newton multigrid scheme and checked for acceptance by the level ℓ monotonicity test. In the general case of more than 2 levels, the multilevel predictor-corrector continuation method consists of a recursive application of the two-level scheme.

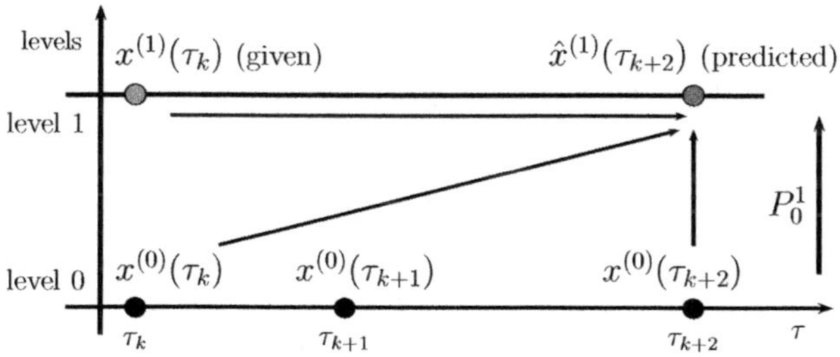

Fig. 3. Two-level predictor-corrector scheme

4 Numerical Results

Microfluidic biochips are used in pharmaceutical, medical and forensic applications as well as in academic research and development for high throughput screening, genotyping and sequencing by hybridization in genomics, protein profiling in proteomics, and cytometry in cell analysis (cf., e.g., Pollard and Castrodale [25]). Recent nanotechnological devices are surface acoustic wave driven biochips with integrated fluidics on top of the chip consisting of a lithographically produced network of channels and reservoirs (see Fig. 4 (left)). The core of the technology are nanopumps featuring surface acoustic waves generated by electric pulses of high frequency. These waves propagate like a miniaturized earthquake, enter the fluid filled channels and thus cause a flow which transports the DNA or protein containing liquid along the network to a reservoir where the chemical analysis is performed (see, e.g., Wixforth et al. [32, 33]. Between the channels and the reservoirs are capillary barriers (cf. Fig. 4 (right)) which have to be designed in such a way that a precise filling of the reservoirs is guaranteed.

Mathematical models for SAW biochips are based on the linearized equations of piezoelectricity in $Q_1 := (0,T_1) \times \Omega_1$

Fig. 4. Microfluidic biochip (left) and capillary barrier (right)

$$\rho_1 \frac{\partial^2 u_i}{\partial t^2} - \frac{\partial}{\partial x_j} c_{ijkl} \frac{\partial u_k}{\partial x_l} - \frac{\partial}{\partial x_j} e_{kij} \frac{\partial \Phi}{\partial x_k} = 0, \tag{13a}$$

$$\frac{\partial}{\partial x_j} e_{jkl} \frac{\partial u_k}{\partial x_l} - \frac{\partial}{\partial x_j} \varepsilon_{jk} \frac{\partial \Phi}{\partial x_k} = 0 \tag{13b}$$

with appropriate initial conditions at $t = 0$ and boundary conditions on $\Gamma_1 := \partial\Omega_1$. Here, ρ_1 and $\mathbf{u} = (u_1, u_2, u_3)^T$ denote the density of the piezoelectric material and the mechanical displacement vector. Moreover, $\varepsilon = (\varepsilon_{ij})$ stands for the permittivity tensor and Φ for the electric potential. The tensors $\mathbf{c} = (c_{ijkl})$ and $\mathbf{e} = (e_{ikl})$ refer to the forth order elasticity tensor and third-order piezoelectric tensor, respectively.

The modeling of the micro-fluidic flow is based on the compressible Navier-Stokes equations in $Q_2 := (0, T_2) \times \Omega_2$

$$\rho_2 \left(\frac{\partial \mathbf{v}}{\partial t} + (\mathbf{v} \cdot \nabla)\mathbf{v} \right) = -\nabla p + \eta \Delta \mathbf{v} + \left(\zeta + \frac{\eta}{3} \right) \nabla(\nabla \cdot \mathbf{v}), \tag{14a}$$

$$\frac{\partial \rho_2}{\partial t} + \nabla \cdot (\rho_2 \mathbf{v}) = 0, \tag{14b}$$

$$\mathbf{v}(x + \mathbf{u}(x,t), t) = \frac{\partial \mathbf{u}}{\partial t}(x,t) \quad \text{on } (0, T_2) \times \Gamma_2 \tag{14c}$$

with suitable initial conditions at $t = 0$. Here, $\rho_2, \mathbf{v} = (v_1, v_2, v_3)^T$ and p are the density of the fluid, the velocity, and the pressure. η and ζ refer to the shear and the bulk viscosity. The boundary conditions include the time derivative $\partial \mathbf{u}/\partial t$ of the displacement of the walls $\Gamma_2 = \partial\Omega_2$ of the microchannels caused by the surface acoustic waves. The induced fluid flow involves extremely different time scales. The damping of the jets created by the SAWs happens on a time scale of nanoseconds, whereas the resulting acoustic streaming reaches an equilibrium on a time scale of milliseconds. We perform a separation of the time-scales by homogenization using an expansion $\mathbf{v} = \mathbf{v}_0 + \varepsilon \mathbf{v}' + \varepsilon^2 \mathbf{v}'' + O(\varepsilon^3)$ of the velocity $\mathbf{v}$ in a scale parameter $\varepsilon > 0$ representing the maximal displacement of the walls and analogous expansions of the pressure p and the density ρ_2. We set $\mathbf{v}_1 := \varepsilon \mathbf{v}', \mathbf{v}_2 := \varepsilon^2 \mathbf{v}''$ and define $p_i, \rho_{2,i}, 1 \le i \le 2$, analogously. Time-averaging the second order (in ε) system according to $\langle w \rangle := T^{-1} \int_{t_0}^{t_0+T} w\,dt, T := 2\pi/\omega$, we arrive at the following Stokes equations in Ω_2

$$-\eta \Delta \mathbf{v}_2 - \left(\zeta + \frac{\eta}{3}\right) \nabla(\nabla \cdot \mathbf{v}_2) + \nabla p_2 = \left\langle -\rho_{2,1} \frac{\partial \mathbf{v}_1}{\partial t} - \rho_{2,0}(\nabla \mathbf{v}_1)\mathbf{v}_1 \right\rangle , \tag{15a}$$

$$\rho_{2,0} \nabla \cdot \mathbf{v}_2 = \langle -\nabla \cdot (\rho_{2,1} \mathbf{v}_1) \rangle , \tag{15b}$$

$$\mathbf{v}_2 = -\langle (\nabla \mathbf{v}_1)\mathbf{u} \rangle \quad \text{on } \Gamma_2 . \tag{15c}$$

which describe the stationary flow pattern, called acoustic streaming, resulting after the relaxation of the high frequency surface acoustic waves (for further details we refer to Gantner et al. [13], Köster [22]).

Table 1. History of the adaptive multilevel predictor-corrector strategy (Capillary barriers, 4 Levels)

level	k	τ	$\Delta\tau$	ΔJ
1	0	2.0E+02		
				2.83E+00
	1	6.3E+02	4.3E+02	1.87E-05
	2	1.1E+03	4.9E+02	3.40E-06
	3	1.6E+03	5.1E+02	1.09E-06
	4	2.3E+03	6.8E+02	5.70E-07
	5	3.5E+03	1.1E+03	3.63E-07
	6	5.3E+03	1.9E+03	1.99E-07
	7	8.8E+03	3.5E+03	1.02E-07
	8	1.6E+04	7.3E+03	4.50E-08
2	2	1.1E+03	9.2E+02	
	4	2.3E+03	1.2E+03	
	6	5.3E+03	3.0E+03	
	8	1.6E+04	1.1E+04	
3	4	2.3E+03	2.1E+03	
	8	1.6E+04	1.4E+04	
4	8	1.6E+04	1.6E+04	

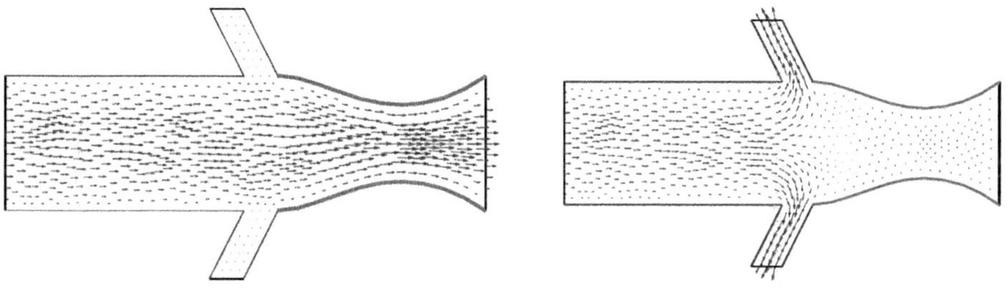

Fig. 5. Optimally designed capillary barrier: Velocity profile in the flow mode (left) and in the stopping mode (right)

We have considered the optimal design of a capillary barrier for a domain consisting of part of a microchannel close to a reservoir with two passive outlet valves to allow for an outflow in case of the stopping mode of the barrier (cf. Fig. 5). The objective functional J has been chosen of tracking type according to

$$J(\mathbf{v}_2, p_2, \alpha) := \frac{1}{2} \int_{\Omega(\alpha)} |\mathbf{v}_2 - \mathbf{v}_2^d|^2 \, dx + \frac{1}{2} \int_{\Omega(\alpha)} |p_2 - p_2^d|^2 \, dx$$

subject to the Stokes system (15a)-(15c) with Signorini type boundary conditions at the junction between the microchannel and the reservoir. We have used $m = 16$ Bézier control points of a Bézier curve representation of the barrier as design variables subject to bilateral constraints. Table 1 contains the history of the multilevel interior-point method described in the previous section in case of four levels $1 \leq \ell \leq 4$ with 2362 degrees of freedom (DOFs) on the coarsest grid (level 1) and 141634 DOFs on the finest grid (level 4). The number k indicates the continuation steps, τ_k and $\Delta\tau_k := \tau_k - \tau_{k-1}$ refer to the inverse of the barrier parameter μ_k and the increment in τ_k, and ΔJ_k is the difference between the corresponding values of the objective functional. We have performed two continuation steps on a coarser grid before proceeding by nested iteration to the next finer grid, and we have used $|\Delta J_k| < TOL$ with $TOL := 1.0E-07$ as a termination criterion for the continuation process. Fig. 5 displays the optimal design of the barrier and the associated velocity profiles in the flow mode (fluid flow into the reservoir) and in the stopping mode (backflow). For further results and a comparison with other continuation methods and update strategies of the barrier parameter we refer to Antil, Hoppe and Linsenmann [2].

Acknowledgement. The second and third authors acknowledge support by the German National Science Foundation within the Priority program SPP 1253 'Optimization with Partial Differential Equations'. The work of the authors has been further supported by the NSF under Grant-No. DMS-0511611, DMS-0707602.

References

[1] Allaire, G.: *Shape Optimization by the Homogenization Method.* Springer, Berlin-Heidelberg-New York, 2002.

[2] Antil, H., Hoppe, R.H.W., Linsenmann, C.: *Optimal design of stationary flow problems by path-following interior-point methods.* to appear in Control Cybernet., 2009.

[3] Armand, P., Benoist, J., Orban, D.: *Dynamic updates of the barrier parameter in primal-dual methods for nonlinear programming.* Comput. Optim. Appl. 41, 1–25, 2008.

[4] Bendsøe, M.P.: *Optimization of Structural Topology, Shape, and Material.* Springer, Berlin-Heidelberg-New York, 1995.

[5] Byrd, R.H., Gilbert, J.C., Nocedal, J.: *A trust region method based on interior point techniques for nonlinear programming.* Math. Program. 89, 149–185, 2000.

[6] Byrd, R.H., Hribar, M.E., Nocedal, J.: *An interior point algorithm for large scale nonlinear programming.* SIAM J. Optim. 9, 877–900, 1999.

[7] Delfour, M.C., Zolesio, J.P.: *Shapes and Geometries: Analysis, Differential Calculus and Optimization.* SIAM, Philadelphia, 2001.

[8] Deuflhard, P.: *Newton Methods for Nonlinear Problems. Affine Invariance and Adaptive Algorithms.* Springer, Berlin-Heidelberg-New York, 2004.

[9] El-Bakry, A.S., Tapia, R.A., Tsuchiya, T., Zhang, Y.: *On the formulation and theory of the Newton interior-point method for nonlinear programming.* J. Optim. Theory Appl. 89, 507–541, 1996.

[10] Eppler, K., Harbrecht, H.: *Second order shape optimization using wavelet BEM.* Optim. Methods Softw. 21, 135–153, 2006.

[11] Fiacco, A.V., McCormick, G.P.: *Nonlinear Programming: Sequential Unconstrained Minimization Techniques.* SIAM, Philadelphia, 1990.

[12] Forsgren, A., Gill, P.E., Wright, M.H.: *Interior methods for nonlinear optimization.* SIAM Rev. 44, 522–597, 2002.

[13] Gantner, A., Hoppe, R.H.W., Köster, D., Siebert, K.G., Wixforth, A.: *Numerical simulation of piezoelectrically agitated surface acoustic waves on microfluidic biochips.* Comput. Vis. Sci. 10, 145–161, 2007.

[14] Gay, M.D., Overton, M.L., Wright, M.H.: *A primal-dual interior method for nonconvex nonlinear programming.* In: Advances in Nonlinear Programming (Y. Yuan; ed.), pp. 31–56, Kluwer, Dordrecht, 1998.

[15] Griewank, A.: *Evaluating Derivatives, Principles and Techniques of Automatic Differentiation.* SIAM, Phildelphia, 2000.

[16] Hackbusch, W.: *Multi-grid solution of continuation problems.* In: Iterative Solution of Nonlinear Systems (T. Meis ans W. Törnig; eds.), Lect. Notes in Math. **953**, pp. 20–45, Springer, Berlin-Heidelberg-New York, 1982.

[17] Haslinger, J., Mäkinen, R.A.E.: *Introduction to Shape Optimization: Theory, Approximation, and Computation.* SIAM, Philadelphia, 2004.

[18] Hoppe, R.H.W., Linsenmann, C., Petrova, S.I.: *Primal-dual Newton methods in structural optimization.* Comput. Vis. Sci. 9, 71–87, 2006.

[19] Hoppe, R.H.W., Mittelmann, H.D.: *A multi-grid continuation strategy for parameter-dependent variational inequalities.* J. Comput. Appl. Math. 26, 35–46, 1989.

[20] Hoppe, R.H.W., Petrova, S.I.: *Primal-dual Newton interior-point methods in shape and topology optimization.* Numer. Linear Algebra Appl. 11, 413–429, 2004.

[21] Hoppe, R.H.W., Petrova, S.I., Schulz, V.: *A primal-dual Newton-type interior-point method for topology optimization.* J. Optim. Theory Appl. 114, 545–571, 2002.

[22] Köster, D.: *Numerical simulation of acoustic streaming on SAW-driven biochips.* SIAM J. Sci. Comput. 29, 2352–2380, 2007.

[23] Mohammadi, B., Pironneau, O.: *Applied Shape Optimization for Fluids.* Oxford University Press, Oxford, 2001.
[24] Nocedal, J., Wächter, A., Waltz, R.A.: *Adaptive barrier update strategies for nonlinear interior methods.* Research Report RC 23563, IBM T.J. Watson Research Center, Yorktown, 2006.
[25] Pollard, J., Castrodale, B.: *Outlook for DNA microarrays: emerging applications and insights on optimizing microarray studies.* Report. Cambridge Health Institute, Cambridge, 2003.
[26] Sokolowski, J., Zolesio, J.P.: *Introduction to Shape Optimization.* Springer, Berlin-Heidelberg-New York, 1992.
[27] Tits, A.L., Wächter, A., Bakhtiari, S., Urban, T.J., Lawrence, C.T.: *A primal-dual interior-point method for nonlinear programming with strong global and local convergence properties.* SIAM J. Optim. 14, 173–199, 2003.
[28] Ulbrich, M., Ulbrich, S., Vicente, L.: *A globally convergent primal-dual interior point filter method for nonconvex nonlinear programming.* Math. Program. 100, 379–410, 2004.
[29] Vanderbei, R.J., Shanno, D.F.: *An interior point algorithm for nonconvex nonlinear programming.* Comput. Optim. Appl. 13, 231–252, 1999.
[30] Wächter, A., Biegler, L.T.: *Line search filter methods for nonlinear programming: motivation and global convergence.* SIAM J. Optim. 16, 1–31, 2005.
[31] Weiser, M., Deuflhard, P.: *Inexact central path following algorithms for optimal control problems.* SIAM J. Control Optim. 46, 792–815, 2007.
[32] Wixforth, A., Scriba, J., Gauer, G.: *Flatland fluidics.* Mst. News 5, 42–43, 2002.
[33] Wixforth, A., Strobl, C., Gauer, C., Toegl, A., Scriba, J., Guttenberg, T.: *Acoustic manipulation of small droplets.* Anal. Bioanal. Chem. 379, 982–991, 2004.
[34] Wright, M.H.: *Interior methods for constrained optimization.* Acta Numer. 1, 341–407,1992.

Numerical Homogeneisation Technique with Domain Decomposition Based a-posteriori Error Estimates

Patrick Le Tallec

Ecole Polytechnique, 91128 Palaiseau, France patrick.letallec@polytechnique.fr

Summary. The purpose of the present work is to review some basic numerical homogeneisation techniques for the simulation of multiscale materials and to introduce an error control strategy at the local level. This error control uses an a posteriori error estimate built on a local problem coupling different representative volume elements. It introduces a weakly coupled adjoint problem to be solved say by a direct Schur complement method. Mortar element techniques as introduced in domain decomposition techniques are used to couple in a weak and cheap form the different representative elements in the error analysis. The strategy is numerically assessed on a model two dimensional problem.

1 Introduction

In many practical situations, there is a significant separation of scales between the global macroscopic problem and the local heterogeneities governing the response of the constitutive materials. Metals, elastomers, construction materials present a microstructure at micronic or submicronic scales which influence their constitutive laws. Dynamic contact problems, cable matrix interactions inside composite materials are similar situations where the physical response of the system occurs at very small scales (millimiter or less) compared to the overall dimensions of the global structure. These scales are out of reach by a direct simulation of the global problem, even when using sophisticated domain decomposition algorithms. Homogeneisation techniques at large propose a general methodology to handle such situations.

The methodology is quite simple [7, 8, 12, 15]. It is based on the notion of representative volume elements (RVE). Each volume element is a microscopic sample of the system under study. Its size H is very small compared to the macroscopic characteristic length L of the global problem, and very large compared to the size ε of the heterogeneities $L >> H >> \varepsilon$. Each sample is solved locally at a microscopic scale taking as boundary conditions uniform displacement data deduced from the pointwise value of the strain tensor observed at macroscopic scale. Once computed at a local scale, the averaged answer defines the macroscopic constitutive response of the material. In such a construction, the ratio between H and ε serves two purposes: a

M. Bercovier et al. (eds.), *Domain Decomposition Methods in Science and Engineering XVIII*, Lecture Notes in Computational Science and Engineering,
DOI: 10.1007/978-3-642-02677-5, © Springer-Verlag Berlin Heidelberg 2009

large ratio reduces the effect of the artificial boundary conditions to be used on the RVE [10], and leaves room to include a statistically representative sample of the local heterogeneities. On the other hand, a smaller ratio reduces the cost of the solution of the local problems. Hence the idea of developing an a posteriori error estimate strategy to assess the choices of the sample size H and of the artificial boundary conditions used at local scale.

The purpose of the present work is therefore to review some basic numerical homogeneisation techniques for the simulation of nonlinear viscoelastic multiscale materials and to introduce a domain decomposition based error control strategy for the local problems. A model mechanical problem is introduced in Section 2. Numerical homogeneisation is reviewed in Section 3. Section 4 is devoted to the introduction of an a posteriori error estimate built on a reference extended local problem coupling different representative volume elements. This introduces a weakly coupled adjoint problem to be transformed into a small interface problem to be solved say by a direct Schur complement method. Mortar element techniques are used at this level to weakly couple the different representative elements in the error analysis. The strategy is assessed on a two dimensional problem in Section 5.

2 Mechanical Problem

Let us consider the quasi static evolution of a given macroscopic solid or structure which occupies the domain Ω at rest, which is fixed on a part $\partial\Omega_\xi$ of its boundary, and which is subjected to a known distribution $\underline{F}$ of specific loads and $\underline{T}$ of surface loads. The problem to solve combines the balance of momentum in reference configuration, a time dependent viscoelastic constitutive law and a time differential equation describing the evolution of the internal variables $\underline{\underline{\varepsilon}}^e_\tau$, each material relaxation time τ corresponding to one specific internal variable $\underline{\underline{\varepsilon}}^e_\tau$. After time discretisation, say by a uniformly stable time implicit Euler scheme, this problem reduces to a sequence of equilibrium problems to be solved at different times t_{n+1}:

Find $\underline{x} - \underline{x}_d \in \mathbf{V}$ and $\underline{\underline{\varepsilon}}^e_\tau$ solution of

$$\int_\Omega \underline{\underline{\sigma}} : \frac{\partial \underline{\hat{U}}^t}{\partial \underline{Y}}\, d\Omega = \int_\Omega \rho \underline{F}(\underline{X}) \cdot \underline{\hat{U}}\, d\Omega + \int_{\partial\Omega_T} \underline{T}(\underline{X}) \cdot \underline{\hat{U}}\, da \quad \forall \underline{\hat{U}} \in \mathbf{V}, \tag{1}$$

$$\underline{\underline{\varepsilon}} = \frac{1}{2}\left(\frac{\partial(\underline{x}-\underline{X})}{\partial \underline{X}} + \frac{\partial(\underline{x}-\underline{X})^t}{\partial \underline{X}} \right), \tag{2}$$

$$\underline{\underline{\sigma}} = \rho \frac{\partial \psi_\infty}{\partial \underline{\underline{\varepsilon}}}(\underline{\underline{\varepsilon}}) + \sum_\tau \underline{\underline{\bar{\sigma}}}_\tau, \qquad \underline{\underline{\bar{\sigma}}}_\tau = \rho \frac{\partial \psi_\tau}{\partial \underline{\underline{\varepsilon}}^e_\tau}(\underline{\underline{\varepsilon}}^e_\tau), \tag{3}$$

$$\underline{\underline{\varepsilon}}^e_\tau = \underline{\underline{\varepsilon}}^{en}_\tau + \underline{\underline{\varepsilon}} - \underline{\underline{\varepsilon}}^n - \Delta t \underline{\underline{\phi}}^{-1}_\tau (\underline{\underline{\bar{\sigma}}}_\tau) \quad \forall \tau. \tag{4}$$

In the above expression, $\mathbf{V}$ denotes the space of kinematically admissible test functions, ψ_∞ and ψ_τ are given free energy potentials characterizing the reversible parts of the stresses, $\underline{\underline{\phi}}^{-1}_\tau$ is a given dissipation function, $\underline{\underline{\varepsilon}}^n$ and $\underline{\underline{\varepsilon}}^{en}_\tau$ are the strain and internal variables values at previous time step. After elimination of the viscoelastic stress $\underline{\underline{\bar{\sigma}}}_\tau$ and linearisation, this problem reduces to a standard elasticity problem

$$\int_\Omega \frac{1}{2}\left(\frac{\partial \delta \underline{x}}{\partial \underline{X}} + \frac{\partial \delta \underline{x}^t}{\partial \underline{X}}\right) : \underline{\underline{\underline{\underline{C}}}} : \frac{\partial \underline{\hat{U}}^t}{\partial \underline{X}} \, d\Omega = R(\underline{\hat{U}}) \quad \forall \underline{\hat{U}} \in \mathbf{V}$$

with branch averaged elasticity tensor

$$\underline{\underline{\underline{\underline{C}}}} = \rho \frac{\partial^2 \psi_\infty}{\partial \underline{\underline{\varepsilon}}^2} + \sum_\tau \left(\frac{1}{\rho}\left(\frac{\partial^2 \psi_\tau}{\partial \underline{\underline{\varepsilon}}_\tau^{e\,2}}\right)^{-1} + \Delta t \frac{\partial \phi_\tau^{-1}}{\partial \underline{\underline{\bar{\sigma}}}_\tau}\right)^{-1}.$$

3 Numerical Homogeneisation

In the original problem, the Cauchy stress $\underline{\underline{\sigma}}$ oscillate rapidly in space at scale ε because the coefficients inside the free energy do so. In theory, one would need to solve this problem at the space scale ε over the whole domain Ω of size L, which is completely out of reach. To overcome this problem, homogeneisation techniques introduce an averaging scale H with $L >> H >> \varepsilon$ and construct around each macroscopic point $\underline{X}$ one sample or a collection of samples $\Omega_H(\underline{X})$ of the material, of size H. Space averages on the local RVE (Representative Volume Elements) $\Omega_H(\underline{X})$ with respect to the local space variable $Y \in \Omega_H(\underline{X})$ will be denoted by $\langle f\rangle_{\Omega_H(\underline{X})} := \frac{1}{|\Omega_H(\underline{X})|}\int_{\Omega_H(\underline{X})} f(\underline{Y})d\Omega_H$. Using smooth test functions $\underline{\hat{U}}$ such that $\frac{\partial \hat{U}}{\partial \underline{X}} \approx \langle \frac{\partial \hat{U}}{\partial \underline{Y}}\rangle_{\Omega_H(\underline{X})}$, the power devoloped by the internal forces in the virtual motion $\underline{\hat{U}}$ can be reduced to

$$\int_\Omega \underline{\underline{\sigma}} : \frac{\partial \underline{\hat{U}}^t}{\partial \underline{X}} \, d\Omega \approx \int_\Omega \langle \underline{\underline{\sigma}}\rangle_{\Omega_H(\underline{X})} : \frac{\partial \underline{\hat{U}}^t}{\partial \underline{X}} \, d\Omega.$$

Compared to the original expression, we may assume that the variations in space of the average tensor $\langle\underline{\underline{\sigma}}\rangle_{\Omega_H(\underline{X})}$ will be very slow, meaning that the second integral can be approximated at macroscopic level by a Gaussian integration rule with few integration points X_G. As explained in [5], the challenge is to identify the averaged stress $\langle\underline{\underline{\sigma}}\rangle_{\Omega_H(\underline{X})}$ at each macroscopic point X. The stress tensor must summarize the local heterogeneous constitutive response of the material as function of the strain field inherited from the macroscopic displacement field. The idea [7, 8, 11] is then to solve the original problem on each local domains Ω_H with imposed strain average

$$\lim_{\underline{Y} \to \partial\Omega_H(\underline{X})} \underline{x}(\underline{Y}) - \underline{Y} - \langle\underline{\underline{\varepsilon}}\rangle_{\Omega_H(\underline{X}_G)} \cdot \underline{Y} = \underline{0}, \tag{5}$$

using Dirichlet or periodic boundary conditions to impose this macroscopic strain field. The choice of boundary conditions does not affect the asymptotic limit of the solution as $H/\varepsilon \to \infty$, but may affect the size of the error for bounded ratios H/ε [10]. In that respect, periodic boundary conditions are usually found to be less intrusive.

The local problem defines a H homogeneized constitutive law

$$\langle \underline{\underline{\sigma}} \rangle (\langle \underline{\underline{\varepsilon}} \rangle_{\Omega_H(\underline{X})}) = \left\langle \rho \frac{\partial \psi_\infty}{\partial \underline{\underline{\varepsilon}}} (\underline{\underline{\varepsilon}}_Y) + \sum_\tau \rho \frac{\partial \psi_\tau}{\partial \underline{\underline{\varepsilon}}^e_\tau} (\underline{\underline{\varepsilon}}^e_\tau) \right\rangle_{\Omega_H(\underline{X})} \tag{6}$$

which may directly be used in a continuous writing of the problem on the whole domain Ω. This leads to a regularized macroscopic problem with unknown $\underline{x}$, which hopefully is the right limit of our original problem when the size of the heterogeneities go to zero and which writes

$$\int_\Omega \langle \underline{\underline{\sigma}} \rangle (\langle \underline{\underline{\varepsilon}}(\underline{x}) \rangle_{\Omega_H(\underline{X})}) : \frac{\partial \underline{\hat{U}}}{\partial \underline{X}}^t \, d\Omega = \int_\Omega \rho \underline{F}(\underline{X}) \cdot \underline{\hat{U}} \, d\Omega + \int_{\partial \Omega_T} \underline{T}(\underline{X}) \cdot \underline{\hat{U}} \, da \quad \forall \underline{\hat{U}} \in \mathbf{V}. \tag{7}$$

The two scales homogeneized formulation of our original problem is obtained by simultaneously writing the global equilibrium problem (7) and the local evolution problems. The downscale coupling comes from the boundary condition (5) used in the local problem which is function of the global solution. The upscale coupling occurs through the averaged constitutive law (6).

We could see in the above numerical homogeneisation a nonlinear domain decomposition technique, with representative volume elements playing the roles of subdomains, and where the restriction and extension operators would be simple averages of strains and stresses respectively. One could also view the global problem as the reduction of the problem to a coarse space built with functions whose restriction on each representative element is linear. The difference is that the decomposition of the original problem as done in the numerical homogeneisation technique does not really build an additive decomposition of our original multiscale problems because of the simplified nonconforming boundary conditions applied to the small scale solutions and because the local domains do not necessarily build a complete partition of the full global domain. Therefore, homogeneisation techniques are inexact in nature, and can only be approximation of the real solutions at the limit of large ratios H/ε and L/H.

4 Error Control

4.1 Motivation and Reference Local Solution

The above methodology is very general. It is arbitrary with respect to the choice of the ratio between the size H of the representative volume element Ω_H and the size ε of the heterogeneities, the choice of the boundary conditions to be imposed on the local problem to be solved on each RVE and the construction of the local geometry and material coefficients inside the RVE.

The theoretical answer is to use as large RVE as possible, the theory proving the asymptotic convergence of the method at the limit $H/\varepsilon \to \infty$ [1]. But such a methodology has a cost, which is the solution of the local problems on the different domains Ω_{mH}. Since calculating over large RVE is costly, compromises must be found using smaller samples and improved boundary conditions. The validity of the resulting approach must then be checked by a posteriori error estimates [16].

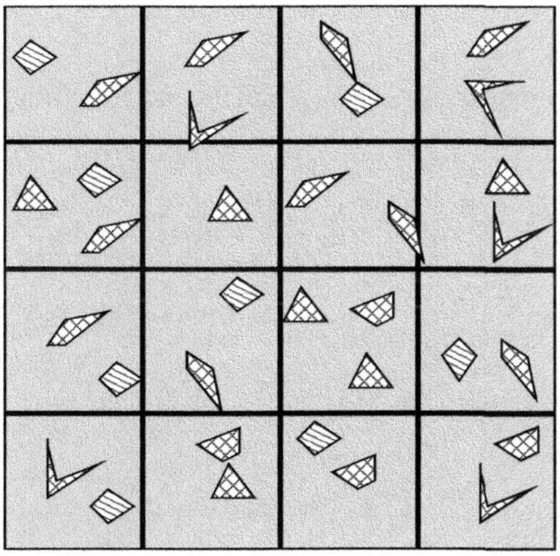

Fig. 1. The reference geometry at local scale, and its decomposition into subcells.

Error estimates must first define a local reference problem. For this purpose, we assume that we can specify a local reference problem at a sufficiently large scale $\bar{H} = NH$, with geometry $\overline{\Omega}_{\bar{H}}$, imposed macroscopic strain, and an adequate choice of boundary conditions (typically periodic boundary conditions in space). This reference geometry is partitioned into subcells (Fig. 1)

$$\overline{\Omega}_{\bar{H}} = \cup_{K=1}^{N^3} \overline{\Omega}_{kH},$$

on which we can introduce a hierarchy of local numerical solutions. The coarsest is the solution of the local problem on a single subcell $\overline{\Omega}_{0H}$ with imposed averaged strain. The second level solves the local problem on each subcell $\overline{\Omega}_{kH}, \forall k = 1, N^3$, with say periodic boundary conditions. This constructs a two scale local solution $(\underline{x}_H, \underline{\underline{\sigma}}_H)$ by juxtaposition of the local fields $\underline{x}_H|_{\Omega_{kH}} = \underline{x}_{kH}$, $\underline{\underline{\sigma}}_H|_{\Omega_{kH}} = \underline{\underline{\sigma}}_{kH}$ and an empirical stress average (and associated variance)

$$\langle \underline{\underline{\sigma}}(\underline{\underline{\varepsilon}}) \rangle = \frac{1}{N^3} \sum_k \langle \underline{\underline{\sigma}}(\underline{\underline{\varepsilon}}) \rangle_{\Omega_{kH}}.$$

The third level would be to compute the full solution $(\underline{x}, \underline{\underline{\sigma}})$ of the nonlinear problem on the large local domain $\overline{\Omega}_{\bar{H}}$.

The problem is then to estimate the distance $(\delta \underline{x}, \delta \underline{\underline{\sigma}})$ between the two scales local solution $(x_H, \underline{\underline{\sigma}}_H)$ and the full local solution $(\underline{x}, \underline{\underline{\sigma}})_{\Omega_{\bar{H}}}$ without computing the full solution. Since the proposed two scales solution may be discontinuous across interdomain boundaries, we must first propose a framework which handles discontinuous fields. This can be achieved by using mortar techniques as in [3, 14] which introduce a finite element notion of weak continuity by constructing local interface finite elements $\mathbf{M}_{kl} = \mathbf{M}_{lk}$ on each interface $\partial\Omega_{kH} \cap \partial\Omega_{lH}$ between neighboring subdomains Ω_{kH} and $\partial\Omega_{lH}$ and by weakly imposing the interface continuity requirement in $\mathbf{M}_{kl}$. In this framework, the full local problem reduces to the following sequence of coupled problems:

Find the displacement $(\underline{x}_k)_k \in \Pi_k \mathbf{V}_{kH}$ and the interface tractions $(\underline{\lambda}_{kl})_{kl} \in \Pi_{kl}\mathbf{M}_{kl}$ such that

$$\int_{\Omega_{kH}} \left(\rho \frac{\partial \psi_\infty}{\partial \underline{\underline{\varepsilon}}}(\underline{\underline{\varepsilon}}_Y) + \sum_\tau \rho \frac{\partial \psi_\tau}{\partial \underline{\underline{\varepsilon}}^e_\tau}(\underline{\underline{\varepsilon}}^e_\tau(\underline{Y}))\right) : \frac{\partial \underline{\hat{U}}}{\partial \underline{Y}}^t d\Omega_H$$
$$+ \sum_l \int_{\partial\Omega_{kH}\cap\partial\Omega_{lH}} \underline{\lambda}_{kl} \cdot \underline{\hat{U}}\, da = 0 \quad \forall \underline{\hat{U}} \in \mathbf{V}_{kH},\ \forall k, \tag{8}$$

$$\underline{\underline{\varepsilon}}_Y(\underline{Y}) = \frac{1}{2}\left(\frac{\partial(\underline{x}_k - \underline{Y})}{\partial \underline{Y}} + \frac{\partial(\underline{x}_k - \underline{Y})}{\partial \underline{Y}}^t\right)(\underline{Y}), \tag{9}$$

$$\underline{\underline{\varepsilon}}^e_\tau = \underline{\underline{\varepsilon}}^{en}_\tau + \underline{\underline{\varepsilon}}_Y - \underline{\underline{\varepsilon}}^n_Y - \Delta t \underline{\underline{\phi}}_\tau^{-1}\left(\rho \frac{\partial \psi_\tau}{\partial \underline{\underline{\varepsilon}}^e_\tau}(\underline{\underline{\varepsilon}}^e_\tau)\right), \tag{10}$$

$$\int_{\partial\Omega_{kH}\cap\partial\Omega_{lH}} \underline{\mu}_{kl} \cdot (\underline{x}_l - \underline{x}_k) da = 0 \quad \forall \underline{\mu}_{kl} \in \mathbf{M}_{kl},\ \forall k < l, \tag{11}$$

$$\underline{\lambda}_{kl} + \underline{\lambda}_{lk} = 0 \quad \forall k < l. \tag{12}$$

In such formulations, the choice of the interface spaces $\mathbf{M}_{kl}$ cannot be completely arbitrary. The initial formulation of [2, 3] uses finite element displacements of degree q without stabilization, and continuous Lagrange multipliers of degree q. Limited modifications of the Lagrange multipliers are necessary on the boundaries of the interfaces. Alternatively, as shown in [13] for second order approximations of the displacements ($q \geq 2$), the formulation of [2, 3] can be used with continuous Lagrange multipliers of degree $q-1$. In order to make the mortar weak continuity constraint diagonal, one can also adopt the dual Lagrange multipliers of Wohlmuth [14]. A last choice advocated in [9] uses finite element displacements of degree q with proper stabilization (bubble additions) together with discontinuous Lagrange multipliers of degree $q-1$ as first developed for three-field matching formulations in [4]. All these choice guarantee an optimal order of convergence between the solution of the above discrete coupled problem and the continuous one [9, 14]. But here we are only interested in error estimates. Thus, for estimating the error, we can use very simple Lagrange multipliers as initially proposed in [6], namely simple polynomials globally defined on each interface and of low order p (typically $4 \leq p \leq 10$).

4.2 Adjoint Equation

The question is now to estimate the distance between the two scales local solution $(x_H, \underline{\underline{\sigma}}_H)$ and the reference solution of the mixed variational system (8-12) without solving the latter system. A first information on the error is given by the residual observed in the formulation (8-12) when plugging our two scales solution. This is useful, but is hard to relate to a meaningful norm. A better strategy would be to solve the variational problem defining the error but its solution is out of reach because it couples all local subdomains together. A compromise must be found. Here, we are only interested in the local averages of the components of the Cauchy stress tensor

$$Q = \int_{\Omega_H} \underline{\underline{\sigma}} : \underline{e}_i \otimes \underline{e}_j \, d\Omega_H,$$

$$\frac{\partial Q_k}{\partial \underline{x}}(\underline{x}_H) \cdot \underline{\hat{U}} = \int_{\Omega_{kH}} \underline{e}_i \otimes \underline{e}_j : \left(\underline{\underline{\underline{\underline{C}}}}_Y : \frac{1}{2}\left(\frac{\partial \underline{\hat{U}}}{\partial \underline{Y}} + \frac{\partial \underline{\hat{U}}}{\partial \underline{Y}}^t \right) \right) d\Omega_H.$$

To estimate the accuracy of $Q(\underline{x})$ as predicted by the two scales local solution, one only needs to obtain an approximate solution of the adjoint equation defined on the collection of subdomains by:

Find the adjoint state $\underline{x}^a$ and the adjoint interface tractions $\underline{\lambda}^a$ such that

$$\int_{\Omega_{kH}} \left(\underline{\underline{\underline{\underline{C}}}}_Y : \frac{1}{2}\left(\frac{\partial \underline{x}^a}{\partial \underline{X}} + \frac{\partial \underline{x}^a}{\partial \underline{X}}^t \right) \right) : \frac{\partial \underline{\hat{U}}}{\partial \underline{X}}^t d\Omega_H + \sum_l \int_{\partial\Omega_{kH} \cap \partial\Omega_{lH}} \underline{\lambda}^a_{kl} \cdot \underline{\hat{U}} \, da$$
$$= \frac{\partial Q}{\partial \underline{x}}(\underline{x}_H) \cdot \underline{\hat{U}} \quad \forall \underline{\hat{U}} \in \mathbf{V}_{kH}, \forall k, \tag{13}$$

$$\int_{\partial\Omega_{kH} \cap \partial\Omega_{lH}} \underline{\mu}_{kl} \cdot (\underline{x}^a_l - \underline{x}^a_k) da = 0 \quad \forall \underline{\mu}_{kl} \in \mathbf{M}_{kl}, \forall k < l, \tag{14}$$

$$\underline{\lambda}^a_{kl} + \underline{\lambda}^a_{lk} = 0 \quad \forall k < l. \tag{15}$$

4.3 Explicit a Posteriori Error Estimate

The adjoint state then allows a direct access to the error on $Q(\underline{x})$. Indeed, writing the adjoint problem (13)–(14) using as test functions $(\underline{\hat{U}}, \underline{\mu})$ the solution $(\delta\underline{x}, \underline{\lambda})$ of the linearized error problem yields

$$\frac{\partial Q}{\partial \underline{x}}(\underline{x}_H) \cdot \delta\underline{x} = \sum_k \int_{\Omega_{kH}} \left(\underline{\underline{\underline{\underline{C}}}}_Y : \frac{1}{2}\left(\frac{\partial \underline{x}^a}{\partial \underline{X}} + \frac{\partial \underline{x}^a}{\partial \underline{X}}^t \right) \right) : \frac{\partial \delta\underline{x}}{\partial \underline{X}}^t d\Omega_H$$
$$+ \sum_{k<l} \int_{\partial\Omega_{kH} \cap \partial\Omega_{lH}} \underline{\lambda}^a_{kl} \cdot (\delta\underline{x}_l - \delta\underline{x}_k) da.$$

Using the symmetry of the elasticity tensor, the continuity of the exact solution, the linearized error equations, and using the weak interface continuity of the adjoint state reduce the above expression to the *explicit error estimate*

$$\frac{\partial Q}{\partial \underline{x}}(\underline{x}_H) \cdot \delta\underline{x} = \sum_k \int_{\Omega_{kH}} (\underline{\underline{\underline{\underline{C}}}}_Y : \delta\underline{\underline{\varepsilon}}) : \frac{\partial \underline{x}^a}{\partial \underline{X}}^t d\Omega_H$$
$$- \sum_{k<l} \int_{\partial\Omega_{kH} \cap \partial\Omega_{lH}} \underline{\lambda}^a_{kl} \cdot (\underline{x}_{lH} - \underline{x}_{kH}) da$$
$$= -\sum_k \int_{\Omega_{kH}} \underline{\underline{\sigma}}_H : \frac{\partial \underline{x}^a}{\partial \underline{X}}^t d\Omega_H - \sum_{k<l} \int_{\partial\Omega_{kH} \cap \partial\Omega_{lH}} \underline{\lambda}^a_{kl} \cdot (\underline{x}_{lH} - \underline{x}_{kH}) da.$$

4.4 Numerical Solution of the Adjoint Problem

.

For solution purposes, the adjoint problem can be rewritten as a Schur complement problem set on the interface with unknown $\bar{X}^a = (Tr_{kl}\underline{x}^a)_{kl} \in \Pi_{k<l} M'_{kl}$. By introducing the local trace

$$Tr_k = \begin{pmatrix} \vdots \\ Tr_{kl} \\ \vdots \end{pmatrix} \quad \text{and restriction} \quad R_k\bar{X} = \begin{pmatrix} \vdots \\ \bar{X}_{kl} \\ \vdots \end{pmatrix},$$

we can immediately rewrite the adjoint problem as the algebraic system

$$\left(\sum_k R_k^t \,(0\ I) \begin{pmatrix} K_k & Tr_k^T \\ Tr_k & 0 \end{pmatrix}^{-1} \begin{pmatrix} 0 \\ I \end{pmatrix} R_k \right)\bar{X}^a = \\ -\sum_k R_k^t \,(0\ I) \begin{pmatrix} K_k & Tr_k^T \\ Tr_k & 0 \end{pmatrix}^{-1} \begin{pmatrix} \frac{\partial Q_k}{\partial \underline{x}}(\underline{x}_H) \\ 0 \end{pmatrix}. \quad (16)$$

This symmetric reduced system is of smalll dimension when one uses low order interface mortars. It can be solved by a direct solver in $\bar{X}^a$. It can also be solved by a few iterations of a domain decomposition algorithm.

5 Numerical Results

The proposed strategy has been tested on a simple two dimensional situation in anisotropic elasticity. The reference local problem uses a periodic geometry at scale $\bar{H}$ made of $n_x \times n_y = 30$ unit subcells of size $l_x \times l_y$ with a non periodic variation of the stiffness coefficients as represented in Fig. 2. In crystal coordinates, the stiffness coefficients C_{1111}, C_{1122} and C_{1212} have a space periodic distribution at the subcell level

$$C_{ijkl} = C^0_{ijkl} * \left(1.1 + \sin\left(\frac{2\pi x}{l_x}\right)\right) * \left(1.1 + \sin\left(\frac{2\pi y}{l_y}\right)\right)$$

with $C^0_{1111} = 3000\,Gpa, C^0_{1122} = 100\,Gpa, C^0_{1212} = 200\,Gpa$ and the crystal direction has a non periodic space variation with a local angle given by

$$\theta = \frac{1}{4}\left(\frac{2\pi x}{n_x l_x} + \frac{2\pi y}{n_y l_y}\right).$$

The global sample (the local geometry built with the 30 subcells) is subjected at infinity to a uniform shear deformation of one percent. A solution computed for the whole local sample cell yields a non periodic shear stress distribution as represented on Fig. 2, with a shear localisation in the weak parts of the sample.

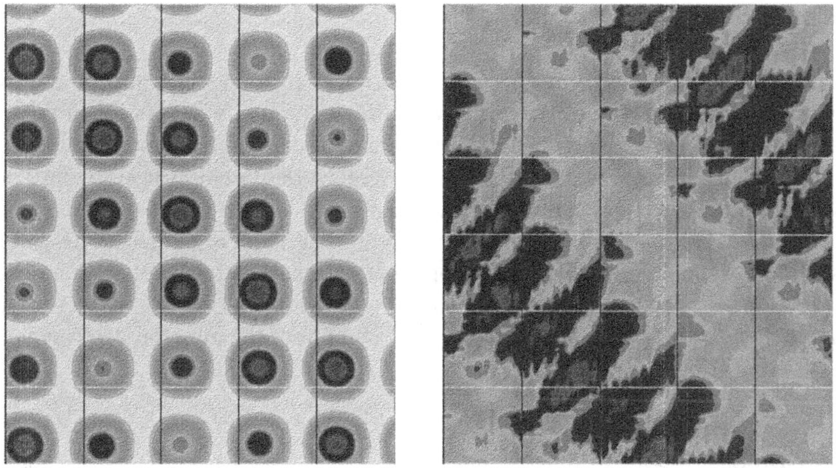

Fig. 2. (Left) Representation of the space variation of the stiffness coefficient C_{xxxx}. The ratio between the minimal and the maximal values is 400. (Right) Representation of the shear stress when computed globally on the full sample.

The two scales solution $\underline{x}_H$ computed independently on each subcell using periodic boundary conditions is represented on Fig. 3. Observe the displacement jumps at the interface in this local construction. The averaged shear stress obtained by this two scales local approach differs from the exact one by $50.6MPa$. The solution of the dual problem is then obtained using polynomial Lagrange multipliers of order 4. In this simple case, the error on the average shear as predicted from the dual solution is equal to $50.3MPa$ to be compared to the real error of $50.6MPa$. This clearly indicates the good accuracy of our error estimate.

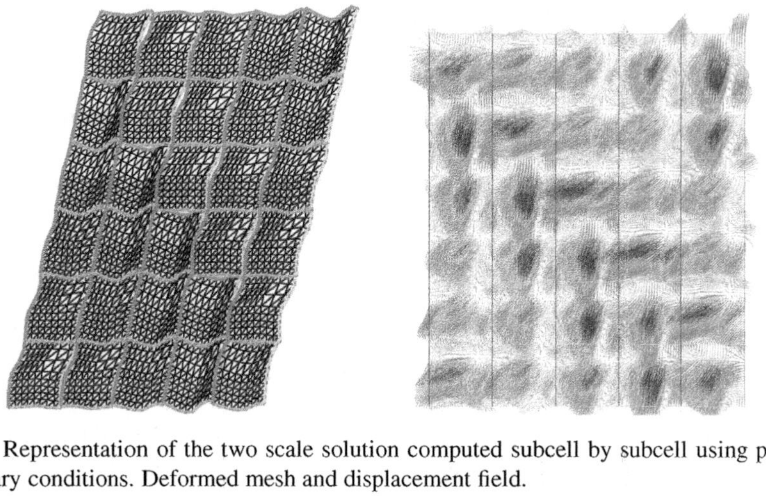

Fig. 3. Representation of the two scale solution computed subcell by subcell using periodic boundary conditions. Deformed mesh and displacement field.

6 Conclusions

In the framework of first order numerical homogeneisation techniques, the present paper has introduced an a-posteriori error estimate to qualify the choice of the local representative volume elements to be used in a two scale finite element method.

The error estimate was developed in the framework of nonlinear viscoelastic materials in small strains but can readily be extended to large strains situations. Its performance was assessed on two dimensional problems. A lot of numerical assessment is still to be done. Recovering microscopic data, implementing mortars in an industrial framework is challenging. Moreover, the local problems are stochastic in nature. What is then the best treatment of the random nature of the material heterogeneities and processes?

References

[1] Alicandro, R., Cicalese, M.: A general integral representation result for the continuum limits of discrete energies with superlinear growth. *SIAM J. Math. Anal.*, 36(1):1–37, 2004.

[2] Bernardi, C., Maday, Y., Patera, A.T.: *Asymptotic and numerical methods for partial differential equations with critical parameters*, chapter Domain decomposition by the mortar element method, 269–286. 1993.

[3] Bernardi, C., Maday, Y., Patera, A.T.: *Nonlinear partial differential equations and their applications*, chapter A new nonconforming approach to domain decomposition: the mortar element method, 13–51. Pitman, 1994.

[4] Brezzi, F., Marini, D.: Error estimates for the three-field formulation with bubble stabilization. *Math. Comp.*, 70:911–934, 2000.

[5] Engquist, B., Li, X., Ren, W., Vanden-Eijnden, E.: Heterogeneous multiscale methods: A review. *Commun. Comput. Phys.*, 2:367–450, 2007.

[6] Farhat, C., Géradin, M.: Using a reduced number of lagrange multipliers for assembling parallel incomplete field finite element approximations. *Comput. Methods Appl. Mech. Engrg.*, 97:330–354, 1992.

[7] Féyel, F.: Multiscale FE^2 elastoviscoplastic analysis of composite structures. *Comp. Mat. Sci.*, 16:344–354, 1999.

[8] Fish, J., Yuan, Z.: Toward realization of computational homogenization in practice. *Internat. J. Numer. Methods Engrg.*, 73:361–380, 2008.

[9] Hauret, P., Le Tallec, P.: A discontinuous stabilized mortar method for general 3d elastic problems. *Comput. Methods Appl. Mech. Engrg.*, 196:4881–4900, 2006.

[10] Hou, T.Y., Wu, X.H., Cai, Z.Q.: Convergence of a multiscale finite element method for elliptic problems with rapidly oscillating coefficients. *Math. Comp.*, 68:913–943, 1999.

[11] Kouznetsova, V.G., Brekelmans, W.A.M., Baaijens, F.P.T.: An approach to micro-macro modeling of heterogeneous materials. *Comp. Mech.*, 27:37–48, 2001.

[12] Moulinec, H., Suquet, P.: A numerical method for computing the overall response of nonlinear composites with complex microstructure. *Comput. Methods Appl. Mech. Engrg.*, 157:69–94, 1998.
[13] Seshaiyer, P.: *Non-conforming hp finite element methods*. PhD thesis, University of Maryland, 1998.
[14] Wohlmuth, B.I.: *Discretization methods and iterative solvers based on domain decomposition*. Springer, 2001.
[15] Zaoui, A.: *Matériaux hétérogènes et composites*. Ecole Polytechnique, 1988.
[16] Zohdi, T., Wriggers, P.: *Numerical Modeling of Heterogeneous Materials*. Springer, 2005.

Multiscale Methods for Multiphase Flow in Porous Media

Jan M. Nordbotten

Department of Mathematics, University of Bergen jan.nordbotten@math.uib.no
Department of Civil and Environmental Engineering, Princeton University

Summary. We aim in this paper to give a unified presentation to some important approaches in multi-phase flow in porous media within the framework of multiscale methods. Thereafter, we will present a modern outlook indicating future research directions in this field.

1 Introduction

Understanding flow in porous media is crucial in applications as diverse as petroleum and geothermal energy recovery, ground water management, waste disposal (including CO_2) in geological formations, and the production of porous materials. Simultaneously, the mathematical and numerical challenges associated with accurate modeling of the strongly non-linear governing equations are profound. Difficulties are characterized by parameters which are anisotropic and discontinuous on all scales of observation, while the solutions are nearly discontinuous and globally coupled even for idealized homogeneous problems.

The equations for porous media flow have an elliptic-hyperbolic structure, where the pressure is governed by an equation which is nearly elliptic, while the fluid saturation is governed by an equation which is nearly hyperbolic. In applications, the principle of mass conservation (which is embedded in both equations), is considered essential. Thus efficient numerical solution techniques are needed which appropriately handle discontinuous coefficients, while honoring mass conservation strictly (see e.g. [3]).

Standard methods are often unsuitable under these circumstances. In terms of spatial discretization, either control volume methods or mixed (or discontinuous) finite element methods are needed to enforce mass conservation in a strong sense [8]. The non-linearities and time dependencies lead to implicit discretizations (in particular for the pressure step). Finally, many of the challenges encountered by multi-grid preconditioners in continuum fluid dynamics (see [6]) are present or even enhanced in porous media (see e.g. [29]). Therefore, domain decomposition preconditioners have become popular.

M. Bercovier et al. (eds.), *Domain Decomposition Methods in Science and Engineering XVIII*, Lecture Notes in Computational Science and Engineering,
DOI: 10.1007/978-3-642-02677-5, © Springer-Verlag Berlin Heidelberg 2009

In industrial petroleum recovery applications, it is common that the parameter field (e.g. permeability), is given at a far higher resolution than what can be resolved with the available computational resources. This has led to a large focus on upscaling methods, and lately this effort has focused on multiscale methods. Frequently, these methods show strong resemblance not only to upscaling, but also to domain decomposition (see e.g. [25]).

We define by "multiscale" in this paper methods which deal with problems defined at a resolution finer than what can be computationally resolved. In reality, this defines two scales, that of the problem and that of the computational resolution, and all the methods herein might well be labeled "twoscale".

We will continue this paper by outlining a framework for classifying various multiscale methods. Thereafter, we will discuss several methods from literature, and identify their proper definition as a multiscale method. Our focus will not be on discussing abstract frameworks, but practical implementations. In particular, we will discuss how permeability upscaling, relative permeability upscaling, and vertical equilibrium formulations all can be seen as multiscale methods. We then give an introduction to state of the art multiscale simulation. Finally, we summarize by giving examples indicating prospects for future developments.

2 A Framework for Discussing Multiscale Methods

There are several frameworks to discuss multiscale methods in. Some useful approaches are Volume Averaging [30], Systematic Upscaling [7] and Variational Multiscale [16]. We will herein use the terminology of the Heterogeneous Multiscale Method (HMM) [12]. Similarities between these frameworks have been discussed elsewhere [11].

Following the presentation of HMM [12], we recall that for a problem

$$f(u,d) = 0, \tag{1}$$

where u is the unknown and d is data, we can postulate the exitence of a "coarse" variable u_D, which satisfies

$$F(u_D, D) = 0. \tag{2}$$

We will assume the functional form of $F(u_D, D)$ to be known, however the coarse data D must be estimated from the fine scale model. The coarse and fine scale models are associated through a compression (also referred to as interpolation) operator $u_D = \mathcal{Q}u$, and some reconstruction (or extrapolation) operator $\mathcal{R}u_D$. Note that while we require $\mathcal{Q}\mathcal{R} = \mathcal{I}$, the reverse does not in general hold.

Of particular interest to the remaing discussion is the finite element formulation of HMM (HMFEM), which consideres minimization problems on the form: Denote by u an element which minimizes

$$\min_{v \in V} A(v) - B(v) \tag{3}$$

for non-linear and linear forms A and B, respectively.

Consider the minimization problem given in Equation (3), and assume it has a unique solution. By introducing a compression operator $\mathcal{Q} : V \to V_D$, where V_D is some coarse scale solution domain, we have the minimization problem equivalent to (3):

$$\min_{v_D \in V_D} \min_{v:\mathcal{Q}v=v_D} A(v) - B(v). \tag{4}$$

We restrict the choice of compression operators under consideration such that also (4) has a unique solution. We note that an 'exact' reconstruction operator with respect to the minimization problem can be defined from (4): $\mathcal{R}_e u_D$ solves

$$\min_{v:\mathcal{Q}v=u_D} A(v) - B(v). \tag{5}$$

We now have the exact coarse scale HMFEM minimization problem

$$\min_{v_D \in V_D} A(\mathcal{R}_e v_D) - B(\mathcal{R}_e v_D). \tag{6}$$

For practical purposes, calulating $\mathcal{R}_e$ is excessively expensive, and an approximation is introduced; $\tilde{\mathcal{R}} \approx \mathcal{R}_e$. It is usually advocated (see e.g. [12]) that since u_D is a macro-scale function, it should vary smoothly, thus it is sufficient to evaluate the integrals appearing in the variational formulation at quadrature points. This allows for great flexibility in localization strategies for approximating $\tilde{\mathcal{R}}$.

3 A Model Problem for Multiphase Flow

The model equation for multiphase flow in porous media is the standard extension of Darcy's law to two phases $\alpha = \{0, 1\}$ (see e.g. [5, 8, 19, 22]):

$$\mathbf{u}_\alpha = -K\lambda_\alpha(\nabla p - \rho_\alpha \mathbf{g}). \tag{7}$$

Here $\mathbf{u}_\alpha$ is the volumetric flux with units [L/T] , K is the intrinsic permeability of the medium [L^2], $\lambda_\alpha = \lambda_\alpha(s_\alpha)$ is the phase mobility as a function of the phase saturation s_α [TLM^{-1}], p is pressure [$ML^{-1}T^{-2}$], $\rho_\alpha = \rho_\alpha(p)$ is phase density [ML^3], and finally $\mathbf{g}$ is the gravitational vector [LT^{-2}], positive downwards. We have neglected the difference between phase pressures, which is a common assumption at reservoir scales [19].

The equations for flow satisfy conservation of mass for each phase

$$\phi \partial_t(\rho_\alpha s_\alpha) = \nabla \cdot (\rho_\alpha \mathbf{u}_\alpha) = b_\alpha, \tag{8}$$

where ϕ denotes the void fraction (porosity) [L^0], which is kept constant is time, but is allowed to vary in space, while b_α represents source and sink terms [$ML^{-3}T^{-1}$].

We close the system by requiring that no more than two phases are present,

$$s_0 + s_1 = 1, \tag{9}$$

and assigning constitutive relationships:

$$K = K(\mathbf{x}), \qquad \phi = \phi(x), \qquad \lambda_\alpha = \lambda_\alpha(s_\alpha). \tag{10}$$

Note that we will frequently use the shorthand $s = s_0$ and $s = 1 - s_1$ in the derivations. In this work we will neglect compressibility.

The key difficulties in (7) and (8) lie firstly in the pronounced heterogeneity in the permeability, which may be discontinuous and contain long-range correlations. Further, the solution may contain shocks due to the non-linear relative permebility functions.

A total pressure - fractional flow formulation is often used for (7)–(10) as this formulation allows for the application of a splitting to exploit the (relatively) weak time dependence of pressure [8]. We obtain a total pressure equation by eliminating $\partial_t s$ from (8):

$$\nabla \cdot \mathbf{u}_T = b_T \tag{11}$$

$$\mathbf{u}_T = -K\lambda_T(\nabla p - \rho_T \mathbf{g}). \tag{12}$$

Equations (11) and (12) are written in two parts to retain the physical fluxes explicitly. This is important to get the mass conservation equation discretized correctly. The flux, source, mobility and density are defined as:

$$b_T = \sum_\alpha b_\alpha \rho_\alpha^{-1}, \qquad \mathbf{u}_T = \sum_\alpha u_\alpha, \tag{13}$$

$$\lambda_T = \sum_\alpha \lambda_\alpha, \qquad \rho_T = \lambda_T^{-1} \sum_\alpha \lambda_\alpha \rho_\alpha.$$

The individual phase fluxes can be recovered from $\mathbf{u}_T$;

$$\mathbf{u}_\alpha = \lambda_\alpha \lambda_T^{-1} \mathbf{u}_T + \lambda_\alpha \lambda_\beta \lambda_T^{-1} K(\rho_\alpha - \rho_\beta)\mathbf{g}, \tag{14}$$

where $\beta = 1 - \alpha$. Equations (11)–(14) together with either of (8) form an equivalent system of equations to (7)–(10).

4 Some Upscaling Methods

In this section, we will investigate upscaling methods aimed at the two main difficulties outlined in our model multiphase flow equations: The heterogeity of the permeability data and the heterogeneity of the saturation solution.

4.1 Permeability Upscaling

Upscaling of permeability by itself is essentially a single phase problem. As such, it is analogous to many other problems in the physical sciences, including most famously heat conduction. Many strategies are applicable to this problem, particularly in the case where scale separation exists. We will herein consider a form of numerical homogenization popular in the porous media community. We recognize that permeability upscaling has previously been discussed within the framework of HMM [2, 12], however the approach taken here is different.

Typically, in porous media, every coarse grid block is assigned a permeability, based on fine scale flow properties. Depending on the level of complexity of the coarse scale numerical solver, this permeability is either isotropic, anisotropic aligned with the grid, or generally anisotropic.

The coarse block permeability is calculated by solving some fine scale problem. The appropriate boundary conditions for this problem can be a cause of debate, but for the sake of the argument, we follow [5] and use a linear potential (we need not consider gravity when upscaling absolute permeability). The coarse permeability is then calculated by postulating the existence of a coarse scale Darcy law for the grid block

$$\langle \mathbf{u} \rangle = -\frac{K_D}{\mu} \langle \nabla p \rangle. \tag{15}$$

Here $< \mathbf{u} >$ is the mean of the calculated velocity, and $< \nabla p >$ is (by Green's theorem) a function of the boundary conditions. By varying the boundary conditions, one can infer the coarse scale permeability K_D.

We will avoid a lengthier discussion of upscaling methods for absolute permeability, and consider how the approach taken above can be seen as within the framework of the HMM.

Consider the following choice of discrete coarse scale variables: a potential vector $\mathbf{p}_D$ and coarse permeability vector (or tensor) $\mathbf{K}_D$. We assume that the coarse scale equations are some appropriate discretization of the elliptic equation on the coarse scale grid, e.g. $F(\mathbf{p}_D, \mathbf{K}_D) = 0$. We define the compression for coarse cell Ω_i as

$$p_{D,i} = \mathcal{Q}p = \frac{1}{L}\int_{\Omega_i} p ds,$$

where L is the arc length of the integral.

The constrained variational form of (5) for the fine scale HMM problem can then be written [24]: Find p' s.t. $p_D = \mathcal{Q}p'$ and

$$(K\nabla p', \nabla q') = 0 \quad \forall\, q' \text{ s.t. } 0 = \mathcal{Q}q'. \tag{16}$$

The data K_D is then obtained from (15), subject to the additional constraint that the anisotropy directions are known (e.g. aligned with the flow or the grid).

We see now that the classical permeability upscaling approach can be seen as a localization of the global fine scale problem in the HMM. Indeed, if we use a piecewise linear potential to impose $p_D = \mathcal{Q}p'$, and solve (16) separately on each subdomain, we obtain exactly the numerical upscaling approach described earlier.

We emphasize that the intuitive, or one might say engineering, approach to upscaling permeability can thus be shown to be related to a multiscale modeling framework. However, this relationship comes at a considerable cost: We have assumed the existence of an equivalent homogeneous coarse scale permeability K_D; defined a highly specialized compression operator; and subsequently made a very crude approximation to what is the known “true” fine scale problem. Indeed, we would encourage interpreting this section not in support of classical permeability upscaling, but rather as a critique pointing out the expected weaknesses.

The observations regarding potential problems with permeability upscaling are not new, and they have previously motivated what is known as transmissibility upscaling for finite volume methods (see e.g. [10, 20]). With this approach, the coarse scale is assumed to satisfy a discrete conservation law of the form

$$\mathbf{f}_D = -B\mathbf{p}_D; \qquad -C\mathbf{f}_D = \mathbf{b}_D. \tag{17}$$

Here, $\mathbf{f}_D$ and $\mathbf{b}_D$ are the coarse fluxes and sources, while B and C are sparse matrices representing conductivity and mass conservation, respectively. The mass conservation matrix C is known, and an upscaling approach is used to determine the coefficients of B.

We note that from a general perspective, permeability upscaling and transmissivity upscaling are closely related, and the previous remarks about the relationship to a HMM framework apply also to transmissibility upscaling. The case of transmissivity upscaling is discussed in more detail in [13]. Our main point of including the transmissivity upscaling, is to show how the macroscale model may be either discrete *a posteriori*, as in the case of permeability upscaling, or *a priori* as in the case of transmissivity upscaling. Note that the advantage of an *a priori* discrete model in this case is that no explicit assumptions are made on a macroscale permeability K_D

4.2 Saturation Upscaling

The second challenge of porous media upscaling is the saturation equation. We will here outline the industry standard perspective.

As with the permeability, we assume the existence of a macroscale extension of Darcy's law. This macroscale extension can as in the previous section be either continuous or discrete; for the sake of the argument we will make assumption that it is continuous, e.g.:

$$\langle \mathbf{u}_\alpha \rangle = -k_{D,\alpha}(s_{D,\alpha})\frac{K_D}{\mu_\alpha}\langle \nabla p_\alpha \rangle. \tag{18}$$

Here s_D is the macroscale saturation, while k_D is the macroscale relative permeability. The compression operator for the macroscale saturation must for mass conservation reasons be defined simply as the cell average saturation.

Applying a similar approach to the permeability upscaling case, one obtains the following rather interesting observations: Firstly, the results are highly sensitive to how one treats the coarse scale potential gradient term [27]. Secondly, the results are (as expected), influenced by imposed boundary conditions. Finally, and more importantly, we note that in analog to permeability upscaling where we saw induced anisotropy at the coarse scale (even without anisotropy at the fine scale), for saturation upscaling we see a strong history dependence on the coarse scale, even when none is present at the fine scale.

From a HMM perspective, we first observe that for the hyperbolic part, these simple approaches are variants of a Godunov method. This allows us to make a more natural interpretation of the hysteresis point: While saturation has a unique compression operator, the problem unfortunately has a strong dependence on the degrees of freedom in the reconstruction operator. Indeed, the problem is that the upscaling method described above aims at being more accurate than a naive first order Godunov method, but the price then becomes selective accuracy, depending on the quality of the reconstruction.

4.3 Vertically Integrated Models

For some porous media applications, such as saltwater intrusion [5] and storage of CO_2 in saline aquifers, vertically integrated models may be applicable [14, 28]. The features allowing a successful application of such a formulation are the dominance of horizontal length scales over vertical length scales, combined by gravity segregation in the systems.

The key concept behind vertically averaged models is to consider equations for an interface between the two fluids resulting from gravity segregation. By integrating over the vertical direction, we obtain governing equations for the interface, which are essentially 2D conservation equations combined by flux functions involving integrals over the vertical direction;

$$F = \int f \mathrm{d}z. \tag{19}$$

These integrals contain subscale information through the explicit dependence on the vertical solution structure. To apply this formulation, effective approximations must be introduced regarding the vertical structure of the pressure field in addition to that of segregated fluids. Common choices are vertical equilibrium (the Dupuit approximation), although more complex choices are possible [26].

Let us consider this approach again within the framework of a multiscale methodology. The scale assumption is that the vertical scales are short, and the associated time scale of equilibration are short. To honor mass conservation, the compression operator taking saturation to interface thickness is vertical integration. A compression operator for the pressure can be taken as the pressure at the bottom of the domain (as in the above references). Following the assumption of short equilibration time in the vertical direction, and reconstruction of initial conditions for a fine scale solver will lead to a vertically segregated, fluid-static system. We will therefore simply assume that the reconstruction operator is the fluid-static distribution. The combined operator $\mathcal{RQ}$ will be exact for problems where the fine scale indeed is vertically segregated.

It is interesting to note how the Dupuit approximation in the vertically integrated model appears immediately with the multiscale framework. Also worth noting is how HMM provides an abstract framework for discussing this approximation beyond the ususal asymptotic arguments.

We conclude this section by reiterating the purpose of these examples. Through relating well established concepts to a common framework, we hope to achieve two aims: Firstly, to provide a unified way of considering physical based (as opposed to strictly numerical) upscaling methods. Secondly, to build support for HMM (or a similar multiscale design) as a general framework to guide upscaling methods.

5 Multiscale Numerical Methods

In this section, we will expand upon the ideas from the previous section, and discuss newer methods gaining interest in porous media. In particular we will discuss a class of numerical methods, which term themselves also multiscale methods, which aim at creating approximate solutions on a coarse scale, retaining a physically plausible fine scale structure. The methods discussed here primarily address the pressure equation, which due to ellipticity is the harder equation, while the saturation equation is resolved on a fine scale [1, 4, 17, 23]. While these methods have much in common with domain decomposition [9, 25], they differ in the focus on fast approximations to the fine scale problem which are physically plausible, rather than the solution to the fine scale problem itself. We will focus in particular on the so-called variational multiscale methods, the general ideas are similar between the formulations.

5.1 The Variational Multiscale Method

The Variational MultiScale (VMS) Method is a general approach to solving partial differential equations [15, 16]. While more specialized in approach than the HMM, we see in VMS a sharper focus on the nature and structure of the fine scale problems. When applied in a similar manner, it can be shown that VMS can be considered a special case of HMM [24].

We therefore consider: Find $u \in U$ such that

$$a(u,v) = b(v) \quad \forall\, v \in V. \tag{20}$$

We take a and b to be bilinear and linear operators, respectively. Although in general the spaces U and V may be different, we will here use $U = V$.

Hughes et. al discuss finite element approximations in terms of the following argument: Let V' be defined such that $V_H \oplus V' = V$, noting that in general V_H and V' need not be orthogonal. Then the following coupled problems are equivalent to (20): Find $u_H \in V_H$ and $v' \in V'$ such that

$$a(u_H, v_H) + a(u', v_H) = b(v_H) \quad \forall\, v_H \in V_H \tag{21}$$

and

$$a(u_H, v') + a(u', v') = b(v') \quad \forall\, v' \in V', \tag{22}$$

The term $a(u', v_H)$ can be quanitified by representing u' in terms of a Green's function for the original problem constrained to the space V' [15]. Thus, we can write the solution of (22) formally as

$$u' = -G'(b - \mathcal{L}u_H), \tag{23}$$

where $b - \mathcal{L}u_H$ is the residual error of the approximate solution u_H to the underlying PDE:

$$\mathcal{L}u - b = 0, \tag{24}$$

and G' is an integral Green's transform, where the kernel is simply the Green's function in V'. By combining (22) and (23), we obtain the finite dimensional variational problem: Find $u_H \in V_H$ such that

$$a(u_H + G'(u_H), v_H) = b(v_H) + a(G'(b), v_H) \quad \forall\, v_H \in V_H. \tag{25}$$

This equation is refered to as a paradigm for multiscale simulation [16].

5.2 A VMS Approach for the Implicit Time-Discretized Pressure Equation

We consider the following coupled partial differential equations, which serve as a prototype for the pressure equation in porous media flow.

$$\nabla \cdot \mathbf{u} = b \quad \text{in } \Omega, \tag{26}$$

$$\mathbf{u} + d(\nabla p - \mathbf{c}) = 0 \quad \text{in } \Omega, \tag{27}$$

$$\mathbf{u} \cdot \nu = 0 \quad \text{on } \partial\Omega. \tag{28}$$

We consider for simplicity only zero Neumann (no-flow) boundaries. These boundary conditions are prevailing in applications. On variational form, the mixed problem can be stated as: Find $p \in W$ and $\mathbf{u} \in \mathbf{V}$ such that

$$(\nabla \cdot \mathbf{u}, w) = (b, w) \quad \forall\, w \in W, \tag{29}$$

$$(d^{-1}\mathbf{u}, \mathbf{v}) - (p, \nabla \cdot \mathbf{v}) = (\mathbf{c}, \mathbf{v}) \quad \forall\, \mathbf{v} \in \mathbf{V}. \tag{30}$$

The permeability is a symmetric and positive definite matrix, justifying the inverse used in (30). The mixed space $\mathbf{V}$ must honor the boundary condition.

Given (29) and (30), we are prepared to introduce our VMS method. Thus, let W and $\mathbf{V}$ be direct sum decompositions $W = W_H \oplus W'$ and $\mathbf{V} = \mathbf{V}_H \oplus \mathbf{V}'$. Our coarse scale variational problem is thus: Find $p_H \in W_H$ and $\mathbf{u}_H \in \mathbf{V}_H$ such that

$$\begin{aligned}&(\nabla \cdot (\mathbf{u}_H + G'_{\mathbf{u}}(\nabla \cdot \mathbf{u}_H, \nabla p_H + d^{-1}\mathbf{u}_H)), w_H)\\ &\quad = (b + \nabla \cdot G'_{\mathbf{u}}(b, \mathbf{c}), w_H) \quad \forall\, w_H \in W_H\end{aligned} \tag{31}$$

and

$$\begin{aligned}&(d^{-1}(\mathbf{u}_H + G'_{\mathbf{u}}(\nabla \cdot \mathbf{u}_H, \nabla p_H + d^{-1}\mathbf{u}_H), \mathbf{v}_H)\\ &\quad - (p_H + G'_p(\nabla \cdot \mathbf{u}_H, \nabla p_H + d^{-1}\mathbf{u}_H), \nabla \cdot \mathbf{v}_H)\\ &\quad = (\mathbf{c} + d^{-1}G'_{\mathbf{u}}(b, \mathbf{c}), \mathbf{v}_H) - (G'_p(b, \mathbf{c}), \nabla \cdot \mathbf{v}_H) \quad \forall\, \mathbf{v} \in \mathbf{V}.\end{aligned} \tag{32}$$

The integral operators $G'_p \in W'$ and $G'_{\mathbf{u}} \in \mathbf{V}'$, which we will refer to as Green's transforms, are the formal solutions to the following (linear) equations:

$$(\nabla \cdot G'_{\mathbf{u}}(g,\mathbf{f}), w') = (g, w') \quad \forall\, w' \in W', \tag{33}$$

$$(d^{-1} G'_{\mathbf{u}}(g,\mathbf{f}), \mathbf{v}') - (G'_p(g,\mathbf{f}), \nabla \cdot \mathbf{v}') = (\mathbf{f}, \mathbf{v}') \quad \forall\, \mathbf{v}' \in \mathbf{V}'. \tag{34}$$

We note that in (31) and (32), we need the Green's transforms evaluated for all members of the spaces W_H and $\mathbf{V}_H$. Since (33) and (34) are linear, it suffices to evaluate (or approximate, as the case will be) G'_p and $G'_{\mathbf{u}}$ for a set of basis functions for W_H and $\mathbf{V}_H$, in addition to the right hand side components b and $\mathbf{c}$.

To proceed further, it is necessary to make an appropriate choice of spaces W_H, $\mathbf{V}_H$, W' and $\mathbf{V}'$. This will not be elaborated here, alternative choices can be found in e.g. [4, 18, 21, 23].

6 Conclusions

In this paper we have made an initial attempt at bringing together diverse approaches to upscaling for multiphase porous media flow problems under the umbrella of multi-scale methods. The goal has not been to complete a comprehensive survey, but rather to illustrate how key ideas can be related.

We have considered upscaling, both static and dynamic; vertically integrated formulations; and modern multiscale simulation. While there is still work to be done before these approaches can be presented seamlessly, we hope that the current exposition will allow the reader to appreciate the subtle similarities.

References

[1] Aarnes, J.E., Krogstad, S., Lie, K.-A.: Multiscale mixed/mimetic methods on corner-point grids. *Comput. Geosci.*, 12(3):297–315, 2008.

[2] Abdulle, A., E, W.: Finite difference heterogeneous multi-scale method for homogenization problems. *J. Comput. Phys.*, 191(1):18–39, 2003. ISSN 0021-9991.

[3] Allen, III, M.B., Behie, G.A., Trangenstein, J.A.: *Multiphase flow in porous media*, volume 34 of *Lecture Notes in Engineering*. Springer-Verlag, Berlin, 1988. ISBN 3-540-96731-1. Mechanics, mathematics, and numerics.

[4] Arbogast, T.: Implementation of a locally conservative numerical subgrid upscaling scheme for two-phase Darcy flow. *Comput. Geosci.*, 6(3-4):453–481, 2002. ISSN 1420-0597. Locally conservative numerical methods for flow in porous media.

[5] Bear, J., Bachmat, Y.: *Introduction to Modeling of Transport Phenomena in Porous Media*. Kluwer Academic, 1991.

[6] Brandt, A.: Barriers to achieving textbook multigrid efficiency (tme) in cfd. `http://hdl.handle.net/2002/14809`, 1998.

[7] Brandt, A.: Principles of systematic upscaling. In J. Fish, ed., *Bridging the Scales in Science and Engineering*. 2008.

[8] Chen, Z., Huan, G., Ma, Y.: *Computational methods for multiphase flows in porous media*. Computational Science & Engineering. Society for Industrial and Applied Mathematics (SIAM), Philadelphia, PA, 2006. ISBN 0-89871-606-3.

[9] Chen, Z., Hou, T.Y.: A mixed multiscale finite element method for elliptic problems with oscillating coefficients. *Math. Comp.*, 72(242):541–576 (electronic), 2003. ISSN 0025-5718.

[10] Durlofsky, L.J.: Coarse scale models of two phase flow in heterogeneous reservoirs: volume averaged equations and their relationship to existing upscaling techniques. *Comput. Geosci.*, 2(2):73–92, 1998. ISSN 1420-0597.

[11] E, W.: The heterogeneous multiscale method and the equation-free approach to multiscale modeling. Preprint.

[12] E, W., Engquist, B.: The heterogeneous multiscale methods. *Commun. Math. Sci.*, 1(1):87–132, 2003. ISSN 1539-6746.

[13] E, W., Engquist, B., Li, X., Ren, W., Vanden-Eijnden, E.: Heterogeneous multiscale methods: a review. *Commun. Comput. Phys.*, 2(3):367–450, 2007. ISSN 1815-2406.

[14] Gasda, S.E., Nordbotten, J.M., Celia, M. A.: Vertical equilibrium with sub-scale analytical methods for geological CO_2 sequestration. *Comput. Geosci.*, In press.

[15] Hughes, T.J.R., Sangalli, G.: Variational multiscale analysis: the fine-scale Green's function, projection, optimization, localization, and stabilized methods. *SIAM J. Numer. Anal.*, 45(2):539–557 (electronic), 2007. ISSN 0036-1429.

[16] Hughes, T.J.R., Feijóo, G.R., Mazzei, L., Quincy, J.-B.: The variational multiscale method—a paradigm for computational mechanics. *Comput. Methods Appl. Mech. Engrg.*, 166(1-2):3–24, 1998. ISSN 0045-7825.

[17] Jenny, P., Lee, S.H., Tchelepi, H.A.:. Adaptive multiscale finite-volume method for multiphase flow and transport in porous media. *Multiscale Model. Simul.*, 3(1):50–64 (electronic), 2004/05. ISSN 1540-3459.

[18] Juanes, R., Dub, F.-X.: A locally conservative variational multiscale method for the simulation of porous media flow with multiscale source terms. *Comput. Geosci.*, 12(3):273–295, 2008.

[19] Lake, L.: *Enhanced Oil Recovery*. Prentice-Hall, 1989.

[20] Lambers, J.V., Gerritsen, M.G., Mallison, B.T.: Accurate local upscaling with variable compact multipoint transmissibility calculations. *Comput. Geosci.*, 12 (3):399–416, 2008.

[21] Larson, M.G., Målqvist, A.: Goal oriented adaptivity for coupled flow and transport problems with applications in oil reservoir simulations. *Comput. Methods Appl. Mech. Engrg.*, 196(37-40):3546–3561, 2007. ISSN 0045-7825.

[22] Muskat, M.: *Physical Principles of Oil Production*. McGraw-Hill, 1949.

[23] Nordbotten, J.M.: Adaptive variational multiscale methods for multiphase flow in porous media. *Multiscale Model. and Simul.*, in press.

[24] Nordbotten, J.M.: Variational and heterogeneous multiscale methods for non-linear problems. *Commun. Math. Sci.*, submitted.

[25] Nordbotten, J.M., Bjørstad, P.E.: On the relationship between the multiscale finite-volume method and domain decomposition preconditioners. *Comput. Geosci.*, 12(3):367–376, 2008.
[26] Nordbotten, J.M., Celia, M.A.: An improved analytical solution for interface upconing around a well. *Wat. Resour. Res.*, 42:W08433, 2006.
[27] Nordbotten, J.M., Nogués, J.P., Celia, M.A.: Appropriate choice of average pressure for upscaling relative permeability in dynamic flow conditions. *SPE J.*, in press.
[28] Nordbotten, J.M., Celia, M.A.: Similarity solutions for fluid injection into confined aquifers. *J. Fluid Mech.*, 561:307–327, 2006. ISSN 0022-1120.
[29] Papadopoulos, A., Tchelepi, H.: Block smoothed aggregation algebraic mutigrid for reservoir simulation systems. In *Proceedings of Copper Mountain Conference on Multigrid Methods*, 2003.
[30] Whitaker, S. *The Method of Volume Averaging* Theory and Applications of Transport in Porous Media, 13. Springer, 1998.

Mixed Plane Wave Discontinuous Galerkin Methods

Ralf Hiptmair[1] and Ilaria Perugia[2]

[1] SAM, ETH Zurich, CH-8092 Zürich; hiptmair@sam.math.ethz.ch
[2] Dipartimento di Matematica, Università di Pavia; ilaria.perugia@unipv.it

Summary. In this paper, we extend the class of plane wave discontinuous Galerkin methods for the two-dimensional inhomogeneous Helmholtz equation presented in Gittelson, Hiptmair, and Perugia [2007]. More precisely, we consider the case of numerical fluxes defined in mixed form, namely, numerical fluxes explicitly defined in terms of both the primal and the flux variable, instead of the primal variable and its gradient. In our error analysis, we rely heavily on the approximation results and inverse estimates for plane waves proved in Gittelson, Hiptmair, and Perugia [2007] and develop a new mixed duality argument.

1 Introduction

The oscillatory behavior of solutions to time harmonic wave problems, along with numerical dispersion, renders standard finite element methods inefficient already in medium-frequency regimes. As an alternative, several ways to incorporate information from the equation into the discretization spaces have been proposed in the literature, giving rise to methods based on shape functions which are solutions to either the primal or the dual problem. The so-called "ultra weak variational formulation" (UWVF) introduced by Després for the Helmholtz equation in the 1990's (see [5, 7]) belongs to this class of methods.

The UWVF was inspired by the domain decomposition approach introduced in [6], where Robin-type transmission conditions were used in order to guarantee well-posedness of the subproblems. The introduction of these impedance interelement traces as unknowns and the use of discontinuous piecewise plane wave basis functions are the basic ingredients of the UWVF. This method, which was numerically proved to be effective, has attracted new interest very recently (see, e.g., [9, 10, 11]).

From a theoretical point of view, the UWVF has been analyzed in [5], where the convergence of discrete solutions to the impedance trace of the analytical solution on the domain boundary was proved. On the other hand, numerical results showed that convergence is achieved not only at the boundary, but in the whole domain. In the recent papers [3, 8], convergence of the h–version of the UWVF was proved by

M. Bercovier et al. (eds.), *Domain Decomposition Methods in Science and Engineering XVIII*, Lecture Notes in Computational Science and Engineering,
DOI: 10.1007/978-3-642-02677-5,

recasting the UWVF in the discontinuous Galerkin (DG) framework: in [3], slightly suboptimal L^2–error estimates were derived for the homogeneous Helmholtz problem by exploiting a result by [5], while in [8], error estimates in a mesh-dependent broken H^1–norm, as well as in the L^2–norm, for the inhomogeneous Helmholtz problem were proved based on duality techniques. All these estimates require a minimal resolution of the mesh to resolve the wavelength which shows that the plane wave discontinuous Galerkin (PWDG) method is not free from the *pollution effect* (see, e.g., [2, 12]).

In order to get the stability properties necessary to develop the theoretical analysis in [8], the numerical fluxes had to be defined in a slightly different way with respect to the original UWVF, introducing mesh and wave number dependent parameters.

In this paper, we extend the class of PWDG methods for the two-dimensional inhomogeneous Helmholtz equation presented in [8] by allowing for numerical fluxes defined in mixed form, namely, numerical fluxes explicitly defined in terms of both the primal and the flux variable, instead of the primal variable and its gradient. For these *mixed* PWDG methods, we essentially prove the same results as in [8] for the *primal* PWDG methods, by exploiting the approximation results and inverse estimates for plane waves proved in [8], and by developing a new mixed duality argument.

2 Mixed Discontinuous Galerkin Approach

Consider the following model boundary value problem for the Helmholtz equation:

$$\begin{aligned} -\Delta u - \omega^2 u &= f && \text{in } \Omega, \\ \nabla u \cdot n + i\omega u &= g && \text{on } \partial\Omega. \end{aligned} \tag{1}$$

Here, Ω is a bounded polygonal/polyhedral Lipschitz domain in $\mathbb{R}^d$, $d = 2,3$, and $\omega > 0$ denotes a fixed wave number (the corresponding wavelength is $\lambda = 2\pi/\omega$). The right hand side f is a source term in $H^{-1}(\Omega)$, n is the outer normal unit vector to $\partial\Omega$, and i is the imaginary unit. Inhomogeneous first order absorbing boundary conditions in the form of impedance boundary conditions are used in (1), with boundary data $g \in H^{-1/2}(\partial\Omega)$.

Denoting by $(\cdot,\cdot)$ the standard complex $L^2(\Omega)$–inner product, namely, $(u,v) = \int_\Omega u\bar{v}\,\mathrm{d}V$, the variational formulation of (1) reads as follows: find $u \in H^1(\Omega)$ such that, for all $v \in H^1(\Omega)$,

$$(\nabla u, \nabla v) - \omega^2 (u,v) + i\omega \int_{\partial\Omega} u\bar{v}\,\mathrm{d}S = (f,v) + \int_{\partial\Omega} g\bar{v}\,\mathrm{d}S\,. \tag{2}$$

Existence and uniqueness of solutions of (2) is well establish, see, e.g., [13, sec. 8.1].

Introduce the auxiliary variable $\sigma := \nabla u / i\omega$ and write problem (1) as a first order system:

$$
\begin{aligned}
i\omega\sigma &= \nabla u && \text{in } \Omega\,, \\
i\omega u - \nabla\cdot\sigma &= \frac{1}{i\omega} f && \text{in } \Omega\,, \\
i\omega\sigma\cdot n + i\omega u &= g && \text{on } \partial\Omega\,.
\end{aligned}
\tag{3}
$$

Now, introduce a partition $\mathcal{T}_h$ of Ω into subdomains K, and proceed as in [4]. By multiplying the first and second equations of (3) by smooth test functions τ and v, respectively, and integrating by parts on each K, we obtain

$$
\begin{aligned}
&\int_K i\omega\,\sigma\cdot\overline{\tau}\,\mathrm{d}V + \int_K u\overline{\nabla\cdot\tau}\,\mathrm{d}V - \int_{\partial K} u\,\overline{\tau\cdot n}\,\mathrm{d}S = 0 && \forall \tau \in \mathbf{H}(\mathrm{div};K) \\
&\int_K i\omega\,u\overline{v}\,\mathrm{d}V + \int_K \sigma\cdot\overline{\nabla v}\,\mathrm{d}V - \int_{\partial K} \sigma\cdot n\overline{v}\,\mathrm{d}S = \frac{1}{i\omega}\int_K f\overline{v}\,\mathrm{d}V && \forall v\in H^1(K)\,.
\end{aligned}
\tag{4}
$$

Introduce discontinuous *discrete* function spaces Σ_h and V_h; replace σ, τ by $\sigma_h, \tau_h \in \Sigma_h$ and u, v by $u_h, v_h \in V_h$. Then, approximate the traces of u and σ across interelement boundaries by the so-called *numerical fluxes* denoted by $\widehat{u}_h$ and $\widehat{\sigma}_h$, respectively (see, e.g., [1] for details) and obtain

$$
\begin{aligned}
&\int_K i\omega\,\sigma_h\cdot\overline{\tau}_h\,\mathrm{d}V + \int_K u_h\overline{\nabla\cdot\tau_h}\,\mathrm{d}V - \int_{\partial K} \widehat{u}_h\,\overline{\tau_h\cdot n}\,\mathrm{d}S = 0 && \forall \tau_h \in \Sigma_h(K) \\
&\int_K i\omega\,u_h\overline{v}_h\,\mathrm{d}V + \int_K \sigma_h\cdot\overline{\nabla v_h}\,\mathrm{d}V - \int_{\partial K} \widehat{\sigma}_h\cdot n\overline{v}_h\,\mathrm{d}S = \frac{1}{i\omega}\int_K f\overline{v}_h\,\mathrm{d}V && \forall v_h\in V_h(K)\,.
\end{aligned}
\tag{5}
$$

At this point, in order to complete the definition of classical DG methods, one "simply" needs to choose the numerical fluxes $\widehat{u}_h$ and $\widehat{\sigma}_h$ (notice that only the normal component of $\widehat{\sigma}_h$ is needed).

In order to define the numerical fluxes, we first introduce the following standard notation (see, e.g., [1]): let u_h and σ_h be a piecewise smooth function and vector field on $\mathcal{T}_h$, respectively. On $\partial K^-\cap\partial K^+$, we define

$$
\begin{aligned}
&\text{the averages:} && \{\{u_h\}\} := \tfrac{1}{2}(u_h^+ + u_h^-), && \{\{\sigma_h\}\} := \tfrac{1}{2}(\sigma_h^+ + \sigma_h^-)\,, \\
&\text{the jumps:} && [\![u_h]\!]_N := u_h^+ n^+ + u_h^- n^-, && [\![\sigma_h]\!]_N := \sigma_h^+\cdot n^+ + \sigma_h^-\cdot n^-\,.
\end{aligned}
$$

Taking a cue from [4], we can now introduce the *mixed* numerical fluxes: on $\partial K^-\cap\partial K^+ \subset \mathcal{F}_h^{\mathcal{I}}$, we define

$$
\begin{aligned}
\widehat{\sigma}_h &= \{\{\sigma_h\}\} - \alpha\,[\![u_h]\!]_N - \gamma[\![\sigma_h]\!]_N, \\
\widehat{u}_h &= \{\{u_h\}\} + \gamma\cdot[\![u_h]\!]_N - \beta\,[\![\sigma_h]\!]_N,
\end{aligned}
\tag{6}
$$

and on $\partial K\cap\partial\Omega \subset \mathcal{F}_h^{\mathcal{B}}$, we define

$$
\begin{aligned}
\widehat{\sigma}_h &= \sigma_h - (1-\delta)\left(\sigma_h + u_h n - \frac{1}{i\omega} g n\right), \\
\widehat{u}_h &= u_h - \delta\left(\sigma_h\cdot n + u_h - \frac{1}{i\omega} g\right).
\end{aligned}
\tag{7}
$$

These numerical fluxes are consistent and therefore the corresponding method is consistent. Moreover, adjoint consistency is guaranteed, due to symmetry; see [1].

We will assume

$$\alpha = \mathrm{a}/\omega \mathrm{h}\,, \qquad \beta = \mathrm{b}\omega \mathrm{h}\,, \qquad \gamma = 0\,, \qquad \delta = \mathrm{d}\omega \mathrm{h}\,, \tag{8}$$

with functions $\mathrm{a} > 0$ on $\mathcal{F}_h^{\mathcal{I}}$, $\mathrm{b} \geq 0$ on $\mathcal{F}_h^{\mathcal{I}}$ and d on $\mathcal{F}_h^{\mathcal{B}}$, all bounded from above (and below in the case of a) independent of the mesh size and ω. Moreover, the choice of d on $\mathcal{F}_h^{\mathcal{B}}$ must ensure that $0 \leq \delta \leq 1$. A further assumption on d will be stated in Sect. 3.

Remark 1. Whenever $\beta = 0$, it is possible to eliminate the auxiliary variable σ_h from the final system and write the mixed DG methods in primal formulation. On the other hand, also in these cases, the PWDG methods defined in the next section differs from the PWDG methods presented in [8] because the fluxes there were defined using $\nabla_h u_h$ instead of σ_h in (6)–(7).

3 Convergence Analysis of the Mixed PWDG Method

We restrict ourselves to the two-dimensional case and assume that Ω is a *convex* polygon and that $\mathcal{T}_h$ is a triangular mesh with possible hanging nodes satisfying the shape regularity assumption.

We carry our analysis of the mixed DG method (5) with numerical fluxes defined by (6) and (7), with parameters satisfying (8). In addition, we opt for a Trefftz type DG method: the local test and trial spaces will be spanned by plane wave functions and their gradients, which belong to the kernel of the Helmholtz operator. In particular, writing

$$PW_\omega^p(\mathbb{R}^2) = \left\{ v \in C^\infty(\mathbb{R}^2) : v(x) = \sum_{j=1}^{p} \alpha_j \exp(i\omega d_j \cdot x),\ \alpha_j \in \mathbb{C} \right\}, \tag{9}$$

with even spaced directions

$$d_j = \begin{pmatrix} \cos(\frac{2\pi}{p}(j-1)) \\ \sin(\frac{2\pi}{p}(j-1)) \end{pmatrix}, \quad j = 1, \ldots, p\,, \tag{10}$$

we set

$$V_h = \{ v \in L^2(\Omega) : v_{|K} \in PW_\omega^p(\mathbb{R}^2)\ \forall K \in \mathcal{T}_h \}\,, \qquad \Sigma_h = V_h^2; \tag{11}$$

notice that $\nabla V_h \subseteq \Sigma_h$ and $\nabla \cdot \Sigma_h \subseteq V_h$.

Let $V \subseteq H^2(\Omega)$ be the space containing all possible solutions u to (1) and $\Sigma = \nabla V$, and denote by $\mathfrak{Q}$ the product space $\mathfrak{Q} = \Sigma \times V$; we set

$$\mathfrak{p} := \begin{bmatrix} \sigma \\ u \end{bmatrix}.$$

Similarly, we define $\mathfrak{Q}_h := \Sigma_h \times V_h$ and denote by $\mathfrak{p}_h$ and $\mathfrak{q}_h$ the vectors containing the discrete solution to (5) and the generic test function in $\mathfrak{Q}_h$, namely,

$$\mathfrak{p}_h := \begin{bmatrix} \sigma_h \\ u_h \end{bmatrix}, \qquad \mathfrak{q}_h := \begin{bmatrix} \tau_h \\ v_h \end{bmatrix}.$$

We define the following seminorm and norms in $\mathfrak{Q} + \mathfrak{Q}_h$:

$$\begin{aligned}
|\mathfrak{q}|_{DG}^2 &= \omega^2 \|\tau\|_{0,\Omega}^2 + \omega \|\beta^{1/2} [\![\tau]\!]_N\|_{0,\mathcal{F}_h^{\mathcal{I}}}^2 + \omega \|\alpha^{1/2} [\![v]\!]_N\|_{0,\mathcal{F}_h^{\mathcal{I}}}^2 \\
&\quad + \omega \|\delta^{1/2} \tau \cdot n\|_{0,\mathcal{F}_h^{\mathcal{B}}}^2 + \omega \|(1-\delta)^{1/2} v\|_{0,\mathcal{F}_h^{\mathcal{B}}}^2, \\
\|\mathfrak{q}\|_{DG}^2 &= |\mathfrak{q}|_{DG}^2 + \omega^2 \|v\|_{0,\Omega}^2, \\
\|\mathfrak{q}\|_{DG^+}^2 &= \|\mathfrak{q}\|_{DG}^2 + \omega \|\beta^{-1/2} \{\{v\}\}\|_{0,\mathcal{F}_h^{\mathcal{I}}}^2 + \omega \|\alpha^{-1/2} \{\{\tau\}\}\|_{0,\mathcal{F}_h^{\mathcal{I}}}^2 + \omega \|\delta^{-1/2} v\|_{0,\mathcal{F}_h^{\mathcal{B}}}^2.
\end{aligned}$$

Multiply the first and second equations in (5) by $-i\omega$ and by $i\omega$, respectively, and add over all $K \in \mathcal{T}_h$, integrating the second term in the second equation by parts. Then, add the conjugate of the first equation to the second equation. Finally, replace $\widehat{u}_h$ and $\widehat{\sigma}_h$ with the numerical fluxes according to (6) and (7), and write the mixed PWDG method as follows: find $\mathfrak{p}_h \in \mathfrak{Q}_h$ such that, for all $\mathfrak{q}_h \in \mathfrak{Q}_h$,

$$\mathcal{A}_h(\mathfrak{p}_h, \mathfrak{q}_h) - \omega^2 (u_h, v_h) = (f, v_h) - \int_{\mathcal{F}_h^{\mathcal{B}}} \delta\, \tau_h \cdot n \overline{g} + \int_{\mathcal{F}_h^{\mathcal{B}}} (1-\delta)\, g \overline{v}_h. \tag{12}$$

Here, $\mathcal{A}_h(\cdot,\cdot)$ is the DG-bilinear form on $(\mathfrak{Q} + \mathfrak{Q}_h) \times (\mathfrak{Q} + \mathfrak{Q}_h)$ defined by

$$\begin{aligned}
\mathcal{A}_h(\mathfrak{p}, \mathfrak{q}) &= \omega^2 (\tau, \sigma) + i\omega \int_{\mathcal{F}_h^{\mathcal{I}}} \beta\, [\![\tau]\!]_N [\![\overline{\sigma}]\!]_N + i\omega \int_{\mathcal{F}_h^{\mathcal{B}}} \delta\, \tau \cdot n \overline{\sigma \cdot n} \\
&\quad + i\omega (\nabla_h \cdot \tau, u) - i\omega \int_{\mathcal{F}_h^{\mathcal{I}}} [\![\tau]\!]_N \{\{\overline{u}\}\} - i\omega \int_{\mathcal{F}_h^{\mathcal{B}}} (1-\delta)\, \tau \cdot n \overline{u} \\
&\quad - i\omega (\nabla_h \cdot \sigma, v) + i\omega \int_{\mathcal{F}_h^{\mathcal{I}}} [\![\sigma]\!]_N \{\{\overline{v}\}\} + i\omega \int_{\mathcal{F}_h^{\mathcal{B}}} (1-\delta)\, \sigma \cdot n \overline{v} \\
&\quad + i\omega \int_{\mathcal{F}_h^{\mathcal{I}}} \alpha\, [\![u]\!]_N \cdot [\![\overline{v}]\!]_N + i\omega \int_{\mathcal{F}_h^{\mathcal{B}}} (1-\delta)\, u \overline{v}.
\end{aligned}$$

Notice that

$$|\mathcal{A}_h(\mathfrak{q}_h, \mathfrak{q}_h)| \geq \frac{1}{\sqrt{2}} |\mathfrak{q}_h|_{DG}^2 \quad \forall \mathfrak{q}_h \in \mathfrak{Q}_h. \tag{13}$$

Moreover, the PWDG method (12) is consistent by construction, and thus

$$\mathcal{A}_h(\mathfrak{p}_h, \mathfrak{q}_h) = \mathcal{A}_h(\mathfrak{p}, \mathfrak{q}_h) - \omega^2 (u - u_h, v_h) \quad \forall \mathfrak{q}_h \in \mathfrak{Q}_h. \tag{14}$$

We develop the theoretical analysis of the method (12) by using Schatz' argument (see [14]). We start by stating the following abstract estimate.

Proposition 1. *Assume $0 < \delta < 1/2$. Denoting by Π_h the L^2–projection onto $\mathfrak{Q}_h$, we have*

$$\|\mathfrak{p} - \mathfrak{p}_h\|_{DG} \leq C_{\text{abs}} \|\mathfrak{p} - \Pi_h \mathfrak{p}\|_{DG^+} + (\sqrt{2}+1) \sup_{0 \neq w_h \in V_h} \frac{\omega |(u - u_h, w_h)|}{\|w_h\|_{0,\Omega}},$$

where $C_{\text{abs}} > 0$ is a constant independent of ω and of the mesh size.

Proof. By the triangle inequality, for all $\mathfrak{q}_h \in \mathfrak{Q}_h$, it holds

$$\|\mathfrak{p}-\mathfrak{p}_h\|_{DG} \leq \|\mathfrak{p}-\mathfrak{q}_h\|_{DG} + \|\mathfrak{q}_h-\mathfrak{p}_h\|_{DG}.$$

From the definition of the DG–norm, (13) and (14), we get

$$\begin{aligned}
\|\mathfrak{q}_h-\mathfrak{p}_h\|_{DG}^2 &= |\mathfrak{q}_h-\mathfrak{p}_h|_{DG}^2 + \omega^2\|v_h-u_h\|_{0,\Omega}^2 \\
&\leq \sqrt{2}\,|\mathcal{A}_h(\mathfrak{q}_h-\mathfrak{p}_h,\mathfrak{q}_h-\mathfrak{p}_h)| + \omega^2(v_h-u_h,v_h-u_h) \\
&\leq \sqrt{2}\,|\mathcal{A}_h(\mathfrak{q}_h-\mathfrak{p},\mathfrak{q}_h-\mathfrak{p}_h)| + \sqrt{2}\omega^2|(u-u_h,v_h-u_h)| \\
&\quad + \omega^2|(v_h-u,v_h-u_h)| + \omega^2|(u-u_h,v_h-u_h)| \\
&= \sqrt{2}\,|\mathcal{A}_h(\mathfrak{q}_h-\mathfrak{p},\mathfrak{q}_h-\mathfrak{p}_h)| + \omega^2|(v_h-u,v_h-u_h)| \\
&\quad + (\sqrt{2}+1)\,\omega^2|(u-u_h,v_h-u_h)|.
\end{aligned}$$

Now, select $\mathfrak{q}_h = \Pi_h\mathfrak{p}$, i.e., $\tau_h = \Pi_{\Sigma_h}\sigma$ and $v_h = \Pi_{V_h}u$, with Π_{Σ_h} and Π_{V_h} denoting the L^2–projections onto Σ_h and V_h, respectively. Since, as consequence of $\nabla\cdot\Sigma_h \subseteq V_h$,

$$\begin{aligned}
(\tau_h-\sigma_h,\Pi_{\Sigma_h}\sigma-\sigma) &= 0\,, \\
\big(\nabla_h\cdot(\tau_h-\sigma_h),\Pi_{V_h}u-u\big) &= 0\,,
\end{aligned}$$

and, integrating by parts and using $\big(\Pi_{\Sigma_h}\sigma-\sigma,\nabla_h(v_h-u_h)\big)=0$,

$$\begin{aligned}
(\nabla_h\cdot(\Pi_{\Sigma_h}\sigma-\sigma),v_h-u_h) = &\int_{\mathcal{F}_h^{\mathcal{I}}} \{\{\Pi_{\Sigma_h}\sigma-\sigma\}\}\cdot[[\overline{v_h-u_h}]]_N \\
&+ \int_{\mathcal{F}_h^{\mathcal{I}}} [[\Pi_{\Sigma_h}\sigma-\sigma]]_N\{\{\overline{v_h-u_h}\}\} \\
&+ \int_{\mathcal{F}_h^{\mathcal{B}}} (\Pi_{\Sigma_h}\sigma-\sigma)\cdot n\,(\overline{v_h-u_h})\,,
\end{aligned}$$

we immediately have

$$|\mathcal{A}_h(\mathfrak{q}_h-\mathfrak{p},\mathfrak{q}_h-\mathfrak{p}_h)| \leq C\,\|\mathfrak{p}-\Pi_h\mathfrak{p}\|_{DG^+}\,\|\mathfrak{q}_h-\mathfrak{p}_h\|_{DG}\,,$$

where $C>0$ is a constant independent of ω and of the mesh size (also independent of α and β; yet it may depend on δ). Moreover,

$$(v_h-u,v_h-u_h) = (\Pi_{V_h}u-u,v_h-u_h) = 0.$$

Therefore,

$$\begin{aligned}
\|\mathfrak{p}-\mathfrak{p}_h\|_{DG} &\leq (C+1)\|\mathfrak{p}-\Pi_h\mathfrak{p}\|_{DG^+} + (\sqrt{2}+1)\,\omega^2\,\frac{|(u-u_h,v_h-u_h)|}{\|\mathfrak{p}-\mathfrak{p}_h\|_{DG}} \\
&\leq (C+1)\|\mathfrak{p}-\Pi_h\mathfrak{p}\|_{DG^+} + (\sqrt{2}+1)\,\omega^2\,\frac{|(u-u_h,v_h-u_h)|}{\omega\,\|v_h-u_h\|_{0,\Omega}},
\end{aligned}$$

from which the result follows.

We bound the term $\sup_{0\neq w_h\in V_h} \frac{\omega|(u-u_h,w_h)|}{\|w_h\|_{0,\Omega}}$ in the estimate of Proposition (1) by a duality argument. To this end, we assume $0<\delta<1/2$.

We will make use of the following theorem proved in [13]. Its original statement makes use of the following weighted norm on $H^1(\Omega)$:

$$\|v\|_{1,\omega,\Omega}^2 = |v|_{1,\Omega}^2 + \omega^2\|v\|_{0,\Omega}^2 \,. \tag{15}$$

Theorem 1. *[13, Propostion 8.1.4] Let Ω be a bounded convex domain (or smooth and star-shaped). Consider the adjoint problem to* (1) *with right-hand side $w \in L^2(\Omega)$:*

$$\begin{aligned} -\Delta\varphi - \omega^2\varphi &= w \qquad && \text{in } \Omega, \\ -\nabla\varphi\cdot n + i\omega\varphi &= 0 && \text{on } \partial\Omega. \end{aligned} \tag{16}$$

Then, $\varphi \in H^2(\Omega)$, and there are constants $C_1(\Omega), C_2(\Omega) > 0$ such that

$$\begin{aligned} \|\varphi\|_{1,\omega,\Omega} &\le C_1(\Omega)\|w\|_{0,\Omega} \,, \\ |\varphi|_{2,\Omega} &\le C_2(\Omega)\,(1+\omega)\,\|w\|_{0,\Omega} \,. \end{aligned} \tag{17}$$

As a consequence of Theorem 1, we have the following bounds:

$$|\varphi|_{2,\Omega} + \omega^2\|\varphi\|_{0,\Omega} \le C(1+\omega)\|w\|_{0,\Omega} \,, \tag{18}$$

and, setting $\Phi = \nabla\varphi/i\omega$,

$$\omega|\Phi|_{1,\Omega} + \omega^2\|\Phi\|_{0,\Omega} \le C(1+\omega)\|w\|_{0,\Omega} \,. \tag{19}$$

The next lemma summarizes the results in Propositions 4.12, 4.13, 4.14 and Lemma 5.6 of [8].

Lemma 1. *Let v be in $H^2(\Omega)$. Then,*

$$\begin{aligned} \|v-\Pi_{V_h}v\|_{0,\Omega} &\le Ch^2(|v|_{2,\Omega}+\omega^2\|v\|_{0,\Omega}) \,, \\ |v-\Pi_{V_h}v|_{1,\Omega} &\le Ch(\omega h+1)(|v|_{2,\Omega}+\omega^2\|v\|_{0,\Omega}) \,, \\ |\Pi_{V_h}v|_{2,\Omega} &\le C(\omega h+1)^2(|v|_{2,\Omega}+\omega^2\|v\|_{0,\Omega}) \,, \\ \|v-\Pi_{V_h}v\|_{0,\mathcal{F}_h} &\le Ch^{3/2}(\omega h+1)^{1/2}(|v|_{2,\Omega}+\omega^2\|v\|_{0,\Omega}) \,, \\ \|\nabla_h(v-\Pi_{V_h}v)\|_{0,\mathcal{F}_h} &\le Ch^{1/2}(\omega h+1)^{3/2}(|v|_{2,\Omega}+\omega^2\|v\|_{0,\Omega}) \,, \end{aligned}$$

with a constant $C>0$ depending only on the bound for the minimal angle of elements and the domain Ω.

For functions which are only in $H^1(\Omega)$, we have the following bounds.

Lemma 2. *Let v be in $H^1(\Omega)$. Then,*

$$\begin{aligned} \|v-\Pi_{V_h}v\|_{0,\Omega} &\le Ch\max\{\omega^{-1},h\}\,(\omega|v|_{1,\Omega}+\omega^2\|v\|_{0,\Omega}) \,, \\ |\Pi_{V_h}v|_{1,\Omega} &\le C(\omega h+1)\max\{\omega^{-1},h\}\,(\omega|v|_{1,\Omega}+\omega^2\|v\|_{0,\Omega}) \,, \\ \|v-\Pi_{V_h}v\|_{0,\mathcal{F}_h} &\le Ch^{1/2}(\omega h+1)^{1/2}\max\{\omega^{-1},h\}\,(\omega|v|_{1,\Omega}+\omega^2\|v\|_{0,\Omega}) \,, \end{aligned}$$

with a constant $C > 0$ depending only on the bound for the minimal angle of elements and the domain Ω. In particular, as soon as $\omega h < 1$,

$$\begin{aligned}
\|v - \Pi_{V_h} v\|_{0,\Omega} &\leq C\omega^{-1} h (\omega |v|_{1,\Omega} + \omega^2 \|v\|_{0,\Omega}) ,\\
|\Pi_{V_h} v|_{1,\Omega} &\leq C\omega^{-1} (\omega h + 1)(\omega |v|_{1,\Omega} + \omega^2 \|v\|_{0,\Omega}) ,\\
\|v - \Pi_{V_h} v\|_{0,\mathcal{F}_h} &\leq C\omega^{-1} h^{1/2} (\omega h + 1)^{1/2} (\omega |v|_{1,\Omega} + \omega^2 \|v\|_{0,\Omega}) .
\end{aligned}$$

Proof. The proof can be carried out by proceeding as in the proofs of Proposition 4.12, Proposition 4.13 and Lemma 5.6 of [8].

Proposition 2. *Let the assumptions of Theorem 1 hold true. With the choice of the flux parameters in* (8) *and* $0 < \delta < 1/2$, *the following estimate holds true:*

$$\sup_{0 \neq w_h \in V_h} \frac{\omega |(u - u_h, w_h)|}{\|w_h\|_{0,\Omega}} \leq C_{\text{dual}}\, \omega h (1 + \omega)(\|\mathfrak{p} - \mathfrak{p}_h\|_{DG} + h \|f - \Pi_{V_h} f\|_{0,\Omega}),$$

with a constant $C_{\text{dual}} > 0$ independent of the mesh and ω.

Proof. Consider the adjoint problem to (1) with right-hand side $w_h \in V_h$:

$$\begin{aligned}
-\Delta \varphi - \omega^2 \varphi &= w_h && \text{in } \Omega,\\
-\nabla \varphi \cdot n + i\omega \varphi &= 0 && \text{on } \partial\Omega.
\end{aligned} \tag{20}$$

Then, from Theorem 1, we have that $\varphi \in H^2(\Omega)$, $\|\varphi\|_{1,\omega,\Omega} \leq C_1(\Omega) \|w_h\|_{0,\Omega}$ and $|\varphi|_{2,\Omega} \leq C_2(\Omega)(1 + \omega) \|w_h\|_{0,\Omega}$, with $C_1(\Omega), C_2(\Omega) > 0$.

Define $\Phi := \nabla \varphi / i\omega$; setting

$$\mathfrak{s} := \begin{bmatrix} \Phi \\ \varphi \end{bmatrix}, \qquad \mathfrak{t} := \begin{bmatrix} \Psi \\ \psi \end{bmatrix}, \qquad \mathfrak{t}_h := \begin{bmatrix} \Psi_h \\ \psi_h \end{bmatrix},$$

the solution $\mathfrak{s}$ satisfies

$$\mathcal{A}_h(\mathfrak{t}, \mathfrak{s}) - \omega^2 (\psi, \varphi) = (\psi, w_h) \quad \forall \mathfrak{t} \in \mathfrak{Q}.$$

The adjoint consistency of the DG method implies that

$$\mathcal{A}_h(\mathfrak{t}_h, \mathfrak{s}) - \omega^2 (\psi_h, \varphi) = (\psi_h, w_h) \quad \forall \mathfrak{t}_h \in \mathfrak{Q}_h.$$

Taking into account adjoint consistency and consistency, we have

$$\begin{aligned}
(u - u_h, w_h) &= (u - w_h) - (u_h, w_h)\\
&= \mathcal{A}_h(\mathfrak{p} - \mathfrak{p}_h, \mathfrak{s}) - \omega^2 (u - u_h, \varphi)\\
&= \mathcal{A}_h(\mathfrak{p} - \mathfrak{p}_h, \mathfrak{s} - \mathfrak{t}_h) + \mathcal{A}_h(\mathfrak{p} - \mathfrak{p}_h, \mathfrak{t}_h) - \omega^2 (u - u_h, \varphi)\\
&= \mathcal{A}_h(\mathfrak{p} - \mathfrak{p}_h, \mathfrak{s} - \mathfrak{t}_h) + \omega^2 (u - u_h, \psi_h) - \omega^2 (u - u_h, \varphi)\\
&= \mathcal{A}_h(\mathfrak{p} - \mathfrak{p}_h, \mathfrak{s} - \mathfrak{t}_h) - \omega^2 (u - u_h, \varphi - \psi_h).
\end{aligned} \tag{21}$$

From $i\omega\nabla\cdot\sigma+\omega^2 u=f$, we have the identity

$$i\omega(\nabla_h\cdot(\sigma-\sigma_h),\varphi-\psi_h)+\omega^2(u-u_h,\varphi-\psi_h)= (f,\varphi-\psi_h)-(i\omega\nabla_h\cdot\sigma_h+\omega^2 u_h,\varphi-\psi_h). \quad (22)$$

Moreover, integrating by parts and using the definition of σ, we obtain the identity

$$\begin{aligned}
& i\omega(\nabla_h\cdot(\Phi-\Psi_h),u-u_h)-i\omega\int_{\mathcal{F}_h^{\mathcal{I}}}[\![\Phi-\Psi_h]\!]_N\{\!\{\overline{u-u_h}\}\!\}\\
&-i\omega\int_{\mathcal{F}_h^{\mathcal{B}}}(\Phi-\Psi_h)\cdot n(\overline{u-u_h})\\
&=-\omega^2(\Phi-\Psi_h,\sigma-\sigma_h)-i\omega(\Phi-\Psi_h,i\omega\sigma_h-\nabla_h u_h)\\
&+i\omega\int_{\mathcal{F}_h^{\mathcal{I}}}\{\!\{\Phi-\Psi_h\}\!\}\cdot[\![\overline{u-u_h}]\!]_N.
\end{aligned} \quad (23)$$

Using (22) and (23), equation (21) becomes

$$\begin{aligned}
(u-u_h,w_h)=&\,i\omega\int_{\mathcal{F}_h^{\mathcal{I}}}\beta\,[\![\Phi-\Psi_h]\!]_N\,[\![\overline{\sigma-\sigma_h}]\!]_N+i\omega\int_{\mathcal{F}_h^{\mathcal{B}}}\delta\,(\Phi-\Psi_h)\cdot n\,(\overline{\sigma-\sigma_h\cdot n})\\
&-i\omega(\Phi-\Psi_h,i\omega\sigma_h-\nabla_h u_h)+i\omega\int_{\mathcal{F}_h^{\mathcal{I}}}\{\!\{\Phi-\Psi_h\}\!\}\cdot[\![\overline{u-u_h}]\!]_N\\
&+i\omega\int_{\mathcal{F}_h^{\mathcal{B}}}\delta\,(\Phi-\Psi_h)\cdot n\,(\overline{u-u_h})+i\omega\int_{\mathcal{F}_h^{\mathcal{I}}}[\![\sigma-\sigma_h]\!]_N\{\!\{\overline{\varphi-\psi_h}\}\!\}\\
&+i\omega\int_{\mathcal{F}_h^{\mathcal{B}}}(1-\delta)\,(\sigma-\sigma_h)\cdot n\,(\overline{\varphi-\psi_h})\\
&+i\omega\int_{\mathcal{F}_h^{\mathcal{I}}}\alpha\,[\![u-u_h]\!]_N\cdot[\![\overline{\varphi-\psi_h}]\!]_N+i\omega\int_{\mathcal{F}_h^{\mathcal{B}}}(1-\delta)\,(u-u_h)(\overline{\varphi-\psi_h})\\
&-(f,\varphi-\psi_h)+(i\omega\nabla_h\cdot\sigma_h+\omega^2 u_h,\varphi-\psi_h).
\end{aligned}$$

Form the Cauchy-Schwarz inequality, since $0<\delta<1/2$, we have

$$\begin{aligned}
\omega|(u-u_h,w_h)|\le&\, C\|\mathfrak{p}-\mathfrak{p}_h\|_{DG}\,\omega^{3/2}\Big(\|\beta^{1/2}[\![\Phi-\Psi_h]\!]_N\|_{0,\mathcal{F}_h^{\mathcal{I}}}\\
&+\|\alpha^{-1/2}\{\!\{\Phi-\Psi_h\}\!\}\|_{0,\mathcal{F}_h^{\mathcal{I}}}+\|\delta^{1/2}(\Phi-\Psi_h)\cdot n\|_{0,\mathcal{F}_h^{\mathcal{B}}}\\
&+\|\alpha^{1/2}[\![\varphi-\psi_h]\!]_N\|_{0,\mathcal{F}_h^{\mathcal{I}}}+\|\beta^{-1/2}\{\!\{\varphi-\psi_h\}\!\}\|_{0,\mathcal{F}_h^{\mathcal{I}}}\\
&+\|\delta^{-1/2}\varphi-\psi_h\|_{0,\mathcal{F}_h^{\mathcal{B}}}\Big)\\
&+\omega|(f,\varphi-\psi_h)|+\omega^2|(\Phi-\Psi_h,i\omega\sigma_h-\nabla_h u_h)|\\
&+\omega|(i\omega\nabla_h\cdot\sigma_h+\omega^2 u_h,\varphi-\psi_h)|.
\end{aligned}$$

We choose $\psi_h=\Pi_{V_h}\varphi$ and $\Psi_h=\Pi_{\Sigma_h}\Phi$. We immediately have

$$\omega^2|(\Phi-\Psi_h,i\omega\sigma_h-\nabla_h u_h)|=0\,,$$
$$\omega|(i\omega\nabla_h\cdot\sigma_h+\omega^2 u_h,\varphi-\psi_h)|=0\,,$$

and, from Lemma 1 and (18),

$$\begin{aligned}\omega\,|(f,\varphi-\psi_h)| &= \|f-\Pi_{V_h}f\|_{0,\Omega}\,\omega\|\varphi-\psi_h\|_{0,\Omega}\\ &\leq C\,\omega\,h^2(1+\omega)\,\|f-\Pi_{V_h}f\|_{0,\Omega}\,\|w_h\|_{0,\Omega}\,.\end{aligned}$$

We estimate all the interelement terms containing $(\varphi-\psi_h)$ and those containing $(\Phi-\Psi_h)$ by using Lemma 1 and (18), and Lemma 2 and (19), respectively. Taking the definitions of the flux parameters into account, we obtain

$$\omega^{3/2}(\text{interelement terms}) \leq C\,\omega\,h\,(1+\omega)\,\|w_h\|_{0,\Omega}\,.$$

The result readily follows.

The following estimate of the L^2–projection error of $\mathfrak{p}$ is a consequence of Lemma 1 and Lemma 2.

Lemma 3. *For any $\mathfrak{p}\in H^2(\Omega)\times H^1(\Omega)$, as soon as $\omega h<1$, we have*

$$\|\mathfrak{p}-\Pi_h\mathfrak{p}\|_{DG^+} \leq C\,h\,\left(|u|_{2,\Omega}+\omega^2\|u\|_{0,\Omega}+\omega|\sigma|_{1,\Omega}+\omega^2\|\sigma\|_{0,\Omega}\right).$$

The complete error estimate is a straightforward consequence of Proposition 1, Proposition 2 and Lemma 3.

Theorem 2. *Let the assumptions of Theorem 1 hold true and assume that the analytical solution to (1) belongs to $H^2(\Omega)$. With the choice of the flux parameters in (8) and $0<\delta<1/2$, provided that $\omega h<1$ and*

$$\omega\,h\,(1+\omega) < \frac{1}{(\sqrt{2}+1)\,C_{\mathrm{dual}}}\,, \tag{24}$$

the following estimate holds true:

$$\begin{aligned}\|\mathfrak{p}-\mathfrak{p}_h\|_{DG} \leq C\,h\,(&|u|_{2,\Omega}+\omega^2\|u\|_{0,\Omega}+\omega|\sigma|_{1,\Omega}+\omega^2\|\sigma\|_{0,\Omega}\\ &+\omega\,h\,(1+\omega)\|f-\Pi_{V_h}f\|_{0,\Omega})\,,\end{aligned}$$

with a constant $C>0$ independent of the mesh and ω.

Remark 2. In the relevant case of $\omega>1$, in order to satisfy the threshold condition (24), we need to require $\omega^2 h$ to be sufficiently small, instead of the milder condition ωh sufficiently small, as required for the best approximation estimates. This reflects the fact that, like for the PWDG methods of [8], the mixed PWDG methods also suffer from a *pollution effect*.

Remark 3. Like for the PWDG methods of [8], the presence of a source term $f\neq 0$ prevents the methods from being higher order convergent when increasing the number of elemental plane waves used in the approximation.

Remark 4. By proceeding like in the proof of Theorem 5.13 of [8], one can prove that, under the threshold conditions of Theorem 2,

$$\begin{aligned}\|u-u_h\|_{0,\Omega} \leq C\,h^{3/2}\,(&|u|_{2,\Omega}+\omega^2\|u\|_{0,\Omega}+\omega|\sigma|_{1,\Omega}+\omega^2\|\sigma\|_{0,\Omega}\\ &+\omega\,h\,(1+\omega)\|f-\Pi_{V_h}f\|_{0,\Omega})\,,\end{aligned}$$

with a constant $C>0$ independent of the mesh and ω.

4 Conclusion

The h-version of the plane wave discontinuous Galerkin method has been shown to converge asymptotically optimally. However, as all other local discretizations of the Helmholtz equation, this method is also affected by numerical dispersion. This is reflected by a threshold condition of the form "$\omega^2 h$ sufficiently small" for the onset of asymptotic convergence. For a primal plane wave DG method, numerical experiments in [8] demonstrate that this condition is essential. There is no reason to believe that the mixed method analyzed in this paper behaves differently.

Yet, in practical applications of the UWVF one rather tries to raise the number of plane waves than to refine the mesh. Hence, it is the p-version of the plane wave DG method that deserves more attention than the h-version.

References

[1] Arnold, D.N., Brezzi, F., Cockburn, B., Marini, L. D.: Unified analysis of discontinuous Galerkin methods for elliptic problems. *SIAM J. Numer. Anal.*, 39 (5):1749–1779, 2002.

[2] Babuška, I.M., Sauter, S.: Is the pollution effect of the FEM avoidable for the Helmholtz equation? *SIAM Rev.*, 42(3):451–484, September 2000.

[3] Buffa, A., Monk, P.: Error estimates for the ultra weak variational formulation of the Helmholtz equation. *M2AN Math. Model Numer. Anal.*, 42(6):925–940, 2008.

[4] Castillo, P., Cockburn, B., Perugia, I., Schötzau, D.: An a priori error analysis of the local discontinuous Galerkin method for elliptic problems. *SIAM J. Numer. Anal.*, 38(5):1676–1706, 2000.

[5] Cessenat, O., Després, B.: Application of an ultra weak variational formulation of elliptic PDEs to the two-dimensional Helmholtz equation. *SIAM J. Numer. Anal.*, 35(1):255–299, 1998.

[6] Després, B.: *Méthodes de décomposition de domaine pour les problèmes de propagation d'ondes en régime harmonique.* PhD thesis, Paris IX Dauphine, 1991.

[7] Després, B.: Sur une formulation variationelle de type ultra-faible. *C. R. Acad. Sci. Paris, Ser. I Math.*, 318:939–944, 1994.

[8] Gittelson, C., Hiptmair, R., Perugia, I.: Plane wave discontinuous Galerkin methods: Analysis of the h–version. *M@AN Math. Model. Numer. Anal.*, 2009. Published online, DOI: 10.1051/m2an/2009002.

[9] Huttunen, T., Malinen, M., Monk, P.: Solving Maxwell's equations using the ultra weak variational formulation. *J. Comput. Phys.*, 2006.

[10] Huttunen, T., Monk, P.: The use of plane waves to approximate wave propagation in anisotropic media. *J. Comput. Math.*, 25(3):350–367, 2007.

[11] Huttunen, T., Monk, P., Kaipio, J.P.: Computational aspects of the ultra-weak variational formulation. *J. Comput. Phys.*, 182(1):27–46, 2002.

[12] Ihlenburg, F.: *Finite Element Analysis of Acoustic Scattering*, Applied Mathematical Sciences, 132. Springer, New York, 1998.
[13] Melenk, J.: *On Generalized Finite Element Methods*. PhD thesis, University of Maryland, USA, 1995.
[14] Schatz, A.H.: An observation concerning Ritz-Galerkin methods with indefinite bilinear forms. *Math. Comp.*, 28:959–962, 1974.

Numerical Zoom and the Schwarz Algorithm

Frédéric Hecht[1], Alexei Lozinski[2], and Olivier Pironneau[1]

[1] Laboratoire Jacques-Louis Lions - Université Pierre et Marie Curie
Frederic.Hecht@upmc.fr and Olivier.Pironneau@upmc.fr
[2] Institut de Mathématiques de Toulouse, Université Paul Sabatier
alexei.lozinski@math.univ-toulouse.fr

Summary. We investigate Schwarz' domain decomposition algorithm as a tool for numerical zoom and compare it with the Subspace Correction Method. Quadrature error is investigated and the convergence of Schwarz' algorithm is sketched for non matching grids. The methods are also compared numerically.

1 Introduction

Often enough engineers do a coarse calculation and then a finer one on a subset (zoom) Λ of the whole domain Ω. We wish here to justify this approach, i.e. to study convergence and errors when the strategy is made into a loop. Obviously one zoom calculation is not enough, unless the problem is nonlinear or time dependent and the iterations for the zoom are seen as part of the nonlinear or time loop.

So the situation is as follows: a coarse calculation is done in Ω, then another one in a zoom $\Lambda \subset \Omega$. The question then is how to set properly the problem in Λ and how to feed in intelligently its solution into the coarse solver to correct it?

Chimera

The chimera method proposed by Steger [11], originally for nonlinear time dependent problems, digs a hole D in the coarse domain strictly inside the zoom region Λ. For example, to compute the hydrostatic pressure u of a porous media flow given its value on $\partial\Omega$ and governed by Darcy's law with porosity K,

$$u - g \in H_0^1(\Omega) : -\nabla \cdot (K \nabla u) = f \quad \text{in } \Omega \tag{1}$$

one chooses a sub-domain D strictly inside Λ and loops on n on the two problems:

Chimera is identical to Schwarz' algorithm for domain decomposition, but the Computational Fluid Dynamic community uses this terminology. In our numerical experiments in the present paper, the hole D will always be chosen as a union of several triangles from the coarse mesh on Ω and $U^n|_{\partial D} = u^{n-1}|_{\partial D}$ will be approximated

M. Bercovier et al. (eds.), *Domain Decomposition Methods in Science and Engineering XVIII*, Lecture Notes in Computational Science and Engineering,
DOI: 10.1007/978-3-642-02677-5,

Require: An initial guess is needed, for instance $u_H^0 = g_H$
1: **for** $n = 1 \dots N$ **do**
2: Solve

$$-\nabla\cdot(K\nabla U^n) = f \text{ in } \Omega\backslash D, \ (U^n - g)|_{\partial\Omega} = 0, \ U^n|_{\partial D} = u^{n-1}|_{\partial D}, \tag{2}$$

3: Solve

$$-\nabla\cdot(K\nabla u^n) = f \text{ in } \Lambda, \quad u^n = U^n \text{ on } \partial\Lambda. \tag{3}$$

4: **end for**

Algorithm 2: Chimera

by taking the values of u^{n-1} at the coarse grid vertices of ∂D. Although this seems to work fine in most cases, the convergence in the natural energy norm is an open problem, as well as the precision because of the unavoidable interpolation of U^n on $\partial\Lambda$ when Λ is not made of elements of the triangulation of Ω. The theoretical analysis of this method will be done here in the maximum norm. Note that an alternative way to impose $U^n|_{\partial D} = u^{n-1}|_{\partial D}$ (even with an arbitrary form of ∂D) would be to use boundary penalty on ∂D or volumic penalty on D. However, such implementations do not perform well in practice as reported in [9].

Hilbert Space Decomposition Method

An alternative idea introduced in [12] and studied in [2, 3] amounts, formally speaking, to finding a subspace correction u to the coarse solution, i.e.

find U, u with $U - g \in H_0^1(\Omega), u \in H_0^1(\Lambda)$ and

$$\int_\Omega (K\nabla(U+u)\cdot\nabla(W+w) - f(W+w)) = 0 \ \forall W \in H_0^1(\Omega), \ \ w \in H_0^1(\Lambda) \tag{4}$$

This equation is easy to discretize, with $u_H \approx U$, $u_h \approx u$, and a simple iterative scheme such as Algorithm 3. We use there the following notations: V_H and V_h are finite element spaces on some regular triangulations $\mathcal{T}_H$ and $\mathcal{T}_h$ of Ω and Λ respectively; $V_{0H} = V_H \cap H_0^1(\Omega)$; V_{gH} is the subspace of V_H consisting of functions equal to g_H on $\partial\Omega$, and $V_{0h} = V_h \cap H_0^1(\Lambda)$.

When $\Lambda \subset \Omega$, Algorithm 3 is also known as the *patch iterator* [7]. The solution $u_{Hh} = \lim_{n\to\infty}(u_H^n + u_h^n)$ thus obtained satisfies the following error estimate even if the two meshes Ω_H and Λ_h do not match:

$$\|u - u_{Hh}\|_{H^1(\Omega)} \le C\big(H^r\|u\|_{H^q(\Omega\backslash\overline{\Lambda})} + h^s\|u\|_{H^q(\Lambda)}\big), \tag{7}$$

where r and s are the maximal degrees of the polynomials used in the construction of V_H and V_h respectively and $q = \max(r,s) + 1$.

Require: an initial guess $u_h^0 \in V_{0h}$ is needed.
1: **for** $n = 1 \dots N$ **do**
2: Find $u_H^n \in V_{gH}$ by

$$\int_\Omega K\nabla u_H^n \cdot \nabla w_H = \int_\Omega f w_H - \int_\Omega K\nabla u_h^{n-1} \cdot \nabla w_H \;\; \forall w_H \in V_{0H} \tag{5}$$

3: Find $u_h^n \in V_{0h}$ by

$$\int_\Lambda K\nabla u_h^n \cdot \nabla w_h = \int_\Lambda f w_h - \int_\Lambda K\nabla u_H^n \cdot \nabla w_h \;\; \forall w_h \in V_{0h} \tag{6}$$

4: **end for**

Algorithm 3: Hilbert Space Decomposition

Harmonic Patch Iterator

The drawback of Hilbert Decomposition method is that its convergence can be very slow when the triangulations $\mathcal{T}_H$, $\mathcal{T}_h$ are not nested. The method needs to be improved; this is the object of Algorithm 4, the Harmonic patch method of [8]. To write it down, we need the following (normally low dimensional) subspace of V_H:

$$V_H^0 = \{v_H \in V_H : \text{supp } v_H \subset \Lambda\}.$$

Require: an initial guess $u_h^0 \in V_{0h}$.
1: **for** $n = 1 \dots N$ **do**
2: Find $\lambda_H^n \in V_H^0$ such that

$$\int_\Omega K\nabla \lambda_H^n \cdot \nabla \mu_H = \int_\Omega f \mu_H - \int_\Omega K\nabla u_h^{n-1} \cdot \nabla \lambda_H \;\; \forall \mu_H \in V_H^0 \tag{8}$$

3: Find $u_H^n \in V_{gH}$ by

$$\int_\Omega K\nabla u_H^n \cdot \nabla w_H = \int_\Omega f w_H - \int_\Omega K\nabla u_h^{n-1} \cdot \nabla w_H - \int_\Omega K\nabla \lambda_H^n \cdot \nabla v_H \;\; \forall w_H \in V_{0H} \tag{9}$$

4: Find $u_h^n \in V_{0h}$ by

$$\int_\Lambda K\nabla u_h^n \cdot \nabla w_h = \int_\Lambda f w_h - \int_\Lambda K\nabla u_H^n \cdot \nabla w_h = 0 \;\; \forall w_h \in V_{0h} \tag{10}$$

5: **end for**

Algorithm 4: Harmonic patch iterator

The new variable λ_H^n is merely auxiliary, and the solution is recovered as $u_{Hh} = \lim_{n\to\infty}(u_H^n + u_h^n)$ exactly as in the case of Algorithm 3. In fact these two algorithms are identical in the case of nested triangulations, in the sense that $u_H^n + u_h^n$ is then rigorously the same in Algorithms 3 and 4 for all $n \geq 1$ although each u_H^n and u_h^n may differ from one algorithm to another. In general, u_{Hh} obtained by the Harmonic

Patch Iterator can be slightly different from the limiting solution of Algorithm 3 but it still satisfies the *a priori* error estimate (7). The additional problem for λ_H^n is normally very cheap to solve permitting a great increase of the convergence rate in comparison with Algorithm 3 as confirmed by the numerical experiments in [8].

One Way Schwarz

If (as will be done in this paper) in order to facilitate the evaluation of $\int_\Lambda K\nabla u_H^n \cdot \nabla w_h$ in (6), one approximates u_H^n there by its interpolation $\gamma_h u_H^n$ on $\mathcal{T}_h$, then the resulting problem for u_h^n can be simplified. Namely, one introduces in each iteration the new unknown $w_h^n \in V_h$ as $w_h^n = \gamma_h u_H^n|_\Lambda + u_h^n$ so that w_h^n solves the same problem (3) as the fine correction u^n in the Schwarz algorithm 2. One can rewrite then Algorithm 3 in terms of u_H^n and w_h^n (without distinguishing between $\gamma_h u_H^{n-1}$ and u_H^{n-1} in the coarse correction step) and this leads to the "One way Schwarz" algorithm 5 proposed in [9]. Note that the successive approximations u_{Hh}^n to u should be defined here as $u_{Hh}^n = \{w_h^n \text{ in } \Lambda,\ u_H^n \text{ outside } \Lambda\}$ just as in the Schwarz algorithm.

Require: 2 initial guesses $u_H^0 \in V_{gH}$ and $w_h^0 \in V_h$ such that $w_h^0 = \gamma_h u_H^0$ on $\partial\Lambda$.
1: **for** $n = 1 \dots N$ **do**
2: Find $u_H^n \in V_{gH}$ by

$$\int_\Omega K\nabla u_H^n \cdot \nabla v_H = \int_\Omega f w_H + \int_\Lambda K\nabla(u_H^{n-1} - w_h^{n-1}) \cdot \nabla v_H \quad \forall v_H \in V_{0H} \tag{11}$$

3: Find $w_h^n \in V_h$ with

$$\int_{\Lambda_h} K\nabla w_h^n \cdot \nabla v_h = \int_{\Lambda_h} f v_h \quad \forall v_h \in V_{0h}, \quad w_h^n = \gamma_h u_H^n \ \text{ on } \partial\Lambda \tag{12}$$

4: **end for**

Algorithm 5: One way Schwarz

The equivalence of Algorithms 3–5 is readily seen in the nested case. In a general situation, the relations between them are rather complicated because of interpolations from one mesh to another. One can see though that our last algorithm is closer to the harmonic version 4, but its performance in practice is situated between those of Algorithms 3 and 4, see [9].

As the subspace correction methods such as Algorithms 3–5 can in principle be prone to instabilities due to the quadrature errors, we wish here to study the Chimera idea more carefully. The convergence of Algorithm 2 and an error estimate for it will be proved in the maximum norm under some natural hypotheses. It will be also numerically compared with Algorithms 3–4.

2 Convergence of Schwarz' Algorithm on Arbitrary Non-Matching Meshes

Convergence of multidomain approximations with overlap of arbitrary finite element meshes is known only in the case of the Mortar method [1]. Convergence of Schwarz' algorithm on arbitrary uniform meshes has been shown by Cai et al. [4] only for finite difference discretization. Their proof relies on the maximum principle and the exponential decay of the solution of elliptic pdes far from the boundaries. The same ideas are used here for triangular first order finite elements.

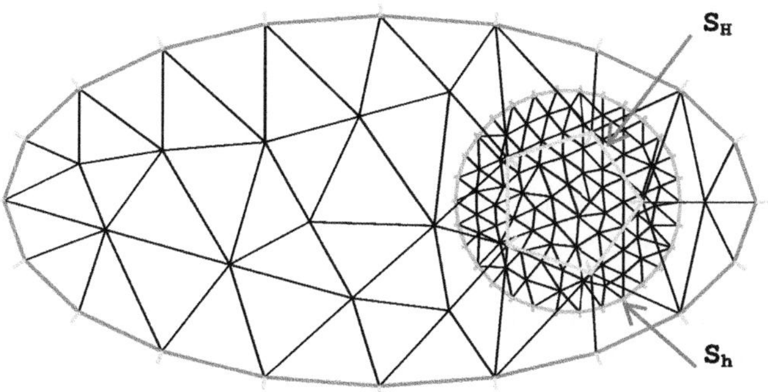

Fig. 1. Triangulations showing Ω_H outside S_H and Ω_h inside S_h.

To solve

$$-\Delta u = f \text{ in } \Omega \quad \text{with} \quad u = g \text{ on } \Gamma = \partial\Omega, \tag{13}$$

we choose two subsets of Ω, Ω_H and Ω_h and two triangulations $\mathcal{T}_H$ of Ω_H, $\mathcal{T}_h$ of Ω_h, such that

$$\Omega_H \cup \Omega_h = \Omega, \quad \partial\Omega_h \subset \Omega_H, \quad \partial\Omega_h \cap \Gamma = \emptyset, \quad \partial\Omega_H \backslash \Gamma \subset \Omega_h.$$

As in Fig. 1 we denote by S_h the boundary of Ω_h and by S_H the part of the boundary of Ω_H different from Γ_H. Next, let

$$V_H = \{v \in C^0(\Omega_H) : v|_K \in P^1, \ \forall K \in \mathcal{T}_H\}, \quad V_{0H} = \{v \in V_H : v|_{\partial\Omega_H} = 0\},$$

and similarly with h. Starting from $u_H^0 = 0$, $u_h^0 = 0$, the discrete Schwarz algorithm (same as Chimera Algorithm 2) finds $u_H^m \in V_H$ and $u_h^m \in V_h$ such that $\forall w_H \in V_{0H}$, $\forall w_h \in V_{0h}$:

$$\begin{aligned} a_H(u_H^m, w_H) &= (f, w_H), \qquad u_H^m|_{S_H} = \gamma_H u_h^{m-1}, \quad u_H^m|_{\Gamma_H} = g_H \\ a_h(u_h^m, w_h) &= (f, w_h), \qquad u_h^m|_{S_h} = \gamma_h u_H^m, \end{aligned} \tag{14}$$

where $a_{H,h}(u,v) = \int_{\Omega_{H,h}} \nabla u \cdot \nabla v$ and γ_H (resp γ_h) is the interpolation operator on V_H (resp V_h).

Hypothesis 1 Assume that the maximum principle holds for each system in (14) independently. Further assume that the solution $v_H \in V_H$ of

$$a_H(v_H, w_H) = 0 \ \forall w_H \in V_{0H}, \quad v_H|_{S_H} = 1, \ v_H|_{\Gamma_H} = 0, \tag{15}$$

satisfies $|v_H|_{\infty,S_h} := \lambda < 1$.

Remark 1. Notice that the maximum principle is known to be true when all the angles of the triangulation are acute [5]. The strict maximum principle of the hypothesis could be checked numerically, a priori. Error estimates in maximum norm of order $h^2 \log \frac{1}{h}$ with respect to the mesh edge size h for linear elements have been obtained by Schatz et al. [13].

Proposition 1. *Assume Hypothesis* 1 *to be satisfied. Then the discrete Schwarz algorithm* (14) *converges to the unique solution* $(u_h^*, u_H^*) \in V_h \times V_H$ *of the following system:*

$$\begin{aligned} a_H(u_H^*, w_H) &= (f, w_H) \ \forall w_H \in V_{0H}, \ u_H^*|_{S_H} = \gamma_H u_h^*, \ u_H^*|_{\Gamma_H} = g_H \\ a_h(u_h^*, w_h) &= (f, w_h) \ \forall w_h \in V_{0h}, \ u_h^*|_{S_h} = \gamma_h u_H^*. \end{aligned} \tag{16}$$

Proof. By the maximum principle and the fact that γ_H and γ_h decrease the L^∞ norms, problems of the type: find $v_H \in V_H$, $v_h \in V_h$

$$\begin{aligned} a_H(v_H, w_H) &= 0 \ \forall w_H \in V_{0H}, \ v_H|_{S_H} = \gamma_H u_h, \ v_H^{m+1}|_{\Gamma_H} = 0 \\ a_h(v_h, w_h) &= 0 \ \forall w_h \in V_{0h}, \ v_h^{m+1}|_{S_h} = \gamma_h v_H, \end{aligned} \tag{17}$$

satisfy

$$\|v_H\|_\infty \le \|u_h\|_{\infty,S_H}, \quad \|v_h\|_\infty \le \|v_H\|_{\infty,S_h}. \tag{18}$$

Combining this with the estimate on the solution of (15) we obtain

$$\|v_h\|_\infty \le \|v_H\|_{\infty,S_h} \le \lambda \|v_H\|_\infty \le \lambda \|u_h\|_\infty. \tag{19}$$

Consider now the mapping $T : V_h \to V_h$ that maps any u_h^{m-1} in (14) to u_h^m. Since T is affine, estimate (19) the problem (17) proves that T is a contraction in the $L^\infty(\Omega_h)$ norm. By Banach contraction theorem we have then that the iterative process $u_h^m = Tu_h^{m-1}$ converges to the unique fixed point u_h^* of T. In other words, u_h^m given by (14) converges to u_h^* in (16), which entails the convergence of u_H^m to u_H^*.

Proposition 2. *Assume Hypothesis* 1 *to be satisfied. Then* (u_h^*, u_H^*) *in* (16) *solves approximately* (13) *with optimal* L^∞ *error. More precisely, we have*

$$\max(\|u_H^* - u\|_{\infty,\Omega_H}, \|u_h^* - u\|_{\infty,\Omega_h}) \le C\left(H^2 \log\frac{1}{H}\|u\|_{H^{2,\infty}(\Omega_H)} + h^2 \log\frac{1}{h}\|u\|_{H^{2,\infty}(\Omega_h)}\right), \tag{20}$$

with a constant C depending only on the domains Ω_H *and* Ω_h.

Proof. Solution u to problem (13) satisfies $u|_\Gamma = g$ and

$$\begin{aligned} a_H(u,w) &= (f,w) \ \forall w \in H_0^1(\Omega_H), \qquad u = \gamma_H u + (u - \gamma_H u) \ \text{ on } S_H, \\ a_h(u,w) &= (f,w) \ \forall w \in H_0^1(\Omega_h), \qquad u = \gamma_h u + (u - \gamma_h u) \ \text{ on } S_h. \end{aligned} \tag{21}$$

Let $e = u_H^* - u$ and $\varepsilon = u_h^* - u$. Setting $w = w_H$ in the first equation and $w = w_h$ in the second, we have

$$\begin{aligned} a_H(e,w_H) &= 0 \ \forall w_H \in V_{0H}, \qquad e = \gamma_H \varepsilon - (u - \gamma_H u) \ \text{ on } S_H, \quad e|_\Gamma = g_H - g \\ a_h(\varepsilon,w_h) &= 0 \ \forall w_h \in V_{0h}, \qquad \varepsilon = \gamma_h e - (u - \gamma_h u) \ \text{ on } S_h. \end{aligned} \tag{22}$$

Let $\Pi_H u \in V_H$ and $\Pi_h u \in V_h$ be the solutions of

$$\begin{aligned} a_H(\Pi_H u, w_H) &= a_H(u, w_H) \quad \forall w_H \in V_{0H}, \quad \Pi_H u = \gamma_H u \ \text{ on } S_H, \ \Pi_H u|_\Gamma = g_H \\ a_h(\Pi_h u, w_h) &= a_h(u, w_h) \quad \forall w_h \in V_{0h}, \quad \Pi_h u = \gamma_h u \ \text{ on } S_h. \end{aligned} \tag{23}$$

By [13], we have

$$\begin{aligned} \|\Pi_H u - u\|_{\infty,\Omega_H} &\le H^2 \log \frac{1}{H} \|u\|_{H^{2,\infty}(\Omega_H)}, \\ \|\Pi_h - u\|_{\infty,\Omega_h} &\le h^2 \log \frac{1}{h} \|u\|_{H^{2,\infty}(\Omega_h)}. \end{aligned} \tag{24}$$

Finally let

$$\varepsilon_H = u_H - \Pi_H u = e + u - \Pi_H u, \quad \varepsilon_h = u_h - \Pi_h u = \varepsilon + u - \Pi_h u.$$

Then $\varepsilon_H \in V_H$, $\varepsilon_h \in V_h$ and

$$\begin{aligned} a_H(\varepsilon_H, w_H) &= 0 \ \forall w_H \in V_{0H}, \qquad \varepsilon_H = \gamma_H(\varepsilon_h + \Pi_h u - u) \ \text{ on } S_H, \ \varepsilon_H|_\Gamma = 0 \\ a_h(\varepsilon_h, w_h) &= 0 \ \forall w_h \in V_{0h}, \qquad \varepsilon_h = \gamma_h(\varepsilon_H + \Pi_H u - u) \ \text{ on } S_h. \end{aligned} \tag{25}$$

The maximum principle (like in (18) and (19)) again yields

$$\|\varepsilon_H\|_\infty \le \|\Pi_h u - u\|_{\infty,S_H} + \|\varepsilon_h\|_{\infty,S_H}, \tag{26}$$

$$\|\varepsilon_h\|_\infty \le \|\Pi_H u - u\|_{\infty,S_h} + \|\varepsilon_H\|_{\infty,S_h}, \tag{27}$$

$$\|\varepsilon_H\|_{\infty,S_h} \le \lambda \|\varepsilon_H\|_\infty. \tag{28}$$

Therefore

$$\max(\|\varepsilon_h\|_\infty, \|\varepsilon_H\|_\infty) \le \frac{1}{1-\lambda}(\|\Pi_H u - u\|_{\infty,\Omega_H} + \|\Pi_h u - u\|_{\infty,\Omega_h}). \tag{29}$$

Combining it with (24) and the triangle inequality we obtain the desired result.

Numerical Tests

We have tested numerically the Schwarz algorithm for the problem $u - \Delta u = xy$, $u|_{\partial\Omega} = xy$ with the solution $u = xy$ and the geometry shown in Fig. 2a. All the computations were done using the integrated environment freefem++ [10]. Convergence of the Schwarz iterations is illustrated in Fig. 2b. The results show that the convergence is linear and $\|u_h^{m+1} - u_h^*\|_0 / \|u_h^{m+1} - u_h^*\|_0 \to 0.67$ while the constant λ in the Hypothesis 1 is $\lambda = 0.75$. When the two meshes are refined by the factor of 2 or 4, these figures do not change much.

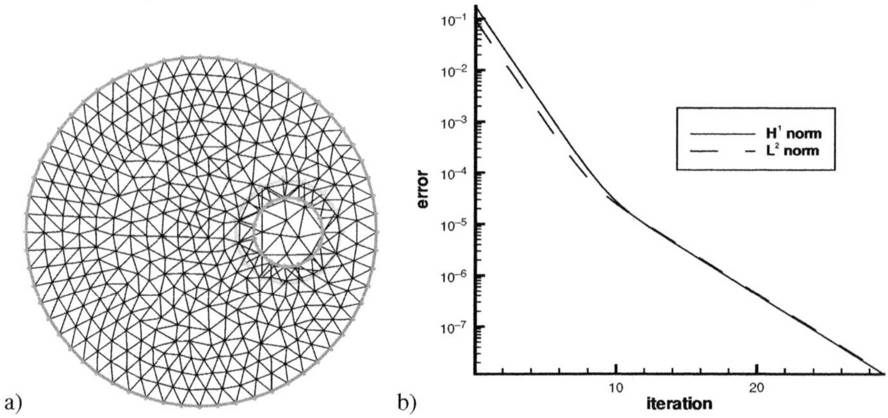

Fig. 2. a) The geometry of Ω_H, Ω_h and the meshes M1; b) Convergence on Schwarz iterations, error with respect to the final discrete solution.

Let us now look at the behavior of the converged Schwarz solution with respect to the mesh refinement. Figure 3c shows it for the mesh M1 as in Fig. 2a, and the error is plotted in Fig. 3d. Table 1 reports the error on three pairs of meshes, namely M1, their twofold refinement M2 and the 4-times refinement M4. We observe the optimal convergence rates

	H^1	L^2	L^∞
M1	$7.2 \cdot 10^{-3}$	$1.4 \cdot 10^{-3}$	$4.0 \cdot 10^{-3}$
M2	$3.6 \cdot 10^{-3}$	$2.9 \cdot 10^{-4}$	$1.2 \cdot 10^{-3}$
M3	$1.6 \cdot 10^{-3}$	$9.8 \cdot 10^{-5}$	$3.8 \cdot 10^{-4}$

Table 1. The relative error in H^1, L^2 and L^∞ norms for the approximated solutions obtained by Schwarz algorithm on meshes M1–M3.

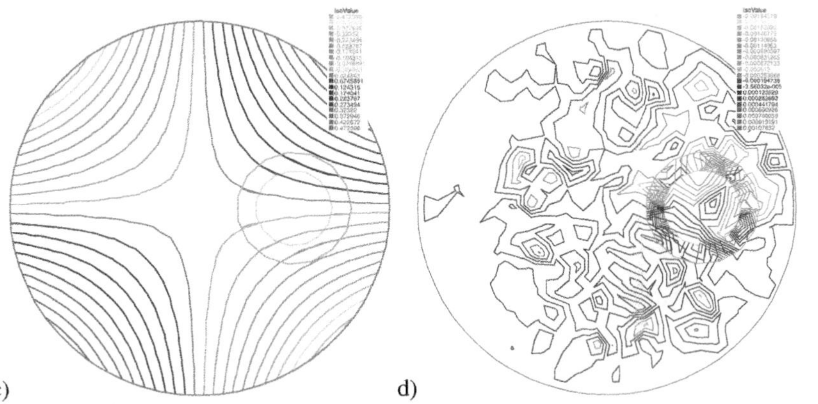

c) d)

Fig. 3. c) The solution obtained by the Schwarz algorithm on M1; d) The error with respect to the exact continuous solution.

3 Numerical Comparison of the Methods

We tested all the Algorithms 2-4 on the benchmark of the Poisson equation $-\Delta u = f$ in $\Omega = (-1,1)^2$ with Dirichlet boundary conditions on $\partial\Omega$ and the exact solution

$$u = \cos\frac{\pi}{2}x\cos\frac{\pi}{2}y + 10\chi(x^2+y^2 < R^2)e^{\frac{1}{R^2}-\frac{1}{R^2-x^2-y^2}} \tag{30}$$

with $R = 0.3$. We choose the patches of the form $\Lambda = (-\varepsilon_\Lambda, \varepsilon_\Lambda)^2$, the holes in Ω of the form $D = (-\varepsilon_D, \varepsilon_D)^2$ and take the triangulations $\mathcal{T}_H$ and $\mathcal{T}_h$ that do not match each other, but the hole D always consists of several triangles from $\mathcal{T}_H$. We used the following options to compute the mixed integrals in (5)–(6) (the same holds for (8)–(10)): the integral $\int_\Omega \nabla u_h^{n-1} \cdot \nabla w_H$ in (5) is evaluated by a numerical quadrature on the fine mesh $\mathcal{T}_h$, the integral $\int_\Lambda \nabla u_H^n \cdot \nabla w_h$ in (6) is approximated by $\int_\Lambda \nabla(\gamma_h u_H^n) \cdot \nabla w_h$, which is easy to evaluate.

Figure 4 presents the convergence history in the H^1 and L^∞ norms of the relative error on iterations for the 4 choices of triangulations. We observe that all the methods converge but that the Harmonic Patch Iterator is in general the most efficient approach. More specifically, it converges to a better approximation in situations a) and b) with the patch of the size $\varepsilon_\Lambda = 0.27$. The gain in accuracy is observed in both H^1 and L^∞ norms. On the contrary, all the methods converge to virtually the same approximated solution in situations c) and d) where we have taken a larger patch with $\varepsilon_\Lambda = 0.4$. However, this choice of meshes does not allow us to compare the relative merits of our algorithms. Indeed, the patch here is so large that the solution outside λ does not feel the spike in f, which lies inside Λ. Therefore, one does not need to iterate here at all: one coarse calculation with one fine correction gives already a fairly good solution. Note also that the errors in Fig. 4 are computed with respect to the exact solution so that these results confirm the convergence of all the methods under mesh refinement.

We do not give here detailed results on our last Algorithm 5 since these results are very close to those of Algorithm 4 for the present benchmark.

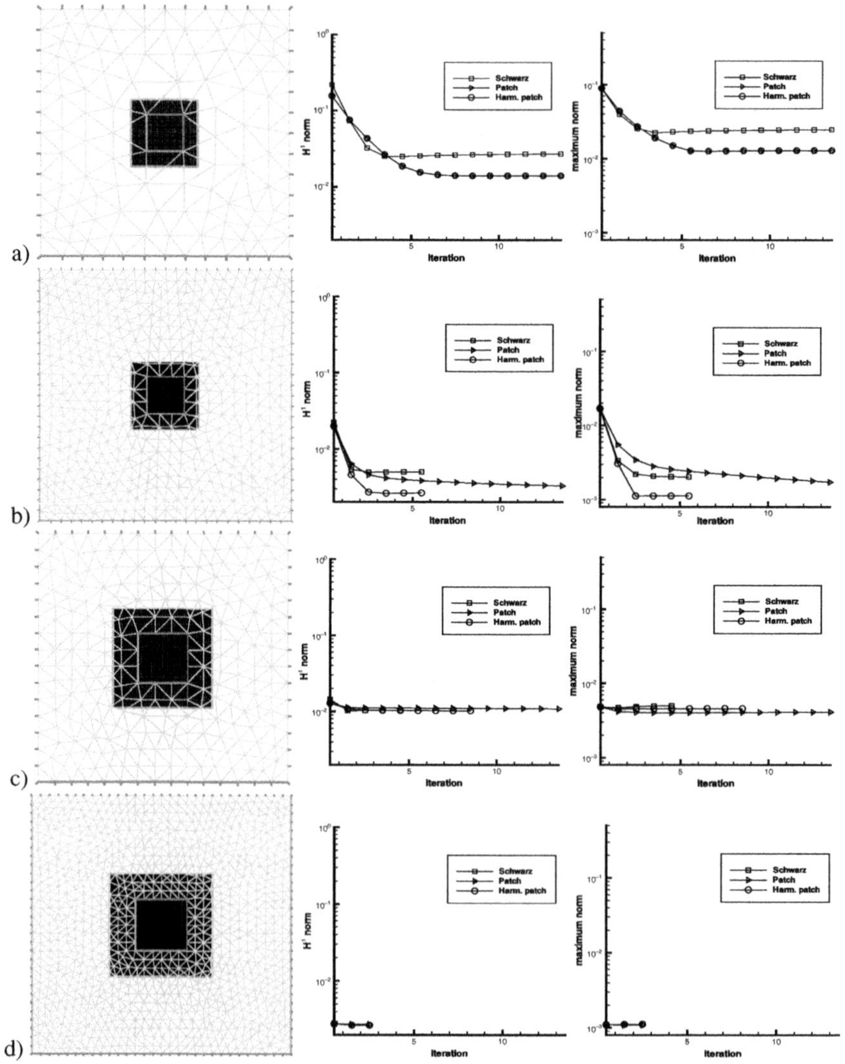

Fig. 4. Results for the benchmark (30). Left – the meshes, middle – relative error on iteration in the H^1 norm, right – relative error in the L^∞ norm. Four pairs of meshes: a) $\varepsilon_\Lambda = 0.27$, $\varepsilon_D = 0.15$, $H = \frac{1}{6}$, $h = \frac{0.27}{15}$; b) the twofold refinement of a); c) $\varepsilon_\Lambda = 0.4$, $\varepsilon_D = 0.2$, $H = \frac{2}{15}$, $h = \frac{1}{50}$; d) the twofold refinement of c).

4 Conclusion

The paper has shown that a standard Schwarz algorithm can be used for a zooming procedure, even with standard interpolations at the boundaries. However it seems that if one needs to iterate between the coarse and fine scales, the Harmonic Patch approach is the most robust way to do it. The numerical quadrature could affect the subspace correction methods in some cases. In order to get rid of the quadrature, one can compute the mixed integrals exactly on the intersection of two triangulations. In the near future we plan to improve on our triangulation intersector by inserting Martin Gander et al 's algorithm [6] into freefem++.

References

[1] Achdou, Y., Maday, Y.: The mortar element method with overlapping subdomains. *SIAM J. Numer. Anal.*, 40(2):601–628, 2002.

[2] Apoung Kamga, J.-B., Pironneau, O.: Numerical zoom for multiscale problems with an application to nuclear waste disposal. *J. Comput. Phys.*, 224(1):403–413, 2007.

[3] Brezzi, F., Lions, J.-L., Pironneau, O.: Analysis of a Chimera method. *C. R. Acad. Sci. Paris Sér. I Math.*, 332(7):655–660, 2001.

[4] Cai, X.-C., Mathew, T.P., Sarkis, M.V.: Maximum norm analysis of overlapping nonmatching grid discretizations of elliptic equations. *SIAM J. Numer. Anal.*, 37(5):1709–1728 (electronic), 2000.

[5] Ciarlet, P.G., Raviart, P.-A.: Maximum principle and uniform convergence for the finite element method. *Comput. Methods Appl. Mech. Engrg.*, 2:17–31, 1973.

[6] Gander, M.J., Japhet, C.: A compact algorithm with optimal complexity for non-matching grid projections. In *Proceedings of the 18th international conference of Domain Decomposition methods*, (these proceedings).

[7] Glowinski, R., He, J., Lozinski, A., Rappaz, J., Wagner, J.: Finite element approximation of multi-scale elliptic problems using patches of elements. *Numer. Math.*, 101(4):663–687, 2005.

[8] He, J., Lozinski, A., Rappaz, J.: Accelerating the method of finite element patches using approximately harmonic functions. *C. R. Math. Acad. Sci. Paris*, 345(2):107–112, 2007.

[9] Hecht, F., Lozinski, A., Perronnet, A., Pironneau, O.: Numerical zoom for multiscale problems with an application to flows through porous media. *Disc. Cont. Dyna. Syst.*, submitted, 2007.

[10] Hecht, F., Pironneau, O., Le Hyaric, A., Ohtsuka, K.: `Freefem++`, ver. 2.23. `http://www.freefem.org`, 2007.

[11] Dougherty, F.C., Steger, J.L., Benek, J.A.: A Chimera grid scheme. In *Advances in Grid Generation*. K.N. Chis and U. Ghia eds., ASME FED-vol. 5. June 1983.

[12] Liions, J.-L., Pironneau, O.: Domain decomposition methods for CAD. *C. R. Acad. Sci. Paris Sér. I Math.*, 328(1):73–80, 1999.

[13] Schatz, A.H., Wahlbin, L.B.: On the quasi-optimality in L_∞ of the $\dot{H}^1$-projection into finite element spaces. *Math. Comp.*, 38(157):1–22, 1982.

BDDC for Nonsymmetric Positive Definite and Symmetric Indefinite Problems

Xuemin Tu[1] and Jing Li[2]

[1] Department of Mathematics, University of California and Lawrence Berkeley National Laboratory, Berkeley, CA 94720-3840 xuemin@math.berkeley.edu
[2] Department of Mathematical Sciences, Kent State University, Kent, OH 44242 li@emath.kent.edu

Summary. The balancing domain decomposition methods by constraints are extended to solving both nonsymmetric, positive definite and symmetric, indefinite linear systems. In both cases, certain nonstandard primal constraints are included in the coarse problems of BDDC algorithms to accelerate the convergence. Under the assumption that the subdomain size is small enough, a convergence rate estimate for the GMRES iteration is established showing that the rate is independent of the number of subdomains and depends only slightly on the subdomain problem size. Numerical experiments for several two-dimensional examples illustrate the fast convergence of the proposed algorithms.

1 Introduction

Domain decomposition methods have been widely used and studied for solving large sparse linear systems arising from finite element discretization of partial differential equations, see [16] and the references therein. The balancing domain decomposition methods by constraints (BDDC) were introduced by Dohrmann [6], see also [9] and [5] for related algorithms. These algorithms originally were designed for the symmetric, positive definite systems. The BDDC methods have also been extended to solving saddle point problems, e.g., Stokes equations [12], nearly incompressible elasticity [7], and flow in porous media [17, 18].

Cai and Widlund [2, 3, 4] studied overlapping Schwarz methods for nonsymmetric and indefinite problems, using a perturbation approach in their analysis, and established that the convergence rates of the two-level overlapping Schwarz methods are independent of the mesh size if the coarse mesh is fine enough.

In this paper, we extend BDDC algorithms to nonsymmetric, positive definite linear systems arising from finite element discretization of advection-diffusion equations, and to symmetric, indefinite systems arising from finite element discretization of Helmholtz equations. A preconditioned GMRES iteration is used. In the preconditioning step of each iteration, a partially sub-assembled finite element problem is solved, for which only the coarse level, primal interface degrees of freedom are

M. Bercovier et al. (eds.), *Domain Decomposition Methods in Science and Engineering XVIII*, Lecture Notes in Computational Science and Engineering,
DOI: 10.1007/978-3-642-02677-5, © Springer-Verlag Berlin Heidelberg 2009

shared by neighboring subdomains. A perturbation approach is used in our analysis to handle the non-symmetry or indefiniteness of the problems. A key point in the analysis is the error bound for a partially sub-assembled finite element problem; we view this partially sub-assembled finite element problem as a non-conforming finite element approximation.

2 Finite Element Discretization

Let $\widehat{W} \subset H_0^1(\Omega)$ be the standard continuous, piecewise linear finite element function space on a shape-regular triangulation of Ω. In this paper, we use the same notation, e.g., u, to denote both a finite element function and the vector of its coefficients with respect to the finite element basis; we will also use the same notation to denote the space of finite element functions and the space of their corresponding vectors, e.g., $\widehat{W}$. In this paper, C always represents a generic positive constant independent of all the parameters and mesh size.

2.1 Nonsymmetric, Positive Definite Problems

We consider the following second order scalar advection-diffusion problem in a bounded polyhedral domain $\Omega \in \mathbf{R}^d$, d = 2, 3,

$$\begin{cases} Lu := -\nu \triangle u + \mathbf{a} \cdot \nabla u + cu = f & \text{in } \Omega, \\ u = 0 & \text{on } \partial\Omega. \end{cases} \tag{1}$$

Here the viscosity ν is a positive constant. The velocity field $\mathbf{a}(x) \in (L^\infty(\Omega))^d$ and $\nabla \cdot \mathbf{a}(x) \in L^\infty(\Omega)$. The reaction coefficient $c(x) \in L^\infty(\Omega)$ and $f(x) \in L^2(\Omega)$. We define $\tilde{c}(x) = c(x) - \frac{1}{2}\nabla \cdot \mathbf{a}(x)$ and assume that there exists a positive constant c_0 such that

$$\tilde{c}(x) \geq c_0 > 0 \quad \forall\, x \in \Omega. \tag{2}$$

We focus on studying the dependence on ν of the performance of our algorithms and assume that all other parameters in the operator L are of order $O(1)$.

The bilinear form associated with the operator L is defined, for functions in the space $H_0^1(\Omega)$, by $a_o(u,v) = \int_\Omega (\nu \nabla u \cdot \nabla v + \mathbf{a} \cdot \nabla u v + cuv)\, dx$, which is positive definite under assumption (2). The weak solution $u \in H_0^1(\Omega)$ of (1) satisfies

$$a_o(u,v) = \int_\Omega fv\, dx \quad \forall\, v \in H_0^1(\Omega). \tag{3}$$

We assume that the weak solution u of the original problem (1), as well as the weak solution of the adjoint problem $L^* u = -\nu \triangle u - \nabla \cdot (\mathbf{a}u) + cu = f$, satisfies the regularity result,

$$\|u\|_{H^2(\Omega)} \leq \frac{C}{\nu} \|f\|_{L^2(\Omega)}, \tag{4}$$

where C is a positive constant independent of ν. Thus, we assume that $\|u\|_{H^2(\Omega)}$ grows proportionally with a decrease of the viscosity ν.

It is well known that the original bilinear form $a_o(\cdot,\cdot)$ has to be stabilized to remove spurious oscillations in the finite element solution for advection-dominated problems. Here, we follow [10, 15] and consider the Galerkin/least-squares method (GALS).

The stabilized finite element problem for solving (3) is: find $u \in \widehat{W}$, such that for all $v \in \widehat{W}$,

$$a(u,v) := a_o(u,v) + \int_\Omega C(x)LuLv\,dx = \int_\Omega fv\,dx + \int_\Omega C(x)fLv\,dx, \tag{5}$$

where $C(x)$ is a positive function which depends on the local element Peclet number; see [19] for details. We note that for all piecewise linear finite element functions u, $Lu = -\nu \triangle u + \mathbf{a}\cdot\nabla u + cu = \mathbf{a}\cdot\nabla u + cu$, on each element. We define $C_s = \max_{x\in\Omega}|C(x)|$ and $C_m = \min_{x\in\Omega}|C(x)|$.

The system of linear equations corresponding to the stabilized finite element problem (5) is denoted by

$$Au = f, \tag{6}$$

where the coefficient matrix A is nonsymmetric but positive definite.

2.2 Symmetric, Indefinite Problems

We consider the solution of the following partial differential equation on a bounded polyhedral domain $\Omega \in \mathbf{R}^d$, d = 2, 3,

$$\begin{cases} -\Delta u - \sigma^2 u = f & \text{in } \Omega, \\ u = 0 & \text{on } \partial\Omega, \end{cases} \tag{7}$$

where σ is a real constant. The weak solution $u \in H_0^1(\Omega)$ of (7) satisfies

$$a(u,v) = (f,v) \quad \forall\, v \in H_0^1(\Omega), \tag{8}$$

where $a(u,v) = \int_\Omega \nabla u\cdot\nabla v - \sigma^2 uv$, and $(f,v) = \int_\Omega fv$. Under the assumption that (8) has a unique solution, we can prove the following regularity result for the weak solution, cf. [11],

$$\|u\|_{H^{1+\gamma}(\Omega)} \le C\left(1 + \frac{\sigma^2}{|\lambda_* - \sigma^2|}\right)\|f\|_{L_2(\Omega)}, \tag{9}$$

where λ_* is the eigenvalue of the corresponding Laplace operator, closest to σ^2. The results hold for $\gamma = 1$, if Ω is convex. In this paper we assume that σ^2 is bounded away from the eigenvalues of the Laplace operator such that the problem is well posed. Therefore we have $\|u\|_{H^{1+\gamma}(\Omega)} \le C(1+\sigma^2)\|f\|_{L_2(\Omega)}$.

The finite element solution for solving (8) is: find $u \in \widehat{W}$, such that

$$a(u,v) = (f,v) \quad \forall v \in \widehat{W}. \tag{10}$$

The resulting system of linear equations has the form

$$Au = (K - \sigma^2 M)u = f, \tag{11}$$

where K is the stiffness matrix, and M is the mass matrix.

3 The BDDC Preconditioners

We decompose the space $\widehat{W}$ into $W_I \oplus \widehat{W}_\Gamma$, where W_I is the product of local subdomain spaces $W_I^{(i)}$, $i = 1,2,\ldots,N$, corresponding to the subdomain interior variables. $\widehat{W}_\Gamma$ is the subspace corresponding to the variables on the subdomain interface Γ. The original discrete problem (6) can be written as: find $u_I \in W_I$ and $u_\Gamma \in \widehat{W}_\Gamma$, such that

$$\begin{bmatrix} A_{II} & A_{I\Gamma} \\ A_{\Gamma I} & A_{\Gamma\Gamma} \end{bmatrix} \begin{bmatrix} u_I \\ u_\Gamma \end{bmatrix} = \begin{bmatrix} f_I \\ f_\Gamma \end{bmatrix}, \tag{12}$$

where A_{II} is block diagonal with one block for each subdomain, and $A_{\Gamma\Gamma}$ corresponds to the subdomain interface variables and is assembled from subdomain matrices across the subdomain interfaces.

Eliminating the subdomain interior variables u_I from (12), we have the Schur complement problem

$$S_\Gamma u_\Gamma = g_\Gamma,$$

where $S_\Gamma = A_{\Gamma\Gamma} - A_{\Gamma I}A_{II}^{-1}A_{I\Gamma}$, and $g_\Gamma = f_\Gamma - A_{\Gamma I}A_{II}^{-1}f_I$.

A partially sub-assembled finite element space $\widetilde{W}$ is defined by $\widetilde{W} = W_I \oplus \widetilde{W}_\Gamma$. Here $\widetilde{W}_\Gamma$ contains the coarse level, continuous primal interface degrees of freedom, in the subspace $\widehat{W}_\Pi$, which are shared by neighboring subdomains, and the remaining dual subdomain interface degrees of freedom which are in general discontinuous across the subdomain interfaces. Then a partially sub-assembled problem matrix $\widetilde{A}$ is defined by

$$\begin{bmatrix} A_{II} & \widetilde{A}_{I\Gamma} \\ \widetilde{A}_{\Gamma I} & \widetilde{A}_{\Gamma\Gamma} \end{bmatrix}, \tag{13}$$

where $\widetilde{A}_{\Gamma\Gamma}$ is assembled only with respect to the coarse level primal degrees of freedom across the interface.

Correspondingly, a partially sub-assembled Schur complement $\widetilde{S}_\Gamma$ is defined by $\widetilde{S}_\Gamma = \widetilde{A}_{\Gamma\Gamma} - \widetilde{A}_{\Gamma I}A_{II}^{-1}\widetilde{A}_{I\Gamma}$. From the definition of S_Γ and $\widetilde{S}_\Gamma$, we see that S_Γ can be obtained from $\widetilde{S}_\Gamma$ by assembling with respect to the dual interface variables, i.e.,

$$S_\Gamma = \widetilde{R}_\Gamma^T \widetilde{S}_\Gamma \widetilde{R}_\Gamma,$$

where $\widetilde{R}_\Gamma$ is the injection operator from the space $\widehat{W}_\Gamma$ into $\widetilde{W}_\Gamma$. We also define $\widetilde{R}_{D,\Gamma} = D\widetilde{R}_\Gamma$, where D is a diagonal scaling matrix. The diagonal elements of D equal 1, for the rows of the primal interface variables, and equal $\delta_i^\dagger(x)$ for the others. Here, for a subdomain interface node x, the inverse counting function $\delta_i^\dagger(x)$ is defined by $\delta_i^\dagger(x) = 1/card(\mathcal{N}_x)$, where $\mathcal{N}_x$ is the set of indices of the subdomains which have x on their boundaries and $card(\mathcal{N}_x)$ is the number of the subdomains in the set $\mathcal{N}_x$.

The preconditioned interface problem in our BDDC algorithm is

$$\widetilde{R}_{D,\Gamma}^T \widetilde{S}_\Gamma^{-1} \widetilde{R}_{D,\Gamma} S_\Gamma u_\Gamma = \widetilde{R}_{D,\Gamma}^T \widetilde{S}_\Gamma^{-1} \widetilde{R}_{D,\Gamma} g_\Gamma. \tag{14}$$

A GMRES iteration is used to solve (14). In each iteration, to multiply S_Γ by a vector, subdomain Dirichlet boundary problems need to be solved; to multiply $\widetilde{S}_\Gamma^{-1}$ by a vector, a partially sub-assembled finite element problem with the coefficient matrix $\widetilde{A}$ needs to be solved, which requires solving subdomain Neumann/Robin boundary problems and a coarse level problem; cf. [13, 19]. After obtaining the interface solution u_Γ, we find u_I by solving subdomain Dirichlet problems.

Another alternative of the BDDC algorithm is to iterate on the full set of variables, instead of on the subdomain interface variables. This alternative preconditioned BDDC operator is of the form

$$(\widetilde{R}_D^T - \mathcal{H} J_D)\widetilde{A}^{-1}(\widetilde{R}_D - J_D^T \mathcal{H}^T)A, \tag{15}$$

where $\widetilde{R}_D$ is a scaled injection operator from $\widehat{W}$ onto $\widetilde{W}$ with the scaling on the subdomain interface defined in the same way as for $\widetilde{R}_{D,\Gamma}$ discussed above. J_D is a map from $\widetilde{W}$ to itself. For any $w \in \widetilde{W}$, the component of $J_D w$, for the subdomain Ω_i, is defined by

$$(J_D w(x))^{(i)} = \sum_{j \in \mathcal{N}_x} \delta_j^\dagger(x)(w^{(i)}(x) - w^{(j)}(x)) \quad \forall x \in \Gamma \cap \partial\Omega_i,$$

where $J_D w$ vanishes in the interior of the subdomain and for the coarse level component. The component of $J_D^T w$ for subdomain Ω_i is then given by

$$(J_D^T w(x))^{(i)} = \sum_{j \in \mathcal{N}_x} (\delta_j^\dagger(x) w^{(i)}(x) - \delta_i^\dagger(x) w^{(j)}(x)) \quad \forall x \in \Gamma \cap \partial\Omega_i.$$

The subdomain interior and the coarse level primal components of $J_D^T w$ also vanish. The operator $\mathcal{H}$ in (15) is a direct sum of the subdomain discrete harmonic extensions $\mathcal{H}^{(i)}$, where $\mathcal{H}^{(i)} = -K_{II}^{(i)-1} K_{I\Gamma}^{(i)}$, $i = 1, 2, \ldots, N$. $\mathcal{H} J_D$ represents the discrete harmonic extension of the jump of the dual interface variables to the interior of the subdomains.

4 Convergence Rate Analysis

The GMRES iteration is used in our BDDC algorithm to solve the preconditioned system of linear equations. To estimate the convergence rate of the GMRES iteration, we use the following result, cf. [8],

Theorem 1. *Let c_1 and C_2 be two positive parameters, $\langle\cdot,\cdot\rangle_\Lambda$ be an inner product defined in a vector space V, T be a linear operator defined on V. If for all $v \in V$,*

$$c_1\langle v,v\rangle_\Lambda \leq \langle v,Tv\rangle_\Lambda, \tag{16}$$
$$\langle Tv,Tv\rangle_\Lambda \leq C_2\langle v,v\rangle_\Lambda, \tag{17}$$

then

$$\frac{\|r_m\|_\Lambda}{\|r_0\|_\Lambda} \leq \left(1-\frac{c_1^2}{C_2}\right)^{m/2},$$

where r_m is the residual at step m of the GMRES iteration applied to T.

Remark 1. The convergence rate of the GMRES iteration using the standard L_2 inner product will not be estimated in this paper. In our numerical experiments, we have found that using the Λ inner product or the standard L_2 inner product gives the same convergence rate. For a study of the convergence rates of the GMRES iteration for an additive Schwarz method in the Euclidean and energy norms, see Sarkis and Szyld [14].

In the following analysis, we focus on obtaining bounds for the two positive constants c_1 and C_2 with respect to appropriate norms.

4.1 Nonsymmetric, Positive Cases

The preconditioned BDDC operator for solving the nonsymmetric, positive definite problem (6) is $T = \widetilde{R}_{D,\Gamma}^T\widetilde{S}_\Gamma^{-1}\widetilde{R}_{D,\Gamma}S_\Gamma$, defined on the subdomain interface variable space $\widehat{W}_\Gamma$. The inner product in the GMRES iteration is defined by $\Lambda = S_\Gamma$. We assume

Assumption 2 *For two-dimensional problems, the coarse level primal subspace $\widehat{W}_\Pi$ contains all subdomain corner degrees of freedom, and for each edge $\mathcal{E}^k$, one edge average degree of freedom and two edge flux average degrees of freedom such that for any $w \in \widetilde{W}$,*

$$\int_{\mathcal{E}^k} w^{(i)}\,ds, \quad \int_{\mathcal{E}^k} \mathbf{a}\cdot\mathbf{n}w^{(i)}\,ds, \quad and \quad \int_{\mathcal{E}^k} \mathbf{a}\cdot\mathbf{n}w^{(i)}s\,ds,$$

respectively, are the same (with a difference of factor -1 corresponding to opposite normal directions) for the two subdomains Ω_i that share $\mathcal{E}^k$.

For three dimensional problems, $\widehat{W}_\Pi$ contains all subdomain corner degrees of freedom, and for each face $\mathcal{F}^l$, one face average degree of freedom and two face flux average degrees of freedom, and for each edge $\mathcal{E}^k$, one edge average degree of freedom, such that for any $w \in \widetilde{W}$,

$$\int_{\mathcal{F}^l} I_h(\vartheta_{\mathcal{F}^l}w^{(i)})\,ds, \quad \int_{\mathcal{F}^l} \mathbf{a}\cdot\mathbf{n}I_h(\vartheta_{\mathcal{F}^l}w^{(i)})\,ds, \quad and \quad \int_{\mathcal{F}^l} \mathbf{a}\cdot\mathbf{n}I_h(\vartheta_{\mathcal{F}^l}w^{(i)})s\,ds,$$

respectively, are the same (with a difference of factor -1 *corresponding to opposite normal directions) for the two subdomains* Ω_i *that share the face* $\mathcal{F}^l$*, and*

$$\int_{\mathcal{E}^k} I_h(\vartheta_{\mathcal{E}^k} w^{(i)})\, ds$$

are the same for all subdomains Ω_i *that share the edge* $\mathcal{E}^k$*. Here* $\vartheta_{\mathcal{F}^l}$ *and* $\vartheta_{\mathcal{E}^k}$ *are the standard finite element face and edge cutoff functions, respectively.*

Theorem 3. *Let Assumption 2 hold. If h is sufficiently small, there then exist positive constants* C_1, C_2, *and* C_3, *which are independent of* H, h, *and* ν, *such that for all* $u_\Gamma \in \widehat{W}_\Gamma$,

$$\langle Tu_\Gamma, Tu_\Gamma\rangle_\Lambda \le C_1 \frac{\Phi^4(H,h)}{\nu^2 \max(\nu, C_m)} \langle u_\Gamma, u_\Gamma\rangle_\Lambda, \tag{18}$$

and

$$c_0 \langle u_\Gamma, u_\Gamma\rangle_\Lambda \le \frac{C_2}{\max(\nu, C_m)} \langle u_\Gamma, Tu_\Gamma\rangle_\Lambda, \tag{19}$$

where $\Phi(H,h) = C(1+\log(H/h))$*. For two dimensions*

$$c_0 = 1 - C_3 \frac{\max(\sqrt{\nu}, \sqrt{C_s}) \max(H\nu, H^2)}{\nu^3 \max(\nu^2\sqrt{\nu}, C_m^2\sqrt{C_m})} \frac{H}{h} \Phi^2(H,h),$$

and for three dimensions

$$c_0 = 1 - C_3 \frac{\max(\sqrt{\nu}, \sqrt{C_s}) \max(H\nu, H^2, \sqrt{Hh})}{\nu^3 \max(\nu^2\sqrt{\nu}, C_m^2\sqrt{C_m})} \frac{H}{h} \Phi^2(H,h)(1+\log(H/h)).$$

4.2 Symmetric, Indefinite Cases

The preconditioned BDDC operator for solving the symmetric, indefinite problem (11) is $T = (\widetilde{R}_D^T - \mathcal{H}J_D)\widetilde{A}^{-1}(\widetilde{R}_D - J_D^T\mathcal{H}^T)A$, defined on $\widehat{W}$. The inner product in the GMRES iteration is defined by $\Lambda = K + \sigma^2 M$. We assume

Assumption 4 *The coarse level primal subspace* $\widehat{W}_\Pi$ *contains all subdomain corner degrees of freedom, one edge average degree of freedom on each edge corresponding to restriction of the plane wave* $\cos(\sigma\theta\cdot x)$ *on the edge with the unit direction vector* θ *chosen orthogonal to the edge, and, for three dimensional problems, one face average degree of freedom on each subdomain boundary face corresponding to restriction of the plane wave* $\cos(\sigma\theta\cdot x)$ *on the face with* θ *chosen orthogonal to the face.*

Theorem 5. *Let Assumptions 4 hold. If* $\sigma(1+\sigma^2)(1+\Phi(H,h))H^\gamma C_L(H,h)$ *is sufficiently small, then for all* $u \in \widehat{W}$,

$$c_1 \langle u,u\rangle_\Lambda \le \langle u, Tu\rangle_\Lambda, \tag{20}$$

$$\langle Tu, Tu\rangle_\Lambda \le C_2(1+\sigma^2H^2)(1+\Phi(H,h)^2)\langle u,u\rangle_\Lambda, \tag{21}$$

where c_1 *and* C_2 *are positive constants independent of* σ, H, *and* h. $\Phi(H,h)$ *is defined in Theorem 3.* $C_L(H,h)$ *equals* $(1+\log(H/h))$ *for three-dimensional problems, and equals* 1 *for two-dimensional problems.*

5 Numerical Experiments

5.1 Nonsymmetric, Positive Definite Cases

We test our BDDC algorithm by solving the advection-diffusion equation (1) on the square domain $\Omega = [-1,1]^2$. The domain Ω is decomposed into square subdomains and each subdomain into uniform triangles. Piecewise linear finite elements are used in our experiments. We take $f = 0$ and $c = 10^{-4}$ in (1) in our example. We choose the most difficult one from the three examples, which were used by Toselli [15] for testing his FETI algorithms. Here the velocity field is $\mathbf{a} = (y, -x)$. The boundary condition is given by:

$$u = 1 \quad \text{for} \begin{cases} y = -1 & 0 < x \leq 1, \\ y = 1, & 0 < x \leq 1, \\ x = 1, & -1 \leq y \leq 1, \end{cases} \quad \text{with} \quad u = 0 \text{ elsewhere on } \partial\Omega.$$

In the GMRES iteration, the L_2 inner product is used and the iteration is stopped when the residual is reduced by 10^{-6}.

In our experiments, we test three different choices of the coarse spaces in the algorithms. In our first test, we test the Robin-Robin algorithms, which is closely related to our BDDC algorithms, see [1]. In our second test, the coarse level primal variables of our BDDC algorithms are only those at the subdomain corners and the subdomain edge averages; no additional continuity constraints corresponding to the flux are enforced across the subdomain edges. This choice of the coarse level primal space does not satisfy Assumption 2. In our last test, in addition to the primal constraints used in the second test, we also include in the coarse level problem two weighted edge average degrees of freedom corresponding to flux continuity constraints for each subdomain edge, as required in Assumption 2. In the following tables, we represent these three different algorithms by RR, BDDC-1, and BDDC-2, respectively.

Table 1 gives the iteration counts of the three algorithms for different number of subdomains with a fixed subdomain problem size. We see that BDDC-2 converges much faster than BDDC-1 and the Robin-Robin algorithm. For the cases where $\nu > 10^{-5}$, the iteration counts are almost independent of the number of subdomains. Even when the viscosity ν goes to zero, the convergence of BDDC-2 is still very fast, while the convergence rates of BDDC-1 and the Robin-Robin algorithm are not satisfactory at all.

From Table 2, we see that the iteration counts of all the algorithms increase with an increase of the subdomain problem size; the increase for BDDC-2 is the smallest.

5.2 Symmetric and Indefinite Cases

Problem (7) is solved on a 2π by 2π square domain with Dirichlet boundary conditions $u = 1$ on the four sides of the square and with $f = 0$. Q1 finite elements are

Table 1. Iteration counts for nonsymmetric, positive definite problems with $H/h = 6$ and changing number of subdomains.

ν	# subdomains	Iteration Count		
		RR	BDDC-1	BDDC-2
$1e-2$	8×8	22	9	3
	16×16	31	7	3
	32×32	49	6	3
$1e-4$	8×8	114	67	12
	16×16	251	111	14
	32×32	475	112	14
$1e-6$	8×8	145	86	14
	16×16	389	199	18
	32×32	> 500	434	26

Table 2. Iteration counts for nonsymmetric, positive definite problems with 4×4 subdomains and changing H/h.

ν	H/h	Iteration Count		
		RR	BDDC-1	BDDC-2
$1e-2$	12	17	11	4
	24	18	12	4
	48	20	13	4
$1e-4$	12	83	72	26
	24	110	104	39
	48	128	122	45
$1e-6$	12	100	87	34
	24	180	165	88
	48	296	290	142

used and the original square domain is decomposed uniformly into square subdomains. In the GMRES iteration, the $\langle \cdot, \cdot \rangle_{K+\sigma^2 M}$ inner product is used; using L_2 inner product gives the same convergence rates. The iteration is stopped when the residual is reduced by 10^{-6}.

In our experiments, we test three different choices of the coarse level primal space in our BDDC algorithm. In our first test, the coarse level primal variables are only those at the subdomain corners. No plane wave continuity constraints are enforced across the subdomain edges; this choice of the coarse level primal space does not satisfy Assumption 4. In our second test, in addition to the subdomain corner variables, we also include one edge average degree of freedom for each subdomain edge, as required in Assumption 4, in the coarse level primal variable space. This

Table 3. Iteration counts for symmetric, indefinite problems with $H/h = 8$ and changing number of subdomains.

σ^2	# subdomains	Iteration Count		
		0-pwa	1-pwa	2-pwa
	16×16	183	37	14
100	24×24	205	20	7
	32×32	> 300	13	6
	16×16	> 300	143	112
200	24×24	> 300	85	39
	32×32	> 300	47	28
	16×16	> 300	> 300	236
400	24×24	> 300	> 300	75
	32×32	> 300	192	49

Table 4. Iteration counts for symmetric, indefinite problems with 24×24 subdomains and changing H/h.

σ^2	H/h	Iteration Count		
		0-pwa	1-pwa	2-pwa
	8	205	20	7
100	12	188	25	8
	16	182	27	8
	8	> 300	85	39
200	12	> 300	108	60
	16	> 300	114	68
	8	> 300	> 300	75
400	12	> 300	> 300	108
	16	> 300	> 300	111

edge average degree of freedom corresponds to the vector determined by the cosine plane wave with the angle θ chosen perpendicular to the edge. In our last test, we further add to the coarse level primal space another plane wave continuity constraint on each edge corresponding to the cosine plane wave with the angle θ chosen tangential to the edge. In the following tables, we represent these three different choices of coarse level primal space by 0-pwa, 1-pwa, and 2-pwa, respectively.

Tables 3 and 4 show the GMRES iteration counts, corresponding to different number of subdomains, different subdomain problem sizes, and the three different choices of the coarse level primal space. With only subdomain corner variables in the coarse level primal space, the convergence cannot be achieved within 300 iterations in most cases. With the inclusion of the edge plane wave augmentations in the coarse level primal space, we see from Table 3 that the iteration counts decrease with an increase of the number of subdomains for a fixed subdomain problem size. We see from Table 4 that when the number of subdomains is fixed and H/h increases, the iteration counts increase slowly, seemingly in a logarithmic pattern of H/h. Tables 3 and 4 also show that the convergence becomes slower with the increase of the shift σ^2 and that the convergence rate is improved by including more plane wave continuity constraints in the coarse level primal subspace.

References

[1] Achdou, Y., Nataf, F.: A Robin-Robin preconditioner for an advection-diffusion problem. *C. R. Acad. Sci. Paris Sér. I Math.*, 325, 1211–1216, 1997.

[2] Cai, X.-C.: Additive Schwarz algorithms for parabolic convection-diffusion equations. *Numer. Math.*, 60(1):41–61, 1991.

[3] Cai, X.-C., Widlund, O.: Domain decomposition algorithms for indefinite elliptic problems. *SIAM J. Sci. Statist. Comput.*, 13(1):243–258, Jan. 1992.

[4] Cai, X.-C., Widlund, O.: Multiplicative Schwarz algorithms for some nonsymmetric and indefinite problems. *SIAM J. Numer. Anal.*, 30(4):936–952, Aug. 1993.

[5] Cros, J.-M.: A preconditioner for the Schur complement domain decomposition method. In *Domain decomposition methods in science and engineering, Proceedings of the 14th International Conference on Domain Decomposition Methods*, 373–380. National Autonomous University of Mexico, 2003.

[6] Dohrmann, C.R.: A preconditioner for substructuring based on constrained energy minimization. *SIAM J. Sci. Comput.*, 25(1):246–258, 2003.

[7] Dohrmann, C.R.: A substructuring preconditioner for nearly incompressible elasticity problems. Technical Report SAND2004-5393, Sandia National Laboratories, Albuquerque, New Mexico, Oct. 2004.

[8] Eisenstat, S.C., Elman, H.C., Schultz, M.H.: Variational iterative methods for nonsymmetric systems of linear equations. *SIAM J. Numer. Anal.*, 20 (2):345–357, 1983.

[9] Fragakis, Y., Papadrakakis, M.: The mosaic of high performance domain decomposition methods for structural mechanics: Formulation, interrelation and numerical efficiency of primal and dual methods. *Comput. Methods Appl. Mech. Engrg.*, 192(35–36):3799–3830, 2003.

[10] Hughes, T.J.R., Franca, L.P., Hulbert, G.M.: A new finite element formulation for computational fluid dynamics. VIII. The Galerkin/least-squares method for advective-diffusive equations. *Comput. Methds Appl. Mech. Engrg.*, 73(2):173–189, 1989.

[11] Li, J., Tu, X.: Convergence analysis of a balancing domain decomposition method for solving interior Helmholtz equations. *Numer. Linear Algebra Appl.*, to appear.

[12] Li, J., Widlund, O.B.: BDDC algorithms for incompressible Stokes equations. *SIAM J. Numer. Anal.*, 44(6):2432–2455, 2006.

[13] Li, J., Widlund, O.B.: FETI–DP, BDDC, and block Cholesky methods. *Internat. J. Numer. Methods Engrg.*, 66:250–271, 2006.
[14] Sarkis, M., Szyld, D.B.: Optimal left and right additive Schwarz preconditioning for minimal residual methods with Euclidean and energy norms. *Comput. Methods Appl. Mech. Engrg.*, 196:1507–1514, 2007.
[15] Toselli, A.: FETI domain decomposition methods for scalar advection-diffusion problems. *Comput. Methods Appl. Mech. Engrg.*, 190(43-44):5759–5776, 2001.
[16] Toselli, A., Widlund, O.B.: *Domain Decomposition Methods—Algorithms and Theory*, Springer Series in Computational Mathematics, 34. Springer, Berlin-Heidelberg-New York, 2005.
[17] Tu, X.: A BDDC algorithm for a mixed formulation of flows in porous media. *Electron. Trans. Numer. Anal.*, 20:164–179, 2005.
[18] Tu, X.: A BDDC algorithm for flow in porous media with a hybrid finite element discretization. *Electron. Trans. Numer. Anal.*, 26:146–160, 2007.
[19] Tu, X., Li, J.: A balancing domain decomposition method by constraints for advection-diffusion problems. *Commun. Appl. Math. Comput. Sci.*, 3:25–60, 2008.

Accomodating Irregular Subdomains in Domain Decomposition Theory

Olof B. Widlund

Courant Institute, 251 Mercer Street, New York, NY 10012, USA widlund@cims.nyu.edu

Summary. In the theory for domain decomposition methods, it has previously often been assumed that each subdomain is the union of a small set of coarse shape-regular triangles or tetrahedra. Recent progress is reported, which makes it possible to analyze cases with irregular subdomains such as those produced by mesh partitioners. The goal is to extend the analytic tools so that they work for problems on subdomains that might not even be Lipschitz and to characterize the rates of convergence of domain decomposition methods in terms of a few, easy to understand, geometric parameters of the subregions. For two dimensions, some best possible results have already been obtained for scalar elliptic and compressible and almost incompressible linear elasticity problems; the subdomains should be John or Jones domains and the rates of convergence are determined by parameters that characterize such domains and that of an isoperimetric inequality. Technical issues for three dimensional problems are also discussed.

1 Introduction

In developing theory for domain decomposition methods of iterative substructuring type, we have typically assumed that each subdomain is quite regular, e.g., the union of a small set of coarse triangles or tetrahedra; see, e.g., [31, Assump. 4.3]; we will call such subdomains *regular*. However, such an assumption is unlikely to hold especially if the subdomains result from using a mesh partitioner, such as METIS, see [18]. Then, the subdomain boundaries might not even be uniformly Lipschitz continuous in the sense that the number of patches required to cover $\partial\Omega$, and in each of which the boundary is the graph of a Lipschitz continuous function, is not uniformly bounded independently of the finite element mesh size. We also note that the shape of the subdomains are likely to change if the mesh size is altered and a mesh partitioner is used several times.

The purpose of this paper is to report on recent development of theory for domain decomposition methods under very weak assumptions on the subdomain partitioning and to categorize the rates of convergence of the algorithms in terms of a few geometric parameters. This work is being carried out in collaboration with C. R. Dohrmann, A. Klawonn, and O. Rheinbach and has so far resulted in four

M. Bercovier et al. (eds.), *Domain Decomposition Methods in Science and Engineering XVIII*, Lecture Notes in Computational Science and Engineering,
DOI: 10.1007/978-3-642-02677-5, © Springer-Verlag Berlin Heidelberg 2009

archival papers, [10, 11, 12, 23]. Results have been obtained for scalar elliptic problems, compressible linear elasticity, and almost incompressible elasticity problems approximated by mixed finite elements with pressure spaces with discontinuous elements.

We will denote a set of nonoverlapping subdomains by $\{\Omega_i\}$. Their closures cover the given domain Ω, and the interface between them is denoted by Γ. We will discuss results for the FETI-DP and BDDC families of algorithms, defined on such a set of nonoverlapping subdomains, as well as results for some two-level Schwarz algorithms based on overlapping subdomains Ω_i'. We will assume that each such subdomain has been obtained from one of the Ω_i by adding one or more layers of finite elements. The FETI-DP and BDDC algorithms are iterative substructuring algorithms, i.e., they provide preconditioners based on nonoverlapping subdomains.

So far, complete results have only been obtained for problems in the plane. We will consider scalar elliptic problems of the following form:

$$-\mathrm{div}\,(\rho(x)\nabla u(x)) = f(x) \quad x \in \Omega, \tag{1}$$

with a homogeneous Dirichlet boundary condition on $\partial\Omega$; we make this choice of boundary condition just to simplify the discussion of our results. The coefficient $\rho(x)$ is strictly positive and assumed to be equal to a constant ρ_i for $x \in \Omega_i$, but is otherwise arbitrary. As is often the case, our results hold equally well for isotropic compressible elasticity problems.

$$-\mathrm{div}\,(2\mu\varepsilon(\mathbf{u}) + \lambda \mathrm{tr}\,(\varepsilon(\mathbf{u}))I) = f \ \text{ in } \Omega \subset \mathbb{R}^n. \tag{2}$$

Here $\varepsilon_{ij}(\mathbf{u}) = (1/2)(\partial u_i/\partial x_j + \partial u_j/\partial x_i)$ and μ and λ the Lamé parameters; in the almost incompressible case, λ/μ takes on very large values.

Older results on domain decomposition methods for linear elasticity are summarized in [31, Chap. 8] and [24] gives more recent results on FETI-DP algorithms. All this work is for regular subdomains. We note that there are extensive and successful numerical results for more general problems; see, e.g., [9, 21, 22].

We use lower order, continuous finite elements and triangulations with shape regular elements, i.e., the diameter of an element is bounded uniformly by a constant times the radius of the largest inscribed circle or sphere and assume that each subdomain is a union of elements. For almost incompressible elasticity, we use an inf-sup stable pair of finite element spaces after introducing the new variable $p = -\lambda \mathrm{div}\,\mathbf{u}$ and assume that the elements of this second finite element space are discontinuous. We can then eliminate this pressure variable element-wise, recover a positive definite problem, and use the same conjugate gradient acceleration as for compressible elasticity.

For a collection of auxiliary results used in the analysis of iterative substructuring algorithms, in the case of regular subdomains, see [31, Sec. 4.6]. Our studies require the generalization of these technical tools to obtain proofs of bounds on the convergence rates of FETI-DP algorithms and on certain overlapping Schwarz methods for less regular subdomains. We also have had to modify some of the reasoning in the main proofs. Four auxiliary results, namely a Poincaré inequality, a Sobolev-type inequality for finite element functions, bounds for certain edge and face terms, and a finite element extension theorem are required in our proofs; see Lemmas 2, 6, 7, and 8. We will work with *John domains* and *Jones domains*, see Definitions 1 and 2; the latter are also known as uniform or (ε,∞)-domains. We will express our bounds on the rate of convergence of our algorithm in terms of the few parameters of Definitions 1, 2, and Lemma 1.

2 A Poincaré Inequality, John and Jones Domains

We will first introduce John domains and then consider a Poincaré inequality for such domains. We will also introduce Jones domains; the latter are needed in order to obtain a finite element extension theorem, required in the analysis of FETI and BDDC algorithms, see [23], but not for the algorithms based on overlapping subdomains, see [8, 9, 10, 11, 12].

We next give a definition of a John domain; see [16] and the references therein. In the proofs of several of our auxiliary results, we will assume that the subdomains belong to this class.

Definition 1 (John Domain). *A domain $\Omega \subset \mathbb{R}^n$, an open, bounded, and connected set, is a John domain if there exists a constant $C_J \geq 1$ and a distinguished central point $x_0 \in \Omega$ such that each $x \in \Omega$ can be joined to it by a rectifiable curve $\gamma : [0,1] \to \Omega$ with $\gamma(0) = x_0$, $\gamma(1) = x$ and $|x - \gamma(t)| \leq C_J \cdot distance(\gamma(t), \partial\Omega)$ for all $t \in [0,1]$.*

This condition can be viewed as a twisted cone condition. We note that certain snowflake curves with fractal boundaries are John domains and that the length of the boundary of a John domain can be arbitrarily much longer than its diameter. We also note that for any choice of the point x_0, there is a point $x \in \Omega$ at a distance of at least $diameter(\Omega)/2$. We find that $diameter(\Omega) \leq 2C_J r_\Omega$, where r_Ω is the radius of the largest ball inscribed in Ω and centered at x_0. Conditions on the boundary are of course also imposed.

In any analysis of any domain decomposition method with a second, coarse level, we need a Poincaré inequality. This inequality is closely related to an isoperimetric inequality. The next lemma is attributed to [28] and [15].

Lemma 1 (Isoperimetric Inequality). *Let $\Omega \subset \mathbb{R}^n$ be a domain and let u be sufficiently smooth. Then,*

$$\inf_{c \in R} \left(\int_\Omega |u - c|^{n/(n-1)} dx \right)^{(n-1)/n} \leq \gamma(\Omega, n) \int_\Omega |\nabla u| \, dx,$$

if and only if,

$$[\min(|A|, |B|)]^{1-1/n} \leq \gamma(\Omega, n) |\partial A \cap \partial B|. \tag{3}$$

Here, $A \subset \Omega$ is an arbitrary open set, and $B = \Omega \setminus \bar{A}$; $\gamma(\Omega, n)$ is the best possible constant and $|A|$ is the measure of the set A, etc.

We note that the domain does not need to be star-shaped or Lipschitz. For $n=2$, the best choice of $c=\bar{u}_\Omega$, the average of u over the domain. A small value of $\gamma(\Omega,n)$ is desirable for our purposes.

It is known that any simply connected plane domain with a finite Poincaré parameter $\gamma(\Omega,2)$ is a John domain; see [7]. It is also known, see [3], that any John domain has a bounded Poincaré parameter $\gamma(\Omega,n)$.

We obtain the standard Poincaré inequality by using the Cauchy-Schwarz inequality, for two dimensions, and the Hölder inequality several times for three.

Lemma 2 (Poincaré's Inequality). *Let Ω be a John domain. Then,*

$$\|u-\bar{u}_\Omega\|^2_{L^2(\Omega)} \le (\gamma(\Omega,n))^2|\Omega|^{2/n}\|\nabla u\|^2_{L^2(\Omega)} \quad \forall u\in H^1(\Omega).$$

Throughout, we will use a weighted $H^1(\Omega_i)-$norm defined by

$$\|u\|^2_{H^1(\Omega_i)} := \int_{\Omega_i}\nabla u\cdot\nabla u\,dx + 1/H_i^2\int_{\Omega_i}|u|^2\,dx = |u|^2_{H^1(\Omega_i)} + 1/H_i^2\int_{\Omega_i}|u|^2\,dx.$$

Here H_i is the diameter of Ω_i. The weight for the L^2-term results from the standard H^1-norm on a domain with diameter one and a dilation. We use Lemma 2 to remove L^2-terms in some of our estimates.

We next consider Jones domains.

Definition 2 (Jones Domains). *A domain $\Omega\subset\mathbb{R}^n$ is a Jones domain if there exists a constant C_U such that any pair of points $x_1\in\Omega$ and $x_2\in\Omega$ can be joined by a rectifiable curve $\gamma(t):[0,1]\to\Omega$ with $\gamma(0)=x_1$, $\gamma(1)=x_2$, and where the Euclidean arc length of $\gamma\le C_U|x_1-x_2|$ and $\min_{i=1,2}|x_i-\gamma(t)|\le C_U\cdot distance(\gamma(t),\partial\Omega)$ for all $t\in[0,1]$.*

It is known, and easy to see, that any Jones domain is a John domain. It is also easy to construct John domains that are not Jones domains. According to [17, Thm. 4], they form the largest class of finitely connected domains for which an extension theorem holds in two dimensions. It is also known that every Jones domain $\mathbb{R}^n$ allows for a bounded extension with respect to the seminorm of H^1, see [17, Thm. 2].

Lemma 3. *Let $\Omega\subset\mathbb{R}^n$ be a Jones domain and let P_0 be the space of constants. There then exists a bounded, linear operator $E_\Omega: H^1(\Omega)/P_0\to H^1(\mathbb{R}^n)/P_0$, which extends any element in $H^1(\Omega)$ to one defined for all of $\mathbb{R}^n$, i.e., $(E_\Omega u)_{|\Omega}=u\,\forall\,u\in H^1(\Omega)/P_0$. The norm of this operator depends only on $C_U(\Omega)$.*

An important tool in any study of elasticity is the second Korn inequality. For a proof for Jones domains, see [14].

Lemma 4 (Korn Inequality for Jones Domains). *Let $\Omega\subset\mathbb{R}^n$ be a bounded Jones domain. Then, there exists a constant C, which depends only on the Jones constant $C_U(\Omega)$ and the dimension n, such that*

$$|\mathbf{u}|^2_{H^1(\Omega)} \leq C \sum_{ij} \|\varepsilon(\mathbf{u})_{ij}\|^2_{L^2(\Omega)}$$

for all $\mathbf{u} \in \left\{\mathbf{u} \in \mathbf{H}^1(\Omega) : \int_\Omega \left(\frac{\partial u_i}{\partial x_j} - \frac{\partial u_j}{\partial x_i}\right) dx = 0, i, j = 1, \ldots, n\right\}$.

Their proof has many details in common with Jones' proof of Lemma 3. In the case of mixed finite element approximations of almost incompressible elasticity, we also need to establish the inf-sup stability of the mixed method. This problem is closely related to the Korn inequality; see, e.g., [4] in which new proofs of both results are given for general Lipschitz domains and the continuous case. There is a proof of the underlying inequality for John domains in [1]; the constant in that estimate depends only on the John parameter $C_J(\Omega)$.

3 FETI-DP and BDDC Algorithms

We first note that these two families of domain decomposition algorithms are closely related. Any such algorithm is characterized by a set of *primal constraints* and it is known that a pair of FETI-DP and BDDC algorithms, with the same primal constraints, have spectra which are almost identical; these spectra determine the rate of convergence of these preconditioned conjugate gradient methods. This result was first established in [27]; see also [26] for a simpler proof and a general discussion and general references on these algorithms.

We denote by $W^h(\Omega_i)$ the standard finite element space of continuous, piecewise linear functions on Ω_i which vanish on $\partial\Omega_i \cap \partial\Omega$. We will denote by h_i the smallest diameter of the finite elements in the subdomain Ω_i. The corresponding finite element trace spaces are denoted by $W^{(i)} := W^h(\partial\Omega_i \cap \Gamma), i = 1, \ldots, N$. The product space of the $W^{(i)}$ is denoted by W and in the context of these iterative methods, we need to consider elements of this space, which are not necessarily continuous across the interface. However, the primal variables are global. Thus, in two dimensions, the values at the subdomain vertices are often chosen to be primal, i.e., to have common values. In the FETI-DP algorithms, the remaining continuity requirements at all the remaining nodes on the interface will only be fully in force when the iteration has converged and certain Lagrange multipliers have reached their correct values. In the BDDC algorithms, on the other hand, continuity across the interface is restored, in each step of the iteration, by replacing the discontinuous values on the interface by a weighted average. The partially assembled subspace, with the primal variables global, will be denoted by $\widetilde{W}$.

We can now formulate our main result, which is also valid for compressible elasticity.

Theorem 1 (Condition Number Estimate). *Let the domain $\Omega \subset \mathbb{R}^2$ be partitioned into subdomains Ω_i, which are partitioned into shape regular elements and which have complements $\complement\Omega_i$ that are Jones domains. Let all values at the subdomain vertices be primal. Then, with M the Dirichlet preconditioner, F the FETI-DP operator, the condition number of the preconditioned conjugate gradient method satisfies*

$$\kappa(M^{-1}F) \le C \max_i (1+\log(H_i/h_i))^2.$$

Here C is a constant which depends only on the parameters $C_J(\Omega_i)$ and $C_U(\mathcal{C}\Omega_i)$ of Definitions 1 and 2, the Poincaré parameters $\gamma(\Omega_i,2)$ of the subdomains, and the shape regularity of the finite elements. The result is also independent of possible jumps in the coefficient ρ_i, or the Lamé parameters, across the interface between the subdomains.

A complete proof of this result is given in [23]. It is as strong a result as those in [24, 25] for regular subdomains. We also note that numerical experiments on very irregular snowflake subdomains have added interesting insight on how best to scale the FETI-DP preconditioners.

We will now indicate what is required to establish the theorem. We denote by $\mathcal{H}$ the discrete harmonic extension operator: $\mathcal{H}(v_\Gamma)$ is the minimal energy extension of the restriction of the finite element function v to the interface Γ. In what follows, $\mathcal{H}(uv)$ will mean the discrete harmonic extension of the finite element function obtained by interpolating the product of u and v. For each edge $\mathcal{E}^{ij}$ (the open set common to $\partial\Omega_i$ and $\partial\Omega_j$ and which does not contain its endpoints) we define an edge cutoff function $\theta_{\mathcal{E}^{ij}}$, which is the discrete harmonic function which equals 1 at all nodes on the edge $\mathcal{E}^{ij}$ and which vanishes at all other interface nodes; cf. [31, Sec. 4.6].

We note that part of the proof of any result on a FETI-DP or BDDC algorithm, such as Theorem 1, is purely algebraic. It is also known that in order to fully prove that theorem, we need to use tools of analysis to establish a result such as Lemma 5; see, e.g., [26] or [24, Sec. 8]. For the set of primal constraints considered in Theorem 1, we need to prove:

Lemma 5. *Let $\mathcal{E}^{ij}$ be an edge common to the boundaries of Ω_i and Ω_j. For all $v \in \widetilde{W}$ and with $v^{(i)} := R^{(i)}v, v^{(j)} := R^{(j)}v$, we have*

$$\begin{aligned}\rho_i|\mathcal{H}(\theta_{\mathcal{E}^{ij}}\delta_i^\dagger(v^{(i)}-v^{(j)}))|^2_{H^1(\Omega_i)} \le\, & C(1+\log(H_i/h_i))^2\rho_i|v^{(i)}|^2_{H^1(\Omega_i)} \\ & + C(1+\log(H_j/h_j))^2\rho_j|v^{(j)}|^2_{H^1(\Omega_j)}.\end{aligned} \tag{4}$$

Here $R^{(i)}$ denotes the restriction operator from $\widetilde{W}$ to $W^{(i)}$. The parameter $\delta_i^\dagger := \rho_i^\gamma/\sum_{j\in\mathcal{N}_x}\rho_j^\gamma$, where $\gamma \in [1/2,\infty)$ and $\mathcal{N}_x$ is the set of indices j of the subregions with x on their boundaries. The constant C in the inequality depends only on the parameters $C_J(\Omega_i)$ and $C_U(\mathcal{C}\Omega_i)$ of Definitions 1 and 2, the Poincaré parameters $\gamma(\Omega_i,2)$ of the subdomains, and the shape regularity of the finite elements.

To prove this lemma, we need three auxiliary results, in addition to Poincaré's inequality. The first is a discrete Sobolev inequality. This inequality, (5), is well known in the theory of iterative substructuring methods. Proofs for domains satisfying an interior cone condition are given in [5] and [6, Sec. 4.9] and a different proof is given in [31, p. 102]. For a proof for John domains, see [10].

Lemma 6 (Discrete Sobolev Inequality). *Let $\Omega_i \subset \mathbb{R}^2$ be a John domain. Then,*

$$\|u - \bar{u}_{\Omega_i}\|^2_{L^\infty(\Omega_i)} \leq C(1 + \log(H_i/h_i))|u|^2_{H^1(\Omega_i)}, \tag{5}$$

for all $u \in W^h(\Omega_i)$. The constant C depends only on the John parameter $C_J(\Omega_i)$ of Ω_i and the shape regularity of the finite elements.

A three-dimensional counterpart of Lemma 6 is given in [31, Subsec. 4.6.2]. This provides an estimate of the L^2−norm of finite element functions over subdomain edges and this result has not yet been extended fully to the case of irregular subdomains.

Another important result provides estimates for different types of edge functions. For regular subdomains in two dimensions, this lemma was first given in [13].

Lemma 7 (Edge Lemma). *Let $\Omega_i \subset \mathbb{R}^2$ be a John domain, let $\mathcal{E}^{ij} \subset \partial\Omega_i$ be an edge, and $\theta_{\mathcal{E}^{ij}} \in W^h(\Omega_i)$ be the finite element function which equals 1 at all nodes of $\mathcal{E}^{ij}$, and which vanishes at all the other nodes on $\partial\Omega_i$, and is discrete harmonic in Ω_i. Then, for any $u \in W^h(\Omega_i)$, we have*

$$|\mathcal{H}(\theta_{\mathcal{E}^{ij}} u)|^2_{H^1(\Omega_i)} \leq C(1 + \log(H_i/h_i))^2 \|u\|^2_{H^1(\Omega_i)}, \tag{6}$$

$$|\theta_{\mathcal{E}^{ij}}|^2_{H^1(\Omega_i)} \leq C(1 + \log(H_i/h_i)), \tag{7}$$

$$\|\theta_{\mathcal{E}^{ij}}\|^2_{L^2(\Omega_i)} \leq CH_i^2(1 + \log(H_i/h_i)). \tag{8}$$

Here, C depends only on the John parameter $C_J(\Omega_i)$ of Ω_i and the shape regularity of the finite elements. The logarithmic factor in (8) can be removed if all angles of the triangulation are acute and W^h is a space of piece-wise linear finite elements.

For a proof, see [23].

In order to advance the work on three dimensional problems, it would be central to develop similar face and edge lemmas under some suitable geometric assumptions; cf. [31, Sec. 4.6] for results in case the subdomains are regular. At this time, we can prove such bounds for a subdomain which contains a Lipschitz domain with edges which are common to those of the subdomain. It is also clear that in the general case, we need a limit on the number of points on each edge since it can easily be seen that the energy of the edge function $\theta_{\mathcal{E}^{ik}}$ will grow in proportion to this number. In a case of many edge nodes, the energy of the face functions must also be large.

We establish inequality (6) by using ideas similar to those of [31, Proofs of Lems. 4.24 and 4.25]. We construct a function $\vartheta_{\mathcal{E}^{ij}}$ which has the same boundary values as $\theta_{\mathcal{E}^{ij}}$ and which satisfies the two inequalities (6) and (7). Since $\theta_{\mathcal{E}^{ij}}$ and $\mathcal{H}(\theta_{\mathcal{E}^{ij}} u)$ are discrete harmonic, the two inequalities (6) and (7) will then hold. We note that in our work on almost incompressible elasticity, described in Section 5, such results are also required for domains with large aspect ratios; see [12, Lem. 5.4]. The bounds as in Lemma 7 will grow linearly with the aspect ratio of the subdomains.

The next lemma was proven for Lipschitz domains and quite general conforming finite elements in [32], using a technique from [2]; see also [31] for a different proof. In [23], we have developed a new proof for more general domains, which uses Lemma 3 and a result by [30]. We note that this result can be viewed as providing an estimate of the rate of convergence of the classical Dirichlet-Neumann algorithm for two subdomains and with a quite irregular interface; see, e.g., [31, Sec. 1.3.3].

Lemma 8 (Extension Lemma). *Let Ω_i and Ω_j, subsets of $\mathbb{R}^n$, be two subdomains with a common $(n-1)$-dimensional interface Γ^{ij}. Furthermore, let Ω_i be a domain with a complement which is a Jones domain, let*

$$W_i^h = \{v_h \in W^h(\Omega_i) : v_h(x) = 0 \text{ at all nodes of } \partial\Omega_i \setminus \Gamma^{ij}\},$$
$$W_j^h = \{v_h \in W^h(\Omega_j) : v_h(x) = 0 \text{ at all nodes of } \partial\Omega_j \setminus \Gamma^{ij}\}.$$

Then, there exists an extension operator

$$E_{ji}^h : W_j^h \longrightarrow W_i^h,$$

with the following properties $\forall u_h \in W_j^h$:

$$(E_{ji}^h u_h)_{|\Omega_j} = u_h \quad \text{and} \quad |E_{ji}^h u_h|_{H^1(\Omega_i)} \leq C|u_h|_{H^1(\Omega_j)}$$

where the constant C depends only on the Jones parameter $C_U(\complement\Omega_i)$ of the complement of Ω_i and the shape regularity of the elements and is otherwise independent of the finite element mesh sizes h_i and h_j and the diameters H_i and H_j.

4 An Overlapping Schwarz Method

When we now turn to another major family of domain decomposition methods, we note that the overlapping Schwarz methods can be used even if the stiffness matrix of the problem is only available in fully assembled form. This is important in many applications. The FETI-DP and BDDC algorithms, in contrast, require access to the stiffness matrices of the subdomains.

In the case of a scalar elliptic problem with constant coefficients in each substructure, a coarse space for a problem in three dimensions can be defined as the range of the interpolation operator

$$I_B^h u(x) = \sum_{i,\ell} u(\mathcal{V}^{i\ell})\theta_{\mathcal{V}^{i\ell}}(x) + \sum_{i,k} \bar{u}_{\mathcal{E}^{ik}}\theta_{\mathcal{E}^{ik}}(x) + \sum_{i,j} \bar{u}_{\mathcal{F}^{ij}}\theta_{\mathcal{F}^{ij}}(x). \tag{9}$$

Here, $\bar{u}_{\mathcal{E}^{ik}}$ and $\bar{u}_{\mathcal{F}^{ij}}$ are averages over edges and faces, respectively. We obtain an analogous expression for two dimensions by dropping the face terms.

With suitable local spaces, it is known that the resulting iterative substructuring algorithm [31, Algo. 5.16] is quite satisfactory with a condition number bound of the form $C\max_i(1+\log(H_i/h_i))$. The constant C is independent of the number of subdomains as well as jumps in the coefficients between the subdomains. By enriching the coarse space by basis functions constructed from the same cutoff functions and the rigid body modes similar results are possible for linear elasticity.

These coarse spaces have also recently been combined with local components based on overlapping subdomains, just as for traditional overlapping Schwarz methods; see [8, 9, 11, 12]. These methods are all additive Schwarz methods and they are therefore definined in terms of a coarse subspace and many local spaces defined by subspaces of finite element functions supported in the overlapping subdomains Ω_i'. We note that there would be no additional technical issues should we choose to work with multiplicative or hybrid Schwarz methods as in [31, Chap. 2]. The resulting algorithms have already proven quite successful for very large problems and an implementation, by Clark Dohrmann, for massively parallel computing systems is now part of the Salinas software system for structural dynamics problems developed at Sandia National Laboratories, Albuquerque, NM.

In the case of Lipschitz subdomains, the weights for the face terms in an interpolation formula such as (9) can easily be bounded by using Cauchy-Schwarz's inequality and an elementary trace theorem such as [29, Thm. 1.2]. For more general subdomains, this argument breaks down but the average can be replaced by any bounded functional, which depends only on the trace of the finite element function on the face, and which reproduces constants. In two dimensions, the average over an edge can simply be replaced by the maximum of the finite element function and we can then use Lemmas 6 and 7 at the expense of an additional logarithmic factor. The same approach would result in a factor H_i/h_i in three dimensions. Instead the face average over $\mathcal{F}^{ij}$ can be replaced by

$$(\nabla \theta_{\mathcal{F}^{ij}}, \nabla \mathcal{H}(\theta_{\mathcal{F}^{ij}} u))_{L^2(\Omega_i)} / (\nabla \theta_{\mathcal{F}^{ij}}, \nabla \theta_{\mathcal{F}^{ij}})_{L^2(\Omega_i)}. \tag{10}$$

and the edge averages by similar expressions. We note that these expressions depend exclusively on the trace of u on the interface Γ. In these formulas, we could equally well integrate over Ω_j or over $\Omega_i \cup \Omega_j$. It is easy to see that this new interpolant also reproduces constants as well as the face and edge terms separately. The energy of the face term with the coefficient, given in (10), can be bounded by $\|\nabla \mathcal{H}(\theta_{\mathcal{F}^{ij}} u)\|_{L^2(\Omega_i)}$. This will result in a bound with two logarithmic factors if we can prove a three dimensional counterpart of Lemma 7.

The analysis of this domain decomposition method is carried out in the framework of the abstract Schwarz theory as in [32, Chap. 2]. If exact solvers are used for the coarse and local problems, each defined on an extended subdomain Ω_i', we primarily need a bound on the energy of the coarse interpolant that we have already discussed. There are essentially no new technical difficulties in obtaining bounds for the local terms in the decomposition of an arbitrary finite element function as in [32, Assump. 2.2].

The coarse space can be enriched so that all rigid body modes are exactly reproduced using formulas similar to that of (10). A result for compressible elasticity can then be obtained for two dimensions and John subdomains. The extension to three dimensions will again essentially require the extension of Lemma 7 to edges and faces in three dimensions.

We have established the following result in [10]. It holds for scalar elliptic problems as well as problems in compressible elasticity.

Theorem 2. *Let $\Omega \subset \mathbb{R}^2$ be partitioned into nonoverlapping subdomains Ω_i, which are John domains, each with a shape regular triangulation. The condition number of our domain decomposition method then satisfies*

$$\kappa(P_{ad}) \leq C \max_i (1 + H_i/\delta_i)(1 + \log(H_i/h_i))^2,$$

where C is a constant which only depends the John and Poincaré parameters of the subdomains, the number of colors required for the overlapping subdomains, and the shape regularity of the elements. The bound is also independent on variations in the coefficients across the interface Γ.

5 Almost Incompressible Elasticity

We also use the same overlapping Schwarz algorithm for almost incompressible elasticity and this is the subject of two papers recently completed; see [11] and [10]. The main emphasis is on regular subdomains, but the result also holds for subdomains that are just John domains and for two dimensions, see [11, Section 6]. As previously pointed out, we have only considered mixed finite element methods, with pressure spaces of discontinuous finite element functions. Our main result is:

Theorem 3 (Condition Number Estimate). *The condition number of our domain decomposition method satisfies*

$$\kappa(P_{ad}) \leq C(1 + (H/\delta))^3(1 + \log(H/h))^2,$$

where C is a constant, independent of the number of subdomains and their diameters and the mesh size and which only depends on the number of colors required for the overlapping subdomains and the shape regularity of the elements and the subdomains. The bound is also independent of the Poisson ratio and of the variations in the coefficients across the interface Γ.

We note that an early application of overlapping Schwarz methods to mixed formulations of linear elasticity and Stokes problems is given in [19]. In that work, the coarse spaces were based on the same mixed finite element methods on coarse meshes and both continuous and discontinuous pressure spaces were considered. An analysis of these methods was not provided, but their performance was shown to be quite competitive with block diagonal and block triangular preconditioners, see [20].

The new algorithm uses a coarse space similar to that of (10). Just as for the compressible elasticity case, it is enriched so as to contain all the rigid body modes. In our work, there are a number of new challenges, in particular, results on subdomains with bad aspect ratios are required. In addition, when applying the abstract Schwarz theory, great care has to be taken when constructing the coarse component of the partitioning of the displacement fields and new ideas are also required when partitioning the remaining part of an arbitrary finite element function into local components. For details, see [11].

References

[1] Acosta, G., Durán, R.G., Muschietti, M.A.: Solutions of the divergence operator on John domains. *Adv. Math.*, 206(2):373–401, 2006.

[2] Astrahancev, G.P.: The method of fictitious domains for a second order elliptic equation with natural boundary conditions. *Ž. Vyvcisl. Mat. i Mat. Fiz.* 18 (1978), no. 1, 118–125; trans. *Comput. Math. Phys.*, 18:114–121, 1978.

[3] Bojarski, B.: Remarks on Sobolev imbedding inequalities. In *Complex analysis, Joensuu 1987*, 52–68. Lecture Notes in Math., 1351. Springer, Berlin, 1988.

[4] Bramble, J.H.: A proof of the inf-sup condition for the Stokes equations on Lipschitz domains. *Math. Models Meth. Appl. Sci.*, 13(3):361–371, 2003.

[5] Bramble, J.H., Pasciak, J.E., Schatz, A.H.: The construction of preconditioners for elliptic problems by substructuring. I. *Math. Comp.*, 47(175):103–134, 1986.

[6] Brenner, S.C., Scott, R.: *The Mathematical Theory of Finite Element Methods*, 2nd ed. Springer, Berlin–Heidelberg–New York, 2002.

[7] Buckley, S., Koskela, P.: Sobolev-Poincaré implies John. *Math. Res. Lett.*, 2:577–593, 1995.

[8] Dohrmann, C.R., Klawonn, A., Widlund, O.B.: Extending theory for domain decomposition algorithms to irregular subdomains. In U. Langer et al. eds., *Domain Decomposition Methods in Science and Engineeering XVII*, 255–261. Lect. Notes Comput. Sci. Eng., 60. Springer, 2007.

[9] Dohrmann, C.R., Klawonn, A., Widlund, O.B.: A family of energy minimizing coarse spaces for overlapping Schwarz preconditioners. In U. Langer et al. eds., *Domain Decomposition Methods in Science and Engineering XVII*, 247–254. Lect. Notes Comput. Sci. and Eng., 60. Springer, 2007.

[10] Dohrmann, C.R., Klawonn, A., Widlund, O.B.: Domain decomposition for less regular subdomains: Overlapping Schwarz in two dimensions. *SIAM J. Numer. Anal.*, 46(4):2153–2168, 2008.

[11] Dohrmann, C.R., Widlund, O.B.: A hybrid domain decomposition method for compressible and almost incompressible elasticity. TR2008-919, Courant Institute of Mathematical Sciences, December 2008.

[12] Dohrmann, C.R., Widlund, O.B.: An overlapping Schwarz algorithm for almost incompressible elasticity. TR2008-912, Dept. Computer Science, Courant Institute of Mathematical Sciences, New York University, May 2008.

[13] Dryja, M., Widlund, O.B.: Some domain decomposition algorithms for elliptic problems. In L. Hayes and D. Kincaid, eds., *Iterative Methods for Large Linear Systems*, 273–291. Academic, 1989.

[14] Durán, R.G., Muschietti, M.A.: The Korn inequality for Jones domains. *Electron. J. Differential Equations*, 2004(127):1–10, 2004.

[15] Federer, H., Fleming, W.H.: Normal and integral currents. *Ann. of Math. (2)*, 72:458–520, 1960.

[16] Hajłasz, P.: Sobolev inequalities, truncation method, and John domains. In *Papers on analysis*, 109–126. *Rep. Univ. Jyväskylä Dep. Math. Stat.*, 83. Univ. Jyväskylä, 2001.

[17] Jones, P.W.: Quasiconformal mappings and extendability of functions in Sobolev space. *Acta Math.*, 147(1-2):71–88, 1981.
[18] Karypis, G., Kumar, V.: *METIS Version 4.0*. University of Minnesota, Department of Computer Science, Minneapolis, MN, 1998.
[19] Klawonn, K., Pavarino, L.F.: Overlapping Schwarz methods for mixed linear elasticity and Stokes problems. *Comput. Methods Appl. Mech. Engrg.*, 165:233–245, 1998.
[20] Klawonn, A., Pavarino, L.F.: A comparison of overlapping Schwarz methods and block preconditioners for saddle point problems. *Numer. Linear Algebra Appl.*, 7:1–25, 2000.
[21] Klawonn, A., Rheinbach, O.: A parallel implementation of Dual-Primal FETI methods for three dimensional linear elasticity using a transformation of basis. *SIAM J. Sci. Comput.*, 28(5):1886–1906, 2006.
[22] Klawonn, A., Rheinbach, O. Robust FETI-DP methods for heterogeneous three dimensional elasticity problems. *Comput. Methods Appl. Mech. Engrg.*, 196(8):1400–1414, 2007.
[23] Klawonn, A., Rheinbach, O., Widlund, O.B.: An analysis of a FETI–DP algorithm on irregular subdomains in the plane. *SIAM J. Numer. Anal.*, 46(5):2484–2504, 2008.
[24] Klawonn, A., Widlund, O.B.: Dual-Primal FETI Methods for Linear Elasticity. *Comm. Pure Appl. Math.*, 59:1523–1572, 2006.
[25] Klawonn, A., Widlund, O.B., Dryja, M.: Dual-primal FETI methods for three-dimensional elliptic problems with heterogeneous coefficients. *SIAM J. Numer. Anal.*, 40(1):159–179, April 2002.
[26] Li, J., Widlund, O.B.: FETI–DP, BDDC, and Block Cholesky Methods. *Internat. J. Numer. Methods Engrg.*, 66(2):250–271, 2006.
[27] Mandel, J., Dohrmann, C.R., Tezaur, R.: An algebraic theory for primal and dual substructuring methods by constraints. *Appl. Numer. Math.*, 54:167–193, 2005.
[28] Maz'ja, V.G.: Classes of domains and imbedding theorems for function spaces. *Soviet Math. Dokl.*, 1: 882–885, 1960.
[29] Nečas, J. *Les méthodes directes en théorie des équations elliptiques*. Academia, Prague, 1967.
[30] Scott, L.R., Zhang, S.: Finite element interpolation of nonsmooth functions satisfying boundary conditions. *Math. Comp.*, 54(190):483–493, 1990.
[31] Toselli, A., Widlund, O.: *Domain Decomposition Methods - Algorithms and Theory*, Springer Series in Computational Mathematics, 34. Springer, Berlin–Heidelberg–New York, 2005.
[32] Widlund, O.B.:. An extension theorem for finite element spaces with three applications. In W. Hackbusch and K. Witsch, eds., *Numerical Techniques in Continuum Mechanics*, 110–122. Braunschweig/Wiesbaden, 1987. Notes on Numerical Fluid Mechanics, v. 16. Vieweg.

Auxiliary Space Preconditioners for Mixed Finite Element Methods

Ray S. Tuminaro[1], Jinchao Xu[2], and Yunrong Zhu[3]

[1] MS 9214, Sandia National Laboratories, Livermore, CA 94551 USA rstumin@sandia.gov
[2] Department of Mathematics, Pennsylvania State University, University Park, PA 16802 USA xu@math.psu.edu
[3] Department of Mathematics, University of California, San Diego, 9500 Gilman Driver, La Jolla, CA 92093 USA zhu@math.ucsd.edu

Summary. This paper is devoted to study of an auxiliary spaces preconditioner for $\mathbf{H}(\text{div})$ systems and its application in the mixed formulation of second order elliptic equations. Extensive numerical results show the efficiency and robustness of the algorithms, even in the presence of large coefficient variations. For the mixed formulation of elliptic equations, we use the augmented Lagrange technique to convert the solution of the saddle point problem into the solution of a nearly singular $\mathbf{H}(\text{div})$ system. Numerical experiments also justify the robustness and efficiency of this scheme.

1 Introduction

In this note, we discuss some implementation details of robust and efficient AMG preconditioners for the $\mathbf{H}(\text{div})$ system:

$$(\lambda \text{div}\, \mathbf{u}, \text{div}\, \mathbf{v}) + (\mu \mathbf{u}, \mathbf{v}) = (\mathbf{f}, \mathbf{v}), \quad \forall \mathbf{v} \in \mathbf{H}(\text{div}), \tag{1}$$

where $\mathbf{f} \in \mathbf{L}^2(\Omega)$ is a vector field and the coefficients $\lambda(x)$ and $\mu(x)$ are assumed to be uniformly positive but may have large variations in the whole domain Ω. Given a triangulation, the finite element problem reads:

$$\text{Find } \mathbf{u}_h \in \mathbf{V}_h(\text{div}): \quad (\lambda \text{div}\, \mathbf{u}_h, \text{div}\, \mathbf{v}_h) + (\mu \mathbf{u}_h, \mathbf{v}_h) = (\mathbf{f}, \mathbf{v}_h), \quad \forall \mathbf{v}_h \in \mathbf{V}_h(\text{div}), \tag{2}$$

where $\mathbf{V}_h(\text{div}) \subset \mathbf{H}(\text{div})$ is a conforming finite element space, e.g. Raviart-Thomas element, or BDM element (c.f. [6]). The finite element discretization (2) gives rise to the following linear system:

$$\mathbf{A}x = b, \tag{3}$$

where $\mathbf{A} = (a_{ij})$ is defined by $a_{ij} = \int_\Omega \lambda \text{div} \varphi_j \text{div} \varphi_i + \mu \varphi_j \cdot \varphi_i dx$ for any basis functions $\varphi_i, \varphi_j \in \mathbf{V}_h(\text{div})$.

M. Bercovier et al. (eds.), *Domain Decomposition Methods in Science and Engineering XVIII*, Lecture Notes in Computational Science and Engineering,
DOI: 10.1007/978-3-642-02677-5,

The importance of $\mathbf{H}(\text{div})$-related problems has promoted vigorous research into efficient multilevel schemes for solving the linear system (3) (see [1, 10, 12, 18, 19] for example). The $\mathbf{H}(\text{div})$ systems (1) arise naturally from numerous problems of practical importance, such as stabilized mixed formulations of the Stokes problem, least squares methods for H^1 systems, and mixed methods for second order elliptic equations, see [1, 19].

Recently, Hiptmair and Xu [12] proposed an innovative approach to solve $\mathbf{H}(\text{curl})$ and $\mathbf{H}(\text{div})$ systems, known as the HX-preconditioner. It relies on a discrete regular decomposition (see Section 2) and the framework of auxiliary space method ([20]). This decomposition links the vector fields in $\mathbf{H}(\text{curl})$ and $\mathbf{H}(\text{div})$ directly with functions in H^1. By using certain grid transfer operators, the evaluation of the preconditioners for $\mathbf{H}(\text{curl})$ and $\mathbf{H}(\text{div})$ systems is essentially reduced to several second-order elliptic operators. Hence, standard (algebraic) multigrid techniques for the H^1 equations can be applied.

In our implementation of the HX-preconditioner for $\mathbf{H}(\text{div})$, we use a "grey-box" multilevel algorithm. More precisely, unlike the standard AMG technique, we rely on certain grid information for construction of the grid-transfer operators, namely the canonical interpolation operators Π_h^{curl}, Π_h^{div} and the discrete curl operator $\mathbf{C}$. The construction of these operators relies solely on coordinates and grid information on the finest level. In particular, we do not need a complete multilevel grid hierarchy which is crucial in standard geometric multilevel algorithms. A similar idea was used in [13] for the AMG implementation of the HX-preconditioner for $\mathbf{H}(\text{curl})$ systems.

The finite element discretization of the mixed problem results in a saddle point problem. There is a significant amount of literature on designing robust preconditioners for the mixed problem, see [2, 7, 11]. Here we use the augmented Lagrangian method to reduce the saddle point problem into a nearly singular $\mathbf{H}(\text{div})$ system, which can be efficiently solved by using the HX-preconditioner.

The remainder of this paper is organized as follows. In Section 2, we discuss some implementation details about the algorithm. In Section 3, we consider solving a mixed formulation of second order elliptic equations. We apply the augmented Lagrange method to reduce the mixed formulation into a nearly singular $\mathbf{H}(\text{div})$ system. Then in Section 4, we present some numerical experiments to justify the robustness and efficiency of the algorithms.

2 HX-Preconditioner for $\mathbf{H}(\text{div})$ Systems

In this section, we summarize the main ingredients used in [12] to derive and analyze the auxiliary space preconditioner. Here, we consider the lowest order Raviart-Thomas space $\mathbf{V}_h(\text{div}) \subset \mathbf{H}(\text{div})$, the lowest order Nédélec space $\mathbf{V}_h(\text{curl}) \subset \mathbf{H}(\text{curl})$ and the standard piecewise linear continuous nodal space $V_h(\text{grad})$. We use Π_h^{grad}, Π_h^{curl} and Π_h^{div} to denote the canonical interpolation operators onto the finite element spaces $V_h(\text{grad})$, $\mathbf{V}_h(\text{curl})$ and $\mathbf{V}_h(\text{div})$, respectively. The HX-preconditioners for $\mathbf{H}(\text{curl})$ and $\mathbf{H}(\text{div})$ systems exploit the following discrete regular decomposition.

Theorem 1. [12, Lemma 5.1] *Let D be the differential operator* curl *or* div, *and D^- be* grad *or* curl *respectively. Then for any $\mathbf{v}_h \in V_h(D)$, we have*

$$\mathbf{v}_h = \widetilde{\mathbf{v}}_h + \Pi_h^D \Phi_h + D^- p_h,$$

where $\widetilde{\mathbf{v}}_h \in V_h(D)$, $\Phi_h \in \mathbf{V}_h(\text{grad})$ and $p_h \in V_h(D^-)$, such that

(1) $\|h^{-1}\widetilde{\mathbf{v}}_h\|_{0,\Omega}^2 + \|\Phi_h\|_{1,\Omega}^2 \lesssim \|D\mathbf{v}_h\|_{0,\Omega}^2$;
(2) $\|p_h\|_{H(D^-)} \lesssim \|\mathbf{v}_h\|_{H(D)}$.

In the above decomposition, when $D = \text{div}$, the discrete potential $p_h \in \mathbf{V}_h(\text{curl})$ is not entirely desirable. In order to avoid solving an $\mathbf{H}(\text{curl})$-elliptic equation for p_h, we apply the decomposition Theorem 1 recursively and replace p_h by a $\Psi_h \in \mathbf{V}_h(\text{grad})$ and some "high frequency" edge element function. More precisely, we obtain a decomposition

$$\begin{aligned}
\mathbf{v}_h &= \sum_{\mathbf{b}\in\mathcal{B}(\text{div})} \mathbf{v}_\mathbf{b} + \Pi_h^{\text{div}} \Phi_h + \text{curl}\, p_h \\
&= \sum_{\mathbf{b}\in\mathcal{B}(\text{div})} \mathbf{v}_\mathbf{b} + \Pi_h^{\text{div}} \Phi_h + \sum_{\mathbf{q}\in\mathcal{B}(\text{curl})} \text{curl}\, p_\mathbf{q} + \text{curl}\, \Pi_h^{\text{curl}} \Psi_h,
\end{aligned}$$

where Φ_h, $\Psi_h \in \mathbf{V}_h(\text{grad})$ and $\mathcal{B}(\text{div})$, $\mathcal{B}(\text{curl})$ are the sets of the basis functions in $\mathbf{V}_h(\text{div})$ and $\mathbf{V}_h(\text{curl})$ respectively. In this decomposition, we have used the fact that curl grad $= 0$. By Theorem 1, this decomposition is stable:

$$\sum_{\mathbf{b}\in\mathcal{B}(\text{div})} \|\mathbf{v}_\mathbf{b}\|_A^2 + \|\Psi_h\|_{1,\Omega}^2 + \sum_{\mathbf{q}\in\mathcal{B}(\text{curl})} \|\text{curl}\, p_\mathbf{q}\|_{0,\Omega}^2 + \|\Phi_h\|_{1,\Omega}^2 \lesssim \|\mathbf{v}_h\|_A^2. \tag{4}$$

Based on this decomposition, the matrix representation of the (additive) auxiliary space preconditioner for the equation (2) is given by

$$\mathbf{B}_h^{\text{div}} := \mathbf{S}_h^{\text{div}} + \mathbf{C}\mathbf{S}_h^{\text{curl}}\mathbf{C}^T + \mathbf{P}_h^{\text{div}}\left(\mathbf{A}_h^{\text{grad}}\right)^{-1}\mathbf{P}_h^{\text{div}T} + \mathbf{C}\mathbf{P}_h^{\text{curl}}\left(\mathbf{A}_h^{\text{grad}}\right)^{-1}\mathbf{P}_h^{\text{curl}T}\mathbf{C}^T, \tag{5}$$

where

- $\mathbf{S}_h^{\text{div}}$ and $\mathbf{S}_h^{\text{curl}}$ are certain smoothers in the Raviart-Thomas and the Nédélec finite element spaces, for example Jacobi or symmetric Gauss-Seidel iterations, which we denote by StandardRelaxation() in the algorithms;
- $\mathbf{C}$ is the discrete curl operator;
- $\mathbf{P}_h^{\text{div}}$ and $\mathbf{P}_h^{\text{curl}}$ are the matrix representation of the canonical interpolation operators Π_h^{div} and Π_h^{curl} respectively;
- $\mathbf{A}_h^{\text{grad}}$ is the related (vectorial) elliptic operator on the finite element spaces $\mathbf{V}_h(\text{grad})$.

Remark 1. We remark that the first two terms in (5) together form an additive *Hiptmair smoother* (see [10]). In a multiplicative version of the preconditioner, we will denote the functions Pre(/Post)FineRelaxation() as multiplicative Hiptmair

Algorithm 1: $u = \mathsf{FineRelaxation}(\mathbf{A}, \mathbf{C}, u, b)$

1. $u \leftarrow \mathsf{StandardRelaxation}(\mathbf{A}, u, b)$;
2. $e \leftarrow \mathsf{StandardRelaxation}(\mathbf{C}^T A\mathbf{C}, 0, \mathbf{C}^T(b - \mathbf{A}u))$;
3. $u \leftarrow u + \mathbf{C}e$;
4. $u \leftarrow \mathsf{StandardRelaxation}(\mathbf{A}, u, b)$;

smoothers, see Algorithm 1. The function PreFineRelaxation() is identical to Algorithm 1 except step 1 is omitted, and the function PostFineRelaxation() is identical to Algorithm 1 except step four is omitted to keep the preconditioner symmetric.

It is important to realize that this special smoother is only needed on the finest level in our implementation, instead of using this smoother on each level as in [10].

When a hierarchy of structured grids is available, standard geometric multigrid can be applied to $\mathbf{A}_h^{\text{grad}}$ in the preconditioner (5). However, in general, the hierarchical information of the grids is not available, for example when the mesh is unstructured. In this case, one may consider using algebraic multigrid (AMG) algorithms. Moreover, instead of assembling the stiffness matrix $\mathbf{A}_h^{\text{grad}}$ explicitly by using the mesh data, we replace it with the following two matrices:

$$\mathbf{A}_1 := \mathbf{P}_h^{\text{div}^T} \mathbf{A} \mathbf{P}_h^{\text{div}},$$
$$\mathbf{A}_2 := \mathbf{P}_h^{\text{curl}^T} \mathbf{C}^T \mathbf{A} \mathbf{C} \mathbf{P}_h^{\text{curl}} = \mathbf{P}_h^{\text{curl}^T} \mathbf{C}^T \mathbf{M}(\mu) \mathbf{C} \mathbf{P}_h^{\text{curl}},$$

where $\mathbf{A}$ is the stiffness matrix defined in (3), and $\mathbf{M}(\mu)$ is the mass matrix defined by $M = (m_{ij})$ with $m_{ij} = \int_\Omega \mu \varphi_i \cdot \varphi_j$. In the formulation of $\mathbf{A}_2$, we used the fact that div curl $= 0$.

We note that $\mathbf{A}_1$ and $\mathbf{A}_2$ are vector Laplacian-like operators defined on the nodal space. Therefore, $\mathbf{A}_1$ and $\mathbf{A}_2$ are amenable to standard AMG algorithms. Here, we make use of the interpolation $\mathbf{P}_h^{\text{div}}$ and $\mathbf{P}_h^{\text{curl}}$, as well as the discrete curl $\mathbf{C}$. All of these three matrices can be constructed using grid information on the fine level. In fact, to compute the matrix $\mathbf{C}$, one needs to expand curlφ_E in terms of the basis of $\mathbf{V}_h(\text{div})$ for any basis function $\varphi_E \in \mathbf{V}_h(\text{curl})$. If $\mathbf{V}_h(\text{div})$ and $\mathbf{V}_h(\text{curl})$ are the lowest order Raviart-Thomas and Nédélec spaces respectively, $\mathbf{C}$ is simply a signed "edge-to-face" incidence matrix. The sign of each entry is determined by the signs of basis functions, i.e. the preset edge and face orientations in the grid. The matrix $\mathbf{P}_h^{\text{div}} = (\mathbf{P}_h^x, \mathbf{P}_h^y, \mathbf{P}_h^z)$ can be computed component-wise, where each of the blocks has the same sparsity pattern as the "face-to-node" incidence matrix. The entries are computed by the surface integral of the nodal basis functions on the normal direction on the face. The computation of $\mathbf{P}_h^{\text{curl}}$ is similar, which can be found in [3, 13].

The operator (5) and the discussion above suggest an additive version of the preconditioner: Algorithm 2.

In Algorithm 2, we may update the solution u after computing each u_1-u_4. By updating the solution and the residual at each step, we define a multiplicative version of the preconditioner. In the multiplicative preconditioner, we replace

Algorithm 2: u =HX_Additive_Preconditioner($\mathbf{A}, b$)

1 %Setup Phase

2 Form $\mathbf{A}_1 \leftarrow \mathbf{P}_h^{\mathrm{div}^T} \mathbf{A} \mathbf{P}_h^{\mathrm{div}}$ efficiently;

3 Standard_AMG_Setup($\mathbf{A}_1$);

4 Form $\mathbf{A}_2 \leftarrow \mathbf{P}_h^{\mathrm{curl}^T} \mathbf{C}^T \mathbf{A} \mathbf{C} \mathbf{P}_h^{\mathrm{curl}}$ efficiently;

5 Standard_AMG_Setup($\mathbf{A}_2$);

6 ———————————————————————;

7 %Solve Phase;

8 $u_1 \leftarrow$ StandardRelaxation($\mathbf{A}, 0, b$);

9 $x \leftarrow$ StandardRelaxation($\mathbf{C}^T \mathbf{A} \mathbf{C}, 0, \mathbf{C}^T b$);

10 $u_2 \leftarrow \mathbf{C}x$;

11 %Perform V-cycles on $\mathbf{A}_1$ and $\mathbf{A}_2$ $a \leftarrow$ Standard_AMG_Vcycle$\left(\mathbf{A}_1, 0, \mathbf{P}_h^{\mathrm{div}^T} b\right)$;

12 $u_3 \leftarrow \mathbf{P}_h^{\mathrm{div}} a$;

13 $p \leftarrow$ Standard_AMG_Vcycle$\left(\mathbf{A}_2, 0, \mathbf{P}_h^{\mathrm{curl}^T} \mathbf{C}^T b\right)$;

14 $u_4 \leftarrow \mathbf{C} \mathbf{P}_h^{\mathrm{curl}} p$;

15 ———————————————————————;

16 $u \leftarrow u_1 + u_2 + u_3 + u_4$;

the additive Hiptmair smoother (line 8-10 in Algorithm 2) by a multiplicative one (PreFineRelaxation). The rest of the algorithm is similar to Algorithm 2, except that after each step, we update the solution u, compute the residual r and replace the b in Line 11 and 13 by the residual r. Finally, in order to guarantee the symmetry of the overall preconditioner, we need to preform a post-smoothing step (PostFineRelaxation) in the end of the algorithm.

3 Application to Mixed Method

As an application, we present the augmented Lagrangian method for solving systems arising from a mixed finite element discretization of the elliptic boundary value problem (see e.g., [6]):

$$\Delta p = f \ \text{ in } \Omega, \quad p|_{\partial\Omega} = 0. \tag{6}$$

The aim is to show that implementing an efficient iterative method for the resulting indefinite linear system reduces to designing an efficient method for the solution of an auxiliary *nearly singular* $\mathbf{H}(\mathrm{div})$ problem. The augmented Lagrangian method has been applied to the mixed formulation of equation (6) in [11].

Given a conforming triangulation $\mathcal{T}_h$, let $\mathbf{V}_h(\mathrm{div}) \subset \mathbf{H}(\mathrm{div})$ and $V_h(0) \subset L^2(\Omega)$ be the corresponding finite element spaces. Then the mixed finite element method for the model problem (6) reads: find $(\mathbf{u}_h, p_h) \in \mathbf{V}_h(\mathrm{div}) \times V_h(0)$ such that

$$\begin{cases} (\mathbf{u}_h, \mathbf{v}_h) + (p_h, \mathrm{div}\mathbf{v}_h) = 0 & \forall \mathbf{v}_h \in V_h(\mathrm{div}) \\ (\mathrm{div}\mathbf{u}_h, q_h) = (f, q_h) & \forall q_h \in V_h(0). \end{cases} \tag{7}$$

A sufficient condition for the well-posedness of the mixed method (7) is the discrete inf-sup condition. Several finite element spaces satisfying the inf-sup condition have been introduced, such as those of Raviart-Thomas [15] and Brezzi-Douglas-Marini [5]. Here we restrict ourselves to the Raviart-Thomas spaces.

The mixed finite element method (7) results in the following linear system:

$$\begin{bmatrix} A & B^* \\ B & 0 \end{bmatrix} \begin{bmatrix} \mathbf{u} \\ p \end{bmatrix} = \begin{bmatrix} 0 \\ f \end{bmatrix}. \tag{8}$$

It is not difficult to see that A is the mass matrix of the Raviart-Thomas element and B is a matrix representation of div*.

The augmented Lagrangian algorithm solves the following equivalent problem to (8) by the Uzawa method:

$$\begin{bmatrix} A+\varepsilon^{-1}B^*B & B^* \\ B & 0 \end{bmatrix} \begin{bmatrix} \mathbf{u} \\ p \end{bmatrix} = \begin{bmatrix} \varepsilon^{-1}B^*f \\ f \end{bmatrix}. \tag{9}$$

Given $(\mathbf{u}^{(k)}, p^{(k)})$, the new iterate $(\mathbf{u}^{(k+1)}, p^{(k+1)})$ is obtained by solving the following system:

$$\begin{cases} (A+\varepsilon^{-1}B^*B)\mathbf{u}^{(k+1)} = \varepsilon^{-1}B^*f - B^*p^{(k)}, \\ p^{(k+1)} = p^{(k)} - \varepsilon^{-1}(f - B\mathbf{u}^{(k+1)}). \end{cases} \tag{10}$$

Convergence of this algorithm has been discussed in many works, see for example [7, 8, 14, 17].

Theorem 2. [14, Lemma 2.1] *Let $(\mathbf{u}^{(0)}, p^{(0)})$ be a given initial guess and for $k \geq 1$, let $(\mathbf{u}^{(k)}, p^{(k)})$ be the iterates obtained via the augmented Lagrangian algorithm* (10). *Then the following estimates hold:*

$$\|p - p^{(k)}\|_{0,\Omega} \leq \left(\frac{\varepsilon}{\varepsilon+\lambda_0}\right)^k \|p - p^{(0)}\|_{0,\Omega},$$

$$\|\mathbf{u} - \mathbf{u}^{(k)}\|_A \leq \sqrt{\varepsilon}\|p - p^{(k)}\|_{0,\Omega} \leq \sqrt{\varepsilon}\left(\frac{\varepsilon}{\varepsilon+\lambda_0}\right)^k \|p - p^{(0)}\|_{0,\Omega},$$

where λ_0 is the minimum eigenvalue of $S = BA^{-1}B^*$.

According to this theorem, the iteration procedure (10) converges rapidly to the solution of (7) for small ε. However, at each iteration one needs to solve a nearly singular $\mathbf{H}(\mathrm{div})$ system

$$(\varepsilon A + B^*B)\mathbf{u}^{(k+1)} = B^*f - \varepsilon B^*p^{(k)}. \tag{11}$$

Thus, an efficient and robust $\mathbf{H}(\mathrm{div})$ solver will result in an optimal iterative method for the saddle point problem (7). We refer to Section 4.3 for the numerical justification.

4 Numerical Results

The proposed solvers are implemented as preconditioners for the conjugate gradient method (CG) in MATLAB. ML's smoothed aggregation solver (c.f. [16]) is used for $\mathbf{A}_1$ and $\mathbf{A}_2$ through the mlmex MATLAB interface [9]. Part of the numerical experiments was done and reported in [4]. Unless otherwise stated, we use two steps of symmetric Gauss-Seidel sub-smoothing on both faces and edges. For all experiments, the convergence is attained when the ℓ^2-norm of the residual is reduced by 1×10^{-10}.

4.1 Constant Coefficients

As the first experiment, we consider the constant coefficient case. We triangulate the domain $\Omega = [0,2]^3$ with an *unstructured* grid. We assume that $\mu > 0$ is a constant in Ω. The following table shows CG-accelerated auxiliary AMG solvers for the $\mathbf{H}(\text{div})$ system:

$$(\text{div}\,\mathbf{u}, \text{div}\,\mathbf{v}) + \mu(\mathbf{u}, \mathbf{v}) = (\mathbf{f}, \mathbf{v}), \quad \forall \mathbf{v} \in \mathbf{H}(\text{div})$$

with respect to different constant values of μ.

Grid		μ							
		10^{-9}	10^{-4}	10^{-3}	10^{-2}	10^{-1}	1	10^{1}	10^{2}
9^3	Additive	15	16	16	17	18	19	21	25
	Multiplicative	5	5	5	6	6	6	6	7
18^3	Additive	15	18	18	19	19	21	23	26
	Multiplicative	5	6	7	7	7	8	8	9
27^3	Additive	15	18	18	19	19	21	24	26
	Multiplicative	5	7	7	7	8	8	9	9

Table 1. Number of CG iterations for AMG $\mathbf{H}(\text{div})$ preconditioners on the unstructured 3D tetrahedral mesh. $\lambda = 1$ and μ is a different constant for each experiment.

From Table 1, we observe that for different mesh sizes, both additive and multiplicative preconditioners result in a uniform and small number of CG iterations. Therefore, the preconditioners in both algorithms are robust with respect to the mesh size, which agrees with the theoretical results in [12]. Also, the iteration numbers are fairly robust with respective to the variation of the coefficient μ. From the table, one may also observe that the multiplicative preconditioner behaves better than the additive ones.

4.2 Variable Coefficients

In this subsection, we consider cases with variable coefficients. We conduct the experiments on the 3D unit cube $[0,1]^3$, triangulated by a uniform tetrahedron mesh (c.f. Fig. 1, each small cube is partitioned into six tetrahedrons).

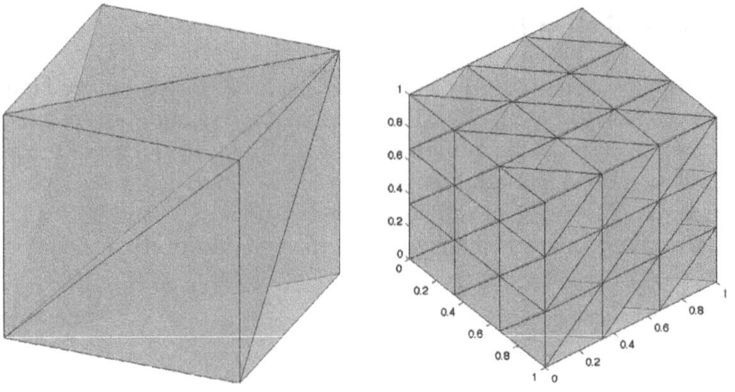

Fig. 1. Uniform Tetrahedra Meshes

First, we experiment with jumps in μ by considering two regions with constant values of μ. Specifically, define

$$\Omega_0 = \left\{ (x,y,z) : \frac{1}{3} \leq x,y,z \leq \frac{2}{3} \right\}, \quad \Omega_1 = \Omega \setminus \Omega_0;$$

let $\mu \equiv 1$ in Ω_1 and choose $\mu = \mu_0$ to be a constant inside Ω_0. λ is fixed to be 1 throughout the whole domain Ω. Table 2 reports the number of iterations on different mesh sizes. Note that the number of iterations are again robust with respect to the variation of the coefficient μ.

Grid		μ_0 10^{-4}	10^{-3}	10^{-2}	10^{-1}	1	10^{1}	10^{2}	10^{3}	10^{4}
9^3	Additive	19	19	19	19	18	19	21	23	23
	Multiplicative	6	5	5	5	5	5	6	6	5
18^3	Additive	19	19	20	18	17	18	20	23	24
	Multiplicative	6	6	6	6	5	5	6	6	6
27^3	Additive	18	19	19	17	17	17	19	22	24
	Multiplicative	6	6	6	5	5	5	6	6	6

Table 2. Number of iterations for CG-accelerated AMG on the 3D tetrahedral mesh problem with jump coefficients. μ_0 is defined inside $[1/3, 2/3]^3$ and is a different constant for each experiment, elsewhere μ is 1, and $\lambda \equiv 1$.

We now consider a jump in λ. As before, we choose $\lambda = \lambda_0$ to be a constant, which varies for different experiments inside the domain Ω_0, and $\lambda = 1$ elsewhere. This time, we fix μ to be 1 in the whole domain Ω. Table 3 reports the number of iterations on different mesh sizes. In this case, the number of iterations varies a little bit. This may be due to some mild deficiencies in the underlying standard AMG solver.

Grid		λ_0								
		10^{-4}	10^{-3}	10^{-2}	10^{-1}	1	10^1	10^2	10^3	10^4
9^3	Additive	31	28	22	19	18	18	18	17	16
	Multiplicative	12	11	8	6	5	5	5	5	5
18^3	Additive	33	29	22	18	17	17	17	16	16
	Multiplicative	11	10	8	6	5	5	5	5	5
27^3	Additive	32	28	21	17	17	16	16	16	16
	Multiplicative	10	9	7	6	5	5	5	5	5

Table 3. Number of iterations for CG-accelerated AMG on the 3D tetrahedral mesh problem with jump coefficients. λ_0 is defined inside $[1/3, 2/3]^3$ and is a different constant for each experiment, elsewhere λ is 1, and $\mu \equiv 1$.

4.3 Augmented Lagrangian Iterations

The augmented Lagrangian algorithm presented in Section 3 requires the solution of a nearly singular $\mathbf{H}(\mathrm{div})$ system (11) at each iteration. This implies that the $\mathbf{H}(\mathrm{div})$ solver should be robust with respect to the (penalty) parameter ε. Table 4 shows the CG-accelerated auxiliary AMG solver for the $\mathbf{H}(\mathrm{div})$ system:

$$(\mathrm{div}\,\mathbf{u}, \mathrm{div}\,\mathbf{v}) + \varepsilon(\mathbf{u}, \mathbf{v}) = (\mathbf{f}, \mathbf{v}), \quad \forall \mathbf{v} \in \mathbf{H}(\mathrm{div})$$

with respect to different ε on structured meshes with different mesh sizes. That is, take $\mu = \varepsilon$ in the example in Subsection 4.1. Here we use ε for consistency with Section 3. As we can see from this table, both additive and multiplicative preconditioners are robust with respect to ε.

Grid		ε							
		10^{-9}	10^{-4}	10^{-3}	10^{-2}	10^{-1}	1	10^1	10^2
9^3	Additive	14	15	15	15	17	18	20	23
	Multiplicative	5	5	5	5	5	5	5	6
18^3	Additive	14	15	15	15	16	17	19	20
	Multiplicative	5	5	5	5	5	5	5	5
27^3	Additive	15	15	15	15	15	17	18	20
	Multiplicative	5	5	5	5	5	5	5	5

Table 4. Number of iterations for CG-accelerated AMG on the 3D tetrahedral mesh $\mathbf{H}(\mathrm{div})$ problem. ε is a different constant for each experiment.

Table 5 shows the number of outer iterations for the augmented Lagrangian method for the mixed formulation of the elliptic equation with respect to different ε, where we used the auxiliary AMG $\mathbf{H}(\mathrm{div})$ solver above to solve the nearly singular system. The tolerance for the augmented Lagrangian iteration is 10^{-8}. In particular, according to the theory, the augmented Lagrangian method converges faster

Grid	ε 10^{-9}	10^{-4}	10^{-3}	10^{-2}	10^{-1}	1	10^{1}	10^{2}
9^3	1	2	3	3	4	7	16	83
18^3	1	2	3	3	4	7	17	87
27^3	1	2	3	3	5	7	17	88

Table 5. Number of iterations for the augmented Lagrangian method for mixed method for elliptic equations on a 3D tetrahedral mesh using the $\mathbf{H}(\mathrm{div})$ solver. ε is a different constant for each experiment.

for smaller ε. We observe this phenomenon in Table 5. Most notably, if we choose $\varepsilon \leq 10^{-9}$ then only one iteration is needed.

5 Conclusions

In this paper, we discuss the implementation of an AMG based HX-preconditioners for the $\mathbf{H}(\mathrm{div})$ systems on unstructured grids. The numerical experiments show the robustness and efficiency of the algorithms even in the presence of large jump coefficients. As an application, we applied these preconditioners to solve the mixed finite element problem by augmented Lagrangian technique. The numerical experiments also show the efficiency of this approach.

Acknowledgement. The first author was supported in part by the DOE Office of Science ASCR Applied Math Research program and by the ASC program at Sandia Laboratories. Sandia is a multiprogram laboratory operated by Sandia Corporation, a Lockheed Martin Company, for the United States Department of Energy's National Nuclear Security Administration under contract DE-AC04-94AL85000. The second author was supported in part by NSF DMS-0609727, NSFC-10528102 and Alexander von Humboldt Research Award for Senior US Scientists. The third author would like to thank Sandia National Lab for the support in summer 2007, and especially thank Bochev Pavel and Chris Siefert for inspiring discussions. He would also like to thank his postdoctoral advisor Professor Michael Holst for his encouragement and support through NSF Awards 0715146 and 0411723.

References

[1] Arnold, D.N., Falk, R.S., Winther, R.: Preconditioning in $H(\mathrm{div})$ and applications. *Math. Comp.*, 66:957–984, 1997.
[2] Bank, R.E., Welfert, B.D., Yserentant, H.: A class of iterative methods for solving mixed finite element equations. *Numer. Math.*, 56:645–666, 1990.
[3] Beck, R.: Graph-based algebraic multigrid for Lagrange-type finite elements on simplicial. Technical Report Preprint SC 99-22, Konrad-Zuse-Zentrum für Informationstechnik Berlin, 1999.

[4] Bochev, P.B., Seifert, C., Tuminaro, R., Xu, J., Zhu, Y.: Compatible gauge approaches for $H(\text{div})$ equations. In *CSRI Summer Proceedings*, 2007.

[5] Brezzi, F., Douglas, J., Marini, L.D.: Two families of mixed finite elements for second order elliptic problems. *Numer. Math.*, 47(2):217–235, 1985.

[6] Brezzi, F., Fortin, M.: *Mixed and hybrid finite element methods*. Springer, 1991.

[7] Chen, Z., Ewing, R.E., Lazarov, R.D., Maliassov, S., Kuznetsov, Y.A.: Multilevel preconditioners for mixed methods for second order elliptic problems. *Numer. Linear Algebra Appl.*, 3(5):427–453, 1996.

[8] Fortin, M., Glowinski, R. *Augmented Lagrangian Methods: Application to the numerical solution of boundary value problems*. North-Holland, Amsterdam, 1983.

[9] Gee, M., Siefert, C., Hu, J., Tuminaro, R., Sala, M.: ML 5.0 smoothed aggregation user's guide. Technical Report SAND2006-2649, Sandia National Laboratories, 2006.

[10] Hiptmair, R.: Multigrid method for $H(\text{div})$ in three dimensions. *Electron. Trans. Numer. Anal.*, 6:133–152, 1997.

[11] Hiptmair, R., Schiekofer, T., Wohlmuth, B. Multilevel preconditioned augmented Lagrangian techniques for 2nd order mixed problems. *Computing*, 57(1):25–48, 1996.

[12] Hiptmair, R., Xu, J.: Nodal Auxiliary Space Preconditioning in $H(\text{curl})$ and $H(\text{div})$ Spaces. *SIAM J. Numer. Anal.*, 45:2483, 2007.

[13] Kolev, T., Vassilevski, P.: Some experience with a H^1-based auxiliary space AMG for $H(\text{curl})$ problems. Technical Report UCRL-TR-221841, Lawrence Livermore Nat. Lab., 2006.

[14] Lee, Y., Wu, J., Xu, J., Zikatanov, L. Robust subspace correction methods for nearly singular systems. *Math. Models Methods Appl. Sci.*, 17(11):1937, 2007.

[15] Raviart, P.A., Thomas, J.: A mixed finite element method fo 2nd order elliptic problems. In I. Galligani and E. Magenes, editors, *Mathematical aspects of the Finite Elements Method*, Lectures Notes in Math. 606, pages 292–315. Springer, Berlin, 1977.

[16] Vaněk, P., Brezina, M., Mandel, J. Convergence of algebraic multigrid based on smoothed aggregation. *Numer. Math.*, 88(3):559–579, 2001.

[17] Vassilevski, P., Lazarov, R.: Preconditioning Mixed Finite Element Saddle-point Elliptic Problems. *Numer. Linear Algebra Appl.*, 3(1):1–20, 1996.

[18] Vassilevski, P.S., Wang, J.: Multilevel iterative methods for mixed finite element discretizations of elliptic problems. *Numer. Math.*, 63(1):503–520, 1992.

[19] Wohlmuth, B.I., Toselli, A., Widlund, O.B.: An iterative substructuring method for Raviart–Thomas vector fields in three dimensions. *SIAM J. Numer. Anal.*, 37(5):1657–1676, 2000.

[20] Xu, J.: The auxiliary space method and optimal multigrid preconditioning techniques for unstructured meshes. *Computing*, 56:215–235, 1996.

Part II

Minisymposia

A Multilevel Domain Decomposition Solver Suited to Nonsmooth Mechanical Problems

Damien Iceta, Pierre Alart, and David Dureisseix

Laboratoire de Mécanique et Génie Civil (LMGC), University Montpellier 2 / CNRS UMR5508, Place E. Bataillon, F-34095 MONTPELLIER CEDEX 5, FRANCE, {iceta,alart,dureisseix}@lmgc.univ-montp2.fr

1 Introduction

A particular class of mechanical systems concerns diffuse non smooth problems for which unilateral conditions may occur within the whole studied domain. For instance, when contact and friction occur as interactions between a large number of bodies, such as for granular media, or with tensegrity structures, [14], when cable slackening may occur on the whole structure.

When such large scale structures are studied, their numerical simulation may take advantage of using domain decomposition (DD) solvers. We do not consider here an outer loop for dealing with non linearities and non smoothness that lead to a series of linear problems, each of them being solved with a classical DD solver as a black box, but we focus on algorithms that allow us to tackle the non smoothness issue at the subdomain level, with a single iterative loop. Such approaches have already be designed for mechanical assemblings with a limited number of unilateral conditions, [5, 6, 7], or for multicontact situations, [1, 2, 3, 4, 12, 15].

We consider in this article approaches suited to multicontact cases, that focus on the non smooth interactions by solving them locally on the one hand, and by solving the global equations on the other hand, iteratively. Among these, one may consider the LArge Time INcrement (LATIN) approaches, [2, 12, 15], that embed a multiscale aspect to derive an scalable DD method, and the Non Smooth Contact Dynamics (NSCD) approaches, genuinely designed for granular media, [10, 13].

Herein, we proposed an approach based on the Gauss-Seidel (or more precisely Jacobi for parallelization purposes) interpretation, [11], of the NSCD method, embedding the same multiscale description used in the LATIN method, and we design the solver in the case of a tensegrity grid steady-state simulation.

2 A Multiscale Description

With a given discretized structure (for instance, with finite element for an elastic problem on a continuum domain, of directly on the equilibrium equations of a

M. Bercovier et al. (eds.), *Domain Decomposition Methods in Science and Engineering XVIII*, Lecture Notes in Computational Science and Engineering,
DOI: 10.1007/978-3-642-02677-5, © Springer-Verlag Berlin Heidelberg 2009

truss-like elastic problem), there are at least two choices for a partitioning into substructures. On the one hand, one can split the nodes into different sets, leading to an interface between two substructures composed by elements linking the nodes of the two corresponding sets. On the other hand, one can split the elements into different sets, leading to an interface defined as shared nodes. The interface behavior is therefore either the behavior of the shared elements (that may be nonlinear), or the behavior of the shared nodes. In this last case, unless if unilateral conditions occur at the interface, [5, 7, 12], the behavior is linear, [3, 15].

This last case is chosen herein. Once the substructuring is performed and the interfaces between each pair of connected substructures are defined, the multiscale description is performed at the interface level. For the quasi-static or steady-state problems we are concerned herein, two dual fields are involved on each interface: the trace of the substructure displacement, and the forces acting on the interface from the neighboring substructures. In the case of discrete structures, the interface is a set of nodes, Figure 1, and the displacement field V on the interface is split into a macro part (denoted with superscript M) and a micro part (denoted with a superscript m): $V = V^M + V^m$. The macro part is chosen as the average translation, rotation and extension in the average plane of the interface, [15]. Therefore, it can be defined with a small number of parameters (9 values per interface in 3D case) stored in a small vector w: $V^M = Pw$, where P stores the basis vectors of macro space. The dual field, i.e. the forces on interface, is split in a similar way: $F = F^M + F^m$. The macro part is also defined with the same small number of parameters, here, the dual macro quantities f, and we chose the basis macro functions such as $F^M = Pf$ and $P^T P$ is the identity matrix.

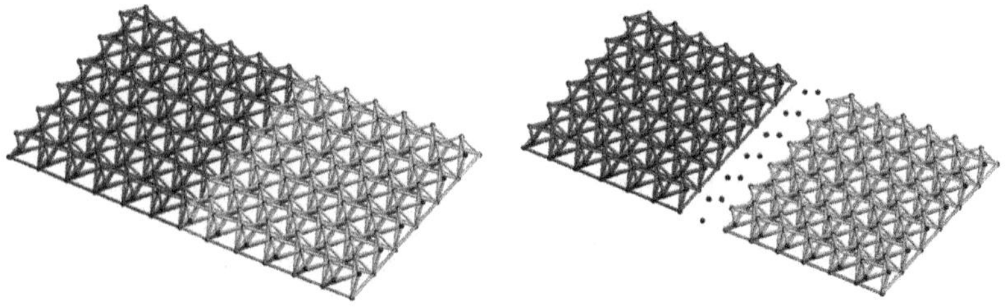

Fig. 1. Element oriented partitioning (left) and perfect discrete interface between the substructures (right).

3 Preliminary: Linear Elastic Case

Substructure Behavior. If we consider a single substructure E, the actions of its neighboring interfaces are the forces $-F_{E\Gamma}$ and the displacement on the boundary $V_{E\Gamma}$. The subscript $E\Gamma$ is used to denote the assembly of the local interfaces of the substructure E. The corresponding balance equation is

$$F_E^d - F_E - C_{E\Gamma}^T F_{E\Gamma} = 0, \tag{1}$$

where $C_{E\Gamma}$ is a boolean mapping matrix that selects the trace on the local interfaces, F_E^d are the prescribed external forces, and F_E are the internal forces. For an elastic media, the internal forces can be expressed with the nodal displacement V_E via the stiffness matrix: $F_E = K_E V_E$.

To prescribe the force equilibrium on each interface, the local forces $F_{E\Gamma}$ can be derived from a single field on the global interface (the gathering of all local interfaces) F_Γ with a signed boolean matrix B_E as for dual approaches, [8, 9]: $F_{E\Gamma} = B_E F_\Gamma$. The dual quantity, the trace of displacement on the local interfaces is: $V_{E\Gamma} = C_{E\Gamma} V_E$.

Interface Behavior. Once the equilibrium of forces on the interfaces is automatically satisfied, their perfect behaviors lead to a displacement continuity:

$$\sum_E B_E^T V_{E\Gamma} = 0 \tag{2}$$

Solution Algorithm. The balance equation (1) for the substructure E can be recast as:

$$K_E V_E + C_{E\Gamma}^T B_E F_\Gamma = F_E^d \tag{3}$$

The first step to design the proposed approach is to condense the information on the interfaces. For sake of simplicity, we consider here that the stiffness matrix K_E is invertible. If this is not the case, for floating substructures for instance, the same procedure can be derived, provided that a suited generalized inverse is used. The gluing condition (2), using (3) to eliminate the internal degrees of freedom, reads:

$$X F_\Gamma = \tilde{F}^d \tag{4}$$

with $\tilde{F}^d = \sum_E B_E^T C_{E\Gamma} K_E^{-1} F_E^d$ and $X = \sum_E B_E^T C_{E\Gamma} K_E^{-1} C_{E\Gamma}{}^T B_E$

In order to solve this problem, we propose a stationary iterative method based on the splitting of the global operator X into $X = X^D - (X^D - X)$. Different choices can be selected to split X (Jacobi, Gauss-Seidel...), that lead to different algorithms. In each case, the iterate number $i+1$ consists in solving: $X^D F_\Gamma^{i+1} = \tilde{F}^d - (X - X^D) F_\Gamma^i$, or when developping $\tilde{F}^d$:

$$X^D (F_\Gamma^{i+1} - F_\Gamma{}^i) = \sum_E B_E^T C_{E\Gamma} V_E^i \tag{5}$$

with $K_E V_E^i = F_E^d - C_{E\Gamma}^T B_E F_\Gamma^i$.

Splitting Choice. X is a dense operator coupling all the degrees of freedom on the global interface. X^D is similar to a preconditioner, or a search direction. Choosing for instance a 'lumped' approximation on the local interfaces on each subdomain leads to $(X^D)^{-1} = \sum_E B_E^T (C_{E\Gamma} K_E C_{E\Gamma}^T) B_E$. An even simpler version uses a constant scalar stiffness d as: $(X^D)^{-1} = \sum_E B_E^T d I_{E\Gamma} B_E$, where $I_{E\Gamma}$ is the identity matrix on the boundary degrees of freedom of the subdomain E. In such a case, due to the fact that the global interface is merely the gathering of all the local interfaces, $\sum_E B_E^T B_E =$

$2I_{E\Gamma}$ and $(X^D)^{-1} = 2d\,I_\Gamma$. Applying $(X^D)^{-1}$ to a vector simply leads to explicit and local computations on each interface independently.

Multiscale Approach. Up to this point, no multilevel feature is involved in the previous algorithm. To do so, the micro-macro description of Section 2 should be used. We propose here to enforce the continuity of the macro displacement and the equilibrium of the macro forces at each iteration, while it is satisfied previously only when the solution has converged. Therefore, for the interfaces connected to each substructure E, the macro generalized forces are supposed to be extracted from a unique macro vector defined on the global interface: $f_{E\Gamma} = c_E f_\Gamma$, where c_E is a signed boolean matrix selecting the entries in f_Γ. The dual quantity is the gap of macro displacements on interfaces $w_\Gamma = \sum_E c_E^T w_{E\Gamma}$. The macro displacement continuity on interfaces gives $w_\Gamma = 0$, which reads:

$$\sum_E c_E^T P_{E\Gamma}^T C_{E\Gamma} V_E = 0 \tag{6}$$

This constraint is therefore to be prescribed at each iteration for the displacement field V_E^i in (5), for which f_Γ is the associated Lagrange multiplier. Therefore, the displacement V_E^i is now obtained by solving:

$$\begin{cases} K_E V_E = F_E^d - C_{E\Gamma}^T B_E F_\Gamma - C_{E\Gamma}^T P_{E\Gamma} c_E f_\Gamma \\ \sum_E c_E^T P_{E\Gamma}^T C_{E\Gamma} V_E = 0 \end{cases} \tag{7}$$

The local condensation on f_Γ for each substructure, and the assembly in (6) leads to the macroscopic (coarse) problem:

$$L_\Gamma f_\Gamma = \sum_E c_E^T P_{E\Gamma}^T C_{E\Gamma} K_E^{-1} (F_E^d - C_{E\Gamma}^T B_E F_\Gamma) \tag{8}$$

with $L_\Gamma = \sum_E c_E^T P_{E\Gamma}^T C_{E\Gamma} K_E^{-1} C_{E\Gamma}^T P_{E\Gamma} c_E$ which is explicitly assembled to maintain the globality of the coarse problem. The size of L_Γ corresponds to the number of macro degrees of freedom involved in the coarse problem.

With a given approximation of the solution $(F_\Gamma^i, V_E^i, f_\Gamma^i)$, one iteration provides the update $(F_\Gamma^{i+1}, V_E^{i+1}, f_\Gamma^{i+1})$:

- During the 'local stage', F_Γ^{i+1} is computed using (5), locally on each interface:

$$F_\Gamma^{i+1} = (X^D)^{-1} \sum_E B_E^T C_{E\Gamma} V_E^i + F_\Gamma{}^i \tag{9}$$

- During the 'coarse step', the macro problem is solved to get f_Γ^{i+1}

$$L_\Gamma f_\Gamma^{i+1} = \sum_E c_E^T P_{E\Gamma}^T C_{E\Gamma} K_E^{-1} (F_E^d - C_{E\Gamma}^T B_E F_\Gamma^{i+1}) \tag{10}$$

- During the 'global step' per subdomain E independently, V_E^{i+1} is updated by solving:

$$K_E V_E^{i+1} = F_E^d - C_{E\Gamma}^T B_E F_\Gamma^{i+1} - C_{E\Gamma}^T P_{E\Gamma} c_E f_\Gamma^{i+1} \tag{11}$$

4 Nonsmooth Case: a Tensegrity Grid

Tensegrity Structures. Tensegrity systems are reticulated spatial structures constituted with rectilinear elements such as 'cables' or 'bars', [14]. Bars are subjected to compression loading, while cables are subjected to traction loading. Joining elements are perfect articulations called 'nodes'. These systems allow for selfstressed states, i.e. stress states that satisfy the equilibrium without external loading. These stress states are mandatory to ensure the overall structure rigidity. The reference problem is herein related to the static behavior of such a structure, with a small perturbation assumption.

Model of a Tensegrity Structure. In the case of a tensegrity structure, the equilibrium (1) remains identical, but the internal forces arise from the internal tension (or compression) r_E in the elements (or the links) that constitute the structure: the cables (with a superscript c) or the bars (with a superscript b)

$$F_E = H_E r_E = H_E^c r_E^c + H_E^b r_E^b \tag{12}$$

where H_E is a mapping from the link set to the node set.

Additionally, the trace of nodal displacement $V_{E\Gamma} = C_{E\Gamma} V_E$ remains identical, but we add the length variation of the links, e_E as the dual quantity of the strength r_E:

$$e_E^b = H_E^{b^T} V_E \quad \text{and} \quad e_E^c = H_E^{c^T} V_E \tag{13}$$

The interface behavior (2) holds again, but the constitutive relations of the links are a linear elastic behavior for the bars,

$$r_E^b + r_E^{b^0} = k_b (e_E^b + e_E^{b^0}), \tag{14}$$

and a nonsmooth complementary condition for the cables, [15],

$$0 \leq \tau_E^c \perp \lambda_E^c \geq 0 \tag{15}$$

where $\tau_E^c = r_E^c + r_E^{c^0}$ and $\lambda_E^c = -e_E^c + k_c^{-1} r_E^c = -(e_E^c + e_E^{c^0}) + k_c^{-1}(r_E^c + r_E^{c^0})$. The superscript 0 denotes the prestress or prestrain that have to be initially prescribed for the structure to exhibit stiffness.

Multiscale Solver. With (12), (13), (14) and (15), the equilibrium reads:

$$K_E V_E + C_{E\Gamma}^T B_E F_\Gamma + H_E^c k_c \lambda_E^c = \tilde{F}_E^d, \tag{16}$$

with the stiffness matrix of the underlying truss $K_E = H_E^b k_b H_E^{b^T} + H_E^c k_c H_E^{c^T}$, and the given right hand side $\tilde{F}_E^d = F_E^d - H_E^b k_b e_E^{b^0}$.

As for the linear case, the first step consists in condensing the problem on the local interface quantities, but keeping the variable λ_E^c traducing the nonsmooth interaction as an unknown, using (16) and (2):

$$X F_\Gamma + \sum_E B_E^T C_{E\Gamma} K_E^{-1} H_E^c k_c \lambda_E^c = \sum_E B_E^T C_{E\Gamma} K_E^{-1} \tilde{F}_E^d, \tag{17}$$

with $X = \sum_E B_E^T C_{E\Gamma} K_E^{-1} C_{E\Gamma}^T B_E$.

Additionally, one must keep the local nonsmooth relationship traducing the behavior of the cables: condensing the equilibrium (1) on cable quantities, while keeping the constitutive equation (15) leads to the Linear Complementary Problem (LCP) with (λ_E, τ_E) as unknowns (and parametrized by F_Γ):

$$\begin{cases} W_E \lambda_E^c - \tau_E^c = -H_E^{c\,T} K_E^{-1} \tilde{F}_E^d + H_E^{c\,T} K_E^{-1} C_{E\Gamma}^T B_E F_\Gamma - r_E^{c\,0} \\ 0 \leq \tau_E^c \perp \lambda_E^c \geq 0 \end{cases} \tag{18}$$

where $W_E = k_c I_E - k_c H_E^{c\,T} K_E^{-1} H_E^c k_c$ (I_E is the identity matrix), is the linear part of the relationship, [15].

As for the linear case, the left hand side X is split, and so is the (local per subdomain) operator W_E: $W_E = W_E^D - (W_E^D - W_E)$. The proposed algorithm iterates both on interface forces F_Γ and on the nonsmooth pair of variables (λ_E, τ_E). Knowing quantities with a superscript i, one iteration will give the update, with superscript $i+1$, such as:

$$\begin{cases} W_E^D \lambda_E^{i+1} - \tau_E^{i+1} = W_E^D \lambda_E^i - k_c \lambda_E^i - k_c H_E^{c\,T} V_E^i - r_E^{c\,0} \\ 0 \leq \tau_E^{i+1} \perp \lambda_E^{i+1} \geq 0 \end{cases} \tag{19}$$

for the nonsmooth part, and (5), that remains unchanged, for the interface part, where V_E^i satisfies $K_E V_E^i = \tilde{F}_E^d - C_{E\Gamma}^T B_E F_\Gamma^i - H_E^c k_c \lambda_E^i$.

Splitting Choice. W_E is a global operator on each subdomain. Choosing a diagonal W_E^D leads to independant LCP on each cable; for instance, one can select $W_E^D = k_c I_E$. Other choices are obviously possible.

This constitutes the monolevel algorithm, whose multilevel version is obtained with a similar procedure as already done for the linear case. With a given approximation of the solution $(F_\Gamma^i, V_E^i, \lambda_E^i, \tau_E^i, f_\Gamma^i)$, one iteration provides the update:

- During 'local stage', F_Γ^{i+1} is computed locally on each interface:

$$F_\Gamma^{i+1} = (X^D)^{-1} \sum_E B_E^T C_{E\Gamma} V_E^i + F_\Gamma^i \tag{20}$$

  and $(\lambda_E^{i+1}, \tau_E^{i+1})$ is computed locally on each subdomain E by solving the nonsmooth but local problem:

$$\begin{cases} W_E^D \lambda_E^{i+1} - \tau_E^{i+1} = W_E^D \lambda_E^i - \tau_E^i \\ 0 \leq \tau_E^{i+1} \perp \lambda_E^{i+1} \geq 0 \end{cases} \tag{21}$$

- During the 'coarse step', the macro multiplier f_Γ^{i+1} is obtained by solving the coarse problem:

$$L_\Gamma f_\Gamma^{i+1} = \sum_E c_E^T P_{E\Gamma}^T C_{E\Gamma} K_E^{-1} (F_E^d - C_{E\Gamma}^T B_E F_\Gamma^{i+1} - H_E^c k_c \lambda_E^{i+1}) \tag{22}$$

- During the 'global step' per subdomain E independently, V_E^{i+1} is updated by solving:

$$K_E V_E^{i+1} = \tilde{F}_E^d - C_{E\Gamma}^T B_E F_\Gamma^{i+1} - C_{E\Gamma}^T P_{E\Gamma} c_E f_\Gamma^{i+1} - H_E^c k_c \lambda_E^{i+1} \tag{23}$$

At to this point, the algorithm is on its way to be tested and compared with previous algorithm based on a LArge Time INcrement approach, [15].

5 Conclusions

The proposed method constitutes a first attempt to combine a multilevel Domain Decomposition technique with a Non Smooth Gauss-Seidel (NSGS) type algorithm. The NSGS algorithm is classically associated with the Non Smooth Contact Dynamics approach and provides a robust solver for the simulation of dense granular media involving not only unilateral contact but also frictional contact or more general localized interactions. The multilevel DD technique ensures the scalability of the solver to deal with large-scale mechanical systems. Moreover the multiscale approach may allow replacing the fine description of some subdomains by their homogenized behavior under additional assumptions. The simulation cost of large-scale systems as granular media may be drastically decreased. For that the previous solver may be extended to dynamical problems without conceptual difficulty. Nevertheless the meaning of the homogenized coarse problem has to be investigated. From a computational viewpoint the chosen formulation, close to the FETI approach, suggests to replace the stationary iterative method by a conjugate gradient algorithm. The projected conjugate gradient method developed in [16] for granular media in sub domains would be usefully combined with the conjugate gradient algorithm of the standard FETI method for solving the interface problem in linear problems. Such a combination may provide a more efficient solver even if the non smoothness does not preserve the conjugating property from one iteration to the following one. A large range of numerical tests have to be performed to validate this strategy.

References

[1] Alart, P., Barboteu, M., Renouf, M.: Parallel computational strategies for multicontact problems: Application to cellular and granular media. *Internat. J. Multiscale Comput. Engrg.*, 1:419–430, 2003.

[2] Alart, P., Dureisseix, D.: A scalable multiscale LATIN method adapted to nonsmooth discrete media. *Comput. Methods Appl. Mech. Engrg.*, 197(5):319–331, 2008.

[3] Barboteu, M., Alart, P., Vidrascu, M.: A domain decomposition strategy for nonclassical frictional multi-contact problems. *Comput. Methods Appl. Mech. Engrg.*, 190:4785–4803, 2001.

[4] Breitkopf, P., Jean, M.: Modélisation parallèle des matériaux granulaires. In *4e Colloque National en Calcul des Structures*, pages 387–392, 1999.

[5] Champaney, L., Cognard, J.-Y., Ladevèze, P.: Modular analysis of assemblages of three-dimensional structures with unilateral contact conditions. *Comput. & Structures*, 73:249–266, 1999.

[6] Dostàl, Z., Gomes Neto, F.A.M., Santos, S.A.: Solution of contact problems by FETI domain decomposition with natural coarse space projection. *Comput. Methods Appl. Mech. Engrg.*, 190(13-14):1611–1627, 2000.

[7] Dureisseix, D., Farhat, C.: A numerically scalable domain decomposition method for the solution of frictionless contact problems. *Internat. J. Numer. Methods Engrg.*, 50(12):2643–2666, 2001.

[8] Farhat, C., Lesoinne, M., LeTallec, P., Pierson, K., Rixen, D.: FETI-DP: a dual-primal unified FETI method. I. A faster alternative to the two-level FETI method. *Internat. J. Numer. Methods Engrg.*, 50(7):1523–1544, 2001.

[9] Farhat, C., Roux, F.-X.: A method of finite element tearing and interconnecting and its parallel solution algorithm. *Internat. J. Numer. Methods Engrg.*, 32(6): 1205–1227, 1991.

[10] Jean, M.: The non-smooth contact dynamics method. *Comput. Methods Appl. Mech. Engrg.*, 177(3-4):235–257, 1999.

[11] Jourdan, F., Alart, P., Jean, M.: A Gauss-Seidel like algorithm to solve frictional contact problems. *Comput. Methods Appl. Mech. Engrg.*, 155(1-2):31–47, 1998.

[12] Ladevèze, P., Nouy, A., Loiseau, O.: A multiscale computational approach for contact problems. *Comput. Methods Appl. Mech. Engrg.*, 191(43):4869–4891, 2002.

[13] Moreau, J.J.: Numerical aspects of the sweeping process. *Comput. Methods Appl. Mech. Engrg.*, 177(3-4):329–349, 1999.

[14] Motro, R.: *Tensegrity*. Hermes Science Publishing, London, 2003.

[15] Nineb, S., Alart, P., Dureisseix, D.: Domain decomposition approach for nonsmooth discrete problems, example of a tensegrity structure. *Comput. & Structures*, 85(9):499–511, 2007.

[16] Renouf, M., Alart, P.: Conjugate gradient type algorithms for frictional multi-contact problems: applications to granular materials. *Comput. Methods Appl. Mech. Engrg.*, 194(18-20):2019–2041, 2005.

A FETI-2LM Method for Non-Matching Grids

François-Xavier Roux

High Performance Computing Unit, ONERA, 29 avenue de la Division Leclerc, 92320, Châtillon, France roux@onera.fr

1 Introduction

In this paper, a new solution methodology based on the FETI-2LM method for non conforming grids is introduced. Thanks to the regularizing properties of the Robin interface matching conditions of the FETI-2LM method, each non conforming condition can be localized inside one subdomain, in such a way that the FETI-2LM method applies exactly in the same way as in the conforming case.

The paper is organized as follows: section 2 recalls the principle of FETI-2LM method, section 3 briefly describes the mortar method for non conforming domains, the new methodology for localizing the multi-point constraints on the non conforming interface derived from the mortar method is introduced in section 4 and section 5 generalizes the methodology in the case of multi-level splitting of a mesh including a non conforming interface.

2 FETI-2LM method

2.1 Discrete Approach

Consider the linear problem $Kx = b$ arising from a finite element discretization of a PDE. The mesh of the entire domain is split in two meshes like in Fig.1, the two subdomains are denoted by Ω_1 and Ω_2, and their interface by Γ_3. Then, the global stiffness matrix and right hand sides have the following block structure:

$$K = \begin{bmatrix} K_{11} & 0 & K_{13} \\ 0 & K_{22} & K_{23} \\ K_{31} & K_{32} & K_{33} \end{bmatrix}, \quad b = \begin{bmatrix} b_1 \\ b_2 \\ b_3 \end{bmatrix} \tag{1}$$

and the subdomain stiffness matrices and right hand sides are:

$$K_1 = \begin{bmatrix} K_{11} & K_{13} \\ K_{31} & K_{33}^{(1)} \end{bmatrix}, \; y_1 = \begin{bmatrix} b_1 \\ b_3^{(1)} \end{bmatrix} \quad K_2 = \begin{bmatrix} K_{22} & K_{23} \\ K_{32} & K_{33}^{(2)} \end{bmatrix}, \; y_2 = \begin{bmatrix} b_2 \\ b_3^{(2)} \end{bmatrix} \tag{2}$$

M. Bercovier et al. (eds.), *Domain Decomposition Methods in Science and Engineering XVIII*, Lecture Notes in Computational Science and Engineering,
DOI: 10.1007/978-3-642-02677-5,

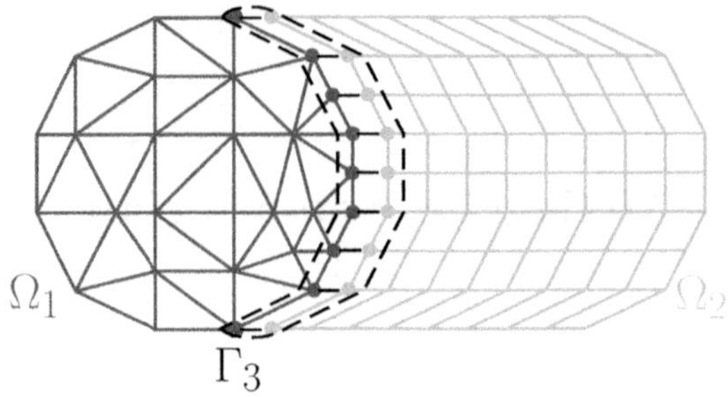

Fig. 1. Two meshes with interface.

with $K_{33}^{(1)} + K_{33}^{(2)} = K_{33}$ and $b_3^{(1)} + b_3^{(2)} = b_3$.

The FETI-2LM method [3] is based on introducing independent generalized Robin boundary conditions on interface Γ_3. Discretized local Robin problem takes the following form:

$$\begin{bmatrix} K_{ii} & K_{i3} \\ K_{3i} & K_{33}^{(i)} + A_{33}^{(i)} \end{bmatrix} \begin{bmatrix} x_i \\ x_3^{(i)} \end{bmatrix} = \begin{bmatrix} b_i \\ b_3^{(i)} + \lambda_3^{(i)} \end{bmatrix} \quad (3)$$

To be the restrictions in the subdomains of the solution of the global problem, the solutions of the local problems must first satisfy the discrete continuity condition:

$$x_3^{(1)} - x_3^{(2)} = 0 \quad (4)$$

The second interface matching condition, that is a discrete condition of equilibrium, is nothing else than the last row of the global discrete system:

$$K_{31}x_1 + K_{32}x_2 + K_{33}x_3 = b_3 \quad (5)$$

Note that these two conditions can be derived by simple algebraic manipulation from any linear system of equations whose matrix has the block form of (1).

Given the splitting of block matrix K_{33} and of vector b_3, the discrete equilibrium equation (5) can be rewritten as:

$$K_{31}x_1 + K_{32}x_2 + K_{33}^{(1)}x_3^{(1)} + K_{33}^{(2)}x_3^{(2)} = b_3^{(1)} + b_3^{(2)} \quad (6)$$

Since the last row of the discrete Robin problem (3) gives:

$$K_{3i}x_i + K_{33}^{(i)}x_3^{(i)} + A_{33}^{(i)}x_3^{(i)} = b_3^{(i)} + \lambda_3^{(i)} \quad (7)$$

equation (6) can be alternatively written as:

$$A_{33}^{(1)}x_3^{(1)} + A_{33}^{(2)}x_3^{(2)} = \lambda_3^{(1)} + \lambda_3^{(2)} \quad (8)$$

Finally, the two discrete interface conditions (4) and (8) can be combined to give the equivalent mixed equations:

$$
\begin{aligned}
A_{33}^{(1)}x_3^{(2)} + A_{33}^{(2)}x_3^{(2)} &= \lambda_3^{(1)} + \lambda_3^{(2)} \\
A_{33}^{(1)}x_3^{(1)} + A_{33}^{(2)}x_3^{(1)} &= \lambda_3^{(1)} + \lambda_3^{(2)}
\end{aligned} \tag{9}
$$

By eliminating inner unknowns in the discrete Robin problem (3), the relation between the trace of the solution on interface $x_3^{(i)}$ and the discrete augmented flux $\lambda_3^{(i)}$ can be explicitly computed:

$$
(K_{33}^{(i)} - K_{3i}K_{ii}^{-1}K_{i3} + A_{33}^{(i)})x_3^{(i)} = \lambda_3^{(i)} + b_3^{(i)} - K_{3i}K_{ii}^{-1}b_i \tag{10}
$$

Denote by $S^{(i)} = (K_{33}^{(i)} - K_{3i}K_{ii}^{-1}K_{i3})$, the Schur complement matrix and by $c_3^{(i)} = b_3^{(i)} - K_{3i}K_{ii}^{-1}b_i$, the condensed right-hand-side.

Replacing $x_3^{(1)}$ and $x_3^{(2)}$ by their values as function of $\lambda_3^{(1)}$ and $\lambda_3^{(2)}$ derived from equation (10) in the mixed interface equations (9), leads to the condensed interface problem associated to the FETI-2LM method:

$$
\begin{bmatrix}
I & I - (A_{33}^{(1)} + A_{33}^{(2)})(S^{(2)} + A_{33}^{(2)})^{-1} \\
I - (A_{33}^{(1)} + A_{33}^{(2)})(S^{(1)} + A_{33}^{(1)})^{-1} & I
\end{bmatrix}
\begin{bmatrix}
\lambda_3^{(1)} \\
\lambda_3^{(2)}
\end{bmatrix}
= \begin{bmatrix}
(S^{(2)} + A_{33}^{(2)})^{-1}c_3^{(2)} \\
(S^{(1)} + A_{33}^{(1)})^{-1}c_3^{(1)}
\end{bmatrix} \tag{11}
$$

2.2 Optimal Interface Operator

FETI-2LM method consists in solving the condensed interface problem (11) via a Krylov space method. Of course, the gradient is not computed using explicit formula (11) but using the implicit one (9) where $x_3^{(1)}$ and $x_3^{(2)}$ are computed by solving the local Robin problems (3).

The main ingredient for the method to be effective is the choice of the operator $A_{33}^{(i)}$ associated with the generalized Robin condition. Analysis of the condensed interface problem (11) clearly shows that the optimal choice consists in taking in each subdomain the Schur complement of the rest of the domain:

$$
A_{33}^{(1)} = S^{(2)} \qquad A_{33}^{(2)} = S^{(1)} \tag{12}
$$

With such a choice, the matrix of the condensed interface problem (11) in the 2-subdomain case is simply the identity matrix and the method is then a direct solver.

In practice the Schur complement is of course too expensive to compute and, since it is a dense matrix, using it would also give a very large bandwidth to the stiffness matrices of the local generalized Robin problems. Sparse approximation of the Schur complement must be used. A purely algebraic methodology has been developed in [5]. It consist in building the approximate Schur complement by assembling exact Schur complements computed on small patches along the interface.

3 Mortar Method

Lagrange multiplier based domain decomposition methods have been extended to the case on non-conforming meshes, especially with the mortar method [1]. At the continuous level, the principle of the method consists in introducing a weak formulation of the interface continuity condition:

$$\int_{\Gamma_3} (u_1 - u_2)\mu = 0 \quad \forall\, \mu \in W \tag{13}$$

where u_i is the solution in subdomain Ω_i of the continuous PDE and W is a suitable set of Lagrange multipliers.

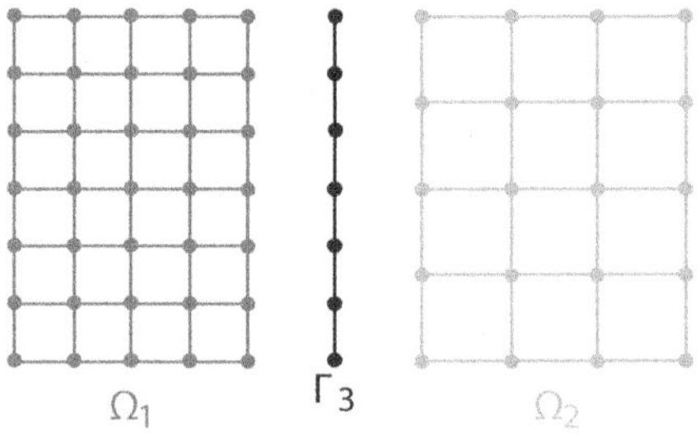

Fig. 2. Non-conforming interfaces.

For the discrete non-conforming case, optimal approximation results have been proved for elliptic second order PDEs when W is the mortar space of one of the two neighboring subdomains.

For instance, for a 2-D problem and linear finite element space V, the mortar space is a subset of the set of the traces on interface Γ of functions belonging to V, consisting of functions that are piecewise constant on the last segments of Γ, like in Fig. 3.

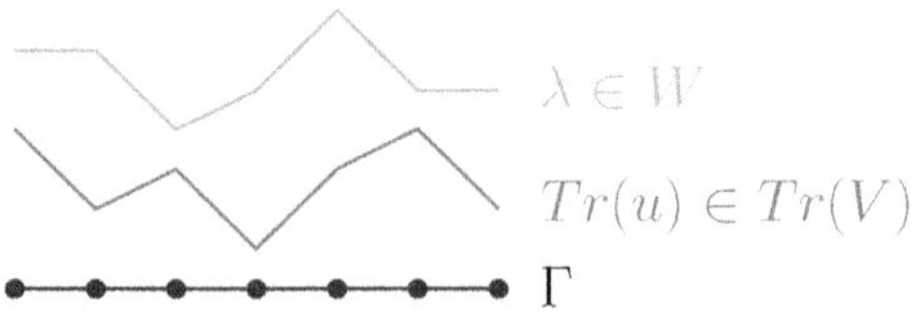

Fig. 3. Mortar space in 2-D for linear elements.

Suppose that, like in Fig. 2, the mortar side is Ω_1, then the global mixed problem associated with the weak formulation (13) of the continuity constraint takes the following discrete form:

$$\begin{bmatrix} K_1 & 0 & B_1^t \\ 0 & K_2 & B_2^t \\ B_1 & B_2 & 0 \end{bmatrix} \begin{bmatrix} \xi_1 \\ \xi_2 \\ \lambda \end{bmatrix} = \begin{bmatrix} y_1 \\ y_2 \\ 0 \end{bmatrix} \quad \text{with } \xi_1 = \begin{bmatrix} x_1 \\ x_3^{(1)} \end{bmatrix},\ \xi_2 = \begin{bmatrix} x_2 \\ x_3^{(2)} \end{bmatrix}, \tag{14}$$

$B_1 = M_{31}R_1$ and $B_2 = -M_{32}R_2$. The R_i matrix is the restriction from subdomain Ω_i to its interface $\partial\Omega_i$, and the M_{3i} matrix is the mass matrix obtained by integration of products of mortar basis functions (living on Γ_3) and traces on $\partial\Omega_i$ of the basis functions of V_i associated with the degrees of freedom of $\partial\Omega_i$.

The FETI method can be applied for solving the global problem (14) [4]. The only difference with the conforming FETI method lies in the fact that, since the B_i matrices are not signed boolean matrices any more, the preconditioning phase must include a scaling taking into account the inhomogeneity induced by the B_i matrices [2].

4 A FETI-2LM Method for Non-conforming Interfaces

In order to avoid dealing with non-conforming interfaces, the multi-point constraints associated with the discrete weak continuity condition:

$$B_1\xi_1 + B_2\xi_2 = 0 \tag{15}$$

may be included inside one subdomain. This means that the targeted subdomain must annex the interface degrees of freedom of the neighboring subdomain. In the case of Fig. 4, it is subdomain Ω_2. The opposite choice could be made as well.

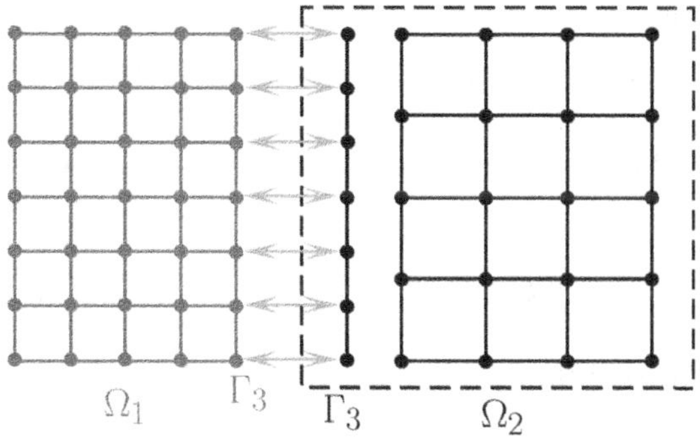

Fig. 4. Inclusion of non-conforming interface in one subdomain.

Then the actual interface between subdomain Ω_1 and extended subdomain Ω_2 is conforming and the weak condition (15) gives multi-point constraints for inner degrees of freedom of extended subdomain Ω_2. This means that the new local stiffness matrices are:

$$[K_1] \qquad \begin{bmatrix} 0 & 0 & M_{31}^t \\ 0 & K_2 & B_2^t \\ M_{31} & B_2 & 0 \end{bmatrix} \tag{16}$$

The restriction matrix R_1 is not present in the mixed matrix of the extended subdomain Ω_2 since only the interface degrees of freedom of Ω_1 have been annexed.

The main apparent issue with this approach is the fact that the matrix of the extended subdomain Ω_2 is highly singular since the annexed degrees of freedom have no stiffness. But, thanks to the generalized Robin conditions on the interface with the FETI-2LM method, each local stiffness matrix is augmented by an approximation of the Schur complement on the interface of the stiffness matrix of the neighboring subdomain:

$$S^{(1)} = K_{33}^{(1)} - K_{31}K_{11}^{-1}K_{13} \qquad S^{(2)} = \begin{bmatrix} 0 & M_{31}^t \end{bmatrix} \begin{bmatrix} K_2 & B_2^t \\ B_2 & 0 \end{bmatrix}^{-1} \begin{bmatrix} 0 \\ M_{31} \end{bmatrix} \tag{17}$$

If $A_{33}^{(1)} \approx S^{(2)}$ and $A_{33}^{(2)} \approx S^{(1)}$ are approximations of the Schur complements defined in equation (17), the augmented stiffness matrices of the generalized Robin problems of the FETI-2LM method are:

$$\begin{bmatrix} K_1 + R_1^t A_{33}^{(1)} R_1 \end{bmatrix} \qquad \begin{bmatrix} A_{33}^{(2)} & 0 & M_{31}^t \\ 0 & K_2 & B_2^t \\ M_{31} & B_2 & 0 \end{bmatrix} \tag{18}$$

None of these matrices is singular any more, in the case where $A_{33}^{(1)} = S^{(2)}$ and $A_{33}^{(2)} = S^{(1)}$, since they are obtained by eliminating unknowns of Ω_2 or inner unknowns of Ω_2, in the well posed global problem (14). The same property holds in general, provided that $A_{33}^{(1)}$ and $A_{33}^{(2)}$ are consistent enough approximations of $S^{(2)}$ and $S^{(1)}$.

The procedure developed in [5] for computing algebraic approximation of the Schur complement applies without any modification to the case where the local matrix has a mixed form, like the matrix of the extended subdomain Ω_2 in (17).

5 Localization of Non-conforming Interface Matching Conditions

In most cases, non-conforming interfaces exist only for engineering or geometrical reason in a limited area of the computational domain. There can be only one non-conforming interface which splits the entire domain into two unbalanced subdomains. In order to get enough subdomains for the domain decomposition solver to be efficient, each initial non-conforming domain must be split into smaller subdomains in such a way that the total number of subdomains is large enough and that the subdomains are balanced.

Therefore, the initial non-conforming interface may be split into several interfaces. Since each non-conforming interface matching conditions couples several degrees of freedom on each side of the interface, it frequently happens that a multi-point constraint associated with a mortar Lagrange multiplier involves degrees of freedom located in more than one subdomain on each side of the non-conforming interface. This leads to a very serious implementation issue since the FETI methods require that each interface connects one subdomain only on each side.

A solution consists in localizing all degrees of freedom associated with a mortar Lagrange multiplier on each side of the non conforming interface to a single subdomain. This means that some degrees of freedom must be annexed by one neighboring subdomain located on the same side of the non-conforming interface, like in Fig. 5, in which Ω_{ij} denotes j^{th} subdomain arising from the splitting of initial domain Ω_i.

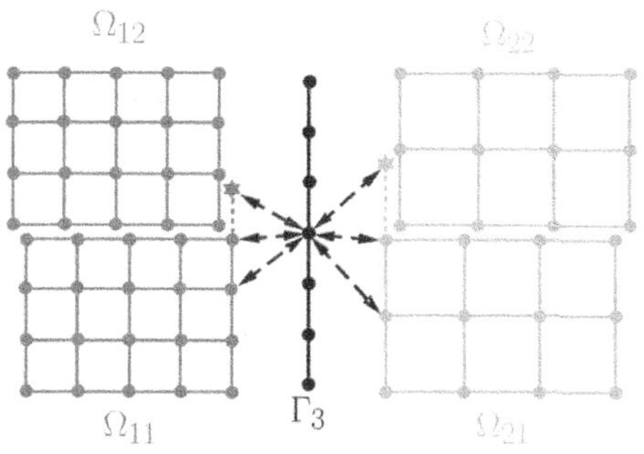

Fig. 5. Localization of non-conforming interface degrees of freedom.

Once again, this procedure adds degrees of freedom with no stiffness, e.g., subdomain Ω_{11}. But the generalized Robin boundary condition on the conforming interface between Ω_{11} and Ω_{12} adds the necessary regularizing terms. And since the FETI-2LM method ensures exact continuity of the solutions along a conforming interface, the value of the solution in Ω_{11} for the annexed degrees of freedom of Ω_{12} is exactly the same as in Ω_{12}.

Thanks to this technique, the initial single non-conforming interface can be split into several non-conforming interfaces, each of them coupling only one subdomain on each side. Therefore, the methodology introduced in section 4 can be applied on each of them.

6 Conclusion

The methodology presented in this paper allows the localization of each multi-point constraint associated with non-conforming interface matching conditions in one subdomain. This localization is to be made in a pre-processing phase. It allows any multi-level splitting of a mesh including a non-conforming interface without any modification in the formulation of the non conforming interface matching conditions.

Thanks to this localization, the FETI-2LM method with the automatic algebraic computation of approximate optimal generalized Robin interface conditions can be implemented without any change from the standard conforming case.

This methodology has been successfully implemented for test problems. Comparison must be made now, in term of convergence speed, between this new non-conforming FETI-2LM method and the classical non-conforming FETI-1LM.

References

[1] Bernardi, C., Maday, Y., Patera, A.T.: Domain decomposition by the mortar element method. In H.G. Kaper ans M. Garbey, editor, *Asymptotic and Numerical Methods for Partial Differential Equations with Critical Parameters*, pages 269–286. N.A.T.O. ASI, Kluwer Academic, 1993.
[2] Farhat, C., Lacour, C., Rixen, D.: Incorporation of linear multipoint constraints in substructure based iterative solvers. part 1 : A numerically scalable algorithm. *Internat. J. Numer. Methods Engrg.*, 43:997–1016, 1998.
[3] Farhat, C., Macedo, A. Lesoinne, M., Roux, F.-X., Magoulès, F., de La Bourdonnaye, A.: Two-level domain decomposition methods with lagrange multipliers for the fast iterative solution of acoustic scattering problems. *Comput. Methods Appl. Mech. Engrg.*, 184:213–240, 2000.
[4] Lacour, C.: Non-conforming domain decomposition method for plate and shell problems. In J. Mandel, C. Farhat, and X.-C. Cai, eds., *Tenth International Conference on Domain Decomposition Methods*, pages 304–310. AMS, Contemporary Mathematics 218, 1998.
[5] Magoulès, F., Roux, F.-X., Series, L.: Algebraic approximation of dirichlet-to-neumann maps for the equations of linear elasticity. *Comput. Methods Appl. Mech. Engrg.*, 195:3742–3759, 2006.

Truncated Nonsmooth Newton Multigrid Methods for Convex Minimization Problems

Carsten Gräser[1,2], Uli Sack[1,2], and Oliver Sander[1,2]

[1] Department of Mathematics, Freie Universität Berlin
[2] DFG Research Center MATHEON {graeser|usack|sander}@math.fu-berlin.de

Summary. We present a new inexact nonsmooth Newton method for the solution of convex minimization problems with piecewise smooth, pointwise nonlinearities. The algorithm consists of a nonlinear smoothing step on the fine level and a linear coarse correction. Suitable postprocessing guarantees global convergence even in the case of a single multigrid step for each linear subproblem. Numerical examples show that the overall efficiency is comparable to multigrid for similar linear problems.

1 Introduction

We consider the minimization problem

$$u \in \mathbb{R}^n: \quad J(u) \leq J(v) \quad \forall v \in \mathbb{R}^n \tag{1}$$

where $J : \mathbb{R}^n \to \mathbb{R} \cup \{\infty\}$ is given by

$$J(v) = \frac{1}{2}\langle Av, v\rangle - \langle b, v\rangle + \varphi(v), \qquad \varphi(v) = \sum_{i=1}^{n} \varphi_i(v_i) \tag{2}$$

for a symmetric positive definite matrix $A \in \mathbb{R}^{n\times n}$ and convex, lower semicontinuous and proper functions $\varphi_i : \mathbb{R} \to \mathbb{R} \cup \{\infty\}$. We will assume that each φ_i is C^2 on a finite number of disjoint intervals $I_i^k \subset \mathbb{R}$ having the property

$$\overline{\operatorname{dom} \varphi_i} = \overline{\{x : \varphi_i(x) < \infty\}} = \bigcup_{k=1}^{m_i} \overline{I_i^k}.$$

Under the above assumptions J is strictly convex, lower semicontinuous, proper, and coercive. Thus (1) has a unique solution [5].

For quadratic obstacle problems the ideas of active-set methods and monotone multigrid have been combined recently to the Truncated Nonsmooth Newton Multigrid (TNNMG) method [6]. Inspired by [7], we generalize this method to non-quadratic nonsmooth energies (2) resulting in a novel globally convergent multigrid

M. Bercovier et al. (eds.), *Domain Decomposition Methods in Science and Engineering XVIII*, Lecture Notes in Computational Science and Engineering,
DOI: 10.1007/978-3-642-02677-5,

method. While our approach is more flexible and significantly easier to implement than the algorithm in [7] the numerical examples indicate that it is comparable to linear multigrid for reasonable initial iterates which can be obtained, e.g., by nested iteration.

2 A Nonsmooth Newton Method

Problem (1) can be equivalently formulated as the following inclusion

$$(A+\partial\varphi)(u) \ni b, \tag{3}$$

where the subdifferential $\partial\varphi$ of φ is the set-valued diagonal operator given by $(\partial\varphi(v))_i = \partial\varphi_i(v_i)$. Similar to the linear case $\varphi = 0$ the nonlinear Gauß-Seidel method $u^{k+1} = u^k + \mathcal{F}u^k$ defined by successive minimization of J in the coordinate directions can be represented by the operator

$$\mathcal{F}(v) = (D+L+\partial\varphi)^{-1}(b-Rv)-v,$$

where we have used the splitting $A = D+L+R$ in the diagonal, left, and right parts. Using a monotonicity argument it can be shown that the nonlinear Gauß-Seidel method converges globally to the solution of (1) [7]. Unfortunately, as in the linear case, the convergence rates deteriorate rapidly if A is a differential operator discretized on finer and finer grids.

It follows from the global convergence of the Gauß-Seidel method that the original problem (3) is equivalent to the fixed-point equation

$$\mathcal{F}(u) = 0 \tag{4}$$

for the operator $\mathcal{F}$ which is single-valued and Lipschitz continuous. This suggests to use a nonsmooth Newton approach for (4) which leads to methods

$$u^{k+1} = u^k - H(u^k)^{-1}\mathcal{F}(u^k) \tag{5}$$

where $H(u^k)$ is a generalized linearization of $\mathcal{F}$. In order to construct such $H(u^k)$ we first derive a linearization of $f_i = (A_{ii}+\partial\varphi_i)^{-1} : \mathbb{R} \to \mathbb{R}$. Since f_i is strictly monotone and Lipschitz continuous it is differentiable almost everywhere by Rademacher's theorem [9]. An element of the generalized Jacobian in the sense of Clarke [3] is given by

$$\partial f_i(x) = \begin{cases} 0 & \text{if } \partial\varphi_i(f_i(x)) \text{ is set-valued,} \\ (a_{ii}+\varphi_i''(f_i(x)))^{-1} & \text{else.} \end{cases} \tag{6}$$

For φ_i'' we use either the derivative from the left or from the right and $(a_{ii} + \varphi_i''(f_i(x)))^{-1}$ is set to zero if both one-sided derivatives tend to infinity.

Given an index set $\mathcal{I} \subset \{1,\dots,n\}$ and a matrix or vector ($n \times 1$ matrix) M we introduce the following notation for truncated versions of M

$$(M_{\mathcal{J}})_{ij} = \begin{cases} M_{ij} & \text{for } i \in \mathcal{J} \\ 0 & \text{else,} \end{cases} \qquad (M_{\mathcal{J}\mathcal{J}})_{ij} = \begin{cases} M_{ij} & \text{for } i,j \in \mathcal{J} \\ 0 & \text{else.} \end{cases}$$

Assuming a chain rule elementary computations lead to a linearization of $\mathcal{F}$ given by

$$\partial\mathcal{F}(v) = -\left(D + L + \varphi''(v + \mathcal{F}v)_{\mathcal{I}(v+\mathcal{F}v)}\right)^{-1} R_{\mathcal{I}(v+\mathcal{F}v)} - I \tag{7}$$

with the index set of inactive components

$$\mathcal{I}(v) = \left\{ i : \partial\varphi_i(v_i) \text{ is single-valued and } \varphi_i''(v_i) \text{ is finite} \right\}.$$

Theorem 1. *If $H(u^k) = \partial\mathcal{F}(u^k)$ is used in a nonsmooth Newton step* (5) *the resulting iteration can be equivalently rewritten as the following two-step method*

$$u^{k+\frac{1}{2}} = u^k + \mathcal{F}(u^k), \tag{8}$$

$$u^{k+1} = u^{k+\frac{1}{2}} + \mathcal{C}(u^{k+\frac{1}{2}}), \tag{9}$$

with the linear correction

$$\mathcal{C}(v) = -\left(J''(v)_{\mathcal{I}(v),\mathcal{I}(v)}\right)^{-1} J'(v)_{\mathcal{I}(v)}. \tag{10}$$

The proof of Theorem 1 is straightforward using the fact that $\mathcal{C}(u^{k+\frac{1}{2}})_i = 0$ for $i \notin \mathcal{I}(u^{k+\frac{1}{2}})$. For obstacle problems it can be found in [6].

Remark 1. By restriction to the $i \in \mathcal{I}(v)$ each φ_i in (9) has a classical first derivative. The second derivatives of φ_i are meant in the sense explained after (6).

Even though linearization of ∂J in (9) is restricted to locally smooth components the derivatives of φ_i might get very large leading to ill-conditioned linear systems and slow multigrid convergence. Therefore we may restrict the linearization further, e.g., to

$$\overline{\mathcal{I}}(v) = \left\{ i \in \mathcal{I}(v) : |\varphi_i''(x) - \varphi_i''(y)| \leq C|x-y| \; \forall x,y \in [v_i - \delta, v_i + \delta] \right\}$$

for a large constant C and a small δ.

Remark 2. Replacing $\mathcal{I}$ by some $\overline{\mathcal{I}} \subset \mathcal{I}$ leads to a truncated linearization $\overline{\partial\mathcal{F}}$ defined analogously to (7). Theorem 1 remains true for $H(u^k) = \overline{\partial\mathcal{F}}(u^k)$ if $\mathcal{I}$ is replaced by $\overline{\mathcal{I}}$.

Due to the leading nonlinear Gauß-Seidel step (8) global convergence can be shown if $J(u^{k+1}) \leq J(u^{k+\frac{1}{2}})$ [7]. Thus the introduction of suitable damping parameters in (9) leads to global convergence. However, very small damping parameters slowing down the convergence may be necessary if $u^{k+\frac{1}{2}} + \mathcal{C}(u^{k+\frac{1}{2}}) \notin \operatorname{dom}(J) = \{v : J(v) < \infty\}$. To overcome this problem we apply damping to a projected correction. If, additionally, we introduce inexact evaluation of $\mathcal{C}$ represented by the error ε^k the algorithm reads

$$u^{k+\frac{1}{2}} = u^k + \mathcal{F}(u^k), \tag{11}$$

$$u^{k+1} = u^{k+\frac{1}{2}} + \rho^k P^k(\mathcal{C}(u^{k+\frac{1}{2}}) + \varepsilon^k), \tag{12}$$

where P^k is the component-wise Euclidean projection onto $\mathrm{dom}(J) - u^{k+\frac{1}{2}}$ and ρ^k is computed by the line search

$$\rho^k = \arg\min_{\rho \in \mathbb{R}} J(u^{k+\frac{1}{2}} + \rho P^k(\mathcal{C}(u^{k+\frac{1}{2}}) + \varepsilon^k)).$$

Since this algorithm satisfies $J(u^{k+1}) \leq J(u^{k+\frac{1}{2}})$ for arbitrary ε^k the following convergence result holds [7].

Theorem 2. *For every $u^0, \varepsilon^k \in \mathbb{R}^n$ the u^k converge to the solution u of* (3).

3 Multigrid

Now we consider the fast and inexact solution of the system (10) by multigrid methods. Since the matrix is symmetric and positive definite on the subspace

$$V^k = \{v \in \mathbb{R}^n : v_i = 0 \ \forall i \notin \bar{\mathcal{I}}(u^{k+\frac{1}{2}})\} \tag{13}$$

standard linear multigrid methods like successive or parallel subspace correction with standard transfer operators for problems in $\mathbb{R}^n$ can be applied if the following two modifications are introduced:

- If a diagonal element in a matrix is zero the subspace correction for the corresponding subspace should be zero as well. This may happen on all levels since the fine matrix has zero rows and columns.
- After the sum of all coarse corrections is prolongated to the fine space $\mathbb{R}^n$ all components $i \notin \mathcal{I}(u^{k+\frac{1}{2}})$ should be set to zero.

With these two modifications each correction in a subspace U of $\mathbb{R}^n$ is now naturally a correction in the Euclidean projection U^k of U onto V^k. Hence the subspace correction method automatically minimizes in suitable subspaces of V^k without explicit construction of these subspaces or their basis functions.

Since there is no need to solve the systems (10) to a certain accuracy applying a single multigrid step is enough to achieve global convergence. The resulting overall algorithm consists of nonlinear smoothing on the fine level and linear multigrid for a reduced linearization. As a generalization of the algorithm in [6] we call it *Truncated Nonsmooth Newton Multigrid* (TNNMG). The algorithm is open to various modifications, e.g.:

- Additional nonlinear smoothing before the linear correction
- Linear smoothing on the fine level can be omitted.
- Alternative smoothers can be applied to the linear correction.

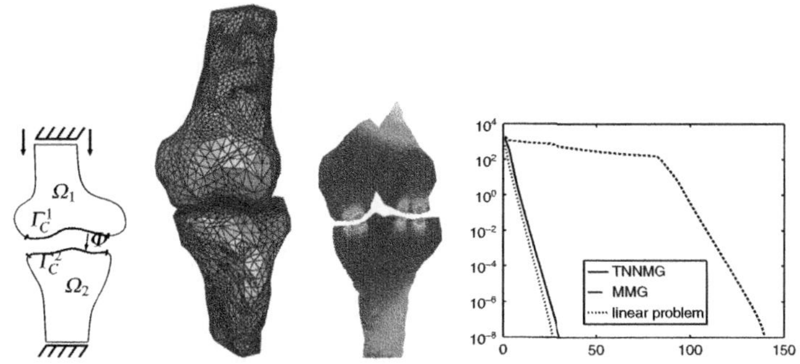

Fig. 1. Two-body contact problem. a) Schematic view; b) solution; c) vertical cut through the von-Mises stress field; d) error per number of iterations.

4 Example I: Two-Body Contact in Linear Elasticity

We will now show how the algorithm can be used to efficiently solve multi-body contact problems in linear elasticity. Consider two disjoint domains Ω_1, Ω_2 in $\mathbb{R}^d, d \in \{2,3\}$, discretized by simplicial grids. The boundary $\Gamma_i = \partial\Omega_i$, $i \in \{1,2\}$, of each domain is decomposed in three disjoint parts $\Gamma_i = \Gamma_{i,D} \cup \Gamma_{i,N} \cup \Gamma_{i,C}$. Let $\mathbf{f}_i \in (L_2(\Omega_i))^d$, $i \in \{1,2\}$, be body force density fields, and $\mathbf{t}_i \in (H^{-1/2}(\Gamma_{i,N}))^d$ be fields of surface traction. The two contact boundaries $\Gamma_{i,C}$ are identified using a homeomorphism $\Phi : \Gamma_{1,C} \to \Gamma_{2,C}$ and the initial distance function $g : \Gamma_{1,C} \to \mathbb{R}$, $g(x) = \|\Phi(x) - x\|$ is defined.

Let $\mathbf{V}_h$ be the space of first-order d-valued Lagrangian finite element functions on $\Omega_1 \cup \Omega_2$ and $\{\mathbf{m}\boldsymbol{\lambda}\}$ the nodal basis in $\mathbf{V}_h$. We denote the basis functions belonging to the n_C^i nodes of the contact boundaries $\Gamma_{i,C}$, $i \in \{1,2\}$ by $\{\mathbf{m}\boldsymbol{\lambda}_C^i\}$, the corresponding coefficients in a vector v by v_C^i and the basis functions for the remaining n^I nodes by $\{\mathbf{m}\boldsymbol{\lambda}_I\}$. The two-body contact problem can then be written as a minimization problem with a quadratic part as in (1) and $n = d(n^I + n_C^1 + n_C^2)$. Here A and b are the stiffness matrix and right-hand-side vector of linear elasticity, respectively. The nonlinearity is the characteristic functional $\varphi = \chi_{\mathcal{K}}$ of the mortar discretized admissible set

$$\mathcal{K} = \{v \in \mathbb{R}^{dn} \mid N\mathsf{D}v_C^1 - N\mathsf{M}v_C^2 \leq \mathsf{g}\} \tag{14}$$

with a sparse mass matrix M, a diagonal mass matrix D, a matrix N which contains the domain normals, and the weak obstacle g. Contrary to (1) $\chi_{\mathcal{K}}$ does not have pointwise structure. To overcome this we introduce the transformed basis [11]

$$\{\widetilde{\mathbf{m}\boldsymbol{\lambda}}\} = OB\{\mathbf{m}\boldsymbol{\lambda}\} = \begin{pmatrix} I & 0 & 0 \\ 0 & O_C & 0 \\ 0 & 0 & I \end{pmatrix} \begin{pmatrix} I & 0 & 0 \\ 0 & I & 0 \\ 0 & (\mathsf{D}^{-1}\mathsf{M})^T & I \end{pmatrix} \begin{pmatrix} \mathbf{m}\boldsymbol{\lambda}_I \\ \mathbf{m}\boldsymbol{\lambda}_C^1 \\ \mathbf{m}\boldsymbol{\lambda}_C^2 \end{pmatrix}.$$

In this basis, we get a minimization problem of the form (1) with a matrix $\widetilde{A} = OBAB^T O^T$, a right-hand side $\tilde{b} = OBb$ and

$$\tilde{\varphi}_{p,0}(v_{p,0}) = \begin{cases} 0 & v_{p,0} \leq (\mathsf{D}^{-1}\mathsf{g})_p,\ p \text{ is vertex on } \Gamma_{1,C} \\ \infty & \text{else.} \end{cases} \tag{15}$$

The matrix O_C is block-diagonal. For each vertex p on $\Gamma_{1,C}$, the $d \times d$ diagonal entry $(O_C)_{pp}$ contains the Householder reflection which maps the first canonical basis vector of $\mathbb{R}^d$ onto the domain normal at p. Due to the point-wise structure (15), a projected block Gauß-Seidel scheme converges. For the coarse grid correction (10) we compute

$$J''(v)_{\mathcal{J}(v),\mathcal{J}(v)} = \widetilde{A}_{\mathcal{J}(v),\mathcal{J}(v)} \qquad \text{and} \qquad J'(v)_{\mathcal{J}(v)} = \left[\widetilde{A}v - \widetilde{b}\right]_{\mathcal{J}(v)},$$

and apply one linear multigrid step to this.

Remark 3. For the transition from the finest to the second finest grid level the standard multigrid prolongation operator P is replaced by $\widetilde{P} = OB^{-1}P$. That way, discretization on the coarser levels is with respect to the nodal basis. Truncation and the transforming prolongation $\widetilde{P}$ can be combined in a single operator. This avoids having to store two fine-grid matrices.

As an example geometry we use the Visible Human data set [1]. We assume bone to be an isotropic, homogeneous, linear elastic material with $E = 17$ GPa and $\nu = 0.3$. The bottom section of the proximal tibia is clamped and a downward displacement of 6 mm is prescribed on the upper section of the femur (see Fig. 1, left). The implementation is based on the DUNE library [2].

We compare the numerical efficiencies of our solver and a monotone multigrid method (MMG), which is currently the fastest known globally convergent solver for two-body contact problems [8, 11]. It is well known that the MMG degenerates to a linear multigrid method once the active set has been found and hence shows linear multigrid convergence asymptotically.

We use nested iteration on two adaptive grids with 44777 vertices in total. Errors are computed by comparing with a precomputed reference solution $\mathbf{u}^*$. The error $e_i = \|\mathbf{u}_i - \mathbf{u}^*\|_A$ is plotted in Fig. 1. As expected, both the TNNMG and the MMG asymptotically show a linear multigrid convergence speed. However, the MMG needs more than 80 iterations to reach the asymptotic phase (see [10] for an explanation), whereas the TNNMG enters the asymptotic phase immediately. Note that iteration counts can be compared directly because both methods do a similar amount of work per iteration.

Remark 4. For two-body contact problems, the TNNMG is considerably easier to implement than the monotone multigrid method. See [10] for details.

5 Example II: The Allen-Cahn Equation

The Allen-Cahn equation is a well established diffuse interface model for phase transition phenomena as, e.g., solidification or crystallographic transformations. It can alternatively be interpreted as a regularization of the sharp interface geometric PDE for mean curvature flow (cf. [4]). The model considers an order parameter $u : \mathbb{R}^d \supset \Omega \to [-1,1]$ where the interval boundaries correspond to the pure phases, and is based on the Ginzburg-Landau energy

$$\mathcal{E}(u) = \int_{\Omega} \left(\frac{\varepsilon}{2} |\nabla u|^2 + \frac{1}{\varepsilon} \psi(u) \right) dx. \tag{16}$$

In the following, the potential ψ is taken to be

$$\psi(u) = \underbrace{\frac{1}{2}\theta \left((1+u)\ln\left(\frac{1+u}{2}\right) + (1-u)\ln\left(\frac{1-u}{2}\right)\right)}_{=:\varphi_\theta(u)} + \frac{1}{2}\theta_c\left(1-u^2\right)$$

which for temperatures θ less than some critical value θ_c takes on the characteristic double-well shape. The Allen-Cahn equation

$$\varepsilon u_t = \varepsilon \Delta u - \varepsilon^{-1}\psi'(u) \tag{17}$$

results as the L^2-gradient flow of (16). For a given triangulation $\mathcal{T}$ of the domain Ω, which for simplicity we will assume to be a polygonal domain in $\mathbb{R}^2$, let $V_\mathcal{T}$ denote the space of continuous piecewise linear finite elements, $\mathcal{N}$ the set of nodes and $\{\lambda_p | p \in \mathcal{N}\} \subset V_\mathcal{T}$ the nodal basis. Time discretization by an unconditionally stable semi-implicit Euler scheme with timestep τ and subsequent finite element discretization of (17) yields the variational problem

$$u_k \in V_\mathcal{T}: \quad a(u_k, v) - \ell_k(v) + \frac{\tau}{\varepsilon^2}(\varphi_\theta'(u_k), v)_\mathcal{T} = 0 \quad \forall v \in V_\mathcal{T} \tag{18}$$

to be solved in the kth time step. Here $(\cdot,\cdot)_\mathcal{T}$ denotes the lumped L^2-product on $V_\mathcal{T}$. Furthermore

$$\ell_k(v) = \left(1 + \frac{\theta_c \tau}{\varepsilon^2}\right)(u_{k-1}, v)_\mathcal{T}, \quad a(v,w) = \tau(\nabla v, \nabla w) + (v,w)_\mathcal{T}$$

are a linear functional and a symmetric, positive definite bilinear form, respectively. Thus (18) is equivalent to the minimization problem in $V_\mathcal{T}$ for

$$J(v) = \frac{1}{2}a(v,v) - \ell_k(v) + \frac{\tau}{\varepsilon^2}\sum_{p\in\mathcal{N}} \varphi_\theta(v(p)) \int_\Omega \lambda_p(x)\,dx.$$

Identifying $V_\mathcal{T}$ with $\mathbb{R}^n$ we are now in the setting described in Section 1.

For the numerical example we choose $\Omega = [-1,1]^2$, $\varepsilon = 10^{-2}$, $\theta_c = 1$, $\tau = 10^{-4}$, and an initial value as shown in Fig. 2. We use one nonlinear smoothing step and a V(3,3)-cycle for the linearized system with nested iteration and compare the averaged convergence rates versus θ. Furthermore the error $e_j = \|u_1^j - u_1^*\|_a$ in the jth TNNMG-step of the first time step is considered where u_1^* is a precomputed reference solution.

As Figs. 2c+d indicate, the TNNMG-method exhibits very fast convergence and robustness wrt. θ which is remarkable as φ_θ is singular for $\theta \to 0$. Note that the convergence rates here never exceed 0.1. Experiments have shown that introducing additional nonlinear smoothing steps does not significantly accelerate convergence any further in this testcase whereas using less linear smoothing results in a considerable slowdown.

Acknowledgement. This work was funded in part by the DFG under contract Ko 1806/3-2.

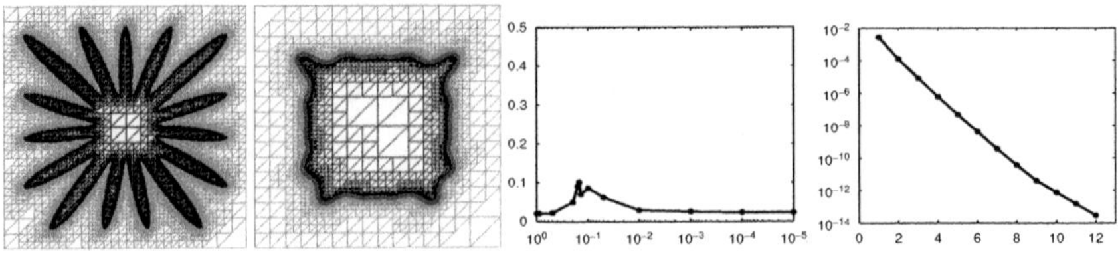

Fig. 2. a) Initial value; b) solution at time step 200; c) averaged convergence rates vs. θ for TNNMG in the first time step; d) error vs. number of TNNMG iterations at $\theta = 0.15$ ($\sim$ 200.000 nodes).

References

[1] The Visible Human Project. `http://www.nlm.nih.gov/research/visible/visible_human.html`.

[2] Bastian, P., Blatt, M., Dedner, A., Engwer, C., Klöfkorn, R., Kornhuber, R., Ohlberger, M., Sander, O.: A generic interface for parallel and adaptive scientific computing. Part II: Implementation and tests in DUNE. *Computing*, accepted.

[3] Clarke, F.H.: *Optimization and Nonsmooth Analysis.* Wiley, New York, 1983.

[4] Deckelnick, K., Dziuk, G., Elliot, C.M.: Computation of geometric partial differential equations and mean curvature flow. *Acta Numer.*, 14, 2005.

[5] Ekeland, I., Temam, R.: *Convex Analysis.* North-Holland, 1976.

[6] Gräser, C., Kornhuber, R.: Multigrid methods for obstacle problems. *J. Comput. Math.*, submitted.

[7] Kornhuber, R.: On constrained Newton linearization and multigrid for variational inequalities. *Numer. Math.*, 91:699–721, 2002.

[8] Kornhuber, R., Krause, R., Sander, O., Deuflhard, P., Ertel, S.: A monotone multigrid solver for two body contact problems in biomechanics. *Comput. Vis. Sci*, 11:3–15, 2008.

[9] Nekvinda, A., Zajíček, L.: A simple proof of the Rademacher theorem. *Časopis Pěst. Mat.*, 113(4):337–341, 1988.

[10] Sander, O.: *Multidimensional Coupling in a Human Knee Model.* PhD thesis, Freie Universität Berlin, 2008.

[11] Wohlmuth, B., Krause, R.: Monotone methods on nonmatching grids for nonlinear contact problems. *SIAM J. Sci. Comput.*, 25(1):324–347, 2003.

A Recursive Trust-Region Method for Non-Convex Constrained Minimization

Christian Groß and Rolf Krause

Institute for Numerical Simulation, University of Bonn.
{gross,krause}@ins.uni-bonn.de

1 Introduction

The mathematical modelling of mechanical or biomechanical problems involving large deformations or biological materials often leads to highly nonlinear and constrained minimization problems. For instance, the simulation of soft-tissues, as the deformation of skin, gives rise to a highly non-linear PDE with constraints, which constitutes the first order condition for a minimizer of the corresponding non-linear energy functional. Besides of the pure deformation of the tissue, bones and muscles have a restricting effect on the deformation of the considered material, leading to additional constraints. Although PDEs are usually formulated in the context of Sobolev spaces, their numerical solution is carried out using discretizations as, e.g., finite elements. Thus, in the present work we consider the following finite dimensional constrained minimization problem:

$$u \in \mathcal{B} : J(u) = \min! \qquad \text{(M)}$$

where $\mathcal{B} = \{v \in \mathbb{R}^n \mid \underline{\varphi} \leq v \leq \overline{\varphi}\}$ and $\underline{\varphi} < \overline{\varphi} \in \mathbb{R}^n$ and the possibly nonconvex, but differentiable, objective function $J : \mathbb{R}^n \to \mathbb{R}$. Here, the occurring inequalities are to be understood pointwise. In the context of discretized PDEs, n corresponds to the dimension of the finite element space and may therefore be very large.

The design of robust and efficient solution methods for problems like (M) is a demanding task. Indeed, globalization strategies, such as trust-region methods (cf., [1, 10]) or line-search algorithms, succeed in computing local minimizers, but are based on the paradigm of Newton's method. This means that a sequence of iterates is computed by solving linear, but potentially large, systems of equations. The drawback is, that due to the utilization of a globalization strategy the computed corrections generally need to be damped. Hence, for two reasons the convergence of such an approach may slow down: often the linear systems of equations are ill-conditioned and therefore iterative linear solvers tend to converge slowly. Moreover, even if the linear system can be solved efficiently, for instance by employing a Krylov-space method in combination with a good preconditioner, globalization

M. Bercovier et al. (eds.), *Domain Decomposition Methods in Science and Engineering XVIII*, Lecture Notes in Computational Science and Engineering,
DOI: 10.1007/978-3-642-02677-5,

strategies tend to reduce the step-size depending on the non-linearity of the objective function.

Therefore, solution strategies which are just based on Newton's method can remain inefficient. In the context of quadratic minimization problems, linear multigrid methods have turned out to be highly efficient since these algorithms are able to resolve also the low frequency contributions of the solution. Similarly, nonlinear multigrids (cf., [4, 6, 8]) aim at a better resolution of the low-frequency contributions of non-linear problems.

Therefore, [3] introduced a class of non-linear multilevel algorithms, called RMTR_∞ (Recursive Multilevel Trust-Region method), to solve problems of the class (M). In the present work, we will introduce a V-cycle variant of the RMTR algorithm presented in [5, 6]. On each level of a given multilevel hierarchy, this algorithm employs a trust-region strategy to solve a constrained nonlinear minimization problem, which arises from level dependent representations of J, $\underline{\varphi}$, $\overline{\varphi}$. An important new feature of the RMTR_∞ algorithm is the L^2-projection of iterates to coarser levels to generate good initial iterates - in contrast to employing the restriction operator to transfer iterates to a coarse level. In fact, the new operator yields significantly better rates of convergence of the RMTR algorithm (for a complete discussion see [6]). To prove first-order convergence, we will state less restrictive assumptions on the smoothness of J than used by [3]. Moreover, we illustrate the efficiency of the RMTR_∞ - algorithm by means of an example from the field of non-linear elasticity in 3D.

2 The Multilevel Setting

The key concept of the RMTR_∞ algorithm, which we will present in Section 3, is to minimize on different levels arbitrary non-convex functions H_k approximating the fine level objective function J. The minimization is carried out employing a trust-region strategy which ensures convergence. Corrections computed on coarser levels will be summed up and interpolated which provide possible corrections on the fine level.

In particular, on each level, m_1 pre-smoothing and m_2 post-smooting trust-region steps are computed yielding trust-region corrections. In between, a recursion is called yielding a coarse level correction which is the interpolated difference between first and last iterate on the next coarser level.

Therefore, we assume that a decomposition of the $\mathbb{R}^n$ into a sequence of nested subspaces is given, such as $\mathbb{R}^n = \mathbb{R}^{n_j} \supsetneq \mathbb{R}^{n_{j-1}} \supsetneq \cdots \supsetneq \mathbb{R}^{n_0}$. The spaces are connected to each other by full-rank linear interpolation, restriction and projection operators, i.e., $I_k : \mathbb{R}^{n_k} \longrightarrow \mathbb{R}^{n_{k+1}}$, $R_k : \mathbb{R}^{n_k} \longrightarrow \mathbb{R}^{n_{k-1}}$ and $P_k : \mathbb{R}^{n_k} \longrightarrow \mathbb{R}^{n_{k-1}}$. Given these operators and a current fine level iterate, $u_{k+1} \in \mathbb{R}^{n_{k+1}}$, the nonlinear coarse level model (cf., [4, 6, 8]) is defined as

$$H_k(u_k) = J_k(u_k) + \langle \delta g_k, u_k - P_{k+1} u_{k+1} \rangle \tag{1}$$

where we assume that a fixed sequence of nonlinear functions $(J_k)_k$ is given, representing J on the coarser levels. Here, the residual $\delta g_k \in \mathbb{R}^{n_k}$ is given by

$$\delta g_k = \begin{cases} R_{k+1}\nabla H_{k+1}(u_{k+1}) - \nabla J_k(P_{k+1}u_{k+1}) & \text{if } j > k \geq 0 \\ 0 & \text{if } k = j \end{cases}$$

In the context of constrained minimization, the fine level obstacles $\underline{\varphi}$, $\overline{\varphi}$ have also to be represented on coarser levels. In our implementation we employ the approach introduced in [2]. Due to the definition of the coarse level obstacles, this ensures that the projection of a fine level iterate and each resulting coarse level correction is admissible.

3 Recursive Trust-Region Methods

The reliable minimization of nonconvex functions H_k depends crucially on the control of the "quality" of the iterates. Line-search algorithms, for instance, scale the length of Newton corrections in order to force convergence to first-order critical points. Similarly, in trust-region algorithms corrections are the solutions of constrained quadratic minimization problems. For a given iterate $u_{k,i} \in \mathbb{R}^{n_k}$, where i denotes the current iteration on level k, a correction $s_{k,i}$ is computed as an approximate solution of

$$\begin{aligned} s_{k,i} \in \mathbb{R}^{n_k} : \psi_{k,i}(s_{k,i}) &= \min! \\ \text{w.r.t. } \|s_{k,i}\|_\infty \leq \Delta_{k,i} \text{ and } &\underline{\varphi}_k \leq u_{k,i} + s_{k,i} \leq \overline{\varphi}_k \end{aligned} \tag{2}$$

Here, $\psi_{k,i}(s) = \langle \nabla H_k(u_{k,i}), s\rangle + \frac{1}{2}\langle B_{k,i}s, s\rangle$ denotes the trust-region model with $B_{k,i}$, a symmetric matrix, possibly approximating the Hessian $\nabla^2 H_k(u_{k,i})$ (if it exists) and $\Delta_{k,i}$ is the trust-region radius.

On the coarse level, the reduction of H_{k-1} starting at the initial iterate $u_{k-1,0} = P_k u_{k,m_1}$ yields a final coarse level iterate u_{k-1}. Therefore, the recursively computed correction is $s_{k,m_1} = I_{k-1}(u_{k-1} - P_k u_{k,m_1})$.

Trust-Region Algorithm, Input: $u_{k,0}, \Delta_{k,0}, \underline{\varphi}_k, \overline{\varphi}_k, H_k, m$

do m times {
 compute $s_{k,i}$ as an approximate solution of (2)
 if $(\rho_{k,i}(s_{k,i}) \geq \eta_1)$
 $u_{k,i+1} = u_{k,i} + s_{k,i}$
 otherwise
 $u_{k,i+1} = u_{k,i}$
 compute a new $\Delta_{k,i+1}$
 }

Algorithm 6: Trust-Region Algorithm

To ensure convergence, corrections are only added to the current iterate, if the *contraction rate* $\rho_{k,i}$ is sufficiently large. The contraction rate compares $H_k(u_{k,i}) - H_k(u_{k,i} + s_{k,i})$ to the reduction predicted by the underlying quadratic model $\psi_{k,i}$. The value of $\psi_{k,i}(s_{k,i})$ prognoses the reduction induced by corrections computed by means of (2). The underlying model for recursively computed corrections s_{k,m_1} is the coarse level objective function H_{k-1}. Thus, we define

$$\rho_{k,i}(s_{k,i}) = \begin{cases} \frac{H_k(u_{k,i})-H_k(u_{k,i}+s_{k,i})}{-\psi_{k,i}(s_{k,i})} & \text{if } s_{k,i} \text{ computed by (2)} \\ \frac{H_k(u_{k,i})-H_k(u_{k,i}+I_{k-1}s_{k-1})}{H_{k-1}(P_k u_{k,i})-H_{k-1}(P_k u_{k,i}+s_{k-1})} & \text{otherwise} \end{cases}$$

Now, a correction is then added to the current iterate if $\rho_{k,i}(s_{k,i}) \geq \eta_1$ where $\eta_1 > 0$. In this case, the next trust-region radius $\Delta_{k,i+1}$ will be chosen larger than the current one, i.e., $\gamma_3 \Delta_{k,i} \geq \Delta_{k,i+1} \geq \gamma_2 \Delta_{k,i} > \Delta_{k,i}$ with $\gamma_3 \geq \gamma_2 > 1$. Otherwise, if $\rho_{k,i}(s_{k,i}) < \eta_1$, the correction will be discarded and $\Delta_{k,i+1}$ chosen smaller than $\Delta_{k,i}$, i.e., $\Delta_{k,i} > \Delta_{k,i+1} \geq \gamma_1 \Delta_{k,i}$, with $0 < \gamma_1 < 1$.

These four steps, computing $s_{k,i}$ by means of (2), computing $\rho_{k,i}$, applying $s_{k,i}$ and the update of the trust-region radius are summarized in Algorithm 6.

Algorithm 7 on Page 141, introduces the RMTR$_\infty$ algorithm, which is a V-cycle algorithm with an embedded trust-region solver.

3.1 Convergence to First-Order Critical Points

To show convergence of the RMTR$_\infty$ algorithm to first-order critical points, we state the following assumptions on H_k, cf. [1].

(A1) For a given initial fine level iterate $u_{j,0} \in \mathbb{R}^{n_j}$, we assume that the level set $\mathcal{L}_j = \{u \in \mathbb{R}^{n_j} \mid \underline{\varphi}_j \leq u \leq \overline{\varphi}_j \text{ and } J_j(u) \leq J_j(u_{j,0})\}$ is compact. Moreover, for all initial coarse level iterates $u_{k,0} \in \mathbb{R}^{n_k}$, we assume that the level sets $\mathcal{L}_k = \{u \in \mathbb{R}^{n_k} \mid \underline{\varphi}_k \leq u \leq \overline{\varphi}_k \text{ and } H_k(u) \leq H_k(u_{k,0})\}$ are compact.

(A2) For all levels $k \in \{0, \ldots, j\}$, we assume that H_k is continuously differentiable on $\mathcal{L}_k$. Moreover, we assume that there exists a constant $c_g > 0$ such that for all iterates $u_{k,i} \in \mathcal{L}_k$ holds $\|\nabla H_k(u_{k,i})\|_2 \leq c_g$.

(A3) For all levels $k \in \{0, \ldots, j\}$, there exists a constant $c_B > 0$ such that for all iterates $u_{k,i} \in \mathcal{L}_k$ and for all symmetric matrices $B_{k,i}$ employed, the inequality $\|B_{k,i}\|_2 \leq c_B$ is satisfied.

Moreover, computations on level $k-1$ are carried out only if

$$\begin{aligned} \|R_k D_{k,m_1} g_{k,m_1}\|_2 &\geq \kappa_g \|D_{k,m_1} g_{k,m_1}\|_2 \\ \varepsilon_\varphi \geq \|D_{k,i}\|_\infty &\geq \kappa_\varphi > 0 \end{aligned} \tag{AC}$$

where $\varepsilon_\varphi, \kappa_\varphi, \kappa_g > 0$, $g_{k,i} = \nabla H_k(u_{k,i})$ and m_1 indexes the recursion. $D_{k,i}$ is a diagonal matrix given by

$$(D_{k,i})_{ll} = \begin{cases} (\overline{\varphi}_k - u_{k,i})_l & \text{if } (g_{k,i})_l < 0 \\ (u_{k,i} - \underline{\varphi}_k)_l & \text{if } (g_{k,i})_l \geq 0 \end{cases}$$

RMTR$_\infty$ Algorithm, Input: $u_{k,0}, \Delta_{k,0}, \underline{\varphi}_k, \overline{\varphi}_k, H_k$
Pre-smoothing call Algorithm 6 with $u_{k,0}, \Delta_{k,0}, \underline{\varphi}_k, \overline{\varphi}_k, H_k, m_1$ receive u_{k,m_1}, Δ_{k,m_1} *Recursion* compute $\underline{\varphi}_{k-1}, \overline{\varphi}_{k-1}, H_{k-1}$ call RMTR$_\infty$ with $u_{k,m_1}, \Delta_{k,m_1}, \underline{\varphi}_{k-1}, \overline{\varphi}_{k-1}, H_{k-1}$, receive s_{k-1} and compute $s_{k,m_1} = I_{k-1}s_{k-1}$ **if** $(\rho_{k,m_1}(s_{k,m_1}) \geq \eta_1)$ $u_{k,m_1+1} = u_{k,m_1} + s_{k,m_1}$ **otherwise** $u_{k,m_1+1} = u_{k,m_1}$ compute a new Δ_{k,m_1+1} *Post-smoothing* call Algorithm 6 with $u_{k,m_1+1}, \Delta_{k,m_1+1}, \underline{\varphi}_k, \overline{\varphi}_k, H_k, m_2$ receive u_{k,m_2}, Δ_{k,m_2} **if** $(k == j)$ **goto** *Pre-smoothing* **else return** $u_{k,m_2} - u_{k,0}$

Algorithm 7: RMTR$_\infty$

In the remainder, we abbreviate $\hat{g}_{k,i} = D_{k,i}g_{k,i}$. Finally, we follow [1] and assume that corrections computed in Algorithm 6 satisfy

$$\psi_{k,i}(s_{k,i}) < \beta_1 \psi_{k,i}(s^C_{k,i}) \tag{CC}$$

where $\beta_1 > 0$ and $s^C_{k,i} \in \mathbb{R}^{n_k}$ solves

$$\psi_{k,i}(s^C_{k,i}) = \min_{t \geq 0:\ s=-tD^2_{k,i}g_{k,i}} \{\psi_{k,i}(s) : \|s\|_\infty \leq \Delta_{k,i} \text{ and } \underline{\varphi}_k \leq u_{k,i}+s \leq \overline{\varphi}_k\} \tag{3}$$

Now, we can now cite Lemma 3.1 from [1].

Lemma 1. *Let* (A1)–(A3) *and* (AC) *hold. Then if* $s_{k,i}$ *in Algorithm* 6 *satisfies* (CC), *we obtain*

$$-\psi_{k,i}(s_{k,i}) \geq c\|\hat{g}_{k,i}\|_2 \min\{\Delta_{k,i}, \|\hat{g}_{k,i}\|_2\} \tag{4}$$

To obtain the next results, the number of applied V-cycles will be indexed by ν, so that $u^\nu_{k,i}$, denotes the i-th iterate on Level k in Cycle ν.

Theorem 1. *Assume that* (A1)–(A3) *and* (AC) *hold. Moreover, assume that in Algorithm 7 at least* $m_1 > 0$ *or* $m_2 > 0$ *holds. We also assume that for each* $s_{k,i}$ *computed* *in Algorithm* 6 (CC) *holds. Then for each sequence of iterates* $(u^\nu_{j,i})_{\nu,i}$, *we obtain* $\liminf_{\nu\to\infty}\|\hat{g}^\nu_{j,i}\|_2 = 0$.

Proof. We will prove the result by contradiction, i.e., we assume that $\exists\varepsilon > 0$ and a sequence $(u^\nu_{j,i})_{\nu,i}$ such that $\liminf_{\nu\to\infty}\|\hat{g}^\nu_{j,i}\|_2 \geq \varepsilon$.

In this case, one can show that $\Delta^\nu_{j,i} \to 0$. Namely, if $\rho^\nu_{j,i} \geq \eta_1$ holds only finitely often, we obtain that $\Delta^\nu_{j,i}$ is increased finitely often but decreased infinitely often. On the other hand, for infinitely many iterations with $\rho^\nu_{j,i} \geq \eta_1$ we obtain

$$H^\nu_k(u^\nu_{k,i}) - H^\nu_k(u^\nu_{k,i} + s^\nu_{k,i}) \geq -\eta_1 \psi^\nu_{k,i}(s^\nu_{k,i})$$

We now exploit Lemma 1, (A1), and $\|\hat{g}^\nu_{j,i}\|_2 \geq \varepsilon$ and obtain for sufficiently large ν that

$$H^\nu_k(u^\nu_{k,i}) - H^\nu_k(u^\nu_{k,i} + s^\nu_{k,i}) \geq c\Delta^\nu_{k,i} \geq c\|s^\nu_{k,i}\|_\infty \to 0$$

Next, we employ the mean value theorem, i.e., $\langle s^\nu_{k,i}, \bar{g}^\nu_{k,i}\rangle = H_k(u^\nu_{k,i} + s^\nu_{k,i}) - H_k(u^\nu_{k,i})$, as well as (A2) and (A3) and obtain for each trust-region correction

$$\begin{aligned}
|\mathrm{pred}^\nu_{k,i}(s^\nu_{k,i})|\,|\rho^\nu_{k,i} - 1| &= \left|H^\nu_k(u^\nu_{k,i} + s^\nu_{k,i}) - H^\nu_k(u^\nu_{k,i}) + \langle s^\nu_{k,i}, g^\nu_{k,i}\rangle + \frac{1}{2}\langle s^\nu_{k,i}, B^\nu_{k,i} s^\nu_{k,i}\rangle\right| \\
&\leq \frac{1}{2}|\langle s^\nu_{k,i}, B^\nu_{k,i} s^\nu_{k,i}\rangle| + |\langle s^\nu_{k,i}, \bar{g}^\nu_{k,i} - g^\nu_k(u^\nu_{k,i})\rangle| \\
&\leq \frac{1}{2}c_B(\Delta^\nu_{k,i})^2 + \|\bar{g}^\nu_{k,i} - g^\nu_k(u^\nu_{k,i})\|_2 \Delta^\nu_{k,i}
\end{aligned}$$

Due to the convergence of $\Delta^\nu_{k,i}$, and, hence, of $(u^\nu_{k,i})_{\nu,i}$, and the continuity of $g^\nu_{k,i}$, we obtain $\rho^\nu_{k,i} \to 1$ for $i \neq m_1$. Hence, on each level, for sufficiently small $\Delta^\nu_{j,i}$, trust-region corrections are successful and applied.

One can also show, that for sufficiently small $\Delta^\nu_{j,i}$ recursively computed corrections will be computed and applied: we find that there exists a $c > 0$ such that $\Delta^\nu_{k,i} \geq c\Delta^\nu_{j,m_1}$ (cf., [6]). In turn, (AC) provides that there exists another constant such that

$$H^\nu_k(u^\nu_{k,m_1}) - H^\nu_k(u^\nu_{k,m_1} + s^\nu_{k,m_1}) \geq c\|\hat{g}^\nu_{k,m_1}\|_2 \min\{c\Delta^\nu_{j,m_1}, \|\hat{g}^\nu_{k,m_1}\|_2\}$$

(cf., the proof of Lemma 4.4, [6]). Now, one can show, cf. Theorem 4.6, [6], that the contraction rates for recursively computed corrections also tend to one, i.e., $\rho^\nu_{j,m_1} \to 1$.

Since $\Delta^\nu_{j,i} \to 0$, we obtain $(\rho^\nu_{j,i})_{\nu,i} \to 1$. But this contradicts $\Delta^\nu_{j,i} \to 0$ and $\liminf_{\nu\to\infty}\|\hat{g}^\nu_{j,i}\|_2 \to 0$ must hold. □

Using exactly the same argumentation as used in Theorem 6.6 of [6], we obtain convergence to first-order critical points, i.e., $\lim_{\nu\to\infty}\|\hat{g}^\nu_{j,i}\|_2 = 0$.

Theorem 2. *Let assumptions* (A1)–(A3), (AC) *hold. Moreover, assume that al least one pre- or post-smoothing step in Algorithm* 7 *will be performed and that* (CC) *holds for each correction computed in Algorithm* 6. *Then, for each sequence of iterates* $(u^\nu_{j,i})_{\nu,i}$ *we obtain* $\lim_{\nu\to\infty}\|\hat{g}^\nu_{j,i}\|_2 = 0$.

4 Numerical Example

In this section, we present an example from the field of non-linear elasticity computed with the RMTR$_\infty$ algorithm which is implemented in OBSLIB++, cf. [7].

R. W. Ogden has introduced a material law for rubber-like materials (cf., [9]). The associated stored energy function is highly non-linear due to a penalty term which prevents the inversion of element volumes:

$$W(\nabla\varphi) = a \cdot \mathrm{tr}E + b \cdot (\mathrm{tr}E)^2 + c \cdot \mathrm{tr}(E^2) + d \cdot \Gamma(\det(\nabla\varphi)) \tag{5}$$

where $\varphi = id + u$. This function is a polyconvex stored energy function depending on the Green - St. Venant strain tensor $E(u)$, i.e., $E(u) = \frac{1}{2}(\nabla u^T + \nabla u + \nabla u^T \nabla u)$, a penalty function $\Gamma(x) = -\ln(x)$ for $x \in \mathbb{R}^+$, and $a = -d, b = \lambda - d, c = \mu + d, d > 0$.

Fig. 1 shows the results of the computation of a contact problem with parameters $\lambda = 34$, $\mu = 136$, and $d = 100$. In particular, a skewed pressure is applied on the top side of the cube, which results in that the cube is pressed towards an obstacle. Problem (M) with $J(u) = \int_\Omega W(\nabla\varphi)dx$ was solved in a finite element framework using both a fine level trust-region strategy and our RMTR$_\infty$ algorithm with $m_1 = m_2 = 2$. Equation (2) was (approximately) solved using 10 projected cg-steps.

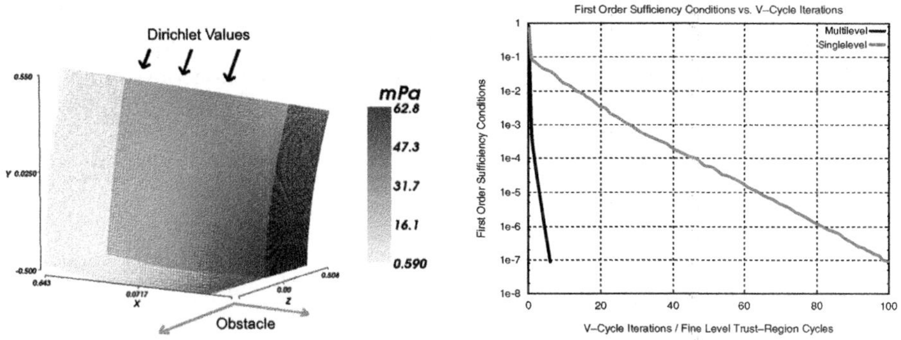

Fig. 1. Nonlinear Elastic Contact Problem 823,875 degrees of freedom. *Left image:* Deformed mesh and von-Mises stress distribution. *Right image:* Comparison of $(\|\hat{g}^V_{j,0}\|_2)_\nu$ computed by our RMTR$_\infty$ algorithm (black line) and by a trust-region strategy, performed only on the finest grid (grey line).

References

[1] Coleman, T.F., Li, Y.: An interior trust region approach for nonlinear minimization subject to bounds. *SIAM J. Optim.*, 6(2):418–445, 1996.

[2] Gelman, E., Mandel, J.: On multilevel iterative methods for optimization problems. *Math. Program.*, 48(1):1–17, 1990.

[3] Gratton, S., Mouffe, M., Toint, P.L., Weber-Mendonca, M.: A recursive trust-region method in infinity norm for bound-constrained nonlinear optimization. *IMA J. Numer. Anal.*, 28(4):827–861, 2008.
[4] Gratton, S., Sartenaer, A., Toint, P.L.: Recursive trust-region methods for multiscale nonlinear optimization. *SIAM J. Optim.*, 19(1):414–444, 2008.
[5] Groß, C.: *A Multiscale Approach for Convex and Non-convex Constrained Minimization*. Diploma thesis, University Bonn, April 2007.
[6] Groß, C., Krause, R.: On the convergence of recursive trust-region methods for multiscale non-linear optimization and applications to non-linear mechanics. Technical Report 710, Institute for Numerical Simulation, University of Bonn, March 2008. Submitted.
[7] Krause, R.: A parallel decomposition approach to non-smooth minimization problems—concepts and implementation. Technical Report 709, Institute for Numerical Simulation, University of Bonn, Germany, 2007.
[8] Nash, S.G.: A multigrid approach to discretized optimization problems. *Optim. Methods Softw.*, 14(1-2):99-116, 2000. ISSN 1055-6788. International Conference on Nonlinear Programming and Variational Inequalities (Hong Kong, 1998).
[9] Ogden, R.W.: *Nonlinear elastic deformations*. Ellis Horwood Series: Mathematics and its Applications. Ellis Horwood, Chichester, 1984. ISBN 0-85312-273-3.
[10] Ulbrich, M., Ulbrich, S., Heinkenschloss, M.: Global convergence of trust-region interior-point algorithms for infinite-dimensional nonconvex minimization subject to pointwise bounds. *SIAM J. Control Optim.*, 37(3):731–764 (electronic), 1999. ISSN 0363-0129.

A Robin Domain Decomposition Algorithm for Contact Problems: Convergence Results

Mohamed Ipopa and Taoufik Sassi

Laboratoire de Mathématiques Nicolas Oresme, LMNO, Université de Caen
Bâtiment Science 3, avenue du Maréchal Juin, 14032, Caen Cedex, France
ipopa@math.unicaen.fr, sassi@math.unicaen.fr

Summary. In this paper, we propose and study a Robin domain decomposition algorithm to approximate a frictionless unilateral problem between two elastic bodies. Indeed this algorithm combines, in the contact zone, the Dirichlet and Neumann boundaries conditions (Robin boundary condition). The primary feature of this algorithm is the resolution on each sub-domain of variational inequality.

1 Introduction

The numerical treatement of nonclassical contact problems leads to very large (due to the large ratio of degrees of freedom concerned by contact conditions) and ill-conditioned systems. Domain decomposition methods are good alternatives to overcome these difficulties (see [2, 3, 13, 14]).

The aim of this paper is to give an idea of the proof for iterative schemes based on domain decomposition techniques for a nonlinear problem modeling the frictionless contact of linear elastic bodies. They were introduced in [11] and can be considered as a generalization to variational inequality of the method described in [7, 15]. In [2, 3, 13, 14], the initial problem is transformed into a unilateral contact problem in the one body and a prescribed displacement problem in the other one. We propose, in this paper, another domain decomposition method in which we solve an unilateral contact problem in each subdomain.

2 Weak Formulation of the Continuous Problem

Let us consider two bodies occupying, in the reference configuration, bounded domains Ω^α, $\alpha = 1, 2$, of the space $\mathbb{R}^2$ with sufficiently smooth boundaries. The boundary $\Gamma^\alpha = \partial\Omega^\alpha$ consists of three measurable, mutually disjoint parts Γ_u^α, Γ_ℓ^α, Γ_c^α so that $\Gamma^\alpha = \overline{\Gamma_u^\alpha} \cup \overline{\Gamma_\ell^\alpha} \cup \overline{\Gamma_c^\alpha}$. The body $\overline{\Omega}^\alpha$ is fixed on the set Γ_u^α. It is subject to surface traction forces $\Phi^\alpha \in (L^2(\Gamma_\ell^\alpha))^2$ and the body forces are denoted by $f^\alpha \in (L^2(\Omega^\alpha))^2$.

M. Bercovier et al. (eds.), *Domain Decomposition Methods in Science and Engineering XVIII*, Lecture Notes in Computational Science and Engineering,
DOI: 10.1007/978-3-642-02677-5, © Springer-Verlag Berlin Heidelberg 2009

On the contact interface determined by Γ_c^1 and Γ_c^2, we consider the contact condition that is characterized by the non-penetration of the bodies and the transmission of forces. To describe the non-penetration of the bodies, we shall use a pre-defined bijective mapping $\chi : \Gamma_c^1 \longrightarrow \Gamma_c^2$, which assigns to each $x \in \Gamma_c^1$ some nearby point $\chi(x) \in \Gamma_c^2$. Let $v^1(x)$ and $v^2(\chi(x))$ denote the displacement vectors at x and $\chi(x)$, respectively. Assuming the small displacements, the *non-penetration condition* reads as follows:

$$v_\nu^1(x) - v_\nu^2(x) = [v_\nu] \leq g(x) \quad \text{with} \quad v_\nu^1(x) \equiv v^1(x) \cdot \nu(x),\ v_\nu^2(x) \equiv v^2(\chi(x)) \cdot \nu(x),$$

where $g(x) = (\chi(x) - x) \cdot \nu(x)$ is the initial gap and $\nu(x)$ is the critical direction defined by $\nu(x) = (\chi(x) - x)/\|\chi(x) - x\|$ or, if $\chi(x) = x$, by the outer unit normal vector to Γ_c^1. We seek the displacement field $u = (u^1, u^2)$ (where the notation u^α stands for $u|_{\Omega^\alpha}$) and the stress tensor field $\sigma = (\sigma(u^1), \sigma(u^2))$ satisfying the following equations and conditions (1)–(2) for $\alpha = 1,2$:

$$\begin{cases} div\, \sigma(u^\alpha) + f^\alpha = 0 & \text{in } \Omega^\alpha, \\ \sigma(u^\alpha) n^\alpha - \Phi_\ell^\alpha = 0 & \text{on } \Gamma_\ell^\alpha, \\ u^\alpha = 0 & \text{on } \Gamma_u^\alpha, \\ \sigma_\nu \leq 0, \sigma_T = 0, [u_\nu] \leq 0, & \text{on } \Gamma_c^\alpha, \\ \sigma_\nu \cdot [u_\nu] = 0 & \text{on } \Gamma_c^\alpha. \end{cases} \tag{1}$$

The symbol div denotes the divergence operator of a tensor function and is defined as

$$div\, \sigma = \left(\frac{\partial \sigma_{ij}}{\partial x_j} \right)_i.$$

The summation convention of repeated indices is adopted. The elastic constitutive law, is given by Hooke's law for homogeneous and isotropic solids:

$$\sigma(u^\alpha) = A^\alpha(x) \varepsilon(u^\alpha), \tag{2}$$

where $A^\alpha(x) = (a_{ijkh}^\alpha(x))_{1 \leq i,j,k,h \leq 2} \in (L^\infty(\Omega^\alpha))^{16}$ is a fourth-order tensor satisfying the usual symmetry and ellipticity conditions in elasticity. The linearized strain tensor $\varepsilon(u^\alpha)$ is given by

$$\varepsilon(u^\alpha) = \frac{1}{2}(\nabla u^\alpha + (\nabla u^\alpha)^T).$$

We will use the usual notations for the stress vector on the contact zone Γ_c^α:

$$\sigma_\nu^\alpha = \sigma_{ij}(u^\alpha) \nu_i^\alpha \nu_j^\alpha, \quad \sigma_T^\alpha = \sigma_{ij}(u^\alpha) \nu_j^\alpha - \sigma_\nu^\alpha \nu_i^\alpha.$$

In order to give the variational formulation corresponding to the problem (1)–(2), let us introduce the following spaces

$$V^\alpha = \{v^\alpha \in (H^1(\Omega^\alpha))^2,\ v = 0 \text{ on } \Gamma_u^\alpha\}, \quad V = V^1 \times V^2.$$

Now, we denote by K the following non-empty closed convex subset of V:

$$K = \{v = (v^1, v^2) \in V,\ [v_\nu] \leq 0 \text{ on } \Gamma_c^1\}.$$

The variational formulation of problem (1)–(2) is

$$\begin{cases} \text{Find } u \in K \text{ such that} \\ a(u, v-u) \geq L(v-u) \quad \forall v \in K, \end{cases} \tag{3}$$

where

$$\begin{aligned} a(u,v) &= a^1(u,v) + a^2(u,v), \\ a^\alpha(u,v) &= \int_{\Omega^\alpha} A^\alpha(x)\varepsilon(u^\alpha)\cdot\varepsilon(v^\alpha)dx, \end{aligned} \tag{4}$$

and

$$L(v) = \sum_{\alpha=1}^{2} \int_{\Omega^\alpha} f^\alpha \cdot v^\alpha\, dx + \int_{\Gamma_\ell^\alpha} \Phi^\alpha \cdot v^\alpha\, d\sigma.$$

There exists a unique solution u to problem (3) (see [5, 6, 12]).

3 The Domain Decomposition Algorithm

In order to split the problem (3) into two subproblems coupled through the contact interface, we first introduce the following spaces and mappings:

$$\begin{aligned} V_0^\alpha &= \{v^\alpha \in V^\alpha,\ v_\nu^\alpha = 0 \text{ on } \Gamma_c^\alpha\}, \quad \alpha = 1,2, \\ \mathcal{H}^{1/2}(\Gamma_c) &= \{\varphi \in (L^2(\Gamma_c))^2,\ \exists v \in V^\alpha,\ \gamma v_{|\Gamma_c} = \varphi\}, \\ H^{1/2}(\Gamma_c) &= \{\varphi \in L^2(\Gamma_c),\ \exists v \in H^1(\Omega^\alpha),\ \gamma v_{|\Gamma_c} = \varphi\}, \end{aligned}$$

where γ is the usual trace operator.

By $P^\alpha : H^{1/2}(\Gamma_c^\alpha) \longrightarrow V^\alpha$, we denote the extension operator from Γ_c^α in Ω^α defined by : $P^\alpha\varphi = v^\alpha, \varphi \in H^{1/2}(\Gamma_c^\alpha)$, where $v^\alpha \in V^\alpha$ satisfies

$$\begin{cases} a^\alpha(v^\alpha, w^\alpha) = 0 & \forall w^\alpha \in V_0^\alpha; \\ v_\nu^\alpha = \varphi & \text{on } \Gamma_c^\alpha. \end{cases}$$

Remark 1. For the sake of simplicity, we shall write $P^1(v_\nu^2)$ and $P^2(v_\nu^1)$ instead of $P^1(v^2 \circ \chi.\nu)$ and $P^2(v^1 \circ \chi^{-1}.\nu)$, respectively.

Let $S_\alpha : H^{1/2}(\Gamma_c^\alpha) \longrightarrow H^{-1/2}(\Gamma_c^\alpha)$ be the following Steklov-Poincaré operator (see [1]), for any $\varphi \in H^{1/2}(\Gamma_c^\alpha)$

$$S_\alpha\varphi = (\sigma(u^\alpha)\nu^\alpha)\nu^\alpha = \sigma_\nu(u^\alpha) \quad \text{on } \Gamma_c^\alpha, \tag{5}$$

where u^α solves the problem

$$\begin{cases} div(\sigma(u^\alpha)) = 0 & \text{in } \Omega^\alpha, \\ \sigma(u^\alpha)\nu^\alpha = 0 & \text{on } \Gamma_\ell^\alpha, \\ u^\alpha = 0 & \text{on } \Gamma_u^\alpha, \\ \sigma_T(u^\alpha) = 0 & \text{on } \Gamma_c^\alpha, \\ u^\alpha\,\nu^\alpha = \varphi & \text{on } \Gamma_c^\alpha. \end{cases} \tag{6}$$

Finally, with any $\varphi \in H^{1/2}(\Gamma_c^\alpha)$, we associate the closed convex set

$$V_-^\alpha(\varphi) = \{v^\alpha \in V^\alpha / v^\alpha \nu^\alpha \leq \varphi \text{ on } \Gamma_c^\alpha\}.$$

The two-body contact problem (3) is approximated by an iterative procedure involving a contact problem for each body Ω^α with a rigid foundation described by:

Given $g_0^\alpha \in H^{1/2}(\Gamma_c)$, $\alpha = 1,2$. For $m \geq 1$, we build the sequence of functions $(u_m^1)_{m\geq 0} \in V^1$ and $(\mathbf{u}_m^2)_{m\geq 0} \in V^2$ by solving the following problems:

$$\textit{Step 1:}\ \begin{cases} \text{Find } u_m^\alpha \in V_-^\alpha(g_{m-1}^\alpha), \\ a^\alpha(u_m^\alpha, w + P^\alpha(g_{m-1}^\alpha) - u_m^\alpha) \geq L^\alpha(w + P^\alpha(g_{m-1}^\alpha) - u_m^\alpha) \quad \forall w \in V_-^\alpha(0). \end{cases} \tag{7}$$

$$\textit{Step 2:}\ \begin{cases} \text{Find } w_m^1 \in V^1, \\ a^1(w_m^1, v) = -a^2(u_m^2, P^2(v_\nu)) + L^2(P^2(v_\nu)) - a^1(u_m^1, v) + L^1(v) \quad \forall v \in V^1. \\ \text{Find } w_m^2 \in V^2, \\ a^2(w_m^2, v) = a^1(u_m^1, P^1(v_\nu)) - L^1(P^1(v_\nu)) + a^2(u_m^2, v) - L^2(v) \quad \forall v \in V^2. \end{cases} \tag{8}$$

$$\textit{Step 3:}\ \begin{cases} g_m^1 = (1-\theta)g_{m-1}^1 + \theta(w_m^2\nu^2 - u_m^2\nu^2) & \text{on } \Gamma_c^1, \\ g_m^2 = (1-\theta)g_{m-1}^2 + \theta(w_m^1\nu^1 - u_m^1\nu^1) & \text{on } \Gamma_c^2. \end{cases} \tag{9}$$

Theorem 1. *The fixed point of the algorithm* (7)–(9) *is the unique solution of the problem* (3).

Proof. We refer to [10] for the proof of this theorem. □

4 Convergence

The convergence of iterative schemes (7)–(9) is proven by the application of Banach's fixed point theorem to a suitable defined operator. In this following, we reformulate (7)–(9) with operators representation.

In order to decouple the influence of exterior forces and boundary data, we define U^α, $\alpha = 1,2$, as solutions of the problems:

$$\begin{cases} -div(\sigma(U^\alpha)) = f^\alpha & \text{in } \Omega^\alpha, \\ \sigma(U^\alpha)\nu^\alpha = \Phi_\ell^\alpha & \text{on } \Gamma_\ell^\alpha, \\ U^\alpha = 0 & \text{on } \Gamma_u^\alpha, \\ \sigma(U^\alpha)\nu^\alpha = 0 & \text{on } \Gamma_c^\alpha. \end{cases} \tag{10}$$

Moreover, we introduce the operator $Q^\alpha : H^{1/2}(\Gamma_c^\alpha) \longrightarrow H^{1/2}(\Gamma_c^\alpha)$ defined by $Q^\alpha g_{m-1}^\alpha = \tilde{u}_{\nu,m}^\alpha$, $\forall g_{m-1}^\alpha \in H^{1/2}(\Gamma_c^\alpha)$, where $\tilde{u}_m^\alpha$ is the solution of

$$\begin{cases} -div(\sigma(\tilde{u}_m^\alpha)) = 0 & \text{in } \Omega^\alpha, \\ \sigma(\tilde{u}_m^\alpha)\nu^\alpha = 0 & \text{on } \Gamma_\ell^\alpha, \\ \tilde{u}_m^\alpha = 0 & \text{on } \Gamma_u^\alpha, \\ \sigma_T(\tilde{u}_m^\alpha) = 0, \sigma_\nu(\tilde{u}_m^\alpha) \leq 0 & \text{on } \Gamma_c^\alpha, \\ \tilde{u}_m^\alpha \nu^\alpha \leq g_{m-1}^\alpha & \text{on } \Gamma_c^\alpha, \\ \sigma_\nu(\tilde{u}_m^\alpha)(\tilde{u}_m^\alpha \nu^\alpha - g_{m-1}^\alpha) = 0 & \text{on } \Gamma_c^\alpha. \end{cases} \tag{11}$$

Then the solution of the problem (7) can be expressed by

$$u_m^\alpha = U^\alpha + P^\alpha(Q^\alpha g_{m-1}^\alpha). \tag{12}$$

Using the Steklov-Poincaré operator, the Step 2 of (7)–(9) can be written as follows:

$$\begin{cases} w_{\nu,m}^1 = S_1^{-1}(\sigma_\nu(u_m^2) - \sigma_\nu(u_m^1)), \\ w_{\nu,m}^2 = S_2^{-1}(\sigma_\nu(u_m^1) - \sigma_\nu(u_m^2)). \end{cases} \tag{13}$$

Then, we have

$$\begin{cases} w_{\nu,m}^1 = a - (Q^1 g_{m-1}^1 + S_1^{-1} S_2 g_{m-1}^2), \\ w_{\nu,m}^2 = b - (Q^2 g_{m-1}^2 + S_2^{-1} S_1 g_{m-1}^1), \end{cases} \tag{14}$$

where $a = -S_1^{-1}S_2U_\nu^2 - U_\nu^1$ and $b = -S_2^{-1}S_1U_\nu^1 - U_\nu^2$.

From (14), we obtain a new expression of (9)

$$\begin{cases} g_m^1 = (1-\theta)g_{m-1}^1 - \theta(2Q^2 g_{m-1}^2 + S_2^{-1}S_1Q^1 g_{m-1}^1) + \theta b_1, \\ g_m^2 = (1-\theta)g_{m-1}^2 - \theta(2Q^1 g_{m-1}^1 + S_1^{-1}S_2Q^2 g_{m-1}^2) + \theta a_1, \end{cases} \tag{15}$$

with $a_1 = -S_1^{-1}S_2U_\nu^2 - 2U_\nu^1$ and $b_1 = -S_2^{-1}S_1U_\nu^1 - 2U_\nu^2$.

Let us introduce, the operator T defined by

$$\begin{aligned} T : (H^{1/2}(\Gamma_c^\alpha))^2 &\longrightarrow (H^{1/2}(\Gamma_c^\alpha))^2 \\ \mathbf{g} \longmapsto T(\mathbf{g}) &= \begin{pmatrix} w_\nu^2 - u_\nu^2 \\ w_\nu^1 - u_\nu^1 \end{pmatrix} = \begin{pmatrix} -2Q^2 g^2 - S_2^{-1}S_1Q^1 g^1 + b_1 \\ -2Q^1 g^1 - S_1^{-1}S_2Q^2 g^2 + a_1 \end{pmatrix} \end{aligned} \tag{16}$$

and T_θ

$$\begin{aligned} T_\theta : (H^{1/2}(\Gamma_c^\alpha))^2 &\longrightarrow (H^{1/2}(\Gamma_c^\alpha))^2 \\ \mathbf{g} &\longmapsto T_\theta(\mathbf{g}) = (1-\theta)\mathbf{g} + \theta T(\mathbf{g}). \end{aligned} \tag{17}$$

Using the definition of the operators T and T_θ, (15) can be expressed by

$$\mathbf{g}_m = T_\theta(\mathbf{g}_{m-1}) = (1-\theta)\mathbf{g}_{m-1} + \theta T(\mathbf{g}_{m-1}). \tag{18}$$

Theorem 2. *The operator T is a Lipschitz operator.*

Theorem 3. *There exists $\theta_0 \in]0,1[$ such that for θ in $]0,\theta_0[$, the operator T_θ is a contraction in a suitable norm equivalent to the $H^{1/2}(\Gamma_c^\alpha)$-norm.*

Remark 2. To prove Theorems 2 and 3, the properties of the operators S_α, Q^α and P^α are very important. Indeed $S_\alpha : H^{1/2}(\Gamma_c^\alpha) \longrightarrow H^{-1/2}(\Gamma_c^\alpha)$ is bounded, bijective, self-adjoint and coercive. The operator $Q^\alpha : H^{1/2}(\Gamma_c^\alpha) \longrightarrow H^{1/2}(\Gamma_c^\alpha)$ is a Lipschitz operator. $S_\alpha Q^\alpha : H^{1/2}(\Gamma_c^\alpha) \longrightarrow H^{-1/2}(\Gamma_c^\alpha)$ is Lipschitz and monotone. The extension operator $P^\alpha : H^{1/2}(\Gamma_c^\alpha) \longrightarrow P^\alpha(H^{1/2}(\Gamma_c^\alpha))$ is continuous and bijective (see [9]).

5 Numerical Experiments

In this section, we describe some numerical results obtained with algorithm (7)–(9) for various values of the parameter θ and various problem sizes. Our implementation uses a recently developed algorithm of quadratic programming with proportioning and gradient projections [4]. The numerical implementations are performed in Scilab 2.7 on a Pentium 4, 1.80 GHz workstation with 256 MB RAM. We set $tol = 10^{-8}$ and we stop the iterations, if their number is greater than eight hundred. For all experiments to be described below, the stopping criterion of algorithm (7)–(9) is

$$\frac{\|g_m^1 - g_{m-1}^1\|}{\|g_m^1\|} + \frac{\|g_m^2 - g_{m-1}^2\|}{\|g_m^2\|} \leq tol,$$

where $\|\cdot\|$ denotes the Euclidean norm. The precision in the inner iterations are adaptively adjusted by the precision achieved in the outer loop.

Let us consider the plane elastic bodies as in Fig. 1,

$$\Omega^1 = (0,3) \times (1,2) \quad \text{and} \quad \Omega^2 = (0,3) \times (0,1)$$

made of an isotropic, homogeneous material characterized by Young's modulus $E_\alpha = 2.1\ 10^{11}$ and Poisson's ratio $\nu_\alpha = 0.277$. The decomposition of Γ^1 and Γ^2 read as:

$$\Gamma_u^1 = \{0\} \times (1,2), \quad \Gamma_c^1 = (0,3) \times \{1\}, \quad \Gamma_\ell^1 = \Gamma^1 \setminus \overline{\Gamma_u^1 \cup \Gamma_c^1},$$
$$\Gamma_u^2 = \{0\} \times (0,1), \quad \Gamma_c^2 = (0,3) \times \{1\}, \quad \Gamma_\ell^2 = \Gamma^2 \setminus \overline{\Gamma_u^2 \cup \Gamma_c^2}.$$

The volume forces vanish for both bodies. The non-vanishing surface traction $\ell^1 = (\ell_1^1, \ell_2^1)$ and $\ell^2 = (\ell_1^2, \ell_2^2)$ on Γ_ℓ^1 and on Γ_ℓ^2, respectively:

$$\begin{aligned}
&\ell_1^1(s,2) = 0, \quad \ell_2^1(s,2) = -100, \quad s \in (0,3),\\
&\ell_1^1(3,s) = 0, \quad \ell_2^1(3,s) = 0, \quad s \in (1,2),\\
&\ell_1^2(s,0) = 0, \quad \ell_2^2(s,0) = 0, \quad s \in (0,3),\\
&\ell_1^2(3,s) = 0, \quad \ell_2^2(3,s) = 0, \quad s \in (0,1).
\end{aligned}$$

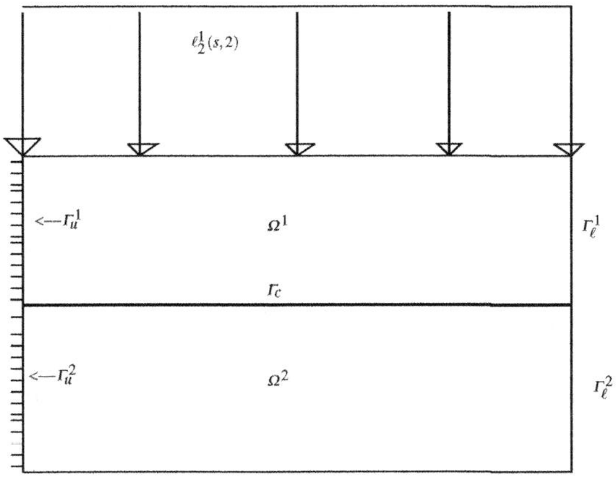

Fig. 1. Setting of the problem.

Fig. 2 illustrates the convergence of the algorithm (7)–(9) for different values of the relaxation parameter θ and various problem sizes with n the number of d.o.f. in $\Omega^1 \cup \Omega^2$ and m the number of d.o.f. on Γ_c^{α}. The results obtained show that the convergence of algorithm (7)–(9) does not depend on the mesh size h. Moreover, this algorithm (7)–9 converges for all $\theta \in]0,1[$.

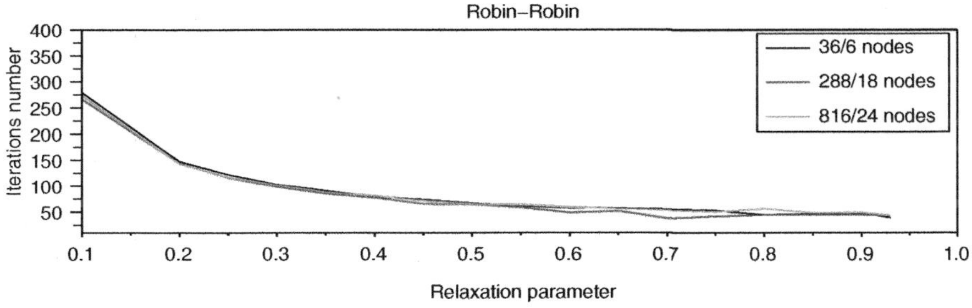

Fig. 2. Convergence rate of the algorithm.

References

[1] Agoshkov, V.I.: *Poincaré-Steklov's operators and domain decomposition methods in finite-dimensional spaces.* First International Symposium on Domain Decomposition Methods for Partial Differential Equations (Paris, 1987), 73–112, SIAM, Philadelphia, PA, 1988.

[2] Bayada, G., Sabil, J., Sassi, T.: *Algorithme de Neumann-Dirichlet pour des problèmes de contact unilatéral: résultat de convergence.* (French)

[Neumann-Dirichlet algorithm for unilateral contact problems: convergence results] C. R. Math. Acad. Sci. Paris 335 (2002), no. 4, 381–386.

[3] Bayada, G., Sabil, J., Sassi, T.: *A Neumann-Neumann domain decomposition algorithm for the Signorini problem.* Appl. Math. Lett. 17 (2004), no. 10, 1153–1159.

[4] Dostal, Z., Schöberl, J.: *Minimizing quadratic functions over non-negative cone with the rate of convergence and finite termination.* Optim. Appl., 30, (2000), no. 1, 23–44.

[5] Duvaut, G., Lions, J.-L.: *Les inéquations en mécanique et en physique*, Dunod, Paris, (1972).

[6] Glowinski, R., Lions, J.-L., Trémolière, R.: *Numerical Analysis of variational Inequalities*, North-Holland, 1981.

[7] Guo, W., Hou, L.S.: *Generalizations and acceleration s of Lions' nonoverlapping domain decomposition method for linear elliptic PDE.* SIAM J. Numer. Anal. 41 (2003), no. 6, 2056–2080

[8] Haslinger, J., Hlavacek, I., Nečas, J.: *Numerical methods for unilateral problems in solid mechanics.* Handbook of numerical analysis, Vol. IV, 313–485, Handb. Numer. Anal., IV, North-Holland, Amsterdam, 1996.

[9] Ipopa, M.A.: *Algorithmes de d'composition de domaine pour les problèmes de contact: Convergence et simulations numériques.* Thesis, Université de Caen, 2008.

[10] Ipopa, M.A., Sassi, T.: *A Robin algorithm for unilateral contact problems.* C.R. Math Acad. Sci. Paris, Ser. I, 346 (2008), 357-362.

[11] Sassi, T., Ipopa, M.A., Roux, F.-X.: *Generalization of Lion's nonoverlapping domain decomposition method for contact problems.* Lect. Notes Comput. Sci. Eng., Vol 60, pp 623-630, 2008.

[12] Kikuchi, N., Oden, J.T.: *Contact problems in elasticity: a study of variational inequalities and finite element methods*, SIAM, Philadelphia, (1988)

[13] Koko, J.: *An Optimization-Based Domain Decomposition Method for a Two-Body Contact Problem.* Num. Funct. Anal. Optim., Vol. 24 , no. 5-6, 587–605, (2003).

[14] Krause, R.H., Wohlmuth, B.I.: *Nonconforming domain decomposition techniques for linear elasticity.* East-West J. Numer. Math. 8 (2000), no. 3, 177–206.

[15] Lions, P.-L.: *On the Schwarz alternating method. III. A variant for nonoverlapping subdomains.* Third International Symposium on Domain Decomposition Methods for Partial Differential Equations (Houston, TX, 1989), 202–223, SIAM, Philadelphia, PA, 1990.

Patch Smoothers for Saddle Point Problems with Applications to PDE-Constrained Optimization Problems

René Simon[1] and Walter Zulehner[2]

[1] SFB 013, Johannes Kepler University Linz, Altenbergerstraße 69, A-4040 Linz, Austria, rene.simon@sfb013.uni-linz.ac.at
[2] Institute of Computational Mathematics, Johannes Kepler University Linz, Altenbergerstraße 69, A-4040 Linz, Austria, zulehner@numa.uni-linz.ac.at

Summary. We consider a multigrid method for solving the discretized optimality system of a PDE-constrained optimization problem. In particular, we discuss the construction of an additive Schwarz-type smoother for a class of elliptic optimal control problems. A rigorous multigrid convergence analysis yields level-independent convergence rates. Numerical experiments indicate that the convergence rates are also independent of the involved regularization parameter.

1 Introduction

In this paper we discuss multigrid methods for solving the discretized optimality system (or Karush-Kuhn-Tucker system, in short KKT system) for optimization problems in function spaces with constraints in form of partial differential equations (PDEs). In particular, we will consider elliptic optimal control problems, see, e.g., [3], and focus on so-called one-shot multigrid methods, see [7], where the multigrid idea is directly applied to the optimality system (instead of a block-wise approach as an alternative).

One of the most important ingredients of such a multigrid method is an appropriate smoother. In this paper we consider patch smoothers: The computational domain is divided into small (overlapping or non-overlapping) sub-domains (patches). One iteration step of the smoothing process consists of solving local problems on each patch one-by-one either in a Jacobi-type or Gauss-Seidel-type manner. This strategy can be seen as an additive or multiplicative Schwarz-type smoother. The technique was successfully used for the Navier-Stokes equations, see [8]. The special case, where each patch consists of a single node of the underlying grid, is usually called a point smoother. Such a smoother was proposed for optimal control problems in [2].

So far, the convergence analysis of multigrid methods with patch smoothers applied to KKT systems of PDE-constrained optimization problems is not as developed as for elliptic PDEs. One line of argument is based on a Fourier analysis, which,

M. Bercovier et al. (eds.), *Domain Decomposition Methods in Science and Engineering XVIII*, Lecture Notes in Computational Science and Engineering,
DOI: 10.1007/978-3-642-02677-5,

strictly speaking, covers only the case of uniform grids with special boundary conditions (and small perturbations of this situation), see [1, 2]. A second and more rigorous strategy exploits the fact that, for certain classes of optimal control problems, the KKT system can be reduced to a compact perturbation of an elliptic system of PDEs. This guarantees the convergence of the multigrid method if the coarse grid is sufficiently fine, see [2]. In [4] the general construction and rigorous analysis of patch smoothers were discussed and applied to the Stokes problem. An extension to KKT systems was presented in [5].

Here we will propose a multigrid method with a patch smoother applied to a reduced system derived from the original KKT system, the same reduced system which was considered in [2]. A rigorous convergence analysis will be presented directly applied to the multigrid method for the reduced system, in contrary what was done in [5]. Compared to the results presented in [5] the numerical experiments show a much better performance of the multigrid method.

In order to keep the notations simple and the strategy transparent the material is presented for a model problem in optimal control only. The extension to more general problems is straight forward.

The paper is organized as follows: In Section 2 the model problem and its discretization are introduced. Section 3 contains the multigrid method, the patch smoother, and the main multigrid convergence result. Finally, in Section 4 some numerical results are presented.

2 An Optimal Control Problem

Let Ω be a bounded convex polygonal domain in $\mathbb{R}^2$. Let $L^2(\Omega)$ and $H^1(\Omega)$ denote the usual Lebesgue space and Sobolev space with norms $\|.\|_{L^2(\Omega)}$ and $\|.\|_{H^1(\Omega)}$, respectively. We consider the following elliptic optimal control problem of tracking type: Find the state $y \in H^1(\Omega)$ and the control $u \in L^2(\Omega)$ such that

$$J(y,u) = \min_{(z,v)\in H^1(\Omega)\times L^2(\Omega)} J(z,v)$$

with cost functional

$$J(z,v) = \frac{1}{2}\|z - y_d\|^2_{L^2(\Omega)} + \frac{\gamma}{2}\|v\|^2_{L^2(\Omega)}$$

subject to the (weak form of the) state equation

$$-\Delta y + y = u \quad \text{in } \Omega, \qquad \frac{\partial y}{\partial n} = 0 \quad \text{on } \Gamma,$$

where Γ denotes the boundary of Ω, $y_d \in L^2(\Omega)$ is the desired state and $\gamma > 0$ is the weight of the cost of the control (or simply a regularization parameter).

By introducing the adjoint state $p \in H^1(\Omega)$ we get the following equivalent optimality system, see e.g., [3]:

1. The adjoint state equation:

$$-\Delta p + p = -(y - y_d) \quad \text{in } \Omega, \qquad \frac{\partial p}{\partial n} = 0 \quad \text{on } \Gamma.$$

2. The control equation:

$$\gamma u - p = 0 \quad \text{in } \Omega.$$

3. The state equation:

$$-\Delta y + y = u \quad \text{in } \Omega, \qquad \frac{\partial y}{\partial n} = 0 \quad \text{on } \Gamma.$$

The control equation yields a simple algebraic relation between the control u and the adjoint state p, which is used to eliminate the control in the state equation. After multiplying by γ we obtain from the state equation:

$$p - \gamma(-\Delta y + y) = 0 \quad \text{in } \Omega, \qquad \frac{\partial y}{\partial n} = 0 \quad \text{on } \Gamma.$$

The weak formulation of the reduced problem in p and y leads to a mixed variational problem: Find $p \in Q = H^1(\Omega)$ and $y \in Y = H^1(\Omega)$ such that

$$\begin{aligned} a(p,q) + b(q,y) &= \langle F,q\rangle \quad \text{for all } q \in Q, \\ b(p,z) - \gamma a(y,z) &= 0 \qquad\quad \text{for all } z \in Y \end{aligned}$$

with

$$a(p,q) = (p,q)_{H^1(\Omega)}, \quad b(q,z) = (q,z)_{L^2(\Omega)}, \quad \langle F,q\rangle = (y_d,q)_{L^2(\Omega)}.$$

Here $(.,.)_H$ denotes the standard scalar product in a Hilbert space H and $\langle .,.\rangle$ is used for the duality product of linear functionals from the dual space H^* and elements in H.

The mixed variational problem can also be written as a variational problem on $Q \times Y$: Find $(p,y) \in Q \times Y$ such that

$$\mathcal{B}((p,y),(q,z)) = \mathcal{F}(q,z) \quad \text{for all } (q,z) \in Q \times Y$$

with the bilinear form

$$\mathcal{B}((p,y),(q,z)) = a(p,q) + b(q,y) + b(p,z) - \gamma a(y,z)$$

and the linear functional

$$\mathcal{F}(q,z) = \langle F,q\rangle.$$

Let $(\mathcal{T}_k)$ be a sequence of triangulations of Ω, where $\mathcal{T}_{k+1}$ is obtained by dividing each triangle into four smaller triangles by connecting the midpoints of the edges of the triangles in $\mathcal{T}_k$. The quantity $\max\{\operatorname{diam} T : T \in \mathcal{T}_k\}$ is denoted by h_k.

We consider the following discretization by continuous and piecewise linear finite elements:

$$Q_k = Y_k = \{w \in C(\bar{\Omega}) : w|_T \in P_1 \text{ for all } T \in \mathcal{T}_k\},$$

where P_1 denotes the polynomials of total degree less or equal to 1. Then we obtain the following discrete variational problem: Find $p_k \in Q_k$ and $y_k \in Y_k$ such that

$$\begin{aligned} a(p_k,q) + b(q,y_k) &= \langle F,q\rangle && \text{for all } q \in Q_k, \\ b(p_k,z) - \gamma a(y_k,z) &= 0 && \text{for all } z \in Y_k. \end{aligned}$$

The discrete mixed variational problem can also be written as a discrete variational problem on $Q_k \times Y_k$: Find $(p_k,y_k) \in Q_k \times Y_k$ such that

$$\mathcal{B}((p_k,y_k),(q,z)) = \mathcal{F}(q,z) \quad \text{for all } (q,z) \in Q_k \times Y_k. \tag{1}$$

By introducing the standard nodal basis for Q_k and Y_k, we finally obtain the following saddle point problem in matrix-vector notation: Find the coefficient vectors $(\underline{p}_k, \underline{y}_k) \in \mathbb{R}^{N_k} \times \mathbb{R}^{N_k}$ such that

$$\mathcal{K}_k \begin{pmatrix} \underline{p}_k \\ \underline{y}_k \end{pmatrix} = \begin{pmatrix} \underline{f}_k \\ 0 \end{pmatrix} \quad \text{with} \quad \mathcal{K}_k = \begin{pmatrix} K_k & M_k \\ M_k & -\gamma K_k \end{pmatrix}. \tag{2}$$

Here N_k denotes the number of nodes of the triangulation $\mathcal{T}_k$, M_k is the mass matrix representing the $L^2(\Omega)$ scalar product on $Y_k = Q_k$, and K_k is the stiffness matrix representing the $H^1(\Omega)$ scalar product on $Y_k = Q_k$.

3 The Multigrid Method

Next we describe the multigrid algorithm: One iteration step for solving (1) at level k is given in the following form:

Let $(p_k^{(0)}, y_k^{(0)}) \in Q_k \times Y_k$ be a given approximation of the exact solution $(p_k,y_k) \in Q_k \times Y_k$ to (1). Then the iteration proceeds in two stages:

1. Smoothing: For $j = 0,1,\ldots,m-1$ compute $(p_k^{(j+1)}, y_k^{(j+1)}) \in Q_k \times Y_k$ by an iterative procedure of the form

$$(p_k^{(j+1)}, y_k^{(j+1)}) = \mathcal{S}_k(p_k^{(j)}, y_k^{(j)}).$$

2. Coarse grid correction: Set

$$\tilde{\mathcal{F}}(q,z) = \mathcal{F}(q,z) - \mathcal{B}((p_k^{(m)}, y_k^{(m)}),(q,z))$$

for $(q,z) \in Q_{k-1} \times Y_{k-1}$ and let $(\tilde{s}_{k-1}, \tilde{r}_{k-1}) \in Q_{k-1} \times Y_{k-1}$ satisfy

$$\mathcal{B}((\tilde{s}_{k-1}, \tilde{r}_{k-1}),(q,z)) = \tilde{\mathcal{F}}(q,z) \quad \text{for all } (q,z) \in Q_{k-1} \times Y_{k-1}. \tag{3}$$

If $k = 1$, compute the exact solution of (3) and set $(s_{k-1}, r_{k-1}) = (\tilde{s}_{k-1}, \tilde{r}_{k-1})$. If $k > 1$, compute approximations (s_{k-1}, r_{k-1}) by applying $\mu \geq 2$ iteration steps of the multigrid algorithm applied to (3) on level $k-1$ with zero starting values. Set

$$(p_k^{(m+1)}, y_k^{(m+1)}) = (p_k^{(m)}, y_k^{(m)}) + (s_{k-1}, r_{k-1}).$$

3.1 The Patch Smoother

We will now define a space decomposition of $\mathbb{R}^{N_k} \times \mathbb{R}^{N_k}$ into N_k subspaces in terms of prolongation matrices $P_{k,i}$ and $Q_{k,i}$, $i = 1, \ldots, N_k$, for the variables p and y, respectively: For each $i \in \{1, \ldots, N_k\}$ representing a node of the triangulation, let $\mathcal{N}_{k,i}$ be the set of all indices consisting of i and the indices of all neighboring nodes (all nodes which are connected to the node with index i by an edge of the triangulation). Then, for each $i \in \{1, \ldots, N_k\}$, the associated local patch consists of all unknowns of $\underline{p}_k$ which are associated to nodes with indices from $\mathcal{N}_{k,i}$ and of the unknown of $\underline{y}_k$ which is associated to the node with index i, see Fig. 1 for an illustration of a local patch. The corresponding canonical embeddings for the variables p and y from the local patches into $\mathbb{R}^{N_k}$ are denoted by $\hat{P}_{k,i}$ and $Q_{k,i}$, respectively.

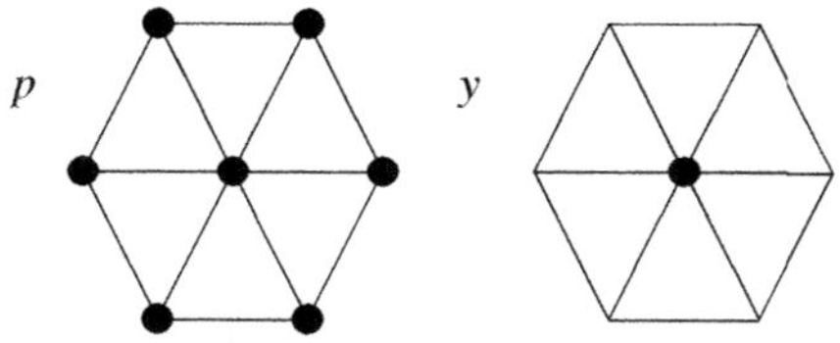

Fig. 1. Local patches

Observe that all entries in $\hat{P}_{k,i}$ and $Q_{k,i}$ are either 0 or 1. A single component of $\underline{y}_k$ belongs to exactly one patch, while a single component of $\underline{p}_k$ belongs, in general, to more than one patch. Let $d_{k,j}$ be the local overlap depth at the node with index j, i.e., the number of all indices l with $j \in \mathcal{N}_{k,l}$, for $j = 1, \ldots, N_k$. Let $\mathcal{D}_k$ be the $N_k \times N_k$ diagonal matrix whose diagonal entries are $d_{k,j}$, $j = 1, \ldots, N_k$. The prolongation matrices $P_{k,i}$ are given by:

$$P_{k,i} = \mathcal{D}_k^{-1/2} \hat{P}_{k,i}.$$

Now we can describe the smoothing procedure: Starting from some approximations $\underline{p}_k^{(j)}$ and $\underline{y}_k^{(j)}$ of the exact solutions $\underline{p}_k$ and $\underline{y}_k$ of (2) we consider iterative methods of form:

$$\underline{p}_k^{(j+1)} = \underline{p}_k^{(j)} + \omega \sum_{i=1}^{N_k} P_{k,i} \underline{s}_{k,i}, \quad \underline{y}_k^{(j+1)} = \underline{y}_k^{(j)} + \omega \sum_{i=1}^{N_k} Q_{k,i} \underline{r}_{k,i},$$

where $(\underline{s}_{k,i}, \underline{r}_{k,i})$ solves a small local saddle point problem of the form

$$\hat{\mathcal{K}}_{k,i} \begin{pmatrix} \underline{s}_{k,i} \\ \underline{r}_{k,i} \end{pmatrix} = \begin{pmatrix} P_{k,i}^T [\underline{f}_k - K_k \underline{p}_k^{(j)} - M_k \underline{y}_k^{(j)}] \\ Q_{k,i}^T [-M_k \underline{p}_k^{(j)} + \gamma K_k \underline{y}_k^{(j)}] \end{pmatrix} \quad \text{for all } i = 1, \ldots, N_k.$$

The local matrix $\hat{\mathcal{K}}_{k,i}$ is given by

$$\hat{\mathcal{K}}_{k,i} = \begin{pmatrix} \hat{K}_{k,i} & M_{k,i}^T \\ M_{k,i} & M_{k,i} \hat{K}_{k,i}^{-1} M_{k,i}^T - \hat{S}_{k,i} \end{pmatrix},$$

where

$$\hat{K}_{k,i} = \hat{P}_{k,i}^T \hat{K}_k \hat{P}_{k,i} \quad \text{with} \quad \hat{K}_k = \frac{1}{\sigma} \operatorname{diag} K_k, \quad M_{k,i} = Q_{k,i}^T M_k \mathcal{D}_k^{1/2} \hat{P}_{k,i}$$

and

$$\hat{S}_{k,i} = \frac{1}{\tau} \left[\gamma Q_{k,i}^T K_k Q_{k,i} + M_{k,i} \hat{K}_{k,i}^{-1} M_{k,i}^T \right].$$

The positive parameters σ and τ have to be chosen such that

$$\hat{K}_k \geq K_k \quad \text{and} \quad \hat{S}_k \geq \gamma K_k + M_k \hat{K}_k^{-1} M_k \tag{4}$$

with

$$\hat{S}_k = \left(\sum_{i=1}^{N_k} Q_{k,i} \hat{S}_{k,i}^{-1} Q_{k,i}^T \right)^{-1}.$$

Observe that there is an additional relaxation factor ω in the smoothing procedure.

For the proposed multigrid method the following convergence result can be shown, see [6]:

Theorem 1. *Let $\omega \in (0,2)$. Then there exists a constant $C > 0$ such that*

$$\|(p_k^{(m+1)} - p_k, y_k^{(m+1)} - y_k)\| \leq C m^{-1/2} \|(p_k^{(0)} - p_k, y_k^{(0)} - y_k)\|,$$

where (p_k, y_k) is the solution of the discrete problem (1), $(p_k^{(0)}, y_k^{(0)})$ is the initial guess, $(p_k^{(m+1)}, y_k^{(m+1)})$ is the result of one multigrid iteration, and the norm is given by

$$\|(q,z)\| = \left(\|q\|_{L^2(\Omega)}^2 + \|z\|_{L^2(\Omega)}^2 \right)^{1/2}.$$

Therefore, the W-cycle multigrid method (i.e. $\mu = 2$) is a contraction with contraction number bounded away from one, independent of the grid level k, if the number m of smoothing steps is sufficiently large.

Remark 1. Observe that the norm used in the last theorem is the L^2-norm, which is weaker than the H^1-norm one would normally expect for the state and the adjoint state.

4 Numerical Experiments

Next we present some numerical results for the domain $\Omega = (0,1) \times (0,1)$ and homogeneous data $y_d = 0$. The initial grid consists of two triangles by connecting the nodes $(0,0)$ and $(1,1)$. For the first series of experiments the regularization parameter γ was set equal to 1. The dependence of the convergence rate on the regularization parameter γ was investigated subsequently.

Randomly chosen starting values for $\underline{p}_k^{(0)}$ and $\underline{y}_k^{(0)}$ for the exact solution $\underline{p}_k = 0$ and $\underline{y}_k = 0$ were used. The discretized problem was solved by a multigrid iteration with a W-cycle ($\mu = 2$) and $m/2$ pre- and $m/2$ post-smoothing steps. The multigrid iteration was performed until the Euclidean norm of the solution was reduced by a factor $\varepsilon = 10^{-8}$. All tests were done with $\sigma = \tau = 0.5$ in order to guarantee (4) and with $\omega = 1.6$, which is motivated by a Fourier analysis on uniform grids.

Table 1 contains the (average) convergence rates q depending on the level k, the total number of unknowns $2N_k$ and the number of smoothing steps, written in the form $m/2$ + $m/2$ (for $m/2$ pre- and $m/2$ post-smoothing steps). It shows a typical multigrid convergence behavior, namely the independence of the grid level and the expected improvement of the rates with an increasing number of smoothing steps.

Table 1. Convergence rates

level k	$2N_k$	1+1	2+2	3+3	5+5
5	2 178	0.301	0.127	0.067	0.023
6	8 450	0.302	0.128	0.066	0.024
7	33 282	0.302	0.135	0.067	0.024
8	132 098	0.302	0.135	0.067	0.024
9	526 338	0.302	0.135	0.068	0.024

Table 2 shows the convergence rates obtained at grid level 7 with 1 pre- and 1 post-smoothing step for values of γ ranging from 1 down to 10^{-6}. Although the analysis presented here does not predict convergence rates that are robust in γ, the numerical experiments indicate robustness with respect to the regularization parameter.

Table 2. Dependence on the regularization parameter γ

γ	1	10^{-2}	10^{-4}	10^{-6}
q	0.302	0.302	0.302	0.302

In summary, the numerical experiments confirm the theoretical results of level-independent convergence rates for the multigrid method with the proposed patch smoother. The convergence rates are much better than in [5] and comparable with the rates presented in [2]. Moreover, they strongly support the conjecture that the convergence rates are also independent of the regularization parameter, as already stated in [2] for the point smoother on the basis of a Fourier analysis on uniform grids.

Acknowledgement. The work was supported by the Austrian Science Fund (FWF) under grant SFB 013/F1309.

References

[1] Arian, E., Ta'asan, S.: Multigrid one-shot methods for optimal control problems: Infinite dimensional control. ICASE-Report 94-52, NASA Langley Research Center, Hampton VA, 1994.

[2] Borzì, A., Kunisch, K., Kwak, D.Y.: Accuracy and convergence properties of the finite difference multigrid solution of an optimal control optimality system. *SIAM J. Control Optim.*, 41(5):1477–1497 (electronic), 2002.

[3] Lions, J.-L.: *Optimal control of systems governed by partial differential equations.* Translated from the French by S. K. Mitter. Die Grundlehren der mathematischen Wissenschaften, 170. Springer, New York, 1971.

[4] Schöberl, J., Zulehner, W.: On Schwarz-type smoothers for saddle point problems. *Numer. Math.*, 95(2):377–399, 2003.

[5] Simon, R., Zulehner, W.: On Schwarz-type Smoothers for Saddle Point Problems with Applications to PDE-constrained Optimization Problems. *Numer. Math.*, 111:445 – 468, 2009.

[6] Simon, R.: *Multigrid solvers for saddle point problems in PDE-constrained optimization.* PhD thesis, Johannes Kepler University Linz, Austria, 2008.

[7] Ta'asan, S.: "One-shot" methods for optimal control of distributed parameter systems I: The finite dimensional case. ICASE-Report 91-2, NASA Langley Research Center, Hampton VA, 1991.

[8] Vanka, S.P.: Block-implicit multigrid solution of Navier-Stokes equations in primitive variables. *J. Comput. Phys.*, 65(1):138–158, 1986.

A Domain Decomposition Preconditioner of Neumann-Neumann Type for the Stokes Equations

Vitorita Dolean[1], Frédéric Nataf[2], and Gerd Rapin[3]

[1] Univ. de Nice Sophia-Antipolis, Laboratoire J.-A. Dieudonné, Nice, France. dolean@math.unice.fr
[2] Laboratoire J. L. Lions, Université Pierre et Marie Curie, 75252 Paris Cedex 05, France, nataf@ann.jussieu.fr
[3] Math. Dep., NAM, University of Göttingen, D-37083, Germany, grapin@math.uni-goettingen.de

Summary. In this paper we recall a new domain decomposition method for the Stokes problem obtained via the Smith factorization. From the theoretical point of view, this domain decomposition method is optimal in the sense that it converges in two iterations for a decomposition into two equal domains. Previous results illustrated the fast convergence of the proposed algorithm in some cases. Our algorithm has shown a more robust behavior than Neumann-Neumann or FETI type methods for particular decompositions; as far as general decompositions are concerned, the performances of the three algorithms are similar. Nevertheless, the computations of the singular values of the interface preconditioned problem have shown that one needs a coarse space whose dimension is less than the one needed for the Neumann-Neumann algorithm. In this work we present a new strategy in order to improve the convergence of the new algorithm in the presence of cross points.

1 Introduction

The last decade has shown that Neumann-Neumann type algorithms, FETI, and BDDC methods are very efficient domain decomposition methods. Most of the early theoretical and numerical work has been carried out for scalar symmetric positive definite second order problems, see for example [5, 10, 11, 17]. Then, the method was extended to different other problems, like the advection-diffusion equations [1, 6], plate and shell problems [20] or the Stokes equations [16, 19]. In the literature one can also find other preconditioners for the Schur complement of the Stokes equations (cf. [2, 19]). Moreover, there exist some Schwarz-type algorithms for non-overlapping decompositions (cf. [13, 14, 15, 18]). Also FETI [8] and BDDC methods [9] are applied to the Stokes problem with success.

Our work is motivated by the fact that in some sense the domain decomposition methods for Stokes are less optimal than the domain decomposition methods for scalar problems. Indeed, in the case of two subdomains consisting of the two

M. Bercovier et al. (eds.), *Domain Decomposition Methods in Science and Engineering XVIII*, Lecture Notes in Computational Science and Engineering,
DOI: 10.1007/978-3-642-02677-5,

half planes it is well known that the Neumann-Neumann preconditioner is an exact preconditioner for the Schur complement equation for scalar equations like the Laplace problem (cf. [17]). A preconditioner is called *exact*, if the preconditioned operator simplifies to the identity. Unfortunately, this does not hold in the vector case. It is shown in [4] that the standard Neumann-Neumann preconditioner for the Stokes equations does not possess this property and the construction of an optimal method is explained. Thus, one can expect a very fast convergence for such an algorithm. Indeed, numerical results clearly support our approach. For an application to the compressible Euler equations see [3].

In Section 2 we recall the domain decomposition method for the Stokes system. Section 3 is dedicated to numerical results for the two-dimensional Stokes problem.

2 DDM for the Stokes Equations

2.1 Stokes Equations

We consider the stationary Stokes problem in a bounded domain $\Omega \subset \mathbb{R}^d$, $d = 2,3$. The Stokes equations are given by a velocity $\mathbf{u}$ and a pressure p satisfying

$$\begin{aligned} -\nu\Delta\mathbf{u} + \nabla p + c\mathbf{u} &= \mathbf{f} \quad \text{in } \Omega, \\ \nabla \cdot \mathbf{u} &= 0 \quad \text{in } \Omega, \end{aligned}$$

and some boundary conditions on $\partial\Omega$. The Stokes problem is a simple model for incompressible flows. The right hand side $\mathbf{f} = (f_1, \dots, f_d)^T \in [L^2(\Omega)]^d$ is a source term, ν is the viscosity and $c \geq 0$ is a constant reaction coefficient. Very often c stems from an implicit time discretization and then c is given by the inverse of the time step size. In the following we denote the d-dimensional Stokes operator by $\mathcal{S}_d(\mathbf{u}, p) := (-\nu\Delta\mathbf{u} + c\mathbf{u} + \nabla p, \nabla \cdot \mathbf{u})$. In the following we will restrict to the two-dimensional case ($d = 2$) but the three-dimensional formulation can be found in [4].

2.2 A New Algorithm for the Stokes Equations

We further introduce the stress depending on a velocity $\mathbf{u} = (u, v)$, a pressure p and the unit normal vector $\mathbf{n}$ on the boundary:

$$\sigma_{\mathbf{n}}(\mathbf{u}, p) := \nu \frac{\partial \mathbf{u}}{\partial \mathbf{n}} - p\mathbf{n}$$

For any vector $\mathbf{u}$, its normal (resp. tangential) component on the interface is $u_{\mathbf{n}_i} = \mathbf{u} \cdot \mathbf{n_i}$ (resp. $u_{\tau_i} = \mathbf{u} \cdot \tau_i$). We denote $\sigma^i_{\mathbf{n}_i}(\mathbf{u}_i, p_i) := \sigma_{\mathbf{n}_i}(\mathbf{u}_i, p_i) \cdot \mathbf{n}_i$ and $\sigma^i_{\tau_i}(\mathbf{u}_i, p_i) := \sigma_{\mathbf{n}_i}(\mathbf{u}_i, p_i) \cdot \tau_i$ as the normal and tangential parts of $\sigma_{\mathbf{n}_i}$, respectively. We now present the new algorithm for the Stokes equations for a general decomposition into non overlapping subdomains $\bar{\Omega} = \cup_{i=1}^N \bar{\Omega}_i$. We denote by Γ_{ij} the interface between subdomains Ω_i and Ω_j, $i \neq j$. The new algorithm for the Stokes system is:

ALGORITHM 1 *Starting with an initial guess* $((\mathbf{u}_i^0, p_i^0))_{i=0}^N$ *satisfying* $\mathbf{u}_{i,\tau_i}^0 = \mathbf{u}_{j,\tau_j}^0$ *and* $\sigma_{\mathbf{n}_i}^i(\mathbf{u}_i^0, p_i^0) = \sigma_{\mathbf{n}_j}^j(\mathbf{u}_j^0, p_j^0)$ *on* Γ_{ij}, $\forall i,j,\ i \neq j$, *the* **correction step** *is expressed as follows for* $1 \leq i \leq N$*:*

$$\begin{cases} \mathcal{S}_2(\tilde{\mathbf{u}}_i^{n+1}, \tilde{p}_i^{n+1}) = 0 & \text{in } \Omega_i \\ \tilde{u}_{i,\mathbf{n}_i}^{n+1} = -\frac{1}{2}(u_{i,\mathbf{n}_i}^n + u_{j,\mathbf{n}_j}^n) & \text{on } \Gamma_{ij} \\ \sigma_{\tau_i}^i(\tilde{\mathbf{u}}_i^{n+1}, \tilde{p}_i^{n+1}) = -\frac{1}{2}(\sigma_{\tau_i}^i(\mathbf{u}_i^n, \tilde{p}_i^n) + \sigma_{\tau_j}^j(\mathbf{u}_j^n, \tilde{p}_j^n)) & \text{on } \Gamma_{ij}, \end{cases} \tag{1}$$

followed by an **updating step** *for* $1 \leq i \leq N$*:*

$$\begin{cases} \mathcal{S}_2(\mathbf{u}_i^{n+1}, p_i^{n+1}) = \mathbf{g} & \text{in } \Omega_i \\ \mathbf{u}_{i,\tau_i}^{n+1} = \mathbf{u}_{i,\tau_i}^n + \frac{1}{2}(\tilde{\mathbf{u}}_{i,\tau_i}^{n+1} + \tilde{\mathbf{u}}_{j,\tau_j}^{n+1}) & \text{on } \Gamma_{ij} \\ \sigma_{\mathbf{n}_i}^i(\mathbf{u}_i^{n+1}, p_i^{n+1}) = \sigma_{\mathbf{n}_i}^i(\mathbf{u}_i^n, p_i^n) & \\ \qquad + \frac{1}{2}(\sigma_{\mathbf{n}_i}^i(\tilde{\mathbf{u}}_i^{n+1}, \tilde{p}_i^{n+1}) + \sigma_{\mathbf{n}_j}^j(\tilde{\mathbf{u}}_j^{n+1}, \tilde{p}_j^{n+1})) & \text{on } \Gamma_{ij}. \end{cases} \tag{2}$$

We have

Proposition 1. *For a domain* $\Omega = \mathbb{R}^2$ *divided into two non overlapping half planes, Algorithm 1 converges in two iterations.*

The new algorithm for the Stokes system is reminiscent of the hybrid approach presented in [7]. Indeed, in both cases, the interface conditions are mixed Dirichlet and Neumann type boundary conditions. However, our approach is different in the sense that it shows the good combination of stress and displacement for the interface conditions in both 2d and 3d (see [4] for details).

3 Numerical Results

In this section we will analyze the performance of the new algorithm in the two-dimensional case. As in [4], we consider the domain $\Omega = (0.2, 1.2) \times (0.1, 1.1)$ decomposed into $N \times N$ subdomains (in the presence of cross points). We choose the right hand side $\mathbf{f}$ such that the exact solution is given by $u(x,y) = \sin(\pi x)^3 \sin(\pi y)^2 \cos(\pi y)$, $v(x,y) = -\sin(\pi x)^2 \sin(\pi y)^3 \cos(\pi x)$ and $p(x,y) = x^2 + y^2$. The viscosity ν is always 1. The interface system is solved by GMRES. In all tables we count the number of iterations needed to reduce the L^∞ norm of the error by the factor $TOL = 10^{-6}$:

$$\max_{i=1,\ldots,N} \|U_k^i - U_h\|_{L^\infty(\Omega_i)} \leq 10^{-6},$$

where $U_k^i = (u_k, v_k, p_k)^i$ is the discrete solution of iteration step k in subdomain Ω_i and $U_h = (u_h, v_h, p_h)$ is the global discrete solution computed by a direct solver applied to the global problem.

A problem in Algorithm 1 is that in the correction step, the local matrices may be singular (the local problems are ill-posed for the pressure, the latter being defined up to an additive constant). To overcome this difficulty we chose to add a penalization term εp with ε sufficiently small to the divergence equation. This penalization term leads however still to ill-conditioned local matrices and an ill-conditioned interface problem. Thus, the reduction of the Euclidean norm of the residual is not a good indicator for the convergence of the algorithm as can be seen in Fig. 1:

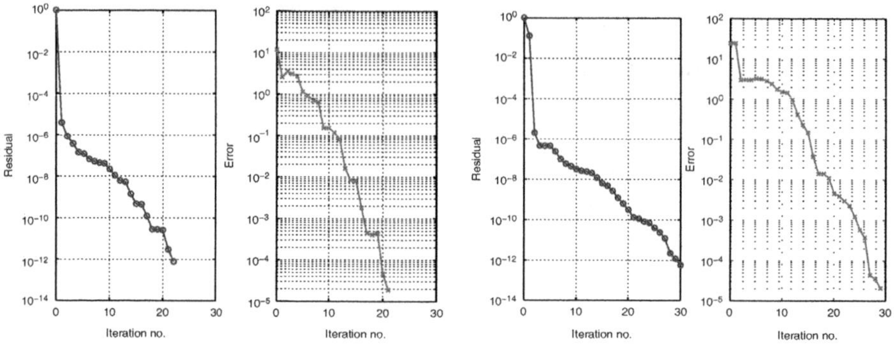

Fig. 1. Convergence of the GMRES algorithm (residual and error) for 3×3 (left) and 4×4 (right) decompositions for $\varepsilon > 0$.

This is also due to the presence of the large eigenvalues in the spectrum, as seen in the Table 1.

$N\times N$	No. of large eigenvalues	$N\times N$	No. of large eigenvalues
2x2	0	6x6	16
3x3	1	7x7	25
4x4	4	8x8	36
5x5	9	9x9	49

Table 1. Number of eigenvalues which are larger than 10 in modulus for a $N\times N$ decomposition.

A very simple way to eliminate the large eigenvalues is to avoid using the penalization term: the local problems are now singular. Consider a local matrix A which corresponds to interior subdomains in the correction (preconditioning) step. It will be singular of co-rank 1. The null space is formed by a vector whose components are constant non-zero only for the pressure componen ts. It can be easily shown that the matrix $B+f\cdot e^t$ is invertible if we choose e (resp. f) to be a vector non-orthogonal to $ker(A)$ (resp. $ker(A^T)$). In our case it is sufficient to take (in order to preserve the sparsity of the matrix A) a vector with null components except for one non-zero component chosen in the right position. Afterward, for any right hand side b in the $Im(A)$, the solution of $Bx=b$ verifies $Ax=b$. In this case no more large eigenvalues will be present in the spectrum and the convergence of the residual will reflect more accurately the convergence of the error as one can see in Fig. 2.

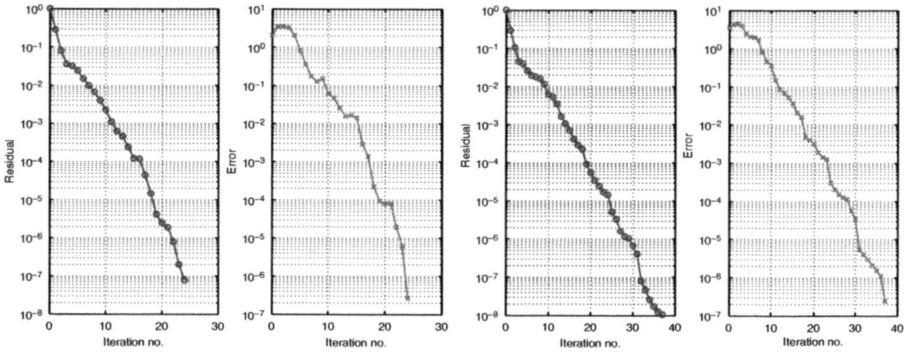

Fig. 2. Convergence of the GMRES algorithm (residual and error) for 3×3 (left) and 4×4 (right) decompositions for $\varepsilon = 0$.

Nevertheless the convergence is still very sensitive to the number of subdomains, which shows the necessity of introducing a coarse space correction in the algorithm. This bad convergence is mainly due to the presence of small eigenvalues in the spectrum of the interface operator (see Fig. 3).

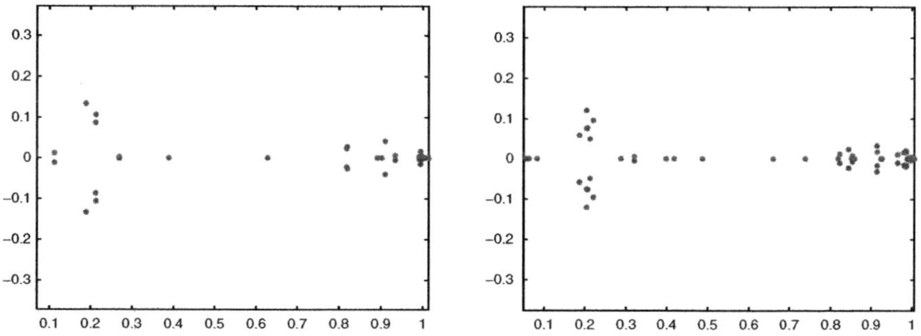

Fig. 3. Eigenvalues of the interface preconditioned operator for 3×3 (left) and 4×4 (right) decompositions.

We need to eliminate the small eigenvalues which can cause bad convergence. In order to do this we will first notice that the error during the iterations of the GMRES method is mainly localized in the corners, as seen in Fig. 4 where the error on component p is visualized.

A solution to this problem could be a deflation method applied to the preconditioning step as seen in [12], where the deflated vectors contain constant non-zero

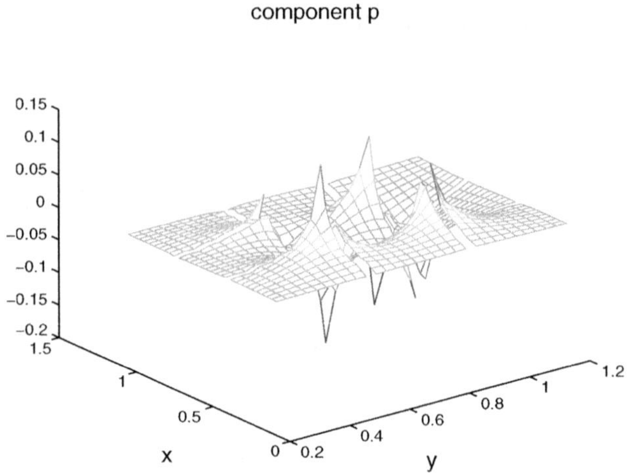

Fig. 4. Error on the component p of the solution.

elements only for the corner components of the solution. As a result, we obtain a better convergence than before. It is however not optimal, since it is dependent on the number of subdomains (see Fig. 5).

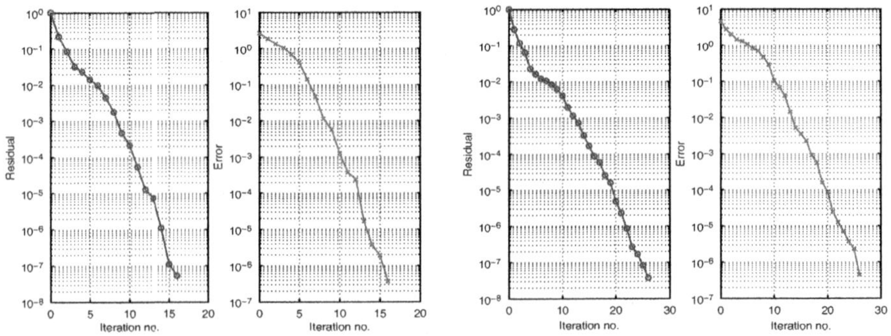

Fig. 5. Convergence of the deflated GMRES algorithm (residual and error) for 3×3 (left) and 4×4 (right) decompositions.

By looking at the spectrum (Fig. 6), we can see that there are still small eigenvalues that have not been taken care of by the deflation method.

As a conclusion we can state that even if the strategy presented is not yet optimal, it leads to an improvement of the previous algorithm, since it eliminates a part of the small eigenvalues, and could thus pave the way to the construction of a more scalable method.

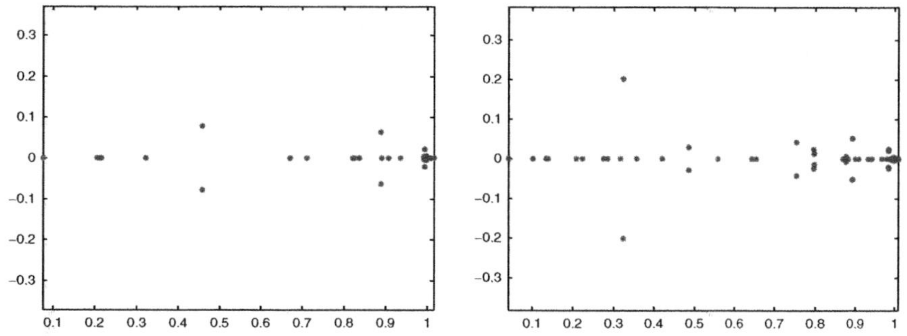

Fig. 6. Eigenvalues of the interface preconditioned operator for 3×3 (left) and 4×4 (right) decompositions.

References

[1] Achdou, Y., Le Tallec, P., Nataf, F., Vidrascu, M.: A domain decomposition preconditioner for an advection-diffusion problem. *Comput. Methods Appl. Mech. Engrg.*, 184:145–170, 2000.

[2] Ainsworth, M., Sherwin, S.: Domain decomposition preconditioners for p and hp finite element approximations of Stokes equations. *Comput. Methods Appl. Mech. Engrg.*, 175:243–266, 1999.

[3] Dolean, V., Nataf, F.: A New Domain Decomposition Method for the Compressible Euler Equations. *M2AN Math. Model. Numer. Anal.*, 40(4):689–703, 2006.

[4] Dolean, V., Nataf, F., Rapin, G.: Deriving a new domain decomposition method for the Stokes equations using the Smith factorization. *Math. Comp.*, 78: 789–814, 2009.

[5] Farhat, C., Roux, F.-X.: A Method of Finite Element Tearing and Interconnecting and its Parallel Solution Algorithm. *Internat. J. Numer. Methods Engrg.*, 32:1205–1227, 1991.

[6] Gerardo-Giorda, L., Le Tallec, P., Nataf, F.: A Robin-Robin preconditioner for advection-diffusion equations with discontinuous coefficients. *Comput. Methods Appl. Mech. Engrg.*, 193:745–764, 2004.

[7] Gosselet, P., Rey, C.: Non-overlapping domain decomposition methods in structural mechanics. *Arch. Comput. Methods Engrg.*, 13(4):515–572, 2006.

[8] Li, J.: A Dual-Primal FETI method for incompressible Stokes equations. *Numer. Math.*, 102:257–275, 2005.

[9] Li, J., Widlund, O.: BDDC algorithms for incompressible Stokes equations. *SIAM J. Numer. Anal.*, 44(6):2432–2455, 2006.

[10] Mandel, J.: Balancing domain decomposition. *Comm. Appl. Numer. Methods*, 9:233–241, 1992.

[11] Mandel, J., Brezina, M.: Balancing domain decomposition: Theory and performance in two and three dimensions. UCD/CCM report 2, 1993.

[12] Nabben, R., Vuik, C.: A comparison of abstract versions of deflation, balancing and additive coarse grid correction preconditioners. *Numer. Linear Algebra Appl.*, 15:355–372, 2008.

[13] Nataf, F.: Interface Conditions for Domain Decomposition Methods for 2D and 3D Oseen equations. *C. R. Acad. Sci. Paris Sér. I Math.* 324:1155–1160, 1997.

[14] Otto, F.-C., Lube, G.: A nonoverlapping domain decomposition method for the Oseen equations. *Math. Models Methods Appl. Sci.*, 8:1091–1117, 1998.

[15] Otto, F.-C., Lube, G., Müller, L.: An iterative substructuring method for div-stable finite element approximations of the Oseen problem. *Computing*, 67: 91–117, 2001.

[16] Pavarino, L.F., Widlund, O.B.: Balancing Neumann-Neumann methods for incompressible Stokes equations. *Comm. Pure Appl. Math.*, 55:302–335, 2002.

[17] De Roeck, Y.H., Le Tallec, P.: Analysis and Test of a Local Domain Decomposition Preconditioner. In R. Glowinski and Y.A. Kuznetsov and G. Meurant and J. Periaux and O.B. Widlund eds., *Fourth International Symposium on Domain Decomposition Methods for Partial Differential Equations*, pages 112–128, 1991.

[18] Rønquist, E.: A Domain Decomposition Solver for the Steady Navier-Stokes Equations. In A. Ilin and L. Scott, eds., *Proc. Internat. Conf. on Spectral and High-Order Methods '95*, pages 469–485.

[19] Le Tallec, P., Patra, A.: Non-overlapping domain decomposition methods for adaptive hp approximations of the Stokes problem with discontinuous pressure fields. *Comput. Methods Appl. Mech. Engrg.*, 145:361–379, 1997.

[20] Le Tallec, P., Mandel, J., Vidrascu, M.: A Neumann-Neumann Domain Decomposition Algorithm for Solving Plate and Shell Problems. *SIAM J. Numer. Anal.*, 35:836–867, 1998.

Non-overlapping Domain Decomposition for the Richards Equation via Superposition Operators

Heiko Berninger

Fachbereich Mathematik und Informatik, Freie Universität Berlin
berninger@math.fu-berlin.de

Summary. Simulations of saturated-unsaturated groundwater flow in heterogeneous soil can be carried out by considering non-overlapping domain decomposition problems for the Richards equation in subdomains with homogeneous soil. By the application of different Kirchhoff transformations in the different subdomains local convex minimization problems can be obtained which are coupled via superposition operators on the interface between the subdomains. The purpose of this article is to provide a rigorous mathematical foundation for this reformulation in a weak sense. In particular, this involves an analysis of the Kirchhoff transformation as a superposition operator on Sobolev and trace spaces.

1 Introduction

The Richards equation, which describes saturated-unsaturated fluid flow in a homogeneous porous medium, reads

$$n\theta(p)_t - \operatorname{div}(K_h kr(\theta(p))(\nabla p - z)) = 0. \tag{1}$$

The unknown water or capillary pressure p, given as the height of a corresponding water column, is a function on $\Omega \times (0,T)$ for a time $T > 0$ and a domain $\Omega \subset \mathbb{R}^d$ ($d = 1,2,3$) inhibited by the porous medium. The function $n : \Omega \to (0,1)$ is the porosity of the soil, $K_h : \Omega \to \mathbb{R}^+$ is the hydraulic conductivity and z is the coordinate in the direction of gravity.

The saturation $\theta : \mathbb{R} \to [\theta_m, \theta_M]$ with $\theta_m, \theta_M \in [0,1]$ is an increasing function of p with $\theta(p) = \theta_M$ (the case of full saturation and ellipticity of (1)) if p is large enough. The relative permeability $kr : [\theta_m, \theta_M] \to [0,1]$ is an increasing function of θ with $kr(\theta_m) = 0$ (degeneracy in (1)) and $kr(\theta_M) = 1$. In this way the Richards equation contains the generalized law of Darcy

$$\mathbf{v} = -K_h kr(\theta(p))(\nabla p - z),$$

for the water flux $\mathbf{v}$. Typical shapes of the nonlinearities θ and kr are depicted in Figs. (a) and (b). However, these functions depend on the soil type so that we

M. Bercovier et al. (eds.), *Domain Decomposition Methods in Science and Engineering XVIII*, Lecture Notes in Computational Science and Engineering,
DOI: 10.1007/978-3-642-02677-5, © Springer-Verlag Berlin Heidelberg 2009

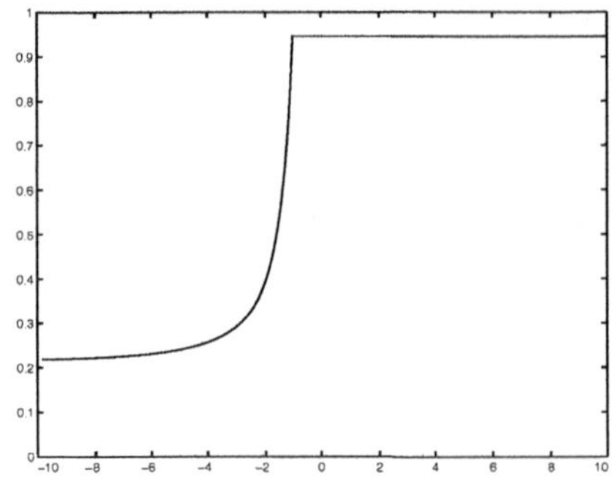

(a) $p \mapsto \theta(p)$

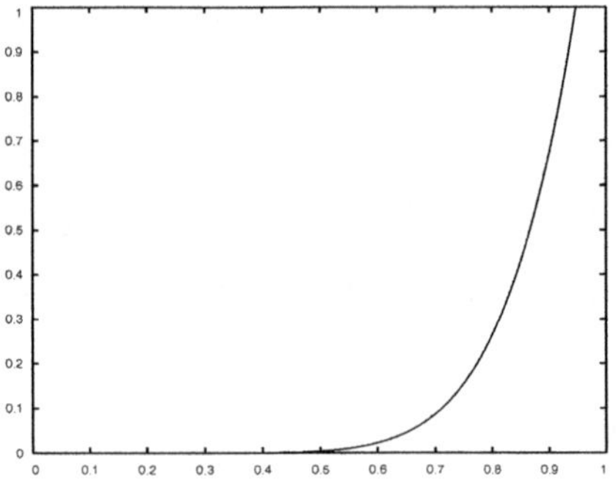

(b) $\theta \mapsto kr(\theta)$

have different nonlinearities θ_i, kr_i on different non-overlapping subdomains Ω_i, $i = 1,\ldots,N \in \mathbb{N}$, constituting a decomposition of Ω.

In the following, we assume $n = K_h = 1$ and $N = 2$ for simplicity. See Figure 1 for a decomposition of Ω into Ω_1 and Ω_2 where $\mathbf{n}$ denotes the outer normal of Ω_1. Moreover, we assume that (1) is discretized implicitly in time but with an explicit

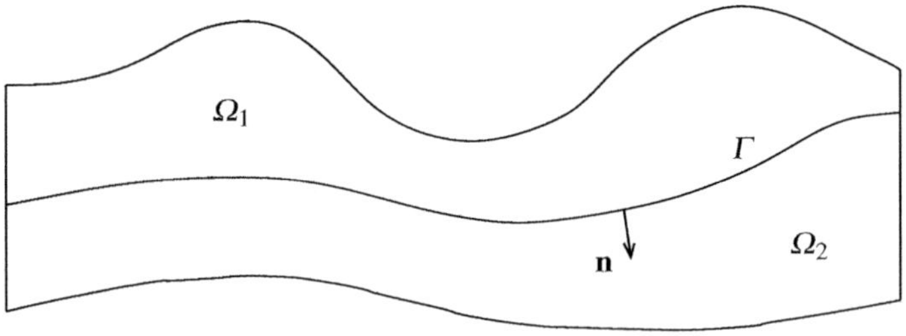

Fig. 1. 2D-domain Ω decomposed into two subdomains.

treatment of the gravitational (convective) term so that with a suitable function f on Ω we arrive at spatial problems of the form

$$\theta_i(p_i) - \operatorname{div}(kr_i(\theta_i(p_i))\nabla p_i) = f \quad \text{on } \Omega_i, \quad i = 1,2. \tag{2}$$

Appropriate interface conditions on $\Gamma := \overline{\Omega}_1 \cap \overline{\Omega}_2$, which are motivated hydrologically, are the continuity of the pressure and the normal water flux $\mathbf{v} \cdot \mathbf{n}$ across Γ. After our implicit–explicit time discretization, this leads to

$$p_1 = p_2 \qquad \text{on } \Gamma, \tag{3}$$

$$kr_1(\theta_1(p_1))\nabla p_1 \cdot \mathbf{n} = kr_2(\theta_2(p_2))\nabla p_2 \cdot \mathbf{n} \quad \text{on } \Gamma. \tag{4}$$

In case of $\theta_1 = \theta_2$ and $kr_1 = kr_2$, these interface conditions can be mathematically derived in a weak sense (and in a very general setting) as a multi-domain formulation for the corresponding global problem, see [2, pp. 131–139].

A powerful tool for the treatment of the Richards equation is Kirchhoff's transformation. It leads to spatial convex minimization problems after time discretization (see [2] for details). Here, we need to apply two different Kirchhoff transformations in the two subdomains. More concretely, we define

$$u_i(x) := \kappa_i(p_i(x)) = \int_0^{p_i(x)} kr_i(\theta_i(q))\,dq \quad \text{a.e. on } \Omega_i\,, \quad i = 1,2\,. \tag{5}$$

Consequently, we obtain

$$kr_i(\theta_i(p_i))\nabla p_i = \nabla u_i\,, \quad i = 1,2, \tag{6}$$

by the chain rule so that with the saturation

$$M_i(u_i) = \theta_i(\kappa_i^{-1}(u_i))\,, \quad i = 1,2, \tag{7}$$

with respect to the new variables the equations (2) are transformed into

$$M_i(u_i) - \Delta u_i = f \quad \text{on } \Omega_i\,, \quad i = 1,2\,. \tag{8}$$

Moreover, the Kirchhoff–transformed interface conditions read

$$\kappa_1^{-1}(u_1) = \kappa_2^{-1}(u_2) \quad \text{on } \Gamma\,, \tag{9}$$

$$\nabla u_1 \cdot \mathbf{n} = \nabla u_2 \cdot \mathbf{n} \quad \text{on } \Gamma\,. \tag{10}$$

Accordingly, boundary conditions on $\partial\Omega$ for (1) and (2) are transformed.

Applying Kirchhoff's transformation is straightforward in the strong formulations above. However, regarding the weak forms, the proof for the equivalence of the physical and the transformed versions is more sophisticated. For example, we need the chain rule (6) in a weak sense in $H^1(\Omega_i)$. Furthermore, $\kappa_i^{-1}(u_i)$, $i = 1,2$, in (9) has to be understood as an element of some trace space. In order to clarify these issues, which already occur in case of a single domain, one has to study the Kirchhoff transformation as a superposition operator in Sobolev and trace spaces. This is the purpose of this paper.

Concretely, we present weak forms of the domain decomposition problems for the time-discretized Richards equation and its transformed version in Section 2. Then we carry out some analysis for the Kirchhoff transformation as a superposition operator in Section 3. Finally, the obtained results are exploited to prove the equivalence of the weak formulations in Section 4.

2 Weak Forms of the Domain Decomposition Problems

In this section we give variational formulations of the domain decomposition problems (2)–(4) and (8)–(10) with homogeneous Dirichlet boundary conditions (compare [3]). We start with some notation and assumptions.

We require $kr_i \in L^\infty(\mathbb{R})$ with $kr_i \geq \alpha$ for some $\alpha > 0$ and $i = 1,2$. (For the general case $\alpha = 0$ as in Figs. 1(a) and 1(b), the results are weaker; see [2, Sec. 1.5.4]). Let θ_i, $i = 1,2$, be bounded Borel–measurable functions on $\mathbb{R}$ and $f \in L^2(\Omega)$. Furthermore, in a decomposition as above, let Ω and Ω_i, $i = 1,2$, be bounded Lipschitz domains in $\mathbb{R}^d$ and Γ a Lipschitz $(d-1)$-dimensional manifold. Now we introduce the spaces

$$V_i := \{v_i \in H^1(\Omega_i) | \, v_{i|\partial\Omega\cap\partial\Omega_i} = 0\}, \quad V_i^0 := H_0^1(\Omega_i), \quad \Lambda := \{v_{|\Gamma} : v \in H_0^1(\Omega)\},$$

and for $w_i, v_i \in V_i$, the forms

$$a_i(w_i, v_i) := (\nabla w_i, \nabla v_i)_{\Omega_i}, \qquad b_i(w_i, v_i) := (kr_i(\theta_i(w_i))\nabla w_i, \nabla v_i)_{\Omega_i},$$

where $(\cdot,\cdot)_{\Omega_i}$ stands for the L^2–scalar product on Ω_i. The norm in $H^1(\Omega)$ will be denoted by $\|\cdot\|_1$. Recall that the trace space Λ is either $H_{00}^{1/2}(\Gamma)$ in case of $\Gamma \cap \partial\Omega \neq \emptyset$ (as in Figure 1) or $H^{1/2}(\Gamma)$ otherwise [8, p. 7]. The restriction $w_{i|\Gamma}$ of a function $w_i \in V_i$ on the interface Γ has to be understood as the application of the corresponding trace operator on w_i.

Finally, let R_i, $i = 1,2$, be any continuous extension operator from Λ to V_i. Then the variational formulation of problem (2)–(4) with homogeneous Dirichlet boundary conditions reads as follows:

Find $p_i \in V_i$, $i = 1,2$, such that

$$(\theta_i(p_i), v_i)_{\Omega_i} + b_i(p_i, v_i) = (f, v_i)_{\Omega_i} \quad \forall v_i \in V_i^0, \; i = 1,2, \tag{11}$$

$$p_{1|\Gamma} = p_{2|\Gamma} \qquad \text{in } \Lambda\,, \tag{12}$$

$$\begin{aligned}(\theta_1(p_1), R_1\mu)_{\Omega_1} + b_1(p_1, R_1\mu) - (f, R_1\mu)_{\Omega_1} = \\ -(\theta_2(p_2), R_2\mu)_{\Omega_2} - b_2(p_2, R_2\mu) + (f, R_2\mu)_{\Omega_2} \quad \forall \mu \in \Lambda\,.\end{aligned} \tag{13}$$

Analogously, the weak formulation of the transformed problem (8)–(10) with homogeneous Dirichlet boundary conditions reads:

Find $u_i \in V_i, i = 1,2$, such that

$$(M_i(u_i), v_i)_{\Omega_i} + a_i(u_i, v_i) = (f, v_i)_{\Omega_i} \quad \forall v_i \in V_i^0, \; i = 1,2, \tag{14}$$

$$\kappa_1^{-1}(u_{1|\Gamma}) = \kappa_2^{-1}(u_{2|\Gamma}) \text{in } \Lambda\,, \tag{15}$$

$$\begin{aligned}(M_1(u_1), R_1\mu)_{\Omega_1} + a_1(u_1, R_1\mu) - (f, R_1\mu)_{\Omega_1} = \\ -(M_2(u_2), R_2\mu)_{\Omega_2} - a_2(u_2, R_2\mu) + (f, R_2\mu)_{\Omega_2} \quad \forall \mu \in \Lambda\,.\end{aligned} \tag{16}$$

The rest of this paper is devoted to prove the equivalence of the variational formulations (11)–(13) and (14)–(16).

3 Kirchhoff Transformation as a Superposition Operator

The difficulties encountered to prove the equivalence of the weak forms in physical and in transformed variables already occur for a single domain. Therefore, we omit the indices $i \in \{1,2\}$ in this section in which we want to address these difficulties. We start with an important definition [1].

Definition 1. *Let p be a real-valued function defined on a subset $S \subset \mathbb{R}^d$, possibly almost everywhere w.r.t. an appropriate measure. Furthermore, let $\kappa : \mathbb{R} \to \mathbb{R}$ be a real function. The* superposition operator *(or* Nemytskij operator*)* $\kappa_S : p \mapsto \kappa(p)$ *is defined by pointwise application*

$$(\kappa_S(p))(x) := \kappa(p(x)),$$

of κ to p (for x almost everywhere) on S. Let X be a normed space consisting of a subset of all measurable functions on S. If the superposition operator satisfies $\kappa_S(p) \in X$ for all $p \in X$, we say that it acts on *the space X. In this case we write $\kappa_X : X \to X$ for the restriction of κ_S on the space X and call κ_X superposition operator on X (induced by κ).*

Here, S will be either Ω or a submanifold Σ of $\partial\Omega$. If not otherwise stated, we assume the conditions listed at the beginning of Section 2 and the Kirchhoff transformation κ given as in (5). We begin by stating the weak chain rule which goes back to J. Serrin (see [5]). Recall that $\kappa' = kr \circ \theta \in L^\infty(\mathbb{R})$ holds for any Lipschitz continuous function $\kappa : \mathbb{R} \to \mathbb{R}$ due to the fundamental theorem of calculus.

Theorem 1. *If $\kappa : \mathbb{R} \to \mathbb{R}$ is Lipschitz continuous then the weak chain rule*

$$\kappa'(p)\nabla p = \nabla(\kappa(p)) \quad \text{a.e. on } \Omega,$$

holds for any $p \in W^{1,1}_{\text{loc}}(\Omega)$ provided $\kappa'(p(x))\nabla p(x)$ is interpreted as 0 whenever $\nabla p(x) = 0$.

We remark that the last condition is an essential part of the theorem since $\kappa'(p(x))$ does not have to be defined for any $x \in \Omega$. Indeed, for $kr \in L^\infty(\mathbb{R})$ the composition $kr \circ \theta(p)$ alone does not make sense for $p \in W^{1,1}_{\text{loc}}(\Omega)$ since it depends on the choice of the representative in the equivalence class kr.

The next lemma is not hard to prove (see [2, Sec. 1.5.4]), however, we must apply the weak chain rule twice in order to obtain (iii).

Lemma 1. *The Kirchhoff transformation κ has the following properties.*

(i) $\kappa : \mathbb{R} \to \mathbb{R}$ *is Lipschitz continuous and has a Lipschitz continuous inverse.*
(ii) $\kappa : \mathbb{R} \to \mathbb{R}$ *and* $\kappa^{-1} : \mathbb{R} \to \mathbb{R}$ *induce Lipschitz continuous superposition operators acting on $L^2(\Omega)$ and on $L^2(\Sigma)$ for any submanifold $\Sigma \subset \partial\Omega$.*
(iii) $\kappa : \mathbb{R} \to \mathbb{R}$ *induces an invertible superposition operator on $H^1(\Omega)$ with*

$$\alpha^{-1}\|p\|_1 \le \|\kappa(p)\|_1 \le \|kr \circ \theta\|_\infty \|p\|_1 \quad \forall p \in H^1(\Omega).$$

By imposing further conditions on the function $kr \circ \theta$, e.g. its boundedness and uniform continuity, the continuity of the superposition operator $\kappa_{H^1(\Omega)}$ can be proved by elementary means (compare [2, Prop. 1.5.14]) — if one assumes $kr \circ \theta$ to be Lipschitz continuous, one even obtains local Lipschitz continuity of $\kappa_{H^1(\Omega)}$ in one space dimension.

The following remarkable characterization of superposition operators acting on $H^1(\Omega)$, however, is a quite profound result, see Marcus and Mizel [6, 7].

Theorem 2. *Let $\Omega \subset \mathbb{R}^d$ be a bounded open set and $\kappa : \mathbb{R} \to \mathbb{R}$ a Borel function. The superposition operator κ_Ω acts on $H^1(\Omega)$ if and only if it is continuous on $H^1(\Omega)$ or, equivalently, if and only if κ is Lipschitz continuous for $d > 1$ or locally Lipschitz in the case $d = 1$, respectively.*

The following proposition contains an important commutativity result. Strangely enough, in order to derive this algebraic property, it seems necessary to assume the continuity of $\kappa_{H^1(\Omega)}$. In the proof we also apply the well-known trace theorem for trace operators $tr_\Sigma : H^1(\Omega) \to H^{1/2}(\Sigma)$ (compare e.g. [4, pp. 1.61, 1.65]).

Proposition 1. *For a submanifold $\Sigma \subset \partial\Omega$ and κ as in Theorem 2, we have the commutativity*

$$\kappa_\Sigma(tr_\Sigma v) = tr_\Sigma(\kappa_\Omega v) \quad \forall v \in H^1(\Omega). \tag{17}$$

Proof. We prove that for any $v \in H^1(\Omega)$

$$\|tr_\Sigma(\kappa_\Omega v) - \kappa_\Sigma(tr_\Sigma v)\|_{L^2(\Omega)} \tag{18}$$

is arbitrarily small by considering a sequence $(v_n)_{n\in\mathbb{N}} \subset C^\infty(\overline{\Omega})$ converging to v in $H^1(\Omega)$. In fact, since Theorem 2 provides the continuity of κ and the trace of a continuous function on Σ coincides with its restriction to Σ, the norm in (18) can be estimated by

$$\|tr_\Sigma(\kappa_\Omega v) - (\kappa_\Omega v_n)_{|\Sigma}\|_{L^2(\Omega)} + \|\kappa_\Sigma(v_{n|\Sigma}) - \kappa_\Sigma(tr_\Sigma v)\|_{L^2(\Omega)}\,. \tag{19}$$

The first term in (19) is at most

$$\|tr_\Sigma\|\,\|\kappa_\Omega v - \kappa_\Omega v_n\|_1\,,$$

due to the trace theorem, and this estimate goes to 0 for $n \to \infty$ by the continuity of $\kappa_{H^1(\Omega)}$. For $d > 1$ where $\kappa : \mathbb{R} \to \mathbb{R}$ is Lipschitz continuous, the second term in (19) can be estimated by

$$L(\kappa_{L^2(\Sigma)})\,\|v_{n|\Sigma} - tr_\Sigma v\|_{L^2(\Sigma)} \le L(\kappa_{L^2(\Sigma)})\,\|tr_\Sigma\|\,\|v_n - v\|_1\,,$$

with Lemma 1 (ii) ($L(\kappa_{L^2(\Sigma)})$ denotes the Lipschitz constant of $\kappa_{L^2(\Sigma)}$) and the trace theorem and, therefore, tends to 0 for $n \to \infty$, too. In one space dimension, (17) is clear since both κ (Theorem 2) and v (Sobolev's embedding theorem) are continuous. □

Note that Proposition 1 does not guarantee $\kappa_\Sigma(tr_\Sigma v) \in H_{00}^{1/2}(\Sigma)$ for $tr_\Sigma v \in H_{00}^{1/2}(\Sigma)$. However, we even have

Proposition 2. *For a submanifold $\Sigma \subset \partial\Omega$ the function κ as in Theorem 2 induces a continuous superposition operator on $H^{1/2}(\Sigma)$ and, if $\kappa(0) = (0)$, on $H_{00}^{1/2}(\Sigma)$, too.*

Proof. With the continuous extension operator $R_\Sigma : H^{1/2}(\Sigma) \to H^1(\Omega)$ given by the trace theorem and using Proposition 1, we can write

$$\kappa_\Sigma = \kappa_\Sigma \circ tr_\Sigma \circ R_\Sigma = tr_\Sigma \circ \kappa_{H^1(\Omega)} \circ R_\Sigma ,$$

and the operator on the right hand side is a composition of continuous operators which obviously acts on $H^{1/2}(\Sigma)$.

Regarding the second case we recall (see [4, p. 1.60]) that $H_{00}^{1/2}(\Sigma)$ is the space of all functions $\mu \in H^{1/2}(\Sigma)$ allowing trivial extensions $\tilde{\mu} \in H^{1/2}(\partial\Omega)$ with the norm

$$\|\mu\|_{H_{00}^{1/2}(\Sigma)} = \|\tilde{\mu}\|_{H^{1/2}(\partial\Omega)} . \tag{20}$$

Now, let $\eta \in H_{00}^{1/2}(\Sigma)$ and $\tilde{\eta}$ be a trivial extension of η in $H^{1/2}(\partial\Omega)$. Then, since $\kappa(0) = 0$ and $\kappa_{\partial\Omega}$ acts on the space $H^{1/2}(\partial\Omega)$, we can conclude $\kappa_{\partial\Omega}(\tilde{\eta}) \in H^{1/2}(\partial\Omega)$ and $\kappa_{\partial\Omega}(\tilde{\eta})_{|\Sigma}$ is a trivial extension of $\kappa_\Sigma(\eta) \in H^{1/2}(\Sigma)$, i.e. by definition $\kappa_\Sigma(\eta) \in H_{00}^{1/2}(\Sigma)$. Moreover, if $\mu \in H_{00}^{1/2}(\Sigma)$ is treated as η, then $\kappa_{\partial\Omega}(\tilde{\eta}) - \kappa_{\partial\Omega}(\tilde{\mu}) \in H^{1/2}(\partial\Omega)$ is a trivial extension of $\kappa_\Sigma(\eta) - \kappa_\Sigma(\mu) \in H_{00}^{1/2}(\Sigma)$. Now, (20) and the continuity of $\kappa_{\partial\Omega}$ provide that, for any $\varepsilon > 0$, we have

$$\|\kappa_\Sigma(\eta) - \kappa_\Sigma(\mu)\|_{H_{00}^{1/2}(\Sigma)} = \|\kappa_{\partial\Omega}(\tilde{\eta}) - \kappa_{\partial\Omega}(\tilde{\mu})\|_{H^{1/2}(\partial\Omega)} \le \varepsilon ,$$

if $\|\tilde{\eta} - \tilde{\mu}\|_{H^{1/2}(\partial\Omega)} = \|\eta - \mu\|_{H_{00}^{1/2}(\Sigma)} \le \delta$ holds with a suitable $\delta > 0$. □

For completeness we remark that Proposition 2 also holds for the trace space $H_0^{1/2}(\Sigma)$, see [2, Prop. 1.5.17].

4 Equivalence of the Weak Formulations

We are now in a position to prove our main result.

Theorem 3. *With the assumptions on θ_i and kr_i, $i = 1,2$, the domain decomposition problem* (11)*–*(13) *is equivalent to its transformed version* (14)*–*(16)*.*

Proof. The following statements are all valid for $i = 1,2$. First, Lemma 1 (iii) provides

$$p_i \in H^1(\Omega_i) \quad \Longleftrightarrow \quad u_i \in H^1(\Omega_i) .$$

Therefore, using (5), by Proposition 1 we can conclude

$$u_{i|\partial\Omega\cap\partial\Omega_i} = \kappa_i(p_i)_{|\partial\Omega\cap\partial\Omega_i} = \kappa_i(p_{i|\partial\Omega\cap\partial\Omega_i}) = \kappa_i(0) = 0 ,$$

i.e. $u_i \in V_i$ if $p_i \in V_i$. In light of Lemma 1 (i), the converse is true, too.

Now, since θ_i are bounded Borel–measurable functions on $\mathbb{R}$ we have

$$\theta_i(p_i(x)) = M_i(u_i(x)) \quad \text{a.e. on } \Omega_i , \tag{21}$$

due to (7) for all $p_i \in V_i$ with $u_i = \kappa_i(p_i)$, and the functions given in (21) are Lebesgue–measurable L^∞–functions on Ω_i. Therefore, the L^2–scalar products, which correspond to each other in (11) and (14) as well as in (13) and (16), respectively, are equivalently reformulated.

Furthermore, the equivalent reformulation of the terms $b_i(\cdot,\cdot)$ in (11) and (13) into the terms $a_i(\cdot,\cdot)$ in (14) and (16), respectively, is provided by the identity

$$kr_i(\theta_i(p_i))\nabla p_i = \kappa_i'(p_i)\nabla p_i = \nabla u_i \quad \text{a.e. on } \Omega_i\,,$$

understood as functions in $(L^2(\Omega_i))^d$. This is a consequence of Theorem 1.

Finally, the equivalence (12)$\Leftrightarrow$(15) requires the commutativity

$$\kappa_i^{-1}(u_i)_{|\Gamma} = \kappa_i^{-1}(u_{i|\Gamma}) \quad \text{in } \Lambda\,,$$

which is obtained by Proposition 1 and 2. □

We close this investigation by noting that, in addition to Dirichlet and Neumann boundary conditions, which have been considered above, boundary conditions of "Signorini-type" can also be suitably Kirchhoff–transformed in a weak sense. However, as in the degenerate case $\alpha = 0$, one can no longer establish the full equivalence result, compare [2, Thm. 1.5.18].

References

[1] Appell, J., Zabrejko, P.P.: *Nonlinear superposition operators.* Cambridge University Press, 1990.

[2] Berninger, H.: *Domain Decomposition Methods for Elliptic Problems with Jumping Nonlinearities and Application to the Richards Equation.* PhD thesis, Freie Universität Berlin, 2007.

[3] Berninger, H., Kornhuber, R., Sander, O.: On nonlinear Dirichlet-Neumann algorithms for jumping nonlinearities. In O.B. Widlund and D.E. Keyes, eds., *Domain Decomposition Methods in Science and Engineering XVI*, volume 55 of *LNCSE*, pages 483–490. Springer, 2007.

[4] Brezzi, F., Gilardi, G.: Functional spaces. In H. Kardestuncer and D.H. Norrie, eds., *Finite Element Handbook*, chapter 2 (part 1), pages 1.29–1.75. Springer, 1987.

[5] Leoni, G., Morini, M.: Necessary and sufficient conditions for the chain rule in $W^{1,1}_{\mathrm{loc}}(\mathbb{R}^N;\mathbb{R}^d)$ and $BV_{\mathrm{loc}}(\mathbb{R}^N;\mathbb{R}^d)$. *J. Eur. Math. Soc. (JEMS)*, 9(2):219–252, 2007.

[6] Marcus, M., Mizel, V.J.: Complete characterization of functions which act, via superposition, on Sobolev spaces. *Trans. Amer. Math. Soc.*, 251:187–218, 1979.

[7] Marcus, M., Mizel, V.J.: Every superposition operator mapping one Sobolev space into another is continuous. *J. Funct. Anal.*, 33:217–229, 1979.

[8] Quarteroni, A., Valli, A.: *Domain Decomposition Methods for Partial Differential Equations.* Oxford Science, 1999.

Convergence Behavior of a Two-Level Optimized Schwarz Preconditioner

Olivier Dubois[1] and Martin J. Gander[2]

[1] IMA, University of Minnesota, 207 Church Street SE, Minneapolis, MN 55455
dubois@ima.umn.edu

[2] Section de mathématiques, Université de Genève, 2-4 Rue du Lièvre, CP 64, 1211 Genève
martin.gander@math.unige.ch

Summary. Optimized Schwarz methods form a class of domain decomposition algorithms in which the transmission conditions are optimized in order to achieve fast convergence. They are usually derived for a model problem with two subdomains, and give efficient transmission conditions for the local coupling between neighboring subdomains. However, when using a large number of subdomains, a coarse space correction is required to achieve parallel scalability. In this paper we demonstrate with a simple model problem that a two-level optimized Schwarz preconditioner is much more effective than a corresponding two-level Restricted Additive Schwarz preconditioner. The weak dependence on the mesh size is retained from the one-level method, while gaining independence on the number of subdomains. Moreover, the best Robin transmission condition is well approximated by using the analysis from the two subdomain case, under Krylov acceleration.

1 Introduction

In the last ten years, a new class of domain decomposition methods has emerged and has been developed: Optimized Schwarz Methods (OSM). The main idea is to replace the Dirichlet transmission conditions of the classical Schwarz iteration by Robin or higher order conditions, and then optimizing the free parameters in these conditions to obtain the best convergence. In addition to providing fast convergence, the optimized transmission conditions allow us to use very small overlapping regions (as well as no overlap), causing only a weak dependence of the convergence on the mesh size. Optimized Schwarz methods were first introduced in [6] for the advection-diffusion equation, and then studied for a variety of problems, for example in [3, 4, 5].

In all of these studies, the analysis of the convergence is done only for a model problem with *two* infinite or rectangular subdomains, for which a Fourier transform can be applied, thus making possible the explicit optimization of the transmission conditions. In more practical situations with many subdomains, numerical experiments show that such optimized transmission conditions lead to efficient local coupling between neighboring subdomains, but there are no theoretical estimates on the convergence rate of the Schwarz iteration in that case.

M. Bercovier et al. (eds.), *Domain Decomposition Methods in Science and Engineering XVIII*, Lecture Notes in Computational Science and Engineering,
DOI: 10.1007/978-3-642-02677-5, © Springer-Verlag Berlin Heidelberg 2009

It is well-known that domain decomposition techniques are not scalable with the number of subdomains, unless a global mean of communication between the subdomains is incorporated. This is often achieved by a coarse space or coarse grid correction. For optimized Schwarz methods, it is often claimed that the same coarse grid corrections applied to "classical" Schwarz methods can be employed to remove the dependence on the number of subdomains, but little numerical evidence of this fact are published, and no theoretical results are yet available. In the early paper of [7], two coarse space preconditioners were proposed that improve considerably the convergence rate of the Schwarz iteration as we increase the number of subdomains; these preconditioners are specific to non-overlapping decompositions, where the problem is first reformulated as an interface problem.

In this paper, we consider the overlapping Optimized Restricted Additive Schwarz (ORAS) preconditioner from [8], and apply a standard coarse grid correction to obtain a two-level method (see [9] and references therein for the two-level Additive Schwarz preconditioner). We verify with experiments that the weak scaling with respect to the mesh size h is preserved, in agreement with the theory on OSM, and that we gain independence on the number of subdomains for generous overlap. Moreover, we investigate whether the formulas for the optimized parameters, derived in the case of two subdomains, still provide good approximations for the best parameters for the two-level ORAS preconditioner when applied to many subdomains.

The paper is organized as follows. In Section 2, we introduce a simple model problem and describe the one-level and two-level preconditioners under consideration. We also discuss some practical implications of an algebraic condition required in the analysis of [8]. In Section 3, we present several numerical results for the two-level preconditioners, in different scaling scenarios. Finally, in Section 4, we find the best Robin parameter numerically and compare it with the values provided by the formulas, which were derived in the two subdomain case.

2 Domain Decomposition Preconditioners

We consider the simple positive definite elliptic problem $-\Delta u = f$, on the unit square, with homogeneous Dirichlet boundary conditions. We use finite differences to discretize this problem on a uniform grid with $n+2$ points in each dimension ($h := 1/(n+1)$). This leads to a linear system $A\mathbf{u} = \mathbf{b}$. The domain of computation is decomposed into $M_x \times M_y$ rectangular subdomains in the natural way, as illustrated by Fig. 1.

2.1 One-Level Preconditioners

Let $\{\tilde{\Omega}_j\}$ denote a non-overlapping partition of the unknowns. By extending these sets with $\frac{C-1}{2}$ layers of unknowns, we get an overlapping decomposition $\{\Omega_j\}$ with a physical overlap of width $L = Ch$. Let $\tilde{R}_j$ and R_j be the restriction operators on the subsets $\{\tilde{\Omega}_j\}$ and $\{\Omega_j\}$ respectively, and $A_j := R_j A R_j^T$ be the induced local matrices.

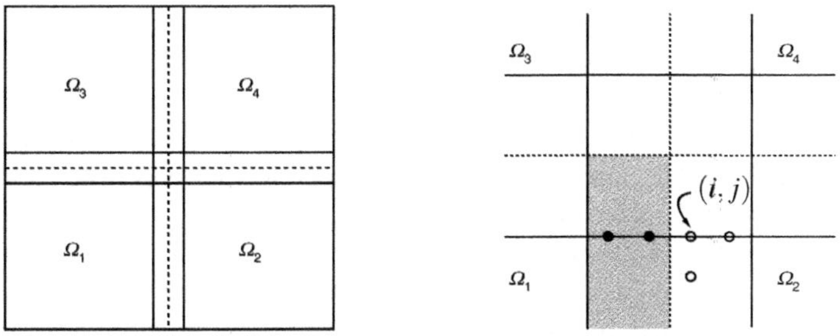

Fig. 1. Example of a uniform domain decomposition into 4 overlapping subdomains. A zoom near the crosspoint is shown on the right.

We consider the following two preconditioners

$$P_{RAS}^{-1} := \sum_{j=1}^{M} \tilde{R}_j^T A_j^{-1} R_j, \qquad P_{ORAS}^{-1} := \sum_{j=1}^{M} \tilde{R}_j^T \tilde{A}_j^{-1} R_j.$$

The first one is the *Restricted Additive Schwarz* (RAS) preconditioner of [2], while the second denotes the *Optimized Restricted Additive Schwarz* (ORAS) preconditioner, recently introduced in [8], where the local matrices are modified to implement optimized Robin interface conditions. Note that with this process, the *physical* overlap is reduced by two mesh layers. For example, a physical overlap of $L = 3h$ for RAS will correspond to an overlap of $\tilde{L} = h$ for ORAS. Here, we will use L to denote exclusively the physical overlap corresponding to the RAS preconditioner as a reference. In the present context, we will refer to the case of *minimal overlap* when choosing the width of the overlapping region to be $L = 3h$. In the case of *generous overlap*, we keep the overlap width proportional to the subdomain size, $L = CH$ (where C is chosen in such a way that we always have $L \geq 3h$).

We often think of an optimized Schwarz method as an iteration-by-subdomain of the form

$$\tilde{A}_j \mathbf{u}_j^{n+1} = \mathbf{f}_j + \sum_{j=1}^{M} \tilde{B}_{jk} \mathbf{u}_k^n, \quad j = 1,2,\ldots,M, \tag{1}$$

whereas in this paper we wish to utilize a stationary iterative method with preconditioner P_{ORAS}^{-1}

$$\mathbf{u}^{n+1} = \mathbf{u}^n + \Big(\sum_{j=1}^{M} \tilde{R}_j^T \tilde{A}_j^{-1} R_j\Big)(\mathbf{f} - A\mathbf{u}^n). \tag{2}$$

It is shown in [8] that the iterations (1) and (2) are ***equivalent*** under two conditions, one of which is an *algebraic condition* given by

$$\tilde{B}_{jk} R_k \tilde{R}_m^T = 0, \quad \text{for any } j, \text{ and } m \neq k. \tag{3}$$

2.2 Interpretation of the Algebraic Condition

The algebraic condition (3) relates the discretization of the transmission conditions with the overlapping (Ω_j) and non-overlapping ($\tilde{\Omega}_j$) domain decompositions. For example, in the case of a decomposition into strips and a 5-point stencil discretization of the Laplacian, condition (3) requires that the overlap be at least three mesh layers wide, i.e $L \geq 3h$.

Now, if the domain decomposition has cross-points, how do we interpret the algebraic condition? To gain more insight, let us consider a simple example with 4 subdomains, labeled as in Fig. 1. In condition (3), let $m = 1$, $k = 2$ and $j = 4$. Fig. 1 also shows a blow-up of the region near the cross-point. After having applied the combination $R_2\tilde{R}_1^T$ to a vector, the only possible nonzero entries are located in the shaded region. Then, condition (3) imposes that the application of $\tilde{B}_{42}$ should not depend on those nodes. Consider in particular the node (i, j) from Fig. 1: it lies on the boundary of Ω_4 and inside $\tilde{\Omega}_2$, hence the operator $\tilde{B}_{42}$ needs to extract a transmission condition there.

If we use a standard finite difference discretization of the normal derivative at (i, j) which is second order accurate ($O(h^2)$), we will need the "illegal node" on the left of (i, j), hence violating the algebraic condition. In practice, we have observed that this causes very slow convergence of iteration (2). To avoid this problem, we have implemented instead a first-order accurate approximation to the normal derivative (using a one-sided finite difference), for which the algebraic condition is satisfied. In that case, modifying the local matrices A_i to $\tilde{A}_i$ also becomes a much simpler task: only diagonal entries for the nodes lying on the boundary of Ω_i need to be modified.

2.3 Two-Level Preconditioners

To introduce a coarse space correction, we proceed as follows. The (non-overlapping) domain decomposition induces a natural coarse mesh with nodes (iH_x, jH_y), where $H_x = 1/M_x$, $H_y = 1/M_y$. We define P_0 to be the *bilinear interpolation* from the coarse to the fine mesh. This induces a coarse matrix using the relation $A_0 := P_0^T A P_0$. We choose to apply the coarse space correction *sequentially*, after the parallel subdomain solves. In the case of a stationary iterative method, preconditioned by a two-level Restricted Additive Schwarz preconditioner (RAS2), we get the iterates

$$\mathbf{u}^{k+\frac{1}{2}} = \mathbf{u}^k + \sum_{j=1}^{M} \tilde{R}_j^T A_j^{-1} R_j(\mathbf{b} - A\mathbf{u}^k),$$

$$\mathbf{u}^{k+1} = \mathbf{u}^{k+\frac{1}{2}} + P_0 A_0^{-1} P_0^T (\mathbf{b} - A\mathbf{u}^{k+\frac{1}{2}}).$$

The same coarse grid component can be added on top of ORAS to get a two-level Optimized Restricted Additive Schwarz preconditioner (ORAS2). To obtain faster convergence, we can apply a GMRES iteration on the corresponding preconditioned linear systems instead.

In this paper, we will experiment only with the optimized one-sided Robin conditions, using the asymptotic formula valid for small values of h (when $L = Ch$), namely $p^* \approx 2^{-1/3} k_{min}^{2/3} L^{-1/3}$ (see [4]). In this formula, the minimum frequency in the min-max problem is chosen to be $k_{min} = \pi$ for the one-level preconditioner, and

$k_{min} = \pi/H$ for the two-level preconditioner, in which case $p^* = O(H^{-2/3})$. The idea behind this choice is that the coarse grid correction should take care of the frequencies below π/H.

3 Numerical Results

In the following results, we solve the preconditioned linear system in two ways. First, we use a stationary iterative method, with right hand side $f \equiv 0$ and random initial guess $\mathbf{u}^{(0)}$, and check the convergence to 0 in the relative ℓ^∞-norm, with tolerance 10^{-6}. Alternatively, we solve the linear system using a preconditioned GMRES (not restarted), with random right-hand side and zero initial guess, with tolerance 10^{-8} on the preconditioned residual.

Our parallel implementation is based on the PETSc library [1]. For the solution of local and coarse problems (i.e. applications of A_j^{-1}), we precompute a full Cholesky factorization.

3.1 Dependence on h

Let us first fix the number of subdomains to $4 \times 4 = 16$ subdomains, use a minimal overlap $L = 3h$, and decrease the fine mesh size. We average the iteration numbers over 25 different random vectors. Fig. 2 illustrates that we obtain the theoretical asymptotic convergence; in fact, with GMRES acceleration, we seem to get a slightly better convergence factor than the expected $1 - O(h^{1/6})$.

More importantly, observe that when using the stationary iterative method, the ORAS2 preconditioner (for which we have chosen $k_{min} = \pi/H$) takes *more* iterations than the one-level ORAS preconditioner (for which $k_{min} = \pi$). This indicates that the Robin parameter with $k_{min} = \pi/H$ does not give the appropriate value; we will confirm this in Section 3. On the other hand, this choice of parameter appears to yield good convergence under GMRES acceleration.

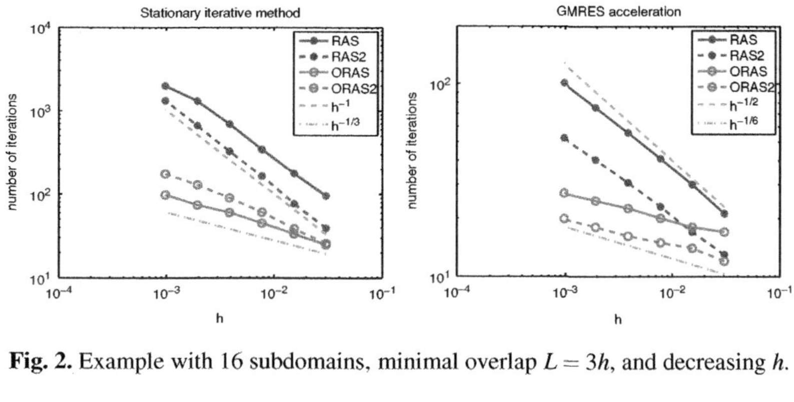

Fig. 2. Example with 16 subdomains, minimal overlap $L = 3h$, and decreasing h.

3.2 Dependence on H, with Generous Overlap

We now fix the fine mesh size with $n = 512$, and increase the number of subdomains while keeping the overlap proportional to H (as much as possible). The following are results obtained for only one instance of a random vector. Table 1 contains the number of iterations; with the GMRES method, the convergence of ORAS2 appears to be independent of the number of subdomains as expected. Also, we can again observe that the performance of ORAS2 with the choice $k_{min} = \pi/H$ is not acceptable for the stationary iterative method, when compared to the ORAS preconditioner.

$\mathbf{M_x \times M_y}$ **(L)**	2×2 $(9h)$	4×4 $(5h)$	8×8 $(3h)$	2×2 $(9h)$	4×4 $(5h)$	8×8 $(3h)$
	Stationary iterative method			Preconditioned GMRES		
RAS	306	739	> 2000	33	60	99
ORAS	27	54	136	16	22	32
RAS2	206	395	533	28	32	31
ORAS2	33	76	174	14	16	17

Table 1. Number of iterations with increasing the number of subdomains, while keeping the overlap proportional to H.

3.3 A Weak Scalability Test

Suppose now that each processor handles a problem of fixed size, in this case 192×192, and let's increase the number of processors. In other words, we keep H/h constant, and always use a minimal overlap $L = 3h$. Table 2 clearly shows that the ORAS2 preconditioner provides significant improvement on the convergence over RAS2 (the difference would become even greater if we increase the ratio H/h).

$\mathbf{M_x \times M_y}$	2×2	4×4	6×6	8×8	9×9
no. of unknowns	147,456	589,824	1,327,104	2,359,296	2,985,984
	Stationary iterative method				
RAS2	439	1082	1528	1557	1798
ORAS2	325	316	323	332	324
	Preconditioned GMRES				
RAS2	40	47	48	48	48
ORAS2	18	20	21	21	21

Table 2. Number of iterations for a weak scaling experiment.

4 Best Robin Parameter

Does the asymptotic formula for the optimized Robin parameter give a good approximation to the best parameter value for the ORAS2 preconditioner? We provide an answer to this question by minimizing the number of iterations taken by the preconditioned iterative method, to obtain the best Robin parameter numerically. Fig. 3 show the results for a fixed problem with mesh size $h = 1/64$ and $4 \times 4 = 16$ subdomains ($H = 1/4$). Fig. 4, on the other hand, plots the behavior of the best Robin parameter as the number of subdomains is increased. We can make two interesting remarks, which are in agreement with our previous experiments:

1. For the stationary iterative method, the best p for ORAS and ORAS2 are very close, and the best convergence appears to be the same in both cases. The asymptotic formula for the optimized Robin parameter with $k_{min} = \pi/H$ gives values far from the best possible.
2. On the other hand, in the case of preconditioned GMRES, the optimized Robin parameters with $k_{min} = \pi$ and $k_{min} = \pi/H$ respectively are very close to the best parameter values, and the convergence of the two-level preconditioner offers significant improvement over the one-level version.

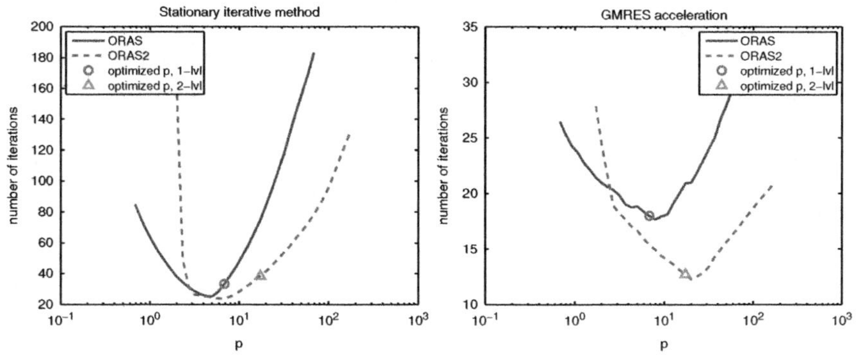

Fig. 3. Convergence for different values of the Robin parameter p, when $h = 1/64$ and $H = 1/4$ (16 subdomains).

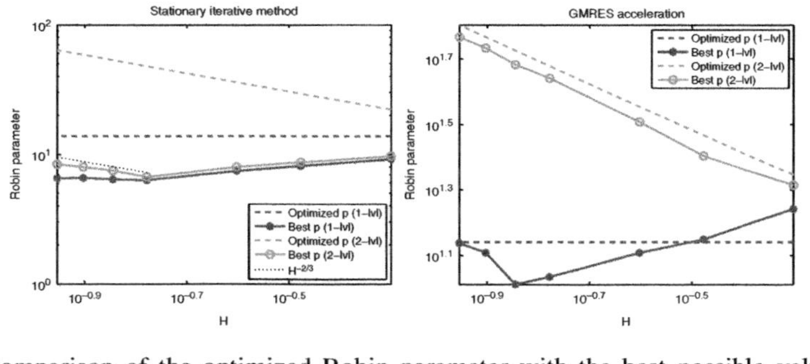

Fig. 4. Comparison of the optimized Robin parameter with the best possible value as we increase the number of subdomains.

References

[1] Balay, S., Buschelman, K., Gropp, W.D., Kaushik, D., Knepley, M.G., Curfman McInnes, L., Smith, B.F., Zhang, H.: PETSc Web page, 2001. `http://www.mcs.anl.gov/petsc`.

[2] Cai, X.-C., Sarkis, M.: A restricted additive Schwarz preconditioner for general sparse linear systems. *SIAM J. Sci. Comput.*, 21:239–247, 1999.

[3] Dubois, O.: *Optimized Schwarz Methods for the Advection-Diffusion Equation and for Problems with Discontinuous Coefficients*. PhD thesis, McGill University, 2007.

[4] Gander, M.J.: Optimized Schwarz methods. *SIAM J. Numer. Anal.*, 44(2):699–731, 2006.

[5] Gander, M.J., Magoulès, F., Nataf, F. Optimized Schwarz methods without overlap for the Helmholtz equation. *SIAM J. Sci. Comput.*, 24(1):38–60, 2002.

[6] Japhet, C.: Optimized Krylov-Ventcell method. Application to convection-diffusion problems. In P.E. Bjørstad, M.S. Espedal, and D.E. Keyes, eds., *Ninth International Conference on Domain Decompositon Methods in Science and Engineering*, 382–389. ddm.org, 1998.

[7] Japhet, C., Nataf, F., Rogier, F.: The Optimized Order 2 Method. Application to convection-diffusion problems. *Future Generation Computer Systems* 18(1):17–30, 2001.

[8] St-Cyr, A., Gander, M.J., Thomas, S.J.: Optimized multiplicative, additive and restricted additive Schwarz preconditioning. *SIAM J. Sci. Comput.*, 29(6):2402–2425, 2007.

[9] Toselli, A., Widlund, O.: *Domain Decomposition Methods—Algorithms and Theory*. Springer Series in Computational Mathematics, 34. Springer, Berlin, 2005.

An Algorithm for Non-Matching Grid Projections with Linear Complexity

Martin J. Gander[1] and Caroline Japhet[2]

[1] Section of Mathematics, University of Geneva, 2-4 Rue du Lièvre, CP 64, 1211 Geneva 4
martin.gander@unige.ch
[2] LAGA, Université Paris 13, 99 Av. J-B Clément, 93430 Villetaneuse, France
japhet@math.univ-paris13.fr

1 Introduction

Non-matching grids are becoming more and more common in scientific computing. Examples are the Chimera methods proposed by [20] and analyzed in [2], the mortar methods in domain decomposition by [1], and the patch method for local refinement by [6], and [17], which is also known under the name 'numerical zoom', see [9]. In the patch method, one has a large scale solver for a particular partial differential equation, and wants to add more precision in certain areas, without having to change the large scale code. One thus introduces refined, possibly non-matching patches in these regions, and uses a residual correction iteration between solutions on the patches and solutions on the entire domain, in order to obtain a more refined solution in the patch regions. The mortar method is a domain decomposition method that permits an entirely parallel grid generation, and local adaptivity independently of neighboring subdomains, because grids do not need to match at interfaces. The Chimera method is also a domain decomposition method, specialized for problems with moving parts, which inevitably leads to non-matching grids, if one wants to avoid regridding at each step. Contact problems in general lead naturally to non-matching grids.

In all these cases, one needs to transfer approximate solutions from one grid to a non-matching second grid by projection. This operation is known in the literature under the name mesh intersection problem in [12], intergrid communication problem in [16], grid transfer problem in [18], and similar algorithms are also needed when one has to interpolate discrete approximations, see [13, Chap. 13].

2 Towards an Optimal Algorithm

There are two problems that need to be addressed in order to obtain an efficient projection algorithm, a combinatorial one and a numerical one: the combinatorial one stems from the fact that in principle, every element of one grid could be intersecting

M. Bercovier et al. (eds.), *Domain Decomposition Methods in Science and Engineering XVIII*, Lecture Notes in Computational Science and Engineering,
DOI: 10.1007/978-3-642-02677-5, © Springer-Verlag Berlin Heidelberg 2009

with every element of the other grid, and hence the naive approach immediately leads to an $O(n^2)$ algorithm, where n is the number of elements. This is well known in the domain decomposition community, see for example [4]. The numerical difficulty is related to the calculation of the intersection of two finite elements, which is numerically difficult, because one needs to take numerical decisions whether two segments intersect or not, and whether one point is in an element or not. For the patch method, [17] state: "Some difficulties remain though since we must compute integrals involving shape functions that are defined on non-compatible meshes". They use as approximation a midpoint rule, computing only in which element the barycenter of the elements of the other grid lies. The influence of the error of a quadrature rule for this problem is studied in [15]. The authors in [4] mention the substantial complexity increase when going from one- to two-dimensional interfaces, and a sophisticated program with many special cases is used to compute the projection, as explained by [3].

If one needs to interpolate values only, the numerical intersection problem is avoided, and an elegant way to reduce the complexity to $O(n\log n)$ was introduced by [10], in form of an additional adaptively refined background Cartesian mesh, called quadtree in 2d and octree in 3d. This approach is currently widely used, for example in contact problems, see [11], where the overall complexity of the simulation process is still dominated by the nonlinear monotone multigrid method. A related approach is to use a binning (or bucket) technique, introduced by [14], see for example [18], and the MpCCI code from the Fraunhofer [8]. Faster algorithms can be obtained, if neighboring information for each element is available: in the case of interpolation, one can use an advancing front technique that starts, for each new point at which one needs to interpolate data, a local search in the neighborhood of the element where the previous point was interpolated. Only if this search is not successful in less than a constant number of steps a brute force search is launched, see [13]. This approach leads to an algorithm with close to linear complexity. A related technique uses a self-avoiding walk, see [12], with a vicinity search. This search can only fail after a boundary element had no intersection with any element of the other mesh, in which case a quad-tree search is employed. This leads to what the authors call approximately linear complexity. Further techniques for treating the boundary are given in [13].

Computing the intersection of elements numerically was first studied in the computer graphics community under the name "polygon clipping", see [21] and references therein. The basic algorithm works as follows: one marches along the edges of one polygon, and whenever an intersection is found, one switches the polygons and marches on the edges of the other one. As soon as one returns to a point already visited, the intersection polygon is obtained. This algorithm is extensively used in computer graphics, and a generalized version, which can also handle self-intersecting polygons can be found in [7], where we also find the quote: "So far we have tacitly assumed that there are no degeneracies, i.e. each vertex of one polygon does not lie on an edge of the other polygon. Degeneracies can be detected in the intersect procedure ... In this case we perturb the vertex slightly. If we take care that the perturbation is less than a pixel width, the output on the screen will be correct."

While for computer graphics, a natural scale for the truncation is the pixel, it is more difficult to determine acceptable perturbations for numerical applications. Since we did not find an elegant and robust solution for degenerate cases in the context of mortar applications, we propose an entirely different algorithm below, which can also easily be generalized to three-dimensional interfaces. A numerically robust way to determine intersections is however presented in [19], who is using adaptive precision floating point arithmetic. The award winning mesh generator "triangle" by the same author computes intersections of two non-matching triangular grids using this approach.

For a problem in two dimensions, the mortar method has one-dimensional interfaces, and a simple algorithm based on the ideas of merge sort computes the projection in $O(n)$ steps, where n is the number of elements touching the interface, see [5]. We show in this paper a generalization of this algorithm to higher dimensions. We use an advancing front technique and neighboring information, which is often available in finite element meshes, in order to obtain an algorithm with linear complexity. Its implementation is surprisingly short, and we give the entire Matlab code. For computing the intersection, we use a new approach, which turns out to be numerically robust and can be generalized to higher dimensions. We show numerical experiments both in 2d and 3d, which illustrate the optimal complexity and negligible overhead of the algorithm.

3 The Algorithm for Computing the Intersection

We now present an algorithm that computes the intersection polygon of two arbitrary triangles. It first computes all edge intersections, and all corners of the triangles that are contained in the other one. Then the algorithm orders the set of points obtained counterclockwise in order to obtain the intersection polygon, see Figure 1. The graphic primitive `EdgeIntersections(X,Y)` computes all intersections of edges of triangle `X` (corner coordinates stored column-wise) with edges of triangle `Y`, including borderline cases by using greater or equal in the decisions. The routine `PointsOfXInY(X,Y)` computes corners of triangle `X` in triangle `Y`, again including borderline cases. The routine `SortAndRemoveDoubles(P)` sorts the points in `P` in counterclockwise order and removes duplicates, which turns out to make the algorithm numerically robust.

In addition to computing the intersection polygon, the algorithm also returns two more results needed later: in `n` which neighboring triangles of `X` will also intersect with `Y`, and in `M` the integrals on the intersection `P` of products of element shape functions of `X` with the ones of `Y`, or any related quantity obtained from the routine `MortarInt`.

This algorithm can be generalized to compute the intersection of tetrahedra in 3d (see also Section 5): one first calculates all points were an edge of one tetrahedron traverses the face of the other, and all corners of one tetrahedron contained in the other. Then one orders the points face by face counterclockwise. Note also that this intersection algorithm can easily be generalized to convex polygons and polyhedra.

```
function [P,n,M]=Intersect(X,Y);
% INTERSECT intersection of two triangles and mortar contribution
%   [P,n,M]=Intersect(X,Y); computes for two given triangles X and Y
%   the points P where they intersect, in n the indices of neighbors
%   of X that also intersect with Y, and the local mortar matrix M
%   of contributions of the element X on the element Y.

[P,n]=EdgeIntersections(X,Y);
Q=PointsOfXInY(X,Y);
if size(Q,2)>1                       % if there are two or more
  n=[1 1 1];                         % interior points, the triangle
end                                  % is candidate for all neighbors
P=[P Q];
P=[P PointsOfXInY(Y,X)];
P=SortAndRemoveDoubles(P);           % sort counterclockwise
M=zeros(3,3);
if size(P,2)>0
  for j=2:size(P,2)-1                % compute local Mortar matrix
    M=M+MortarInt(P(:,[1 j j+1]),X,Y);
  end;
end;
```

Fig. 1. Algorithm for computing the intersection polygon of two triangles.

4 The Projection Algorithm with Linear Complexity

We now show an algorithm that computes, for two non-matching triangular meshes representing the same planar geometry, the associated mortar projection matrix, see [1], or any other similar quantity on each intersection polygon defined by `MortarInt` in the `Intersect` procedure. The algorithm is using advancing fronts and the fact that each triangle knows which are its neighbors, see Fig. 2. The input of the algorithm are two triangular grids. The grid node coordinates are stored column wise in `N`. The triangles are stored row wise in `T`, the first three numbers referring to the nodal coordinates of the triangle in `N`, and the next three to the neighboring triangles in `T`, both ordered counterclockwise. The algorithm then works as follows: it starts with a pair of intersecting triangles (assumed to be the first ones in `Ta` and `Tb`), which are often trivially available at a corner, but otherwise could also be found by one direct search. We then compute first the intersection of these two triangles using the intersection routine from Section 3. We then add the neighbors of the triangle from mesh a as candidates in a list `al`, since they could intersect with our triangle from mesh b. Picking triangles from list `al` one by one, we compute their intersection with the current triangle from mesh b and add non treated neighbors to the list `al` until all triangles in `al` have been treated. This implies that the starting triangle from mesh b cannot intersect any triangles from mesh a any more. Now we put all the neighbors of the starting triangle of mesh b into a list `bl`, and perform the same steps as for the first triangle on each one in the list `bl`, until it becomes empty, and the algorithm terminates.

We now address the complexity of our algorithm: the key step is that we stored a starting candidate from list `al` for each of the triangles added to list `bl` in list `bil`. This information is obtained without extra calculation in the computation of the intersection. Thus there is never a search needed for a candidate triangle of mesh a

```
function M=InterfaceMatrix(Na,Ta,Nb,Tb);
% INTERFACEMATRIX projection matrix for non-matching triangular grids
%   M=InterfaceMatrix(Na,Ta,Nb,Tb); takes two triangular meshes Ta
%   and Tb with associated nodal coordinates in Na and Nb and
%   computes the associated mortar projection matrix M

bl=[1];                          % bl: list of triangles of Tb to treat
bil=[1];                         % bil: list of triangles Ta to start with
bd=zeros(size(Tb,1)+1,1);        % bd: flag for triangles in Tb treated
bd(end)=1;                       % guard, to treat boundaries
bd(1)=1;                         % mark first triangle in b list.
M=sparse(size(Nb,2),size(Na,2));
while length(bl)>0
  bc=bl(1); bl=bl(2:end);        % bc: current triangle of Tb
  al=bil(1); bil=bil(2:end);     % triangle of Ta to start with
  ad=zeros(size(Ta,1)+1,1);      % same as for bd
  ad(end)=1;
  ad(al)=1;
  n=[0 0 0];                     % triangles intersecting with neighbors
  while length(al)>0
    ac=al(1); al=al(2:end);      % take next candidate
    [P,nc,Mc]=Intersect(Nb(:,Tb(bc,1:3)),Na(:,Ta(ac,1:3)));
    if ~isempty(P)               % intersection found
      M(Tb(bc,1:3),Ta(ac,1:3))=M(Tb(bc,1:3),Ta(ac,1:3))+Mc;
      t=Ta(ac,3+find(ad(Ta(ac,4:6))==0));
      al=[al t];                 % add neighbors
      ad(t)=1;
      n(find(nc>0))=ac;          % ac is starting candidate for neighbor
    end
  end
  tmp=find(bd(Tb(bc,4:6))==0); % find non-treated neighbors
  idx=find(n(tmp)>0);            % take those which intersect
  t=Tb(bc,3+tmp(idx));
  bl=[bl t];                     % and add them
  bil=[bil n(tmp(idx))];         % with starting candidates Ta
  bd(t)=1;
end
```

Fig. 2. Algorithm with linear complexity for computing the intersection of two non-matching triangular grids and the associated mortar projection matrix.

that could intersect the currently treated triangle from mesh b. The algorithm treats triangles of mesh b one by one, and checks for each triangle at most a constant number of triangles in mesh a, which shows that the average complexity is linear. The worst-case complexity however is quadratic, namely when the constant equals the total number of triangles in mesh a. This situation arises when every triangle of mesh a intersects every triangle of mesh b, and quadratic complexity is unavoidable in this case.

Note that our algorithm does not depend on the number of dimensions; it only uses the fact that each element has a given number of neighbors, which in the implementation shown is three. We however assumed that the two meshes are connected, and also that the intersection of one element with the elements of the other mesh are simply connected. Otherwise the algorithm would need extra starting points in order to find the complete intersections.

5 Numerical Experiments

We show in Fig. 3 a comparison of our algorithm with the brute force search,

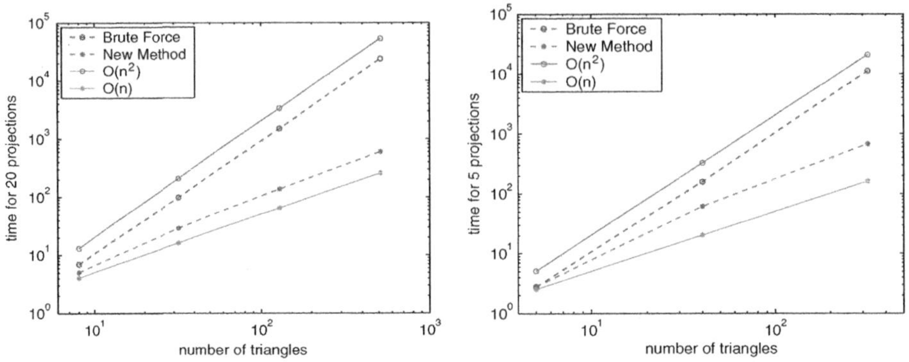

Fig. 3. Comparison in computing time for two-dimensional meshes on the left, and three-dimensional meshes on the right.

where for every element in the first mesh the intersection with every element in the second mesh is computed. On the left, we show the average computing time for twenty projection calculations in two dimensions, each time with two random triangular meshes, and on the right a similar comparison for the three-dimensional case, where we show the average computing time for five projection calculations. In addition to the asymptotic superiority, we also see that the new algorithm is already competitive for small meshes, i.e. the algorithmic overhead is negligible.

6 Conclusions

The intersection algorithm we presented for two triangles can be made slightly faster by first using an inexact range test to quickly exclude non-intersecting triangles, before starting the actual computation of the intersection.

The projection algorithm itself has also been extended to contact problems, where the interfaces of the two neighboring domains do not quite lie in the same physical manifold, and an additional projection "normal" to the interface is necessary. All codes and a demo are available at `www.unige.ch/~gander`.

Acknowledgement. We thank a referee for detailed comments. This research was supported in part by the Swiss National Science Foundation Grant 200020-1 17577/1.

References

[1] Bernardi, C., Maday, Y., Patera, A.T.: Domain decomposition by the mortar element method. In H. G. Kaper and M. Garbey, eds., *Asymptotic and Numerical Methods for Partial Differential Equations with Critical Parameters*, vol. 384, 269–286. N.A.T.O. ASI, Kluwer Academic, 1993.

[2] Brezzi, F., Lions, J.-L., Pironneau, O.: Analysis of a Chimera method. *C. R. Acad. Sci. Paris Sér. I Math.*, 332(7):655–660, 2001.

[3] Flemisch, B., Hager, C.: Mortar projection algorithms. Private Communication, 2007.

[4] Flemisch, B., Kaltenbacher, M., Wohlmuth, B.I.: Elasto-acoustic and acoustic-acoustic coupling on nonmatching grids. *Internat. J. Numer. Methods Engrg.*, 67(13):1791–1810, 2006.

[5] Gander, M.J., Japhet, C., Maday, Y., Nataf, F.: A new cement to glue nonconforming grids with Robin interface conditions: The finite element case. In R. Kornhuber et al. eds., *Proceedings of the 15th international domain decomposition conference*, volume 40, 259–266. Springer LNCSE, 2005.

[6] Glowinski, R., He, J., Lozinski, A., Rappaz, J., Wagner, J.: Finite element approximation of multi-scale elliptic problems using patches of elements. *Numer. Math.*, 101(4):663–687, 2005.

[7] Greiner, G., Hormann, K.: Efficient clipping of arbitrary polygons. *ACM Trans. Graph.*, 17(2):71–83, 1998.

[8] Fraunhofer Institute: *MpCCI 3.0.6-12 Documentation*. Fraunhofer Institute for Algorithms and Scientific Computing, 2007.

[9] Apoung Kamga, J.-B., Pironneau, O.: Numerical zoom for multiscale problems with an application to nuclear waste disposal. *J. Comput. Phys*, 224:403–413, 2007.

[10] Knuth, D.N.: *The Art of Computer Programming*, volume 3. Addison-Wesley, 1973.

[11] Krause, R., Sander, O.: Fast solving of contact problems on complicated geometries. In R. Kornhuber et al. eds., *Proceedings of the 15th international domain decomposition conference*, vol. 40, 495–502. Springer LNCSE, 2005.

[12] Lee, P., Yang, C.-H., Yang, J.-R.: Fast algorithms for computing self-avoiding walks and mesh intersections over unstructured meshes. *Adv. Engrg. Softw.*, 35: 61–73, 2004.

[13] Löhner, R.: *Applied CFD Techniques: An Introduction Based on Finite Element Methods*. Wiley, 2001.

[14] Löhner, R., Morgan, K.: An unstructured multigrid method for elliptic problems. *Internat. J. Numer. Methods Engrg.*, 24:101–115, 1987.

[15] Maday, Y., Rapetti, F., Wohlmuth, B.I.: The influence of quadrature formulas in 2d and 3d mortar element methods. In *Recent developments in domain decomposition methods (Zürich, 2001) Lect. Notes Comput. Sci.*, vol. 23, 203–221. Springer, 2002.

[16] Meakin, R.L.: A new method for establishing intergrid communication among systems of overset grids. *10th AIAA Comput. Fluid Dyn. Conf. (Honolulu, HI),*, 91(1586), 1991.
[17] Picasso, M., Rappaz, J., Rezzonico, V.: Multiscale algorithm with patches of finite elements. *Comm. Numer. Methods Engrg.*, 24(6):477–491, 2007.
[18] Plimpton, S., Hendrickson, B., Stewart, J.: A parallel rendezvous algorithm for interpolation between multiple grids. *J. Parallel Distr. Comput.*, 64(2), 2004.
[19] Shewchuk. J.R.: Adaptive precision floating-point arithmetic and fast robust geometric predicates. *Comput. Geom.*, 18(3):305–363, 1997.
[20] Steger, J.L., Dougherty, F.C., Benek, J.A.: A chimera grid scheme, advances in grid generation. In K.N. Ghia and U. Ghia, eds., *ASME FED*, vol. 5, 1983.
[21] Weiler, K., Atherton, P.: Hidden surface removal using polygon area sorting. *Siggraph* 11(2):214–222, 1977.

A Maximum Principle for L^2-Trace Norms with an Application to Optimized Schwarz Methods

Sébastien Loisel and Daniel B. Szyld

Temple University {loisel,szyld}@temple.edu

Summary. Harmonic functions attain their pointwise maximum on the boundary of the domain. In this article, we analyze the relationship between various norms of nearly harmonic functions and we show that the trace norm is maximized on the boundary of the domain. One application is that the Optimized Schwarz Method with two subdomains converges for all Robin parameters $\alpha > 0$.

1 Introduction

Given a domain $\Omega \subset \mathbb{R}^2$, consider the model problem

$$-\nabla \cdot (a\nabla u) + cu = f \text{ in } \Omega \quad \text{and} \quad u = 0 \text{ on } \partial\Omega, \tag{1}$$

where $a : \Omega \to M_{2\times 2}$ is a symmetric and coercive 2×2 matrix valued function of $x \in \Omega$ and c is a non-negative function of $x \in \Omega$. If we have a domain decomposition $\Omega = \Omega_1 \cup \Omega_2$ and given functions v_0, w_0 on Ω_1, Ω_2, respectively, typical domain decomposition algorithms iteratively solve problems of the type $(-\nabla \cdot (a\nabla) + c)v_k = f$ in Ω_1 and $(-\nabla \cdot (a\nabla) + c)w_k = f$ in Ω_2, $k \geq 1$, with some boundary conditions. In the classical Schwarz algorithm, the local problems use Dirichlet data. Optimized Schwarz Methods replace the Dirichlet subproblems by Robin subproblems; see [3] for a detailed discussion and bibliography. The analysis of the convergence of Optimized Schwarz Methods turned out to be more complicated than that of classical Schwarz methods; see [4, 5, 7, 8, 9].

Schwarz's idea [11] to prove the convergence was to use the maximum principle. This is based on the observation that the *error* iterates $u_k^{(i)} - u$ are (a,c)-harmonic on each subdomain, for every $k \geq 1$; i.e., they solve the PDE $-\Delta \cdot (a\Delta u_k^{(i)}) + cu_k^{(i)} = 0$. In a recent paper [10], we have shown that trace norms can be used in a similar way to show the convergence of Optimized Schwarz Methods, which use Robin boundary conditions on the interfaces between the subdomains.

Let Ω be a domain and $\Gamma_1 \subset \Omega$, $\Gamma_2 \subset \partial\Omega$ be curves. Our goal is to give some conditions under which there is a positive $\omega < 1$ such that the inequality

M. Bercovier et al. (eds.), *Domain Decomposition Methods in Science and Engineering XVIII*, Lecture Notes in Computational Science and Engineering,
DOI: 10.1007/978-3-642-02677-5,

$$\int_{\Gamma_1} v^2 \leq \omega \int_{\Gamma_2} v^2, \tag{2}$$

is satisfied for every (nearly) (a,c)-harmonic function v on Ω satisfying suitable boundary conditions. This result is more general than the one we presented in [9], where it is assumed that v is exactly (a,c)-harmonic.

The structure of this article is as follows. In Section 2, we introduce the notion of (ε,a,c)-harmonicity, and ε-relative uniformity. With these definitions, we are able to prove our main result, which is that the maximum trace L^2 norm is attained on the boundary. In Section 3, we discuss some applications to Schwarz methods.

2 The Maximum Principle for L^2-Trace Norms

To describe our main result, we must first discuss certain Sobolev estimates; we refer the reader to [1, 2, 7], and references therein for details.

2.1 Preliminaries on the Domain and the Interfaces

Let ρ be a nonnegative function on Ω. Let $H^1(\Omega,\rho)$ be the space of functions v of finite weighted Sobolev norm

$$\|v\|^2_{H^1(\Omega,\rho)} = \int_\Omega (|\nabla v|^2 + |v|^2)\rho.$$

If $\rho(x)$ goes to zero linearly as x approaches the boundary $\partial\Omega$, then the trace map $u \to u|_{\partial\Omega}$ is discontinuous, i.e., there is no trace space [6].

Let Ω be parametrized by a function $\Phi(x,y)$. The domain of Φ is $Z = \{(x,y) | x_0 \leq x \leq x_2 \text{ and } p(x) \leq y \leq q(x)\}$, where $x_0 < x_2$ are real numbers. For simplicity, all domains in this paper are Lipschitz.[1] We assume that p and q are C^1 and that Φ is C^2. Because of the parametrization, Ω is furthermore piecewise C^1 and connected.[2] For each fixed x, we define Γ_x to be the curve parametrized by $y \to \Phi(x,y)$. In the context of domain decompositions, Ω is one of the two overlapping subdomains. Choosing x_1 between x_0 and x_2, Γ_{x_1} and Γ_{x_2} are the interfaces defined by the boundary of the overlap. We define $U_x = \cup_{x_0 \leq \xi \leq x} \Gamma_\xi$, the part of Ω "to the left" of Γ_x; see Fig. 1. If v is in $H^1(\Omega)$, we define

$$e(x) = e(x,v) = \int_{\Gamma_x} v^2. \tag{3}$$

Thus, our goal is to find a positive $\omega < 1$, and conditions on Ω, Φ, v so that $e(x_1) < \omega e(x_2)$, i.e., that (2) holds.

[1] We do not assume that Ω is convex.

[2] We could relax the connectedness hypothesis by using one chart per component.

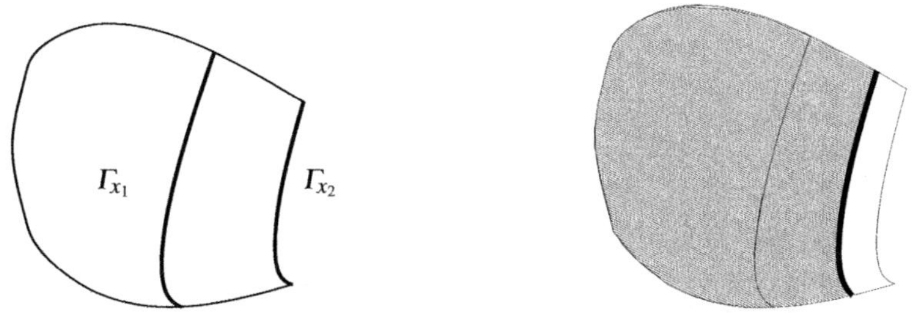

Fig. 1. Left: the domain Ω and the two interfaces Γ_{x_1} and Γ_{x_2} (in bold). Right: U_x (shaded) and Γ_x (bold).

2.2 (ε, a, c)-Harmonicity

A function v in $H^1(\Omega)$ is said to be (a,c)-*harmonic* if $(-\nabla \cdot a\nabla + c)v = 0$. Such functions obey the maximum principle: the L^∞ norm of an (a,c)-harmonic nonconstant function is attained on the boundary, and not on the interior. We want to find a notion of near-(a,c)-harmonicity, which we will call (ε,a,c)-harmonicity, such that a related maximum principle holds. To that end, let v be in $H^1(\Omega)$ with $v = 0$ on $\partial\Omega \setminus \Gamma_{x_2}$ and let ν be the outer normal to U_x. Consider the quantity

$$S = \int_{x_1}^{x_2} \int_{p(x)}^{q(x)} v(a\nu) \cdot \nabla v |\Phi_y| \, dy \, dx$$

$$= \int_{x_1}^{x_2} \int_{U_x} -v(-\nabla \cdot a\nabla + c)v + (\nabla v)^T a(\nabla v) + cv^2 \, ds \, dx \tag{4}$$

$$= \int_{\Omega} [-v(-\nabla \cdot a\nabla + c)v + (\nabla v)^T a(\nabla v) + cv^2]\rho, \tag{5}$$

where we have used Green's integration by parts to obtain (4), and Fubini's Theorem to obtain (5). The function $\rho(\mathbf{x})$ is therefore the Lebesgue measure of the set $\{\xi \in (x_1,x_2) : \mathbf{x} \in U_\xi\}$, and hence $\rho(\mathbf{x}) = O(dist(\mathbf{x}, \Gamma_{x_2}))$. We want to be able to compare S with

$$\|v\|_L^2 = \int_{\Omega} [(\nabla v)^T a(\nabla v) + cv^2]\rho, \tag{6}$$

which is a norm that is equivalent to $\|v\|_{H^1(\Omega,\rho)}$.[3] To that end, for $\varepsilon \geq 0$, we say that $v \in H^1(\Omega)$ is (ε,a,c)-*harmonic* if

$$-\varepsilon\|v\|_L^2 \leq \int_{\Omega} v(-\nabla \cdot a\nabla + c)v\rho \leq \varepsilon\|v\|_L^2. \tag{7}$$

Note that an (a,c)-harmonic function is $(0,a,c)$-harmonic. If v is (ε,a,c)-harmonic, and if $v = 0$ on $\partial\Omega \setminus \Gamma_{x_2}$, then we have

[3] The equivalence of norms is a variant of the standard argument that the bilinear form of the elliptic operator is equivalent to the H^1 norm, but with the added weight ρ; see [2] for details. In the case $c = 0$, a variant of the Friedrichs' inequality is used.

$$(1-\varepsilon)\|v\|_L^2 \leq S \leq (1+\varepsilon)\|v\|_L^2.$$

We define

$$s(x) = \frac{\int_{\Gamma_x} \Phi_x^T av}{\int_{\Gamma_x} (av)^2}, \tag{8}$$

that is, for any $x \in (x_1, x_2)$, $s(x)a(\Phi(x,\cdot))v(\Phi(x,\cdot))$ is the orthogonal projection[4] of $\Phi_x(x,\cdot)$ onto the span of av in $L^2(\Gamma_x)$. Let g be a C^0 vector field. Then for $v \in H^1(\Omega)$, $D_g v = g \cdot \nabla v \in L^2(\Omega)$. We will use the field $g = \Phi_x - sav$.

2.3 ε-Relative Uniformity

We now turn to the notion of *relative uniformity*. We first want to impose a condition so that Φ_x is not too tangent to Γ_x in the sense that there are constants $C_1 > 0$ and $C_2 < \infty$, such that $C_1 \leq s(x) \leq C_2$ for all x.

If a is the identity and Φ is conformal, then $\Phi_x \cdot \Phi_y = 0$, i.e., v is parallel to Φ_y and $s(x)$ is strictly positive and bounded, cf. (8). If Φ is not conformal, but still $0 < C_1 \leq s \leq C_2 < \infty$, we say that Φ is "nearly conformal". Using (6), if v is (ε, a, c)-harmonic, for a fixed $\varepsilon > 0$, and $v = 0$ on $\partial\Omega \setminus \Gamma_{x_2}$, then there are constants $C_a = C_a(\varepsilon)$ and $C'_a = C'_a(\varepsilon)$ such that

$$C_a(\varepsilon)\|v\|_L^2 \leq \int_{x_1}^{x_2} s \int_{p(x)}^{q(x)} 2v(av) \cdot \nabla v |\Phi_y| \, dy\, dx \leq C'_a(\varepsilon)\|v\|_L^2. \tag{9}$$

Specifically, one may use $C_a(\varepsilon) = 2(C_1 - \varepsilon C_2)$ and $C'_a(\varepsilon) = 2(C_2 + \varepsilon C_2)$. Further assume that there are constants $C_v < \infty$ and $C_0 < \infty$ such that

$$\left| \int_{x_1}^{x_2} \int_{p(x)}^{q(x)} 2v(\Phi_x - sav) \cdot \nabla v |\Phi_y| \, dy\, dx \right| \leq C_v \|v\|_L^2, \tag{10}$$

$$\left| \int_{x_1}^{x_2} \int_{p(x)}^{q(x)} v^2 \frac{\Phi_y \cdot \Phi_{xy}}{|\Phi_y|} \, dy\, dx \right| \leq C_0 \|v\|_L^2. \tag{11}$$

If a is the identity and Φ is conformal, then $C_v = C_0 = 0$. Our allowances for $C_v, C_0 > 0$ means that we can use a Φ which is "nearly conformal".

Furthermore, if there is a diffeomorphism that turns a into the identity and Φ into a conformal map, we also have $C_v = C_0 = 0$. Thus, if the interfaces Γ_{x_1} and Γ_{x_2} are "nearly parallel" in the metric induced by a, these constants will be small. In our Definition 1 we want C_v and C_0 to be small in the sense that their sum is smaller than $C_a(\varepsilon)$.

Definition 1. *Let Ω be a domain, fix $\varepsilon > 0$ and consider the elliptic problem (1). Let Φ be a parametrization of Ω as above. Let $s(x)$ be as in (8), and let the positive constants $C_a(\varepsilon)$, C_v, and C_0 be such that (9), (10), and (11) hold. We say that the parametrization is* ε-relatively uniform *if the inequality $C_a(\varepsilon) - C_v - C_0 > 0$ holds, for every v which is (ε, a, c)-harmonic with $v = 0$ on $\partial\Omega \setminus \Gamma_{x_2}$.*

[4] There is no reason to prefer this particular choice of $s(x)$, but we hope that it makes $\Phi_x - sav$ small in a useful way.

2.4 Maximum Principle for L^2-Trace Norms

We present our main Theorem which depends on the elliptic operator (as parametrized by a and c), on the domain Ω and its parametrization Φ, as well as on $\varepsilon > 0$.

Theorem 1. (Maximum principle for a trace norm) *Fix $\varepsilon > 0$ and let Φ be an ε-relatively uniform parametrization of Ω. Then, there exists a positive $\omega < 1$ such that, for every (ε, a, c)-harmonic $v \in H^1(\Omega)$ with $v = 0$ on $\partial\Omega \setminus \Gamma_{x_2}$, then the estimate $e(x_1) \leq \omega e(x_2)$ is satisfied.*

Proof. This proof proceeds in two steps. First, we justify differentiating under the integral sign, then we use Green's Theorem and various trace estimates to show that $e' \geq 0$. Let K be an upper bound for $|p(x)|$ and $|q(x)|$. Let $w \in H^1(Z)$ with $w(x, p(x)) = w(x, q(x)) = 0$ for $x \in (x_1, x_2)$. Let

$$\varphi(x) = \int_{p(x)}^{q(x)} w^2(x,y)\,dy. \tag{12}$$

Let $\varphi \in C_c^\infty(x_1, x_2)$, i.e., an infinitely differentiable, compactly supported test function on the interval (x_1, x_2), and consider the number

$$\eta = \eta(w) = \int_{x_1}^{x_2} \varphi(x)\varphi'(x)\,dx = \int_{x_1}^{x_2} \int_{p(x)}^{q(x)} w^2(x,y)\varphi'(x)\,dy\,dx.$$

We can extend w, first by zero to the strip $S = \{(x,y) | x_1 < x < x_2\}$ (since $p, q \in C^1$), then using a continuous extension operator to $H^1(\mathbb{R}^2)$ to obtain a new function $\tilde{w}$; see, e.g., [1]. We have $\tilde{w}|_Z = w$ and $\tilde{w}|_{S \setminus Z} = 0$. Likewise, by trivial extension of φ, we can consider $\tilde{\varphi}(x,y) = \pi(y)\varphi(x) \in C_c^\infty(\mathbb{R}^2)$ where $\pi(y)$ a smooth function which is uniformly one on $[-K, K]$, and zero outside of $[-K-1, K+1]$. Then we have

$$\begin{aligned}
\eta = \int_{\mathbb{R}^2} \tilde{w}^2(x,y) D_x \tilde{\varphi}(x,y)\,dx\,dy &= -\int_{\mathbb{R}^2} D_x(\tilde{w}^2(x,y))\tilde{\varphi}(x,y)\,dx\,dy \\
&= -\int_Z 2w(x,y)w_x(x,y)\varphi(x)\,dx\,dy \\
&= -\int_{x_1}^{x_2} \int_{p(x)}^{q(x)} 2w(x,y)w_x(x,y)\,dy\,\varphi(x)\,dx.
\end{aligned}$$

Hence, φ has a weak derivative and it is given by $\varphi'(x) = \int_{p(x)}^{q(x)} 2ww_x\,dy$. If we use $w(x,y) = v(\Phi(x,y))\sqrt{|\Phi_y|}$ in (12), we recover $e(x)$ as in (3). We thus obtain that $e(x)$ is weakly differentiable and that its weak derivative is

$$(D^w e)(x) = e'(x) = \int_{p(x)}^{q(x)} \left(2v\Phi_x \cdot \nabla v |\Phi_y| + v^2 \frac{\Phi_y \cdot \Phi_{xy}}{|\Phi_y|} \right) dy.$$

Therefore, by adding and subtracting the appropriate term and using (9), (10), and (11), we have

$$\begin{aligned} e(x_2) - e(x_1) &= \int_{x_1}^{x_2} e'(x)\,dx \\ &= \int_{x_1}^{x_2}\int_{p(x)}^{q(x)} 2v\Phi_x \cdot \nabla v |\Phi_y|\,dy\,dx + \int_{x_1}^{x_2}\int_{p(x)}^{q(x)} v^2 \frac{\Phi_y \cdot \Phi_{xy}}{|\Phi_y|}\,dy\,dx \\ &= \int_{x_1}^{x_2} s \int_{p(x)}^{q(x)} 2v(av) \cdot \nabla v |\Phi_y|\,dy\,dx \\ &\quad + \int_{x_1}^{x_2}\int_{p(x)}^{q(x)} 2v(\Phi_x - sav) \cdot \nabla v |\Phi_y|\,dy\,dx \\ &\quad + \int_{x_1}^{x_2}\int_{p(x)}^{q(x)} v^2 \frac{\Phi_y \cdot \Phi_{xy}}{|\Phi_y|}\,dy\,dx, \end{aligned} \tag{13}$$

and thus

$$e(x_2) - e(x_1) \geq (C_a - C_v - C_0)\|v\|_L^2. \tag{14}$$

Similarly, one obtains

$$\begin{aligned} e(x_2) - e(x_1) &\leq (C_a' + C_v + C_0)\|v\|_L^2 \\ e(x_2) &\leq (C_a' + C_v + C_0 + C_T)\|v\|_L^2 \end{aligned} \tag{15}$$

where C_T is the constant of the trace inequality from $(H^1(\Omega), \|\cdot\|_L)$ to $L^2(\Gamma_{x_1})$.[5] Combining (14) and (15), one obtains the desired inequality

$$e(x_1) \leq \left(1 - \frac{C_a - C_v - C_0}{C_a' + C_v + C_0 + C_T}\right) e(x_2). \qquad \square$$

If we use the estimates $C_a = C_1 - \varepsilon C_2$ and $C_a' = C_2 + \varepsilon C_2$, we can make the dependence of ω on ε explicit:

$$\omega \leq 1 - \frac{C_1 - \varepsilon C_2 - C_v - C_0}{C_2 + \varepsilon C_2 + C_v + C_0 + C_T} < 1,$$

so long as the numerator $C_1 - \varepsilon C_2 - C_v - C_0 > 0$.

We mention that in [9] a version of our Theorem 1 for a block Gauss-Seidel algorithm is proved in the special case of a rectangular domain with $\Phi(x,y) = (x,y)$. We also mention that, while we proved Theorem 1 in the plane, it also holds in higher dimensions and on manifolds, under suitable hypotheses; see [9]. This is important because one cannot rely, e.g., on conformal maps in dimensions higher than 2 to prove results for general domains.

---

[5] This trace inequality follows from the following argument: since $\rho \gg 0$ on Γ_{x_1}, there is some neighborhood U of Γ_{x_1} such that $\rho(\mathbf{x}) > a > 0$ everywhere in U. Then, $\int_U (\nabla v)^2 + v^2 < \frac{1}{a}\int_\Omega ((\nabla v)^2 + v^2)\rho$, and we use the Trace Theorem [2] for $H^1(U)$.

3 Applications to Schwarz Methods

This maximum principle can be used to prove convergence of the classical Schwarz iteration. If Σ is a domain in the plane and $\Sigma = \Sigma_1 \cup \Sigma_2$ is an overlapping domain decomposition, and given $u_0 \in H_0^1(\Sigma)$, the alternating Schwarz method is

$$\begin{aligned} (-\nabla \cdot a\nabla + c)u_{k+j/2} &= f && \text{in } \Sigma_j, \\ u_{k+j/2} &= 0 && \text{on } \partial\Sigma, \\ u_{k+j/2} &= u_{k+(j-1)/2} && \text{on } \partial\Sigma_j \cap \Sigma_{3-j}; \end{aligned} \tag{16}$$

with $k = 0,1,2,\dots$ and $j = 1,2$. Obviously, the iteration converges if the Dirichlet data (16) converge to zero in $L^2(\partial\Sigma_j)$. Let $v_{k+j/2}^{(\ell)} := u - u_{k+j/2}^{(\ell)}$ be the error terms. By setting $\Omega = \Sigma_1$ in Theorem 1, we see that $\|v_{k+1/2}\|_{L^2(\partial\Sigma_2)} < \sqrt{\omega}\|v_k\|_{L^2(\partial\Sigma_1)}$. Similarly, if Theorem 1 holds with $\Omega = \Sigma_2$, we obtain that $\|v_{k+1}\|_{L^2(\partial\Sigma_1)} < \sqrt{\omega}$ $\|v_{k+1/2}\|_{L^2(\partial\Sigma_2)}$. Chaining these together, one obtains

$$\|v_{k+1}\|_{L^2(\partial\Sigma_1)} < \omega\|v_k\|_{L^2(\partial\Sigma_1)}$$

and so the classical Schwarz iteration converges, and the error is multiplied by ω at every full iteration.

It is commonplace to use inexact solvers for the local problems, e.g., the multigrid algorithm. Such methods generate *inner iterates*: for each j,k, one obtains a sequence $u_{k+j/2}^{(\ell)}$, $\ell = 1,2,\dots$ which converges to $u_{k+j/2}$ in the limit. However, the iteration is typically stopped before the residual is zero. The inequality (7) is a condition on the size of the residual. If the residual $f - (-\nabla \cdot a\nabla + c)u_{k+j/2}^{(\ell)}$ is small, then for the error term, we have that $(-\nabla \cdot a\nabla + c)v_{k+j/2}^{(\ell)}$ is small, and so $v_{k+j/2}^{(\ell)}$ is (ε, a, c)-harmonic for some small ε.[6] Hence, the Schwarz iteration is robust in the sense that it will tolerate inexact local solvers.

Less obviously, a consequence of Theorem 1 is that the Optimized Schwarz Method converges. To that end, we say that a domain decomposition for which Theorem 1 holds for $\Omega = \Sigma_1$ as well as $\Omega = \Sigma_2$ is said to be ε-relatively uniform.

Theorem 2. *Let $\varepsilon > 0$ and assume that the domain decomposition is ε-relatively uniform. Then the Optimized Schwarz Method for the general elliptic problem (1) converges geometrically for any Robin parameter $\alpha > 0$.*

We prove this theorem in [9], but without the benefit of the $\varepsilon > 0$ parameter. Although we have not proved it, we hope that the robustness of the ε-harmonicity condition can be used to show that the Optimized Schwarz Method can also use inexact local solvers.

[6] If exact solvers are used, then (7) is verified with $\varepsilon = 0$ and $v = v_k$. If an inexact solver is used, then (7) can be used as a stopping criterion for $v = v_{k+j/2}^{(\ell)}$, assuming that the inexact solver can reach (7). If the inexact solver stops and (7) is not satisfied, it is possible for the outer iteration to stagnate, never reaching an error of 0.

Acknowledgement. We thank a referee for helpful comments. This research was supported in part by the U.S. Department of Energy under grant DE-FG02-05ER25672.

References

[1] Adams, R.A., Fournier, J.J.F.: *Sobolev Spaces.* Academic (Elsevier), Oxford, 2003.

[2] Evans, L.C.: *Partial Differential Equations.* AMS, Providence, RI, 1998.

[3] Gander, M.J.: Optimized Schwarz Methods. *SIAM J. Numer. Anal.*, 44:699-731, 2006.

[4] Kimn, J-H.: Overlapping Schwarz Algorithms using Discontinuous Iterates for Poisson's Equation. Ph.D. Thesis, Department of Mathematics, New York University, New York, 2001.

[5] Kimn, J.-H.: A Convergence Theory for An Overlapping Schwarz Algorithm using Discontinuous Iterates. *Numer. Math.*, 100:117-139, 2005.

[6] Kufner, A.: *Weighted Sobolev Spaces.* Wiley, Chistester, U.K., 1985.

[7] Lions, J.L., Magenes, E.: *Non-Homogeneous Boundary Value Problems and Applications I.* Springer, 1970.

[8] Lions, P.L.: On the Schwarz alternating method. III: A variant for nonoverlapping subdomains. In T. F. Chan, R. Glowinski, J. Périaux, O. Widlund, eds. *Third International Symposium on Domain Decomposition Methods for Partial Differential Equations*, SIAM, 1990, Philadelphia, pp. 202–223.

[9] Loisel, S., Szyld, D.B.: On the convergence of Algebraic Optimizable Schwarz Methods with applications to elliptic problems. Research Report 07-11-16, Department of Mathematics, Temple University, November 2007.

[10] Nataf, F.: Absorbing boundary conditions in block Gauss-Seidel methods for convection problems. *Math. Models Methods Appl. Sci.*, 6:481–502, 1996.

[11] Schwarz, H.: Über einige *Abbildungsaufgaben. J. Reine Angew. Math.*, vol. 70, 105–120, 1869.

An Extended Mathematical Framework for Barrier Methods in Function Space

Anton Schiela

Zuse Institute Berlin, Takustr. 7, 14195 Berlin, Germany, schiela@zib.de

1 Convex State Constrained Optimal Control

In this note, we extend the mathematical framework in [7] of barrier methods for state constrained optimal control problems with PDEs to a more general setting. In [7] we modelled the state equation by $Ly = u$ with L a closed, densely defined, surjective operator. This restricts the applicability of our theory mainly to certain distributed control problems. Motivated by the discussion in [6], we consider in this work operator equations of the more general form $Ay - Bu = 0$, where A is closed, densely defined and with closed range and B is continuous. While this change in framework only neccessitates minor modificatios in the theory, it extends its applicability to large additional classes of control problems, such as boundary control and finite dimensional control.

To make this paper as self contained as possible, assumptions and results of [7] are recapitulated, but for brevity proofs and more detailed information are only given when there are differences to [7]. This is possible, because our extension has only a very local effect.

Let Ω be an open and bounded Lipschitz domain in $\mathbb{R}^d$ and $\overline{\Omega}$ its closure. Let $Y := C(\overline{\Omega})$ and $U := L_2(Q)$ for a measurable set Q, equipped with an appropriate measure. Standard examples are $Q = \Omega$ with the Lebesgue measure for distributed control, $Q = \partial\Omega$ with the boundary measure for boundary control and $Q = \{1, 2, \ldots, n\}$ with the counting measure for finite dimensional controls.

Define $X := Y \times U$ with $x := (y, u)$ and consider the following convex minimization problem, the details of which are fixed in the rest of Section 1.

$$\begin{aligned} \min_{x \in X} J(x) \quad & \text{s.t. } Ay - Bu = 0 \\ & \underline{u} \le u \le \overline{u}, \qquad \underline{y} \le y \le \overline{y}. \end{aligned} \tag{1}$$

We will now specify our abstract theoretical framework and collect a couple of basic results about this class of problems.

M. Bercovier et al. (eds.), *Domain Decomposition Methods in Science and Engineering XVIII*, Lecture Notes in Computational Science and Engineering,
DOI: 10.1007/978-3-642-02677-5, © Springer-Verlag Berlin Heidelberg 2009

1.1 Linear Equality Constraints

By the equality constraint $Ay - Bu = 0$ we model a partial differential equation (cf. Section 1.3 below).

Assumption 1 *Let R be a Banach space. Assume that $B : U \to R$ is a continuous linear operator and that $A : Y \supset \operatorname{dom} A \to R$ is a densely defined and closed linear operator with a closed range.*

Assume that there is a finite dimensional subspace $V \subset U$ of essentially bounded functions on Q, such that $R = \operatorname{ran} A \oplus B(V)$, i.e., for each $r \in R$ there are unique $r_Y \in \operatorname{ran} A$ and $r_V \in B(V)$ with $r = r_Y + r_V$.

Closed operators are a classical concept of functional analysis. For basic results we refer to [9, Kapitel IV.4] for more details, see [5]. In many applications A is bijective, i.e., the equation $Ay = r$ has a unique solution y for all $r \in R$. However, there are several important cases (such as pure Neumann problems), where only a Fredholm alternative holds while the corresponding optimal control problems are still well posed. Introduction of V includes these cases. If A is surjective, then $V = \{0\}$. Consider now the operator

$$\begin{aligned} T : Y \times U \supset \operatorname{dom} A \times U &\to R \\ (y,u) &\mapsto Ay - Bu. \end{aligned} \tag{2}$$

From our assumptions it can be shown easily that T is densely defined, closed and surjective. Since T is closed, $E := \ker T$ is a closed subspace of X.

By density of $\operatorname{dom} A$ in Y, we can define an adjoint operator A^*. For every $l \in R^*$ the mapping $y \to \langle l, Ay \rangle$ is a linear functional on $\operatorname{dom} A$. We define $\operatorname{dom} A^*$ as the subspace of all $l \in R^*$ for which $y \to \langle l, Ay \rangle$ is continuous on $\operatorname{dom} A$ and can thus by density be extended uniquely to a continuous functional on Y. Hence, for all $l \in \operatorname{dom} A^*$ there is a unique $A^* l \in Y^*$ for which $\langle l, Ay \rangle = \langle A^* l, y \rangle \ \forall y \in \operatorname{dom} A$. This defines $A^* : R^* \supset \operatorname{dom} A^* \to Y^*$.

1.2 Inequality Constraints and Convex Functionals

The inequality constraints in (1) are interpreted to hold pointwise almost everywhere and define a closed convex set of $G \subset X$. Some of the inequality constraints may not be present.

Assumption 2 *Assume that $E = \ker T$ is weakly sequentially compact. Assume that there is a strictly feasible point $\check{x} = (\check{y}, \check{u}) \in E$, which satisfies*

$$0 < d_{\min} := \operatorname*{ess\,inf}_{t \in \Omega} \min\{\check{u}(t) - \underline{u}(t), \overline{u}(t) - \check{u}(t), \check{y}(t) - \underline{y}(t), \overline{y}(t) - \check{y}(t)\}. \tag{3}$$

Assume that $J : X \to \overline{\mathbb{R}} := \mathbb{R} \cup \{+\infty\}$ is lower semi-continuous, convex, and coercive on the feasible set $E \cap G$, that J is continuous at $\check{x}$ (cf. (3)) and that its subdifferential ∂J is uniformly bounded in X^* on bounded sets of X.

Weak sequential compactness of E can usually shown by taking into account slightly stronger regularity properties of A. Often $\operatorname{dom} A$ is contained in a reflexive (Sobolev)-Space.

Denote by $\chi_C(x)$ the indicator function of a set $C \subset X$, which vanishes on C and is $+\infty$ otherwise. Then we can rewrite (1) as an unconstrained minimization problem defined by the functional:

$$\begin{aligned} F &: X \to \overline{\mathbb{R}} := \mathbb{R} \cup \{+\infty\} \\ F &:= J + \chi_E + \chi_G. \end{aligned} \tag{4}$$

By our assumptions F is a lower semi-continuous, convex, and coercive functional with a non-empty domain and does thus admit a minimizer by weak compactness of E (cf. e.g. [4, Prop. II.1.2]).

Assumption 3 *Assume that F is strongly convex (w.r.t. some norm $\|\cdot\|$):*

$$\exists \alpha > 0 : \alpha \|x-y\|^2 \le F(x) + F(y) - 2F\left(\frac{x+y}{2}\right) \quad \forall x,y \in \operatorname{dom} F \tag{5}$$

Usually, optimal control problems with Tychonov regularization satisfy (5).

1.3 Example: A class of Elliptic PDEs

To illustrate our theoretical framework, we consider a class of elliptic PDEs, which was analysed by Amann [1] in an even more general framework.

Let Ω be a bounded domain in $\mathbb{R}^d$ with a smooth boundary Γ. Let $\mathbf{a} \in C(\overline{\Omega}, \mathbb{R}^{d\times d})$, $\mathbf{b}, \mathbf{c} \in C(\overline{\Omega}, \mathbb{R}^d)$, $a_0 \in L_\infty(\Omega)$, $b_0 \in C(\Gamma)$. Assume that $\mathbf{a}$ is symmetric positive definite, uniformly in Ω. Denote by $\gamma(\cdot) : W^{1,s} \to L_2(\Gamma)$ the boundary trace operator, which exists continuously if $s > 3/2$. For $1 < q < \infty$ and $1/q + 1/q' = 1$ consider the following continuous elliptic differential operator in the weak formulation:

$$\begin{aligned} A &: W^{1,q}(\Omega) \to (W^{1,q'}(\Omega))^* \\ \langle Ay, p\rangle &:= \int_\Omega \langle \mathbf{a}\nabla y + \mathbf{b}y, \nabla p\rangle + \langle \nabla y, \mathbf{c}p\rangle + a_0 yp\, dt + \int_\Gamma b_0 \gamma(y)\gamma(p)\, ds. \end{aligned} \tag{6}$$

Let $f \in (W^{1,q'}(\Omega))^*$. By [1, Theorem 9.2] a Fredholm alternative holds for the solvability of the equation $Ay = f$. This means that either it is uniquely solvable, or the homogenous problem has a finite dimensional space of nontrivial solutions with basis vectors $w_i \in W^{1,q}(\Omega)$. Then there is a finite number of conditions $\langle w_i, f\rangle = 0$ under which $Ay = f$ is non-uniquely solvable. This implies that A has a closed range with finite codimension and a kernel of the same dimension. In case of solvability, we have (cf. [1, 9.3(d)]):

$$\|y\|_{W^{1,q}} \le C(\|f\|_{(W^{1,q'})^*} + \|y\|_{(W^{1,q'})^*}). \tag{7}$$

If $q > d$, then by the Sobolev embedding theorems $W^{1,q}(\Omega) \hookrightarrow C(\overline{\Omega})$ and we may redefine A as an unbounded operator

$$A : C(\overline{\Omega}) \supset W^{1,q}(\Omega) \to (W^{1,q'}(\Omega))^*.$$

Since $C^\infty(\overline{\Omega})$ is dense in $C(\overline{\Omega})$, A is densely defined, and closedness of A follows easily from (7), continuity of the embedding $W^{1,q}(\Omega) \hookrightarrow C(\overline{\Omega})$, and closedness of $\operatorname{ran} A$. Hence, setting $Y := C(\overline{\Omega})$, $R := (W^{1,q'}(\Omega))^*$, and $\operatorname{dom} A := W^{1,q}(\Omega)$, A fits into our framework. Its adjoint operator

$$A^* : W^{1,q'}(\Omega) \supset \operatorname{dom} A^* \to C(\overline{\Omega})^*.$$

is defined by $\langle y, A^* p\rangle = \langle Ay, p\rangle$ via the right hand side in (6). This expression is well defined for all $y \in \operatorname{dom} A = W^{1,q}(\Omega)$, and $\operatorname{dom} A^*$ is the set of all p, for which $\langle Ay, p\rangle$ is continuous on $\operatorname{dom} A$ with respect to $\|y\|_\infty$ and have thus a unique continuous extension to an element of $C(\overline{\Omega})^*$.

By the choice of B we select how the control acts on the state. Two examples are distributed control

$$B_\Omega : L_2(\Omega) \to (W^{1,q'}(\Omega))^* \qquad \langle B_\Omega u, p\rangle := \int_\Omega u \cdot p\, dt,$$

and Neumann or Robin boundary control

$$B_\Gamma : L_2(\Gamma) \to (W^{1,q'}(\Omega))^* \qquad \langle B_\Gamma u, p\rangle := \int_\Gamma u \cdot \gamma(p)\, ds.$$

If $q' < d/(d-1)$ is chosen sufficiently large, B_Ω is continuous by the Sobolev embedding theorem for $d \le 3$ and B_Γ is continuous by the trace theorem for $d \le 2$. If $d = 3$, then $\gamma : W^{1,q'} \to L_2(\Gamma)$ is not continuous and thus the case $d = 3$ is not included in our framework for B_Γ. This has been a principal problem for the analysis (not only for barrier methods) of state constrained optimal control problems (cf. e.g. [3]). However, in [8] new techniques have been developed to overcome this restriction, which are likely to carry over to the analysis of barrier methods.

If $Ay = f$ is not uniquely solvable, then we have to assert that $u \in U$ can be split into $u = u_Y + u_V$, such that $\langle w_i, Bu_Y\rangle = 0$ and $u_V \in L_\infty$. Since all $w_i \in W^{1,q}$ are bounded, such an u_V can easily be constructed from these w_i in our cases $B = B_\Omega$ and $B = B_\Gamma$.

2 The Homotopy Path and its Properties

We analyse the main properties of the homotopy path of barrier regularizations. For brevity, we give only proofs here, when they differ from [7].

Definition 1. *For all $q \ge 1$ and $\mu > 0$ the functions $l(z;\mu) : \mathbb{R}_+ \to \overline{\mathbb{R}}$*

$$l(z;\mu) := \begin{cases} -\mu \ln(z) : & q = 1 \\ \dfrac{\mu^q}{(q-1)z^{q-1}} : & q > 1 \end{cases}$$

are called barrier functions of order q. *We extend their domain of definition to $\mathbb{R}$ by setting $l(z;\mu) = \infty$ for $z \le 0$.*

Their derivatives can be computed as $l'(z;\mu) = -\mu^q z^{-q}$. Bounds like $z \geq \bar{z}$ and $z \leq \underline{z}$, are incorporated by shifting the arguments.

Using these barrier *functions* $l(z;\mu)$, we construct barrier *functionals* $b(z;\mu)$ on suitable spaces Z to implement constraints of the form $z \geq 0$ on a measurable set $B \subset \overline{\Omega}$ by computing the integral over l:

$$\begin{aligned} b(\cdot;\mu) : Z &\to \overline{\mathbb{R}} \\ z &\mapsto \int_B l(z(t);\mu)\,dt. \end{aligned}$$

By $b'(z;\mu)$ we denote the formal derivative of $b(z;\mu)$, defined by

$$\langle b'(z;\mu), \delta z\rangle := \int_Q l'(z;\mu)\delta z\,dt,$$

if the right hand side is well defined. The following result connects these formal derivatives to the subifferentials of convex analysis (cf. e.g. [4, Section I.5]).

Proposition 1. *Consider $b : L_p(Q) \to \overline{\mathbb{R}}$, $1 \leq p < \infty$ on a measurable set Q. Then either $\partial b(z;\mu) = \emptyset$, or $\partial b(z;\mu) = \{b'(z;\mu)\}$.*

Consider $b : C(Q) \to \overline{\mathbb{R}}$ on a compact set Q and assume $\emptyset \neq \partial b(z;\mu) \subset M(Q) \cong C(Q)^*$. Then on the set of strictly feasible points $S := \{t \in Q : z(t) > 0\}$ we have

$$m|_S = b'(z;\mu)|_S \qquad \forall m \in \partial b(z;\mu). \tag{8}$$

In particular, $\partial b(z;\mu) \cap L_1(Q) = \{b'(z;\mu)\}$. Moreover,

$$\langle m, \delta z\rangle \leq \langle b'(z;\mu), \delta z\rangle \leq 0 \qquad \forall 0 \leq \delta z \in C(Q) \tag{9}$$

and

$$\left\|b'(z;\mu)\right\|_{L_1(Q)} = \min_{m \in \partial b(z;\mu)} \|m\|_{M(Q)}. \tag{10}$$

Adding barrier functionals to F, we obtain another convex functional F_μ defined by

$$\begin{aligned} F_\mu(x) := F(x) + b(x;\mu) &= J(x) + \chi_E(x) + \chi_G(x) + b(x;\mu) \\ &= J(x) + \chi_E(x) + b(x;\mu). \end{aligned} \tag{11}$$

Our definition implies $F_0 = F$, which means that the original state constrained problem is included in our analysis.

Theorem 4 (Existence of Minimizers). *Let $F : X \to \overline{\mathbb{R}}$ be defined by* (4) *and suppose that Assumptions 1–2 hold. Assume that F_{μ_0} is coercive for some $\mu_0 > 0$.*

Then (11) *admits a unique minimizer $x(\mu) = (u(\mu), y(\mu))$ for each $\mu \in]0;\mu_0]$. Moreover, $x(\mu)$ is strictly feasible almost everywhere in Ω and bounded in X uniformly in $\mu \in [0,\mu_0]$.*

Next we study first order optimality conditions for barrier problems. For this purpose, we first have to study the subdifferential of χ_E, the characteristic function for the equality constraints $Ay - Bu = 0$, which can by (2) be written as $Tx = 0$. It is at this point, where our theory differs from [7].

Lemma 1. *If Assumption 1 holds, then there is a constant M, such that for each $u \in U$ there are $y \in Y$, $u_Y \in U$ and $u_V \in V$ with $Ay - Bu_Y = 0$ and*

$$u = u_Y + u_V \qquad \|y\|_\infty + \|u_V\|_\infty \le M \|u\|_U . \tag{12}$$

Proof. For $u \in U$ let $Bu = r$ and $r = r_Y + r_V$ as in Assumption 1. Since $\operatorname{ran} A$ and $B(V)$ ($\dim V < \infty$) are closed, [9, Satz IV.6.3] yields a constant c independent of r, such that $\|r_Y\| + \|r_V\| \le c\|r\| \le c\|B\|\,\|u\|_U$.

By closedness of $B(V)$, the mapping $B : V \to B(V)$ is open, which yields a constant C such that for each $r_V \in B(V)$ there is $u_V \in V$ with $Bu_V = r_V$ and $\|u_V\|_U \le C\|r_V\|$. Since all norms are equivalent on finite dimensional spaces, and V is a space of bounded functions, we even have $\|u_V\|_\infty \le C\|r_V\|$.

Similarly, because $\operatorname{ran} A$ is closed, $A : Y \supset \operatorname{dom} A \to \operatorname{ran} A$ is an open mapping by [9, Satz IV.4.4] and for each $r_Y \in \operatorname{ran} A$ there is $y \in \operatorname{dom} A$ with $Ay = r_Y$ and $\|y\|_\infty \le C\|r_Y\|$.

This altogether yields $\|y\|_\infty + \|u_V\|_\infty \le C(\|r_Y\| + \|r_V\|) \le M\|u\|_U$ and thus (12).

Proposition 2. *Let X, R be Banach spaces and $T : X \supset \operatorname{dom} T \to R$ a closed, densely defined, linear operator with closed range. Denote by χ_E the indicator function of $E := \mathbf{ker} T$. Then*

$$\partial \chi_E(x) = \operatorname{ran} T^* \quad \forall x \in E. \tag{13}$$

Proof. Since, by definition of the subdifferential, $\partial\chi_E(x) = (\ker T)^\perp$, (13) is a consequence of the *closed range theorem* for closed operators on Banach spaces [5, Theorem IV.1.2], which asserts $(\ker T)^\perp = \operatorname{ran} T^*$.

Theorem 5 (First Order Optimality Conditions). *Suppose that the Assumptions 1–2 hold. For $\mu \ge 0$ let x be the unique minimizer of F_μ.*

Then there are $(j_y, j_u) = j \in \partial J(x)$, $m \in \partial b(y;\mu) \subset Y^*$ and $p \in \operatorname{dom} A^*$ such that

$$\begin{aligned} j_y + m + A^* p &= 0 \\ j_u + b'(u;\mu) - B^* p &= 0 \end{aligned} \tag{14}$$

holds. If y is strictly feasible, then $\partial b(y;\mu) = \{b'(y;\mu)\}$ and m is unique.

Proof. Let x be a minimizer of F_μ. Then $0 \in \partial F_\mu(x) = \partial(J + \chi_E + b)(x)$.

To show that (14) has a solution, we have to apply the sum-rule of convex analysis twice:

$$0 \in \partial(J + \chi_E + b) = \partial J + \partial(\chi_E + b) = \partial J + \partial\chi_E + \partial b.$$

To be able to apply the sum-rule to a sum $f + g$ of convex, lower semi-continuous functions, they have to satisfy an additional regularity condition, such as the following (cf. e.g. [2, Theorem 4.3.3]):

$$0 \in \mathrm{int}(\mathrm{dom} f - \mathrm{dom} g). \tag{15}$$

Let now B_X be the unit ball in a normed space X. We observe that showing (15) is equivalent to showing that there is $\varepsilon > 0$ such that each $x \in \varepsilon B_X$ can be written as a difference $x_1 - x_2$ with $x_1 \in \mathrm{dom} f$ and $x_2 \in \mathrm{dom} g$.

By (3) there exists a strictly feasible point $\check{x} = (\check{y}, \check{u})$, which implies $\check{x} \in \mathrm{dom}(\chi_E + b)$. Our assumptions on J include continuity at $\check{x}$ and hence boundedness in some ball $\check{x} + \varepsilon B_X$. Thus,

$$\varepsilon B_X = (\check{x} + \varepsilon B_X) - \check{x} \subset \mathrm{dom} J - \mathrm{dom}(b + \chi_E),$$

and we conclude that (15) is fulfilled for $f = J$ and $g = \chi_E + b$. Therefore the sum-rule can be applied and yields $\partial(J + \chi_E + b) = \partial J + \partial(\chi_E + b)$.

Next we show that $\partial(\chi_E + b) = \partial \chi_E + \partial b$ by verifying (15) for b and χ_E. Here $Y = C(\overline{\Omega})$ is crucial because it guarantees that $(\check{u}, \check{y} + rB_Y) \in \mathrm{dom} b$ for $r < d_{\min}$ via (3). By (12) there is $\delta > 0$ such that for each $u \in \delta B_U$ we find an $y \in (r/2)B_Y$ with $Ay - Bu_Y = 0$ and u_V with $\|u_V\|_\infty \le r$, such that $u = u_Y + u_V$.

Thus $(\check{y} + y, \check{u} + u_Y) \in \mathrm{dom} \chi_E$ and $(\check{y} + y - w, \check{u} - u_V) \in \mathrm{dom} b$ for all $w \in (r/2)B_Y$ by (3). Consequently, for sufficiently small ε and arbitrary $(w, u) \in \varepsilon B_X$ we have

$$\begin{aligned} w &= (\check{y} + y) - (\check{y} + y - w) \\ u &= \underbrace{(\check{u} + u_Y)}_{\in \mathrm{dom} \chi_E} - \underbrace{(\check{u} - u_V)}_{\in \mathrm{dom} b}. \end{aligned}$$

This finally shows (15) and the sum-rule yields $0 \in \partial J + \partial \chi_E + \partial b$.

This is an inclusion in $Y^* \times U^*$. It implies that there are $(j_y, j_u) \in \partial J(x)$, $(\nu, p) \in \partial \chi_E(x)$, $m \in \partial b(y; \mu)$, and $l \in \partial b(u; \mu)$, such that

$$\begin{aligned} j_y + \nu + m &= 0 \\ j_u + \lambda + l &= 0. \end{aligned}$$

Proposition 2 applied to T as defined in (2) yields $(\nu, \lambda) \in \mathrm{ran}\, T^*$. Hence there is $p \in \mathrm{dom}\, T^*$ with $\nu = A^* p$ and $\lambda = B^* p$. Proposition 1 characterizes m and l in terms of barrier gradients. This yields (14). If y is strictly feasible, then $m = b'(y; \mu)$ by Proposition 1.

Once, existence of the barrier gradients is established, their uniform boundedness for $\mu \to 0$ can again be shown as in [7].

Proposition 3. *Suppose that the Assumptions 1–2 hold. Then for each $\mu_0 > 0$*

$$\sup_{\mu \in [0; \mu_0[} \|m\|_{Y^*} \le C.$$

Just as in [7] this result allows also to derive uniform bounds on the adjoint state $p(\mu)$ in some suitable Sobolev space. The results on the analytic properties of the central path carry over literally from [7].

Theorem 6. *Suppose that that the Assumptions 1–3 hold. Let $x(\mu)$ be a barrier minimizer for $\mu \geq 0$ and x_* be minimizer of F. Then*

$$F(x(\mu)) \leq F(x_*) + C\mu_0 \tag{16}$$

$$\|x(\mu) - x_*\| \leq C\sqrt{\frac{\mu}{\alpha}}. \tag{17}$$

$$\|x(\mu) - x(\tilde{\mu})\| \leq \frac{c}{\sqrt{\alpha\mu}}|\mu - \tilde{\mu}| \quad \forall \tilde{\mu} \geq 0. \tag{18}$$

Finally, we remark that the results on strict feasibility of the homotopy path, which depend on the regularity of $y(\mu)$, carry over from [7].

Acknowledgement. Supported by the DFG Research Center MATHEON "Mathematics for key technologies."

References

[1] Amann, H.: Nonhomogeneous linear and quasilinear elliptic and parabolic boundary value problems. In H.J. Schmeisser and H. Triebel, eds., *Function Spaces, Differential Operators and Nonlinear Analysis.*, pages 9–126. Teubner, Stuttgart, Leipzig, 1993.

[2] Borwein, J.M., Zhu, Q.J.: *Techniques of Variational Analysis.* CMS Books in Mathematics. Springer, 2005.

[3] Casas, E.: Boundary control of semilinear elliptic equations with pointwise state constraints. *SIAM J. Control Optim.*, 31:993–1006, 1993.

[4] Ekeland, I., Témam, R.: *Convex Analysis and Variational Problems.* Classics in Applied Mathematics, 28. SIAM, 1999.

[5] Goldberg, S.: *Unbounded Linear Operators.* Dover, 1966.

[6] Hinze, M., Schiela, A.: Discretization of interior point methods for state constrained elliptic optimal control problems: Optimal error estimates and parameter adjustment. Technical Report SPP1253-08-03, Priority Program 1253, German Research Foundation, 2007.

[7] Schiela, A.: Barrier methods for optimal control problems with state constraints. ZIB Report 07-07, Zuse Institute Berlin, 2007.

[8] Schiela, A.: Optimality conditions for convex state constrained optimal control problems with discontinuous states. ZIB Report 07-35, Zuse Institute Berlin, 2007.

[9] Werner, D.: *Funktionalanalysis.* Springer, 3rd ed., 2000.

Optimized Schwarz Preconditioning for SEM Based Magnetohydrodynamics

Amik St-Cyr[1], Duane Rosenberg[1], and Sang Dong Kim[2]

[1] Institute for Mathematics Applied to Geosciences, National Center for Atmospheric Research, Boulder, CO, USA: {amik,duaner}@ucar.edu
[2] Department of Mathematics, Kyungpook National University, Daegu 702-701, South Korea: skim@knu.ac.kr.

Summary. A recent theoretical result on optimized Schwarz algorithms, demonstrated at the algebraic level, enables the modification of an existing Schwarz procedure to its optimized counterpart. In this work, it is shown how to modify a bilinear finite-element method based Schwarz preconditioning strategy originally presented in [6] to its optimized version. The latter is employed to precondition the pseudo-Laplacian operator arising from the spectral element discretization of the magnetohydrodynamic equations in Elsässer form.

1 Introduction

This work concerns the preconditioning of a pseudo-Laplacian operator[3] associated with the saddle point problem arising at each time-step in a spectral element based adaptive MHD solver. The approach proposed is a modification of the method developed in [6] where an overlapping Schwarz preconditioner for the pseudo-Laplacian was constructed using a low order discretization of the weak Laplacian. The finite-element blocks, representing the additive Schwarz, are replaced by so called optimized Schwarz blocks [12]. Two types of overlapping subdomains, employed to construct the finite-element block preconditioning are investigated. The first one is cross shaped and shows good behavior for additive Schwarz (AS) and restricted additive Schwarz (RAS). Improved convergence rates of the optimized RAS (ORAS) version are completely dominated by the corner effects [2]. Opting for a second grid that includes the corners seems to correct this issue. For the zeroth order optimized transmission condition (OO0) an exact tensor product form is available while for the second order version (OO2) version a slight error is introduced in order to preserve the properties of the operators and enable the use of fast diagonalization techniques (FDM) [3].

[3] A.k.a: consistent Laplacian or approximate pressure Schur complement.

M. Bercovier et al. (eds.), *Domain Decomposition Methods in Science and Engineering XVIII*, Lecture Notes in Computational Science and Engineering,
DOI: 10.1007/978-3-642-02677-5,

2 Governing Equations and Discretization

For an incompressible fluid with constant mass density ρ_0, in two spatial dimensions, the magnetohydrodynamic (MHD) equations are:

$$\partial_t \mathbf{u} + \mathbf{u} \cdot \nabla \mathbf{u} = -\nabla p + \nabla \times \mathbf{b} \times \mathbf{b} + \nu \nabla^2 \mathbf{u}, \tag{1}$$

$$\partial_t \mathbf{b} = \nabla \times (\mathbf{u} \times \mathbf{b}) + \xi \nabla^2 \mathbf{b} \tag{2}$$

$$\nabla \cdot \mathbf{u} = 0, \qquad \nabla \cdot \mathbf{b} = 0 \tag{3}$$

$$\mathbf{u}(\mathbf{x}, t=0) = \mathbf{u}_i, \;\; \mathbf{b}(\mathbf{x}, t=0) = \mathbf{b}_i; \;\; \mathbf{u}(\mathbf{x}, t)|_{\partial \mathbb{D}} = \mathbf{u}^b, \;\; \mathbf{b}(\mathbf{x}, t)|_{\partial \mathbb{D}} = \mathbf{b}^b \tag{4}$$

where $\mathbf{u}$ and $\mathbf{b}$ are the velocity and magnetic field (in Alfvén velocity units, $\mathbf{b} = \mathbf{B}/\sqrt{\mu_0 \rho_0}$ with $\mathbf{B}$ the induction and μ_0 the permeability); p is the pressure divided by the (constant) mass density, ρ_0, and ν and ξ are the kinematic viscosity and the magnetic resistivity. In the closed domain $\mathbb{D}$, these equations are solved in Elsässer form [5]:

$$\partial_t \mathbf{Z}^{\pm} + \mathbf{Z}^{\mp} \cdot \nabla \mathbf{Z}^{\pm} + \nabla p - \nu^{\pm} \nabla^2 \mathbf{Z}^{\pm} - \nu^{\mp} \nabla^2 \mathbf{Z}^{\mp} = 0 \tag{5}$$

$$\nabla \cdot \mathbf{Z}^{\pm} = 0, \tag{6}$$

with $\mathbf{Z}^{\pm} = \mathbf{u} \pm \mathbf{b}$ and $\nu^{\pm} = \frac{1}{2}(\nu \pm \eta)$. The initial and boundary conditions for $\mathbf{Z}^{\pm}$ are trivially specified in terms of (4); we do not provide them here. The velocity $\mathbf{u}$ and magnetic field $\mathbf{b}$ can be recovered by expressing them in terms of $\mathbf{Z}^{\pm}$. For spatial discretization of (5)-(6), a $\mathbb{P}_N - \mathbb{P}_{N-2}$ spectral element formulation is chosen to prevent the excitation of spurious pressure modes. In the latter formalism, the domain $\mathbb{D}$ is composed of a union of non-overlapping quadrangles, $\mathbf{E}_k$: $\mathbb{D} \supseteq \bigcup_{k=1}^{K} \mathbf{E}_k =: \mathcal{T}_h$ where $\mathbf{P}_N = \{v_h \in L^2(\mathbb{D}) \mid v_h|_{\mathbf{E}_k} \circ T_{\mathbf{E}_k} \in (\mathbb{P}_N \otimes \mathbb{P}_N)(\mathbf{E}_k) \; \forall \; \mathbf{E}_k \in \mathcal{T}_h \}$ and $T_{\mathbf{E}_k}$ is the image of the reference element $[-1,1] \times [-1,1]$. Finally $\mathbf{P}_N$ and the space $\mathbf{U}_\gamma := \{\mathbf{w} \in (H^1(\mathbb{D}))^2 | \; \mathbf{w} = \gamma \text{ on } \partial \mathbb{D}\}$, with the usual definition for $H^1(\mathbb{D})$, are employed to define finite dimensional representations of $\mathbf{Z}^{\pm}$ (and $\mathbf{u}$, $\mathbf{b}$), p and test functions, $\zeta^{\pm}$ and q:

$$\mathbf{Z}_h^{\pm} \in \mathbf{U}^N = \mathbf{U}_{\mathbf{Z}^b} \bigcap (\mathbf{P}_N \cap C^0(\mathbb{D}))^2, \quad \zeta_h^{\pm} \in \mathbf{U}_0^N = \mathbf{U}_\mathbf{0} \bigcap (\mathbf{P}_N \cap C^0(\mathbb{D}))^2,$$
$$p_h, q_h \in \mathbf{Y}_0^{N-2} = L_0^2(\mathbb{D}) \bigcap \mathbf{P}_{N-2}, \tag{7}$$

see for instance [9][4]. The basis for the velocity expansion in $\mathbf{P}_N \cap C^0(\mathbb{D})$ is the set of Lagrange interpolating polynomials on the Gauss-Lobatto-Legendre (GL) quadrature nodes, and the basis for the pressure is the set of Lagrange interpolants on the Gauss-Legendre (G) quadrature nodes $\{\eta_l\}_{l=0}^{N-2}$. The functions in $\mathbf{U}^N$ and $\mathbf{Y}_0^{N-2}$ are represented as expansions in terms of tensor products of basis functions within each subdomain $\mathbf{E}_k$. Substituting these (Galerkin) truncations into the variational form of (5)-(6), and using appropriate quadrature rules, we arrive at a set of semi-discrete equations written in terms of spectral element operators:

[4] $L_0^2(\mathbb{D}) := \{p \in L^2(\mathbb{D}) \mid \int_{\mathbb{D}} p = 0\}$ which fixes the null space for the pressure.

$$\mathbf{M}\frac{d\hat{\mathbf{Z}}_j^{\pm}}{dt} = -\mathbf{M}\mathbf{C}^{\mp}\hat{\mathbf{Z}}_j^{\pm} + \mathbf{D}_j^{\mathrm{T}}\hat{\boldsymbol{p}}^{\pm} - \nu_{\pm}\mathbf{L}\hat{\mathbf{Z}}_j^{\pm} - \nu_{\mp}\mathbf{L}\hat{\mathbf{Z}}_j^{\mp} \tag{8}$$

$$\mathbf{D}^j\hat{\mathbf{Z}}_j^{\pm} = 0, \tag{9}$$

for the jth ($\in [1,\cdots,d]$) component. The variables $\hat{\mathbf{Z}}^{\pm}$ represent the time-dependent coefficients of the polynomial expansions of $\mathbf{Z}_h^{\pm}$ collocated at the GL node points, and $\hat{\boldsymbol{p}}^{\pm}$ are values of the pressure coefficients at the G node points. Hence, in this discretization vector quantities reside on a different mesh than the pressures leading to a staggered formulation. Note that because the constraints (6), are enforced separately on $\mathbf{Z}^{\pm}$, Eq. (8) contains a different pressure for *each* Elsässer vector. This essentially adds a pressure force, $-\frac{1}{2}\nabla(p^+ + p^-)$, to the momentum equation, and an electromotive force, $-\frac{1}{2}\nabla(p^+ - p^-)$, to the induction equation. In effect, we add a Lagrange variable to the induction equation, in the same way that it already exists for the velocity, such that $\nabla\cdot\mathbf{B} = 0$ in its discrete form–a form of *divergence cleaning* that renders the gradient (curl) and divergence operators consistent numerically. Note that if $\hat{\boldsymbol{p}}^+ = \hat{\boldsymbol{p}}^-$, the discrete approximation faithfully reproduces the continuous equations; in practice we find good agreement between these fields. The operators $\mathbf{M}$, $\mathbf{L}$, and $\mathbf{C}$, are the well-known mass matrix, weak Laplacian and advection operators, respectively (*c.f.* [4]), and $\mathbf{D}_j$ represent the Stokes derivative operators, in which the GL basis function and its derivative operator are interpolated to the G node points, and multiplied by the G quadrature weights. All two dimensional operators are computed as tensor products of their component 1D operators. We note the effect of $\mathbf{D}_j$, on (vector) quantities in $\mathbf{U}^N$: they take a derivative that itself resides on the G nodes; hence, the discrete divergence (9) is collocated on the same grid as the discrete pressure. The effect of the transposed Stokes operator $\mathbf{D}_j^{\mathrm{T}}$, on the other hand, is to compute a derivative of a $\mathbf{Y}^{N-2}$ quantity, which will be collocated with the $\mathbf{Z}^{\pm}$, $\mathbf{u}$, and $\mathbf{b}$. For time marching, we employ a simple second-order Runge-Kutta scheme (RK2) [1, p. 109]. The complete time discretization at each stage is (from 8):

$$\hat{\mathbf{Z}}_j^{\pm,k} = \hat{\mathbf{Z}}_j^{\pm,n} - \frac{k}{2}\Delta t\mathbf{M}^{-1}\left(\mathbf{M}\mathbf{C}^{\mp}\hat{\mathbf{Z}}_j^{\pm,k-1} - \mathbf{D}_j^{\mathrm{T}}\hat{\boldsymbol{p}}^{\pm,k-1} + \nu_{\pm}\mathbf{L}\hat{\mathbf{Z}}_j^{\pm,k-1} + \nu_{\mp}\mathbf{L}\hat{\mathbf{Z}}_j^{\mp,k-1}\right) \tag{10}$$

where $k = 1$ for the first stage and $k = 2$ for the last one[5]. We require that each stage satisfy (9) in its discrete form, so multiplying (10) by $\mathbf{D}^j$, summing over j, and setting the term $\mathbf{D}^j\hat{\mathbf{Z}}_j^{\pm} = 0$, we arrive at the following pseudo-Poisson equation for the pressures, $\hat{\boldsymbol{p}}^{\pm,k-1}$:

$$E\hat{\boldsymbol{p}}^{\pm,k-1} := \mathbf{D}^j\mathbf{M}^{-1}\mathbf{D}_j^{\mathrm{T}}\hat{\boldsymbol{p}}^{\pm,k-1} = \mathbf{D}^j\hat{\boldsymbol{g}}_j^{\pm,k-1}, \tag{11}$$

where, for completeness, the quantity

---

5 $\hat{\mathbf{Z}}_j^{\pm,k=0} := \hat{\mathbf{Z}}_j^{\pm,n}$ and $\hat{\mathbf{Z}}_j^{\pm,n+1} := \hat{\mathbf{Z}}_j^{\pm,k=2}$

$$\hat{g}_j^{\pm,k-1} = \frac{1}{k}\Delta t\, \mathbf{M}^{-1}\big(\mathbf{M}\mathbf{C}^{\mp}\hat{\mathbf{Z}}_j^{\pm,k-1} + \nu_{\pm}\mathbf{L}\hat{\mathbf{Z}}_j^{\pm,k-1} + \nu_{\mp}\mathbf{L}\hat{\mathbf{Z}}_j^{\mp,k-1}\big) - \hat{\mathbf{Z}}_j^{\pm,n}$$

is the remaining inhomogeneous contribution (see [11]). In general, we are interested in high Reynolds number—where ν and η tend to zero—solutions of (5)-(6) (or (8)-(9)), for which the nonlinear terms $\mathbf{Z}^{\mp}\cdot\nabla\mathbf{Z}^{\pm}$ (or $\mathbf{C}^{\mp}\hat{\mathbf{Z}}_j^{\pm}$) dominate the viscous terms. We note that explicit time-stepping presents no problem if the Courant restriction is not violated. Equation (11) is solved using a preconditioned iterative Krylov method, and our focus in the remainder of this paper concerns the preconditioning of this system.

3 From Classical to Optimized Schwarz

The principle behind optimized Schwarz methods consists of replacing the Dirichlet *transmission* condition present in the classical Schwarz approach by a more general Robin boundary condition [8]. The latter contains a positive parameter that can be used to enhance convergence. Optimized Schwarz methods find the best parameter through analytical techniques. For instance, a Fourier analysis of certain continuous elliptic partial differential equations, is performed in [7] (and references therein). In [6], there is numerical evidence that the weak Laplacian is spectrally close to the pseudo-Laplacian (11). Consequently, the construction of the various Schwarz preconditioners are based on a weak formulation of the Poisson problem.

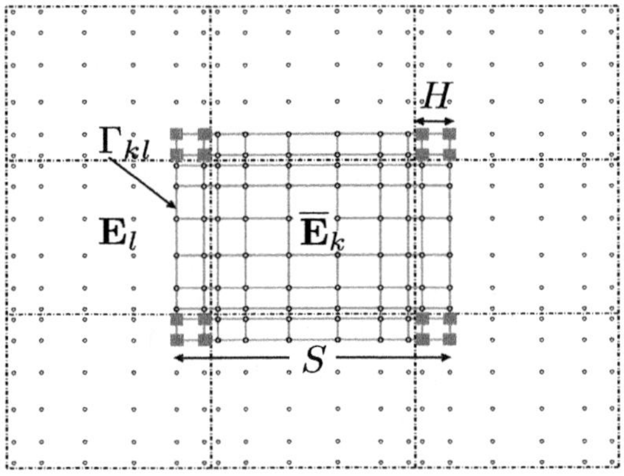

Fig. 1. One overlapping subdomain, $\bar{\mathbf{E}}_k$. Overlapping corner nodes are represented as red squares. $\mathbf{E}_l$, is a nonoverlapping neighboring element.

Suppose that the linear elliptic operator $\mathcal{L} := -\Delta$ with forcing f and boundary conditions $\mathcal{P} := \frac{\partial}{\partial \mathbf{n}}$ needs to be solved on $\mathbb{D}$. Then, an iterative algorithm that can be employed to solve the global problem $\mathcal{L}p = f$ is

$$\begin{aligned} &\mathcal{L}p_k^{n+1} = f \quad \text{in } \bar{\mathbf{E}}_k \\ &\mathcal{P}(p_k^{n+1}) = 0 \quad \text{on } \partial\mathbb{D}\cap\bar{\mathbf{E}}_k \text{ and } p_k^{n+1} = p_l^n \text{ on } \Gamma_{kl},\ \forall l \text{ s.t. } \partial\bar{\mathbf{E}}_k\cap\mathbf{E}_l \neq \emptyset \end{aligned} \tag{12}$$

where the sequence with respect to n will be convergent for any initial guess u^0. This is none other than the classical Schwarz algorithm at the continuous level corresponding to RAS at the matrix level. The optimized version of the above algorithm replaces the Dirichlet transmission conditions between subdomains by

$$\left[\frac{\partial p_k}{\partial \mathbf{n}} + T(p_k, r, q, \tau)\right]_{\Gamma_{kl}}^{n+1} = \left[\frac{\partial p_l}{\partial \mathbf{n}} + T(p_l, r, q, \tau)\right]_{\Gamma_{kl}}^{n} \tag{13}$$

where $T(p_k, r, q, \tau) \equiv rp_k - \frac{\partial}{\partial \tau}(q\frac{\partial p_k}{\partial \tau})$, defines a transmission condition of order 2 with two parameters, $r = r(x,y)$ and $q = q(x,y)$, with $r, q \geq 0$ on Γ_{kl} and $q = 0$ at $\partial\Gamma_{kl}$ as specified in [10]. The algorithm, like in the classical case, converges to the solution of $\mathcal{L}p = f$ with $\mathcal{P}(p) = 0$ on $\partial\mathbb{D}$ [8]. Its discrete algebraic version is

$$\tilde{A}_k \hat{\boldsymbol{p}}_k^{n+1} = \begin{pmatrix} A_k^{ii} & A_k^{i\Gamma} \\ C_k^{\Gamma i} & C_k^{\Gamma\Gamma} \end{pmatrix} \begin{pmatrix} \hat{\boldsymbol{p}}_k^i \\ \hat{\boldsymbol{p}}_k^\Gamma \end{pmatrix}^{n+1} = \begin{pmatrix} f_k^i \\ f_k^\Gamma + C_k \hat{\boldsymbol{p}}^n \end{pmatrix}$$

with $C_k^{\Gamma i}$, $C_k^{\Gamma\Gamma}$ and C_k corresponding to the discrete expressions of the optimized transmission conditions. At this point notice that A_k^{ii} is exactly the same block as in the original Schwarz algorithm. A simple manipulation leads to the following preconditioned system

$$\left\{I - \sum_{k=1}^{K} R_{\mathbf{E}_k}^T \tilde{A}_k^{-1} \begin{pmatrix} 0 & 0 \\ 0 & C_k \end{pmatrix} R_{\bar{\mathbf{E}}_k}^T\right\} \hat{\boldsymbol{p}} = \left\{\sum_{k=1}^{K} R_{\mathbf{E}_k}^T \tilde{A}_k^{-1} R_{\bar{\mathbf{E}}_k}\right\} f \tag{14}$$

where $R_{\bar{\mathbf{E}}_k}$ and $R_{\mathbf{E}_k}^T$ are Boolean restriction and extension (by zero) matrices to the G quadrature points of element $\bar{\mathbf{E}}_k$ and $\mathbf{E}_k$, respectively. As seen in Fig. 1, two different overlapping domains can be considered. The first is cross-shaped (without corner nodes) and imposes optimized boundary conditions on $\partial\bar{\mathbf{E}}_k \backslash \{4 \text{ corner elements}\}$ and Dirichlet ones on the 4 corner elements represented by the square G quadrature points in Fig. 1. The second is the square (with corner elements) domain having optimized conditions on $\partial\bar{\mathbf{E}}_k$. The above results are completely algebraic and independent of the underlying space discretization method. The complete proof in the additive and multiplicative case with and without overlap can be found in [12]. Finally the one level optimized Schwarz preconditioned linear system (11) is

$$P_{ORAS}^{-1} E \hat{\boldsymbol{p}}^{\pm,k-1} = P_{ORAS}^{-1} \mathbf{D}^j \hat{\boldsymbol{g}}_j^{\pm,k-1} \tag{15}$$

where $P_{ORAS}^{-1} \equiv \{\sum_{i=1}^{K} R_{\mathbf{E}_i}^T \tilde{A}_i^{-1} R_{\bar{\mathbf{E}}_i}\}$ and k is the RK2 stage number.

4 Discretization of the Optimized Schwarz

In order to obtain the optimized preconditioner, it suffices to compute the matrices $\tilde{A}_k^{-1}$ in equation (14) from the model problem (12) at convergence ($p_k = p_k^n$) with the boundary condition (13). For simplicity, the *rhs* of the latter is set to g. Therefore the weak formulation of the problem is to find $u_k \in H^1(\bar{\mathbf{E}}_k)$ *such that*

$$\int_{\bar{\mathbf{E}}_k} \nabla\phi \cdot \nabla p + \sum_{l \in nei_k} \int_{\Gamma_{kl}} \left(r\phi p + q\frac{\partial \phi}{\partial \tau_{kl}}\frac{\partial p}{\partial \tau_{kl}}\right) = \int_{\bar{\mathbf{E}}_k} \phi f_k + \sum_{l \in nei_k} \int_{\Gamma_{kl}} \phi g \tag{16}$$

for all $\phi \in H^1(\bar{\mathbf{E}}_k)$. We introduce the tiling of the Gauss-Legendre quadrature points in element $\bar{\mathbf{E}}_k$ by $\mathcal{Q}_h^k = \cup_{l=1}^{m_k} \mathbf{Q}_l$, and the finite dimensional space $V_k := \{v_h \in C^0(\bar{\mathbf{E}}_k) \mid v_h|_{\mathbf{Q}_l} \circ T_{\mathbf{Q}_l} \in (\mathbb{P}_1 \otimes \mathbb{P}_1)(\mathbf{Q}_l),\ \forall\, \mathbf{Q}_l \in \mathcal{Q}_h^k\} \cap H^1(\bar{\mathbf{E}}_k)$. The basis used to represent polynomials in V_k are tensor products of the one dimensional linear hat functions $\phi_i(\eta_j) = \delta_{ij}$ depicted in Fig. 2 at each Gauss-Legendre quadrature point $\{\eta_l\}_{l=-2}^{N}$. Using the one dimensional definition for the stiffness and lumped mass matrices,

$$K_{ij}^k := \int_{\eta_{-2}}^{\eta_N} \frac{d\phi_i}{d\eta}\frac{d\phi_j}{d\eta} d\eta \quad \text{and} \quad M_{ij}^k := \int_{\eta_{-2}}^{\eta_N} \phi_i(\eta)\phi_j(\eta) d\eta,$$

respectively, leads to the following tensor product representation of (16):

$$\begin{aligned} \tilde{A}_k := (K^k + T^k_{r_b,r_t}) \otimes (M^k + T^k_{q_l,q_r}) + (M^k + T^k_{q_b,q_t}) \otimes (K^k + T^k_{r_l,r_r}) \\ - T^k_{r_b,r_t} \otimes T^k_{q_l,q_r} - T^k_{q_b,q_t} \otimes T^k_{r_l,r_r}. \end{aligned} \tag{17}$$

In the last expression, $T^k_{a,b}$ is a matrix with only two non-zero entries at $(1,1)$ and (N,N), which are set to a and b respectively. The notation $q_{r,l,b,t}$ stands for the q optimized parameter at either the right, left, bottom or top boundaries (*idem* for parameter r).

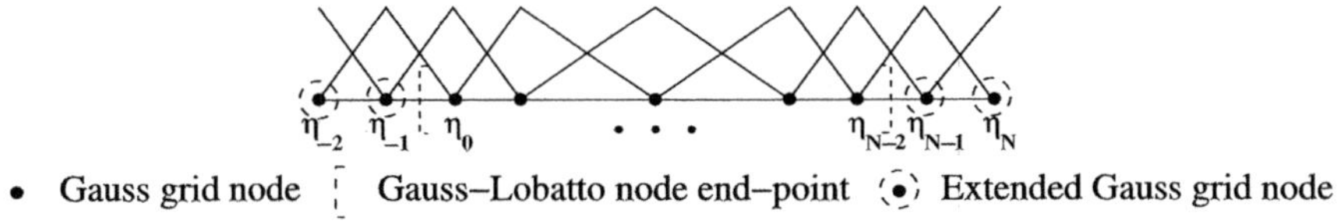

Fig. 2. Schematic of the assembly procedure.

	r	q
OO0, overlap H	$2^{-1/3}(k_{\min}^2)^{1/3}H^{-1/3}$	0
OO2, overlap H	$2^{-3/5}(k_{\min}^2)^{2/5}H^{-1/5}$	$2^{-1/5}(k_{\min}^2)^{-1/5}H^{3/5}$

Table 1. Parameters r and q used in the transmission blocks. $k_{\min} = \pi/S$ with S the characteristic size of the element normal to the face where the parameters are required.

When rectangular elements are considered the FDM (e.g., [3]) can be used to invert the optimized blocks. The number of operations required to invert $N^d \times N^d$ matrix using such a technique is $O(N^{d+1})$ and the application of the inverse is performed using efficient tensor products in $O(N^{d+1})$ operations. We propose the form derived in (17) with the r and q parameters constant on their respective faces but eliminating the last two terms of the form $T^k_{\cdot,\cdot} \otimes T^k_{\cdot,\cdot}$. This is done in order for the fast diagonalization technique to be applicable. Indeed, the modified mass matrix $M^k + T^k_{\cdot,\cdot}$ is still symmetric and positive definite while the matrix $K^k + T^k_{\cdot,\cdot}$ is still symmetric. This enables the use of the modified mass matrix in an inner product and the simultaneous diagonalization of both tensors. When $q = 0$, the proposed formula is exact. Formulas for r and q are given in Table 1.

5 Numerical Experiments

The RAS preconditioner described above was implemented in the MHD code. This version allows for variable overlap of the extended grid. The ORAS counterpart has also been implemented as described, and for comparison, we use a high-order block Jacobi (BJ). We consider first tests of a single pseudo-Poisson solve on a $[0,1]^2$ bi-periodic domain with exact solution $p = \cos(2\pi x)\cos(2\pi y)$. In the first experiment, we use a grid of $E = 8 \times 8$ elements, and iterate using BiCGStab until the residual is 10^{-8} times that of the initial residual. The extended grid overlap is 2, and the initial starting guess for the Krylov method is composed of random noise.

The first test uses non-FDM preconditioners to investigate the effect of including corner transfers on the optimization. The results are presented in Fig. 3, in which we consider only the OO0 optimization. Note that even though the RAS is much less sensitive to the subdomains without corners, especially at higher N_v, the OO0 with subdomains including corners requires fewer iterations. Clearly, including the corners is crucial to the proper functioning of the optimized methods.

In the second experiment, all the parameters are maintained except we use a grid of $E = 16 \times 16$ elements together with the FDM version of the preconditioners to investigate performance. These results are presented on the right-most figure of Fig. 3.

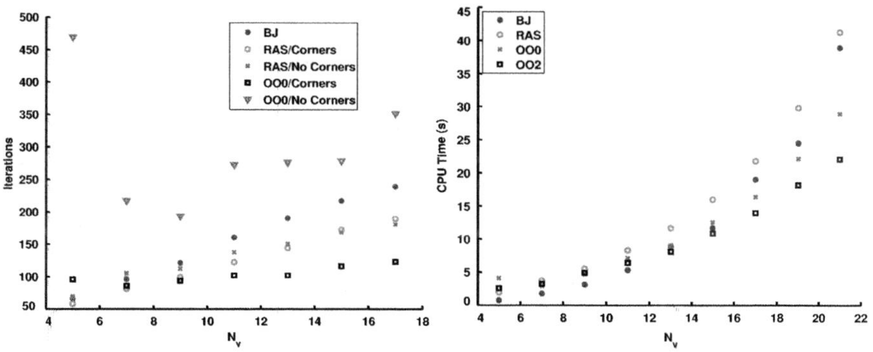

Fig. 3. *Left:* Plot of iteration count vs GLL-expansion node number for different preconditioners using subdomains with and without corners on an 8×8 element grid. *Right:* Comparison of of CPU time vs. GLL-expansion node number of FDM-based preconditioners with subdomain including corners on a 16×16 element grid.

Acknowledgement. NCAR (National Center for Atmosphereic Research) is supported by the National Science Foundation. S.D. Kim was supported by award KRF-2008-313-C00094.

References

[1] Canuto, C., Hussaini, M.Y., Quarteroni, A., Zang, T.A.: *Spectral methods in fluid dynamics*. Springer, 1988.

[2] Chniti, C., Nataf, F., Nier, F.: Improved interface condition for 2D domain decomposition with corner : a theoretical determination. Technical Report hal-00018965, Hyper articles en ligne, 2006.

[3] Couzy, W.: *Spectral element discretization of the unsteady Navier-Stokes equations and its iterative solution on parallel computers*. PhD thesis, École Polytechnique Fédérale de Lausanne, 1995.

[4] Deville, M.O., Fischer, P.F., Mund, E.H.: *High-Order Methods for Incompressible Fluid Flow*. Cambridge University Press, 2002.

[5] Elsässer, W.M.: The hydromagnetic equations. *Phys. Rev.*, 79:183, 1950.

[6] Fischer, P.F.: An overlapping Schwarz method for spectral element solution of the incompressible Navier-Stokes equations. *J. Comput. Phys.*, 133:84–101, 1997.

[7] Gander, M.J.: Optimized Schwarz methods. *SIAM J. Numer. Anal.*, 44(2):699–731, 2006.

[8] Lions, P.-L.: On the Schwarz alternating method. I. In R. Glowinski, G.H. Golub, G.A. Meurant, and J. Périaux, eds., *First International Symposium on Domain Decomposition Methods for Partial Differential Equations*, pages 1–42, Philadelphia, PA, 1988. SIAM.

[9] Maday, Y., Patera, A.T., Rønquist, E.M.: The $\mathbb{P}_N$-$\mathbb{P}_{N-2}$ method for the approximation of the Stokes problem. Technical report, Pulications du Laboratoire d'Analyse Numérique, Université Pierre et Marie Curie, 1992.

[10] Nataf, F.: Interface connections in domain decomposition methods. In A. Bourlioux and M.J. Gander, eds., *Modern Methods in Scientific Computing and Applications*, vol. 75 of *NATO Science Series II*, pages 323–364. Kluwer Academic, 2001.

[11] Rosenberg, D., Pouquet, A., Mininni, P.D. Adaptive mesh refinement with spectral accuracy for magnetohydrodynamics in two space dimensions. *New J. Phys.*, 9(304), 2007.

[12] St-Cyr, A., Gander, M.J., Thomas, S.J. Optimized multiplicative, additive, and restricted additive Schwarz preconditioning. *SIAM J. Sci. Comput.*, 29(6):2402–2425, 2007.

Nonlinear Overlapping Domain Decomposition Methods

Xiao-Chuan Cai

Department of Computer Science, University of Colorado at Boulder, Boulder, CO 80309, cai@cs.colorado.edu

Summary. We discuss some overlapping domain decomposition algorithms for solving sparse nonlinear system of equations arising from the discretization of partial differential equations. All algorithms are derived using the three basic algorithms: *Newton* for local or global nonlinear systems, *Krylov* for the linear Jacobian system inside Newton, and *Schwarz* for linear and/or nonlinear preconditioning. The two key issues with nonlinear solvers are robustness and parallel scalability. Both issues can be addressed if a good combination of Newton, Krylov and Schwarz is selected, and the right selection is often dependent on the particular type of nonlinearity and the computing platform.

1 Introduction

For solving partial differential equations on large scale parallel computers, domain decomposition is a natural choice. Overlapping Schwarz methods and non-overlapping iterative substructuring methods are the two major classes of domain decomposition methods [13, 14, 15]. In this paper we only consider overlapping methods for solving large sparse nonlinear system of equations arising from the discretization of nonlinear partial differential equations, i.e., for a given nonlinear function $F : R^n \to R^n$, we compute a vector $u \in R^n$, such that

$$F(u) = 0, \tag{1}$$

starting from an initial guess $u^{(0)} \in R^n$. Here $F = (F_1, \ldots, F_n)^T$, $F_i = F_i(u_1, \ldots, u_n)$, and $u = (u_1, \ldots, u_n)^T$. One of the popularly used techniques for solving (1) is the so-called *inexact Newton* algorithms (IN) which are described briefly here. Suppose $u^{(k)}$ is the current approximate solution and $J = F'(u^{(k)})$, a new approximate solution $u^{(k+1)}$ can be computed through the following steps: first find an inexact Newton direction $p^{(k)}$ by solving the Jacobian system

$$Jp^{(k)} = F(u^{(k)}) \tag{2}$$

such that $\|F(u^{(k)}) - Jp^{(k)}\| \leq \eta_k \|F(u^{(k)})\|$, then compute the new approximate solution

M. Bercovier et al. (eds.), *Domain Decomposition Methods in Science and Engineering XVIII*, Lecture Notes in Computational Science and Engineering,
DOI: 10.1007/978-3-642-02677-5,

$$u^{(k+1)} = u^{(k)} - \lambda^{(k)} p^{(k)}. \tag{3}$$

Here $\eta_k \in [0,1)$ is a scalar that determines how accurately the Jacobian system needs to be solved using, for example, Krylov subspace methods. $\lambda^{(k)}$ is another scalar that determines the step length in the selected inexact Newton direction. Sometimes when J is not explicitly available, one can use the matrix-free version [11]. IN has several well-known features.

(a) Fast convergence. If the initial guess is close enough to the desired solution then the convergence is very fast (quadratic) provided that the η_k's are sufficiently small.
(b) Non-robustness. The convergence, or fast convergence, happens only if a good initial guess is available. Generally it is difficult to obtain such an initial guess especially for nonlinear equations that have unbalanced nonlinearities [12]. The step length $\lambda^{(k)}$ is often determined by the components with the strongest nonlinearities, and this may lead to an extended period of stagnation in the nonlinear residual curve. We say that the nonlinearities are "unbalanced" when $\lambda^{(k)}$, in effect, is determined by a subset of the overall degrees of freedom.
(c) Scalability. The parallel scalability of the method is mostly determined by how the Jacobian system (2) is solved.

There are a number of strategies [7, 8, 10], such as linesearch, trust region, continuation or better ways to choose the forcing term, to make the algorithm more robust or converge faster, however, these strategies are all based on certain global knowledge of F or J. In other words, all equations in the system are treated equally as if they were some of the worst equations in the system. Other ways to look at the global nature of IN are

(d) To advance from $u^{(k)}$ to $u^{(k+1)}$, all n variables and equations need to be updated even though in many situations n can be very large, but only a small number of components of $u^{(k)}$ receive significant updates.
(e) If a small number of components of the initial guess $u^{(0)}$ are not acceptable, the entire $u^{(0)}$ is declared bad.
(f) There are two global control variables η_k and $\lambda^{(k)}$. Any slight change of F may result in the change of η_k or $\lambda^{(k)}$, and any slight change of η_k or $\lambda^{(k)}$ may result in some global function evaluations and/or the solving of global Jacobian systems. For example, if the search direction $p^{(k)}$ has one unacceptable component, then the entire steplength is reduced.

Note that these global operations can be expensive when n is large and when the number of processors is large. Using domain decomposition methods, more localized treatments can be applied based on the location or the physical nature of the nonlinearities, and the number of global operations can be made small in some situations.

We should point out that the words "local" and "global" have different meanings in the context of domain decomposition methods [15] than in the context of nonlinear equation solvers [7], among others. In nonlinear solvers, "local" means a small neighborhood of the exact solution of the nonlinear system, and "global" means a relatively large neighborhood of the exact solution of the nonlinear system. In domain decomposition, "local" means some subregions in the computational domain and "global" means the whole computational domain.

All the algorithms to be discussed in the paper are constructed with a combination of the three basic techniques: Newton, Krylov and Schwarz. Newton is the basic nonlinear solver that is used for either the system defined on the whole space or some subspaces (subdomain subspace or coarse subspace). Krylov is the basic linear solver that is used inside a Newton solver. Schwarz is a preconditioner for either the linear or the nonlinear solver. Many algorithms can be derived with different combinations of the three basic algorithms. For a given class of problems and computing platform, a special combination might be necessary in order to obtain the best performance. The three basic algorithms are all well understood individually, however, the construction of the best combination remains a challenge. The same can be said for the software. All software components are readily available in PETSc [1], but some of the advanced combinations have to be programmed by the user.

We next define (informally) some notations for describing domain decomposition methods. u is understood as a discrete (or coefficients of a finite element) function defined on the computational domain Ω which is already partitioned into a set of subdomains $\{\Omega_1^\delta, \cdots, \Omega_N^\delta\}$. Here Ω_i^δ is an δ-extension of Ω_i, and the collection of $\{\Omega_i\}$ is a non-overlapping partition of Ω. We define R_i^δ as a restriction operator associated with Ω_i^δ and R_i^0 as the restriction operator associated with Ω_i. We denote $u_{\Omega_i^\delta}$ as the restriction of u on Ω_i^δ, and $u_{\partial\Omega_i^\delta}$ as the restriction of u on the "boundary" of Ω_i^δ. Here we use the word "domain" to denote the mesh points in the interior of the domain and "boundary" to denote the mesh points on the boundary of the domain. Similarly, we may restrict the nonlinear function to a subdomain, such as $F_{\Omega_i^\delta}$. For boundary value problems considered in this paper, we assume

$$F_{\Omega_i^\delta}(u) = F_{\Omega_i^\delta}(u_{\Omega_i^\delta}, u_{\partial\Omega_i^\delta}).$$

That is to say that there are no "global equations" in the system that may couple the equations defined at a mesh point to equations defined outside a small neighborhood.

The rest of the paper is organized as follows. In Section 2, we discuss the most popular overlapping nonlinear domain decomposition method, Newton-Krylov-Schwarz algorithm, and in Sections 3–6, we discuss some more advanced nonlinear methods. Some final remarks are given in Section 7.

2 Newton-Krylov-Schwarz Algorithms

Newton-Krylov-Schwarz (NKS) is simply the application of a linear Schwarz preconditioner for solving the Jacobian equation (2) in the inexact Newton algorithm [2, 3]. Depending on what type of Schwarz preconditioner is used (additive, multiplicative, restricted, one-level, two-level, etc), there are several NKS algorithms. Let us define the subdomain preconditioners as

$$J_i = R_i^\delta J (R_i^\delta)^T,\ i = 1,\ldots,N,$$

then the additive Schwarz preconditioner can be written as

$$M_{AS}^{-1} = \sum_{i=1}^{N} (R_i^\delta)^T J_i^{-1} R_i^\delta.$$

Because of its simplicity, NKS has become one of the most popular domain decomposition methods for solving nonlinear PDEs and is the default nonlinear solver in PETSc [1]. The nonlinear properties of NKS are exactly the same as that of inexact Newton. For example, the initial guess has to be sufficiently close to the solution in order to obtain convergence, and fast convergence can be achieved when the nonlinearity is well balanced. NKS addresses the scalability issue (c) of IN well, but not the other issues (a, b, d–f).

3 Classical Schwarz Alternating Algorithms

Let $(u^{(0)}_{\Omega_1^\delta}, \ldots, u^{(0)}_{\Omega_N^\delta})$ be the initial guess for all subdomains. The classical Schwarz alternating algorithm (SA) can be described as follows:

$$\begin{aligned}
&k = 1,\ldots,\ \textit{till convergence condition is satisfied}\\
&\quad i = 1,\ldots,N\\
&\qquad \textit{define } u^{(k)}_{\partial\Omega_i^\delta} \textit{ using } \{u^{(k-1)}_{\Omega_j^\delta}, 1 \le j \le N\} \textit{ or } \{u^{(k)}_{\Omega_j^\delta}, 1 \le j < i\}\\
&\qquad \textit{compute } u^{(k)}_{\Omega_i^\delta} \textit{ by solving } F_{\Omega_i^\delta}(u^{(k)}_{\Omega_i^\delta}, u^{(k)}_{\partial\Omega_i^\delta}) = 0.
\end{aligned} \tag{SA}$$

The algorithm doesn't belong to the class of IN algorithms and, in general, not share properties (a–f). The method is usually not used by itself as a nonlinear solver because of its slow convergence, but in some cases when the nonlinearities are isolated within some of the subdomains, the method can be a good alternative to IN. Note that SA doesn't involve any global operations.

4 Nonlinear Additive Schwarz Preconditioned Inexact Newton Algorithms

The basic idea of nonlinearly preconditioned inexact Newton algorithms [4, 9] is to find the solution $u \in R^n$ of (1) by solving an equivalent system

$$\mathcal{F}(u) = 0 \tag{4}$$

using IN. Systems (1) and (4) are said to be equivalent if they have the same solution. For any given $v \in R^n$, we define a subdomain projection $T_i(v)$, which is a function with support in Ω_i^δ, as the solution of the following subspace nonlinear system

$$F_{\Omega_i^\delta}(v - T_i(v)) = 0,$$

for $i = 1,\dots,N$. Then a nonlinearly preconditioned function is defined as

$$\mathcal{F}(u) = \sum_{i=1}^{N} T_i(u).$$

It can be shown that, under certain conditions, for this particular $\mathcal{F}$, (1) and (4) offer the same solution subject to the error due to different stopping conditions and preconditioners. This algorithm is often referred to as the additive Schwarz preconditioned inexact Newton algorithm (ASPIN). Sometimes we call it a left preconditioned IN because in the linear case (i.e., $F(u) = Ju - b$) $\mathcal{F}(u) = (\sum_{i=1}^{N}(R_i^\delta)^T J_i^{-1} R_i^\delta)(Ju - b)$.

When using IN to solve (4), the Jacobian of $\mathcal{F}$, or its approximation, is needed. Because of the special definition of the function $\mathcal{F}$, its Jacobian can only be given as the sum of matrix-vector products and the explicit elements of $\mathcal{F}'$ are not available.

It is known that for left preconditioned linear iterative methods, the stopping condition is often influenced by the preconditioner. The impact of the preconditioner on the stopping condition can be removed if the preconditioner is applied to the right. Unlike linear preconditioning, the switch from left to right is not trivial in the nonlinear case. A right nonlinear preconditioner will be discuss in a later section of the paper.

5 Nonlinear Elimination Algorithms

The nonlinear elimination algorithm (NE) was introduced in [12] for nonlinear algebraic systems with local high nonlinearities. It was not introduced as a domain decomposition method, but we include it in the paper because it is the main motivation for the algorithm to be discussed in the next section. Suppose that the function F is more nonlinear in the subdomain Ω_i^δ, then we can eliminate all unknowns in this particular subdomain and let Newton work on the rest of the variables and equations. Let $y = u|_{\Omega_i^\delta}$ and $x = u|_{\Omega\backslash\Omega_i^\delta}$, then using the implicit function theorem, under some assumptions, we can solve for y in terms of x; i.e., solve

$$F_{\Omega_i^\delta}(x,y) = 0$$

for y, which symbolically equals to $y = F_{\Omega_i^\delta}^{-1}(x)$. After the elimination, we can use the regular Newton method for the rest of the system which is more balanced, at least in theory,

$$F_{\Omega\backslash\Omega_i^\delta}(x, F_{\Omega_i^\delta}^{-1}(x)) = 0.$$

The algorithm has some obvious advantages. We mention some of its disadvantages as a motivation for the algorithm to be discussed in the next section. In practice, it is often difficult to tell which components are more nonlinear than the others, and the situation may change from iteration to iteration. The algorithm may introduce sharp jumps in the residual function near the interface of x and y. Such jumps may lead to slow convergence or divergence. Some improved versions are given in [6]. In the next section, we combine the ideas of ASPIN and NE into a right preconditioned Newton method.

6 Nonlinear Restricted Additive Schwarz Algorithms

In [5], a right preconditioned inexact Newton algorithm was introduced as follows: Find the solution $u \in R^n$ of (1) by first solving a preconditioned nonlinear system

$$F(G(v)) = 0$$

for v, and then obtain $u = G(v)$. For any given $v \in R^n$, we define a subdomain projection $T_i(v)$, which is a function with support in Ω_i^δ, as the solution of the following subspace nonlinear system

$$F_{\Omega_i^\delta}(v + T_i(v)) = 0,$$

for $i = 1, \ldots, N$. Then the nonlinear preconditioning function is defined as

$$G(v) = v + \sum_{i=1}^{N} R_i^0 T_i(v).$$

Here the non-overlapping restriction operator R_i^0 effectively removes the sharp jumps on the interfaces of the overlapping subdomains. In the linear case

$$G(v) = v - \left(\sum_{i=1}^{N} (R_i^0)^T J_i^{-1} R_i^\delta \right) (Jv - b),$$

which can be regarded as a restricted additive Schwarz preconditioned Richardson method.

This preconditioner doesn't have to be applied at every outer Newton iteration. It is used only when some local high nonlinearities are sensed, somehow. Below we describe the overall algorithm (NKS-RAS). The goal is to solve equation (1) with a given initial guess $u^{(0)}$. Suppose $u^{(k)}$ is the current solution.

Step 1 (Nonlinearity Checking): *Check local and global stopping conditions.*
- *If the global condition is satisfied, stop.*
- *If local conditions indicate that nonlinearities are not balanced, go to Step* 2.
- *If local conditions indicate that nonlinearities are balanced, set* $\tilde{u}^{(k)} = u^{(k)}$, *go to Step* 3.

Step 2 (RAS): *Solve local nonlinear problems on the overlapping subdomains to obtain the subdomain corrections* $T_i(u^{(k)})$

$$F_{\Omega_i^\delta}(u^{(k)} + T_i(u^{(k)})) = 0 \quad \textit{for } i = 1, \ldots, N.$$

Drop the solution in the overlapping part of the subdomain and compute the global function $G(u^{(k)})$ and set

$$\tilde{u}^{(k)} = G(u^{(k)}).$$

Go to Step 3.

Step 3 (NKS): *Compute the next approximate solution $u^{(k+1)}$ by solving the following system*

$$F(u) = 0$$

with one step of NKS using $\tilde{u}^{(k)}$ as the initial guess.

Go to Step 1.

The nonlinearity checking step is important. However, we only have a few ad hoc techniques such as computing the residual norm subdomain by subdomain (or field by field in the case of multi-physics applications). If some of the subdomain (or subfield) norms are much larger than for other subdomains, we label these subdomains as highly nonlinear subdomains and proceed with the RAS elimination step. Otherwise, when the nonlinearity is more or less balanced we bypass the RAS step and go directly to the global NKS step. The subdomain nonlinear systems in Step 2 do not need to be solved very accurately since the solutions are used only to construct an initial guess for Step 3. In NKS-RAS, a nonlinear system is set up on each subdomain, but in practice, not all subdomain nonlinear problem needs to be solved. In the not-too-nonlinear regions, the solver may declare to have converged in 0 iteration.

7 Concluding Remarks

In this paper, we have given a quick overview of overlapping domain decomposition methods for solving nonlinear partial differential equations. The two key issues of nonlinear methods are robustness and scalability. Both issues can be addressed by using some combinations of the three basic algorithms: Newton, Krylov and Schwarz. Several algorithms are presented in the paper together with some of their advantages and disadvantages. Depending on the particular types of nonlinearities and the computing platform, different combinations of the three basic algorithms may be needed in order to obtain the best performance and robustness. Due to page limit, applications have not been discussed in the paper. Some of them can be found in the references.

Acknowledgement. This research was supported in part by DOE under DE-FC02-01ER25479 and DE-FC02-04ER25595, and in part by NSF under grants ACI-0305666, CNS-0420873, CCF-0634894, and CNS-0722023.

References

[1] Balay, S., Buschelman, K., Gropp, W.D., Kaushik, D., Knepley, M., McInnes, L.C., Smith, B.F., Zhang, H.: *PETSc Users Manual*, ANL, 2008.

[2] Cai, X.-C., Gropp, W., Keyes,D., Melvin, R., Young, D.P.: *Parallel Newton-Krylov-Schwarz algorithms for the transonic full potential equation*, SIAM J. Sci. Comput., 19 (1998), 246–265.

[3] Cai, X.-C., Gropp, W.D., Keyes, D.E., Tidriri, M.D.: *Newton-Krylov-Schwarz methods in CFD*, Proceedings of the International Workshop on the Navier-Stokes Equations, Notes in Numerical Fluid Mechanics, R. Rannacher, eds. Vieweg, Braunschweig, 1994.

[4] Cai, X.-C., Keyes, D.E.: *Nonlinearly preconditioned inexact Newton algorithms*, SIAM J. Sci. Comput., 24 (2002), 183–200.

[5] Cai, X.-C., Li, X.: *Inexact Newton methods with nonlinear restricted additive Schwarz preconditioning for problems with high local nonlinearities*, in preparation.

[6] Cai, X.-C., Li, X.: *A domain decomposition based parallel inexact Newton's method with subspace correction for incompressible Navier-Stokes equations*, Lecture Notes in Computer Science, Springer, 2009.

[7] Dennis, J.E., Schnabel, R.B.: *Numerical Methods for Unconstrained Optimization and Nonlinear Equations*, SIAM, 1996.

[8] Eisenstat, S.C., Walker, H.F.: *Choosing the forcing terms in an inexact Newton method*, SIAM J. Sci. Comput., 17 (1996), 16–32.

[9] Hwang, F.-N., Cai, X.-C.: *A parallel nonlinear additive Schwarz preconditioned inexact Newton algorithm for incompressible Navier-Stokes equations*, J. Comput. Phys., 204 (2005), 666–691.

[10] Kelley, C.T., Keyes, D.E.: *Convergence analysis of pseudo-transient continuation*, SIAM J. Numer. Anal., 35 (1998), 508–523.

[11] Knoll, D., Keyes, D.E.: *Jacobian-free Newton-Krylov methods: a survey of approaches and applications*, J. Comput. Phys., 193 (2004), 357–397.

[12] Lanzkron, P.J., Rose, D.J., Wilkes, J.T.: *An analysis of approximate nonlinear elimination*, SIAM J. Sci. Comput., 17 (1996), 538–559.

[13] Smith, B., Bjørstad, P., Gropp, W.: *Domain Decomposition. Parallel Multilevel Methods for Eliptic Partial Differential Equations*, Cambridge University Press, New York, 1996.

[14] Quarteroni, A., Valli, A.: *Domain Decomposition Methods for Partial Differential Equations*, Oxford University Press, Oxford, 1999.

[15] Toselli, A., Widlund, O.: *Domain Decomposition Methods—Algorithms and Theory*, Springer, Berlin, 2005.

Optimized Schwarz Waveform Relaxation: Roots, Blossoms and Fruits

Laurence Halpern

LAGA, Université Paris XIII, 99 Avenue J-B Clément, 93430 Villetaneuse, France, halpern@math.univ-paris13.fr

1 Introduction: Parallel Processing of Evolution Problems

There are several ways to solve in parallel the evolution problem

$$P(\partial_t, \partial_1, \cdots, \partial_d)u = f.$$

- *Explicit time discretization* is naturally parallel.
- *Implicit time discretization + spatial domain decomposition.* For the heat equation for instance, with an implicit Euler scheme in time, this amounts to solving at each step the linear problem

  $$\frac{u}{\Delta t} - \Delta u = f.$$

  This gives a very well conditioned problem, and multigrid or domain decomposition algorithms can be used without coarse grid preconditioner (see the reference book of [21] and references therein). Thereafter, improved algorithms were designed (see presentation by F. Nataf in the same minisymposium). However a uniform time-step is needed.

Both procedures imply an exchange of information between processors at every time-step, which can be very penalizing when using a large number of processors. Quoting [1] "*A major obstacle to achieving significant speed-up on parallel machines is the overhead associated with synchronizing the concurrent processes*". One way to overcome this delay problem was invented in the seventies with the concept of **asyncronous algorithms**, [4]. An excellent review can be found in [1]. In that context, amounts of information are sent without waiting for the request. However convergence is weakened, if not destroyed.

- *Time or Space-time multigrid.* Quoting [14], in the previous approach, "*the potential for parallelism is limited to the parallelism of the elliptic solver, since the time dimension is treated strictly sequentially. ... Thus it seems natural to ask whether ... a parallelization strategy for the time-dependent part of the problem*

M. Bercovier et al. (eds.), *Domain Decomposition Methods in Science and Engineering XVIII*, Lecture Notes in Computational Science and Engineering,
DOI: 10.1007/978-3-642-02677-5,

[can] be found ". The problem has been studied mainly for parabolic operators, leading to the parabolic multigrid method of [11], or the space-time multigrid in [14], or the parareal algorithm of [16].
- *Waveform relaxation*, and more particularly the Schwarz waveform relaxation algorithms, were designed independently in [7, 10]. The main feature of these algorithms is their flexibility. It permits to choose the space and time meshes independently in the subdomains, leading to local space-time refinement with time windows. Different numerical schemes can be used in the subdomains, or even different models can be coupled, and the method adjusts to underlying computing hardware.

2 Roots: Waveform Relaxation for ODEs

The ancestor is Emile Picard, which in the *Journal de mathématiques* introduced the "Méthode des approximations successives" to prove the existence of solutions to ordinary differential equations. The algorithm was reinvented in practical implementations in [15]. We describe below the Picard algorithm for functions y_j in different spaces $\mathbb{R}^{N_j}$:

SYSTEM OF ODE'S	PICARD ITERATES
$\dfrac{dy_1}{dt} = f_1(t, y_1, y_2, \cdots, y_p),$	$\dfrac{dy_1^{(k+1)}}{dt} = f_1(t, y_1^{(k)}, y_2^{(k)}, \cdots, y_p^{(k)}),$
	$\vdots$
$\dfrac{dy_j}{dt} = f_j(t, y_1, y_2, \cdots, y_p),$	$\dfrac{dy_j^{(k+1)}}{dt} = f_j(t, y_1^{(k)}, y_2^{(k)}, \cdots, y_p^{(k)}),$
	$\vdots$
$\dfrac{dy_p}{dt} = f_p(t, y_1, y_2, \cdots, y_p),$	$\dfrac{dy_p^{(k+1)}}{dt} = f_p(t, y_1^{(k)}, y_2^{(k)}, \cdots, y_p^{(k)}),$
	$\vdots$

Note that it is naturally parallel. The error at step $k+1$ on the time interval $[t_0, T]$ is given by:

$$\|y^{(k+1)} - y\|_\infty \leq \frac{L^k (T - t_0)^k}{k!} \|y^{(0)} - y\|_\infty. \tag{1}$$

At first sight, this tends to zero rapidly with k. However, it can be seen using the Stirling formula $n! \approx \sqrt{2\pi n}\,(\frac{n}{e})^n$ that $\frac{eL(T-t_0)}{k}$ has to be smaller than 1 before the error will start to decrease, which implies many iterations in case of large time intervals, or large Lipschitz constants.

Waveform relaxation algorithms are extensions both of the Picard's "approximations successives" and relaxation methods for algebraic systems. The parallel formulation, of Jacobi type, is obtained by replacing in the "j-" line, $y_j^{(k)}$ by $y_j^{(k+1)}$ in the function f. For the analysis, see [20], [22], for a review [3]. The principal results are:

1. Linear convergence on unbounded time intervals for linear systems with dissipation.
2. Superlinear convergence for finite time.
3. When the ordinary differential equation stems from a discretization in space of a partial differential equation, the convergence rate depends on the discretization parameters, and deteriorates as one refines the mesh.

Later on continuous versions have been developed, like the Schwarz waveform relaxation for partial differential equations.

3 Blossoms: Classical Schwarz Waveform Relaxation for Parabolic Equations

To solve $\mathcal{L}u = f$ in $\Omega \times (0,T)$, with initial condition u_0, with $\mathcal{L}$ the heat operator, it was proposed in [7] to introduce the overlapping domain decomposition algorithm in space-time:

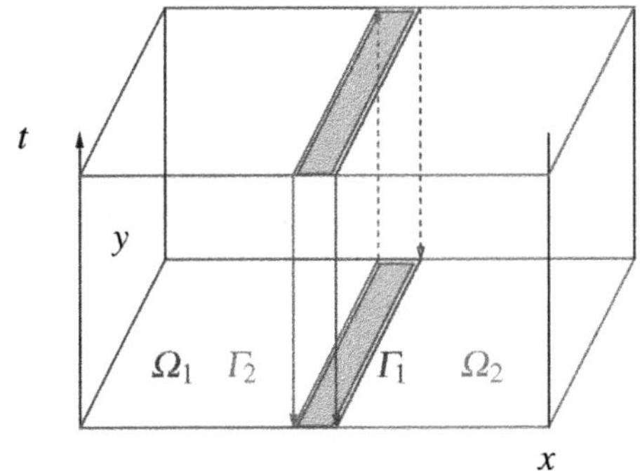

$$\begin{cases} \mathcal{L}u_1^{k+1} = f & \text{in } \Omega_1 \times (0,T), \\ u_1^{k+1}(\cdot,0) = u_0 & \text{in } \Omega_1, \\ u_1^{k+1} = u_2^k & \text{on } \Gamma_1 \times (0,T), \end{cases}$$

$$\begin{cases} \mathcal{L}u_2^{k+1} = f & \text{in } \Omega_2 \times (0,T), \\ u_2^{k+1}(\cdot,0) = u_0 & \text{in } \Omega_2, \\ u_2^{k+1} = u_1^k & \text{on } \Gamma_2 \times (0,T), \end{cases}$$

corresponding to an infinite block Jacobi waveform relaxation. It can be compared to the parallel version of the Schwarz algorithm, introduced in [17], and that is where the name comes from. The algorithm has the same convergence properties 1,2 as the waveform relaxation algorithm. Consider the advection-diffusion-reaction equation in two dimensions,

$$\partial_t u + \mathbf{a} \cdot \nabla u - \nu \Delta u + cu = f.$$

The convergence factor for two semi-infinite subdomains on the Fourier side (ω is the dual variable of time, η is the dual variable of y), is given by $\rho = e^{-\frac{L}{\nu}\sqrt{a_x^2+4\nu(b+i(\omega+a_y\eta)+\nu\eta^2)}}$, and we have for a small overlap L,

$$\sup_{(\omega,\eta)\in\mathbb{R}^{1+1}} |\rho| = e^{-\frac{L}{\nu}\sqrt{a_x^2+4\nu b}} = 1 - \mathcal{O}(L).$$

4 Fruits: Optimized Schwarz Waveform Relaxation for Parabolic Equations

Such algorithms, relying on the theory of absorbing boundary conditions, were announced in [8], for one-dimensional wave and heat equations. For the advection-diffusion-reaction equation in two dimensions, the first thorough analysis was performed in [18]. We proposed in [2] to use a new type of transmission conditions between the subdomains, with or without overlap,

$$\mathcal{B}u := \nu\partial_n u - \frac{a_x}{2}u + \frac{p}{2}u + 2q(\partial_t + a_y u - \nu\Delta_y u + bu), \tag{2}$$

The case $q = 0$ corresponds to Robin transmission conditions, and was explored in [19], with extension to first order conditions (*i.e.* without the Laplace-Beltrami operator in (2)). The convergence factor for two semi-infinite subdomains is given by

$$\begin{aligned} \rho(z,p,q) &= \left(\frac{p+qz-\sqrt{a_x^2+4\nu b+z}}{p+qz+\sqrt{a_x^2+4\nu b+z}}\right)^2 e^{-\frac{L}{\nu}\sqrt{a_x^2+4\nu b+z}}, \\ z &= 4\nu\big(i(\omega+a_y\eta)+\nu|\eta|^2\big). \end{aligned} \tag{3}$$

The terms $O0$ and $O2$ correspond to Robin and second order transmission condition (2), with coefficients determined for two subdomains by minimizing the convergence factor. Note that the use of such transmission conditions leads to problems of the same complexity as the Dirichlet transmission conditions.

Convincing comparisons between Dirichlet and optimized transmission conditions where given in [18], the coefficients being computed numerically. Choosing the best coefficients p and q in (2) leads to a new best approximation problem, that was solved by hand for Robin conditions (q=0) in [6] and studied in depth in [1]. Note that various asymptotics come into play: the size of the overlap in terms of the mesh-size in space Δx, the frequencies actually supported by the grid: $\omega \in [\pi/T, \pi/\Delta T]$, $\eta \in [\pi/Y, \pi/\Delta y]$. Asymptotic values of the optimal parameter were given in [6] for Robin transmission conditions in one dimension. We give in Table 1 the asymptotic values in dimension $n > 1$, which were first included in our manuscript for [6], but discarded by the referee who thought it was too complicated. In [1] we give asymptotic values of the parameters of the order 2 method for two subdomains in any dimension. These asymptotics make the algorithm easy to use and fast. Table 2 shows the asymptotic values of the convergence factor for various algorithms.

Note that the value of the optimal parameters depends on the rate between Δt and Δx, while the convergence factor does not. The convergence factor of the optimized of order 2 method is almost independent of the mesh-size, unlike classical methods. Furthermore, the convergence is even better with a small overlap.

We now discuss numerically the dependence with respect to the number of subdomains. In Fig. 1, we consider the advection-diffusion equation in one-dimension on the domain $[0,6]$, with $a = 1$ and $\nu = 1/5$. The final time is $T = 2.5$. It is discretized with implicit Euler, $\Delta x = 0.01$, $\Delta t = 0.0025$. The overlap is 4 gridpoints.

method	parameter p
O0 overlap Δx	$\begin{cases} (\nu(a^2+4\nu b))^{1/3}\Delta x^{-1/3} & \text{if } \beta=1 \\ (2\nu(a^2+4\nu b))^{1/3}\Delta x^{-1/3} & \text{if } \beta=2 \end{cases}$
O0 no overlap	$\begin{cases} (2\pi\nu\sqrt{(n-1)(a^2+4\nu b)})^{1/2}\Delta x^{-1/2} & \text{if } \beta=1 \\ (8\pi\nu(a^2+4\nu b))^{1/4}\Delta x^{-1/2} & \text{if } \beta=2,\ \nu\leq\bar{\nu} \\ \left(\frac{8\pi\nu(a^2+4\nu b)((n-1)^2\nu^2\pi^2+1)}{\sqrt{(n-1)^2\nu^2\pi^2+1}+(n-1)\nu\pi}\right)^{1/4}\Delta x^{-1/2} & \text{if } \beta=2,\ \nu>\bar{\nu} \end{cases}$

Table 1. Summary of the asymptotic parameters p in dimension $n>1$ in the Robin transmission conditions, for $\Delta t=\Delta x^{\beta}$, $\beta=1,2$, Δx the discretization step in all spatial directions. $\bar{\nu}\approx 1.5437/(\pi(n-1))$, and $a=a_x$.

	overlap	no overlap
Dirichlet	$1-O(\Delta x)$	
Optimized order 0	$1-O(\Delta x^{1/3})$	$1-O(\Delta x^{1/2})$
Optimized order 2	$1-O(\Delta x^{1/5})$	$1-O(\Delta x^{1/4})$

Table 2. Asymptotic expansion of the convergence factor in dimension $n>1$.

For the optimized Robin (resp. Order 2 optimized) algorithm, the same coefficient is used for all subdomains, computed by the explicit formulas given in [2], that is $p=2.054275607$ (resp. $p=1.366061845$, $q=0.1363805228$). We compute the number of iterations necessary to reach an error of 10^{-6} (the error is measured as the discrete L^2 error in time on the interface), as a function of the number of subdomains. The convergence history depends strongly on the number of subdomains. However optimized Schwarz waveform relaxation beats the classical Schwarz in two ways: the convergence is much faster (there is a factor 8 to 10), second the influence of the increase of the number of subdomains is much weaker.

Systematic testing is undergoing to see how relevant these values are for more general geometries and any number of subdomains.

5 Other Fruits: Optimized Schwarz Waveform Relaxation for Other Types of PDEs

The strategy of Schwarz waveform relaxation applies to every type of partial differential equations. We have in each case proposed new algorithms, and compared them to classical Schwarz waveform relaxation, with Dirichlet transmission conditions.

- ♣ We have designed algorithms for the wave equation in one or two dimensions. In two dimensions, due to the presence of evanescent waves, we have been led to choose overlapping algorithms, with optimized of order 2 transmission conditions. The coefficients are given by an exact formula as functions of the overlap and the desired tolerance on the error in [5].
- ♣ For the Schrödinger equation in one dimension, we have tested (complex) Robin conditions, which behave much better with overlap. The coefficient is computed

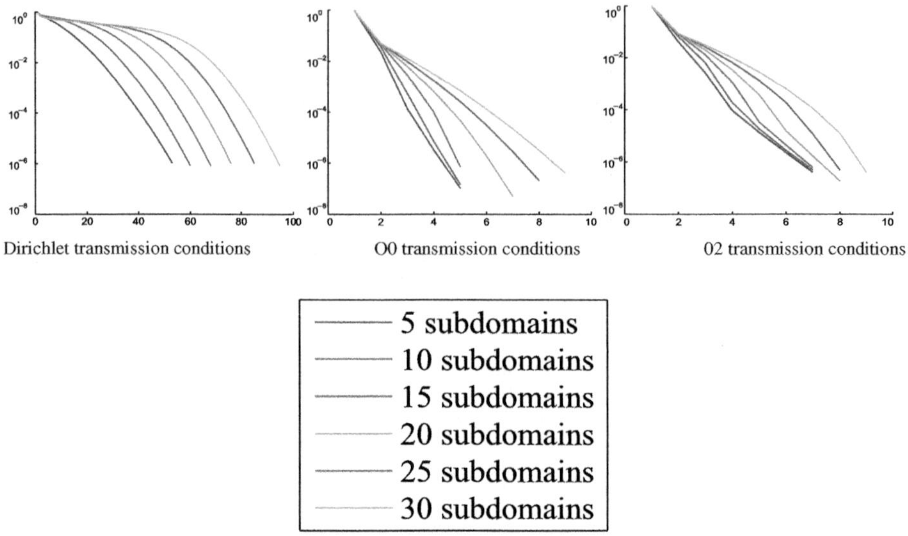

Fig. 1. Convergence history for the three main algorithms for 5 to 30 subdomains. Logaritmic scale.

again by an explicit formula. Moreover, we have suggested to use the exact Dirichlet to Neumann map, designed for the Crank-Nicolson scheme. The convergence is achieved in 2 or 3 iterations, even with non constant potential (linear or parabolic), for which the exact Dirichlet to Neumann map is not known, see [12].

♣ For nonlinear waves, we have designed nonlinear Schwarz waveform relaxation algorithms, which again converge in a few iterations, see [13].

6 New Blossoms: Space-Time Coupling and Refinements

In principle, nothing can prevent us from using different time and space meshes in different subdomains, in any of our algorithms. The question is how to match the discrete approximations on the interfaces of the subdomains. These interfaces have dimension 1 (time) + $N-1$ (space). We designed in [9] an optimal algorithm in one dimension, that we used for the wave equation and space-time refinement. Using a leapfrog scheme in each subdomain, which is of order two in space and time, we were able to keep a CFL number equal to 1 in each subdomain, which minimizes the dispersion and gives convergence in two iterations on time windows, and proved the overall solution to be an order 2 approximation of the wave equation. The extension to the 2-D wave equation is ongoing.

For parabolic problems, we use Discontinuous Galerkin methods in time with Optimized Order 2 Schwarz waveform relaxation (see presentation by Caroline

Japhet in DD18). The projection algorithm in time is coupled with an optimal projection algorithm in space (see contribution by Gander and Japhet in DD18), and mortar projection. Extension to non linear problems is ongoing.

References

[1] Amitai, D., Averbuch, A., Israeli, M., Itzikowitz, S., Turkel, E.: A survey of asynchronous finite-difference methods for parabolic PDEs on multiprocessors. *Appl. Numer. Math.*, 12:27–45, 1993.

[2] Bennequin, D., Gander, M.J., Halpern, L.: A homographic best approximation problem with application to optimized Schwarz waveform relaxation. *Math. Comp.*, 78:185–223, 2009.

[3] Burrage, K., Dyke, C., Pohl, B.: On the performance of parallel waveform relaxations for differential systems. *Appl. Numer. Math.*, 20:39–55, 1996.

[4] Chazan, D., Miranker, W.: Chaotic relaxation. *J. Linear Algebra Appl.*, 2:192–222, 1969.

[5] Gander, M.J., Halpern, L.: Absorbing boundary conditions for the wave equation and parallel computing. *Math. Comp.*, 74(249):153–176, 2004.

[6] Gander, M.J., Halpern, L.: Optimized Schwarz waveform relaxation methods for advection reaction diffusion problems. *SIAM J. Numer. Anal.*, 45(2):666–697, 2007.

[7] Gander, M.J., Stuart, A.M.: Space time continuous analysis of waveform relaxation for the heat equation. *SIAM J. Sci. Comput.*, 19:2014–2031, 1998.

[8] Gander, M.J., Halpern, L., Nataf, F.: Optimal convergence for overlapping and non-overlapping Schwarz waveform relaxation. In C-H. Lai, P. Bjørstad, M. Cross, and O. Widlund, eds., *Eleventh international Conference of Domain Decomposition Methods*, pages 27–36, 1999.

[9] Gander, M.J., Halpern, L., Nataf, F.: Optimal Schwarz waveform relaxation for the one dimensional wave equation. *SIAM J. Numer. Anal.*, 41(5):1643–1681, 2003.

[10] Giladi, E., Keller, H.B.: Space time domain decomposition for parabolic problems. *Numer. Math.*, 93(2):279–313, 2002.

[11] Hackbusch, W.: Parabolic multi-grid methods. In R. Glowinski and J.-L. Lions, eds., *Computing Methods in Applied Sciences and Engineering, VI*, pages 189–197. North-Holland, 1984.

[12] Halpern, L., Szeftel, J.: Optimized and quasi-optimal Schwarz waveform relaxation for the one dimensional Schrödinger equation. In U. Langer, M. Discacciati, D.E. Keyes, O.B. Widlund, and W. Zulehner, eds., *Domain Decomposition Methods in Science and Engineering XVII*, pages 221–228. Springer, 2008.

[13] Halpern, L., Szeftel, J.: Nonlinear Schwarz waveform relaxation for semilinear wave propagation. *Math. Comp.*, 2009.

[14] Horton, G., Vandewalle, S. A space-time multigrid method for parabolic partial differential equations. *SIAM J. Sci. Comput.*, 16(4):848–864, 1995.

[15] Lelarasmee, E., Ruehli, A.E., Sangiovanni-Vincentelli, A.L.: The waveform relaxation method for time-domain analysis of large scale integrated circuits. *IEEE Trans. CAD Integrated Circuits Syst.*, 1:131–145, 1982.

[16] Lions, J.-L., Maday, Y., Turinici, G.: A parareal in time discretization of PDE's. *C. R. Acad. Sci. Paris Sér. I Math.*, 332:661–668, 2001.

[17] Lions, P.-L.: On the Schwarz alternating method. I. In R. Glowinski, G.H. Golub, G.A. Meurant, and J. Périaux, eds., *First International Symposium on Domain Decomposition Methods for Partial Differential Equations*, pages 1–42. SIAM, 1988.

[18] Martin, V.: *Méthodes de décomposition de domaines de type relaxation d'ondes pour des équations de l'océnographie*. PhD thesis, Université Paris 13, 2003.

[19] Martin, V.: An optimized Schwarz waveform relaxation method for the unsteady convection diffusion equation. *Appl. Numer. Math.*, 52(4):401–428, 2005.

[20] Miekkala, U., Nevanlinna, O.: Convergence of dynamic iteration methods for initial value problems. *SIAM J. Sci. Statist. Comput.*, 8:459–482, 1987.

[21] Quarteroni, A., Valli, A.: *Domain Decomposition Methods for Partial Differential Equations*. Oxford Science Publications, 1999.

[22] Vandewalle, S., Horton, G.: Fourier mode analysis of the multigrid waveform relaxation and time-parallel multigrid methods. *Computing*, 54(4):317–330, 1995.

Optimized Schwarz Methods

F. Nataf

Laboratoire J. L. Lions, CNRS UMR7598, Université Pierre et Marie Curie, 75252 Paris Cedex 05, France, nataf@ann.jussieu.fr

Summary. The strategy of domain decomposition methods is to decompose the computational domain into smaller subdomains. Each subdomain is assigned to one processor. The equations are solved on each subdomain. In order to enforce the matching of the local solutions, interface conditions have to be written on the boundary between subdomains. These conditions are imposed iteratively. The convergence rate is very sensitive to these interface conditions. The Schwarz method is based on the use of Dirichlet boundary conditions. It can be slow and requires overlapping decompositions. In order to improve the convergence and to be able to use non-overlapping decompositions, it has been proposed to use more general boundary conditions. It is even possible to optimize them with respect to the efficiency of the method. Theoretical and numerical results are given along with open problems.

1 Introduction: Original Schwarz Method (1870)

The first domain decomposition method was developed at the end of the 19th century by the mathematician H. A. Schwarz. His goal was to study the Laplace operator. At that time, the main tool for this purpose was Fourier analysis and more generally the use of special functions. Geometries of the domain were essentially restricted to simple configurations: rectangles and disks, see Fig. 1. His idea was to study the case of a domain that is the union of simple domains. For example, let $\Omega = \Omega_1 \cup \Omega_2$ with $\Omega_1 \cap \Omega_2 \neq \emptyset$. We want to solve

$$\begin{aligned} -\Delta u &= f \quad \text{in } \Omega \\ u &= 0 \quad \text{on } \partial\Omega. \end{aligned} \tag{1}$$

Schwarz proposed the following algorithm (Alternating Schwarz Method):

Let (u_1^n, u_2^n) be an approximation to $(u_{|\Omega_1}, u_{|\Omega_2})$ at step n of the algorithm, (u_1^{n+1}, u_2^{n+1}) is defined by

$$\begin{aligned} -\Delta u_1^{n+1} &= f && \text{in } \Omega_1 & \qquad -\Delta u_2^{n+1} &= f && \text{in } \Omega_2 \\ u_1^{n+1} &= 0 && \text{on } \partial\Omega_1 \cap \partial\Omega & u_2^{n+1} &= 0 && \text{on } \partial\Omega_2 \cap \partial\Omega \\ u_1^{n+1} &= u_2^n && \text{on } \partial\Omega_1 \cap \overline{\Omega_2}. & u_2^{n+1} &= u_1^{n+1} && \text{on } \partial\Omega_2 \cap \overline{\Omega_1}. \end{aligned}$$

M. Bercovier et al. (eds.), *Domain Decomposition Methods in Science and Engineering XVIII*, Lecture Notes in Computational Science and Engineering,
DOI: 10.1007/978-3-642-02677-5,

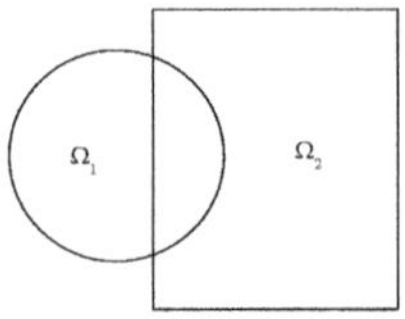

Fig. 1. Overlapping domain decomposition

The problem in domain Ω_1 has to be solved before the problem in domain Ω_2. This algorithm is sequential. Schwarz proved linear convergence of (u_1^n, u_2^n) to $(u_{|\Omega_1}, u_{|\Omega_2})$ as n tends to infinity.

A slight modification of the algorithm is

$$\begin{array}{llll} -\Delta u_1^{n+1} = f & \text{in } \Omega_1 & -\Delta u_2^{n+1} = f & \text{in } \Omega_2 \\ u_1^{n+1} = 0 & \text{on } \partial\Omega_1 \cap \partial\Omega & u_2^{n+1} = 0 & \text{on } \partial\Omega_2 \cap \partial\Omega \\ u_1^{n+1} = u_2^n & \text{on } \partial\Omega_1 \cap \overline{\Omega_2}. & u_2^{n+1} = u_1^n & \text{on } \partial\Omega_2 \cap \overline{\Omega_1}. \end{array} \tag{2}$$

Problems in domains Ω_1 and Ω_2 may be solved concurrently. The algorithm is parallel and is adapted to parallel computers.

The discrete version of (2) is the **RAS** algorithm, see [2, 7].

1.1 Towards Faster Methods: Two Families of Methods

The benefit of the above Schwarz algorithms is the saving in memory requirements. Indeed, if the problems are solved by direct methods, the cost of the storage is non-linear with respect to the number of unknowns. By dividing the original problem into smaller pieces the amount of storage can be significantly reduced. As far as CPU is concerned, the original Schwarz algorithms work fine for some problems but may be very slow for others. Roughly speaking for time dependent problems with relatively small time steps, the methods will perform well (e.g. transient compressible flow computations). But for steady state problems (e.g. Helmholtz or harmonic Maxwell's equations), it can be very slow. Another weakness is the need of overlapping subdomains. Indeed, only the continuity of the solution is imposed and nothing is imposed on the matching of the fluxes. When there is no overlap convergence is thus impossible.

The slowness of the method and the need for overlapping subdomains are linked. Indeed, it can be proved that the convergence rate of the Schwarz method is a continuous function of the size of the overlap denoted δ. For small overlaps the convergence rate is close to one. Actually it can be proved that for small overlaps the convergence rate varies as $1 - C^t\delta$.

In order to remedy the drawbacks of the original Schwarz method, two families of methods have been developed. They both work in the non-overlapping case and consist of introducing the normal derivative of the solution, but in two very different ways:

- write a substructured formulation of the domain decomposition problem where the matching of the solution and of its normal derivative along the interface are imposed explicitly.
- Modify the original Schwarz method by replacing the Dirichlet interface conditions on $\partial\Omega_i \setminus \partial\Omega$, $i = 1,2$, by Robin interface conditions ($\partial_{n_i} + \alpha$, where n is the outward normal to subdomain Ω_i), see [17].

The first approach corresponds to "Neumann-Neumann or FETI Methods". The second approach is developped in what follows.

More generally, a complete overview of various domain decomposition methods may be found in a few books [4, 22, 23, 24] or in the proceedings of various conferences on domain decomposition methods, see e.g. [1, 3, 16] and references therein.

2 Modified Schwarz Method

The Restrictive Additive Schwarz Method presents the drawback of needing overlapping subdomains in order to converge. In this chapter, we consider several improvements:

- replacement of the Dirichlet interface conditions by mixed interface conditions which yield convergence for non overlapping domain decompositions, see section 2.1;
- optimization of the interface conditions for faster convergence, see section 2.2;
- replacement of the fixed point iterative strategy of (2) by Krylov type methods, see [10].

2.1 Generalized Schwarz Methods

A major improvement of the Schwarz method comes from the use of other interface conditions. It has first been proposed by P.L. Lions to replace the Dirichlet interface conditions by Robin interface conditions, see [17]. Let α be a positive number; the modified algorithm is:

$$\begin{aligned}
-\Delta u_1^{n+1} &= f \quad \text{in } \Omega_1,\\
u_1^{n+1} &= 0 \quad \text{on } \partial\Omega_1 \cap \partial\Omega,\\
\left(\frac{\partial}{\partial n_1} + \alpha\right) u_1^{n+1} &= \left(-\frac{\partial}{\partial n_2} + \alpha\right) u_2^n \quad \text{on } \partial\Omega_1 \cap \overline{\Omega_2}
\end{aligned}$$

($\mathbf{n}_1$ and $\mathbf{n}_2$ are the outward normals on the boundary of the subdomains),

$$\begin{aligned}
-\Delta u_2^{n+1} &= f \quad \text{in } \Omega_2,\\
u_2^{n+1} &= 0 \quad \text{on } \partial\Omega_2 \cap \partial\Omega\\
\left(\frac{\partial}{\partial n_2} + \alpha\right) u_2^{n+1} &= \left(-\frac{\partial}{\partial n_1} + \alpha\right) u_1^n \quad \text{on } \partial\Omega_2 \cap \overline{\Omega_1}.
\end{aligned}$$

The convergence proof given by P. L. Lions in the elliptic case was extended by B. Després [6] to the Helmholtz equation. A general presentation is given in [5]. It can also be extended to more general interface conditions with second order tangential derivatives in the interface conditions, see [19].

2.2 Optimal Interface Conditions

In the preceding section, we have seen that a general convergence result holds for interface conditions with Robin or second order tangential derivatives. Actually these conditions are not the most general. Rather than giving the general conditions in an *a priori* form, we shall derive them in this section so as to have the fastest convergence. We establish the existence of interface conditions which are optimal in terms of iteration counts. The corresponding interface conditions are pseudo-differential and are not practical. Nevertheless, this result is a guide for the choice of partial differential interface conditions. Moreover, this result establishes a link between the optimal interface conditions and artificial boundary conditions. This is also a help when dealing with the design of interface conditions since it gives the possibility to use the numerous papers and books published on the subject of artificial boundary conditions, see e.g. [8, 12].

We consider a general linear second order elliptic partial differential operator $\mathcal{L}$ and the problem:

$$\text{Find } u \text{ such that } \mathcal{L}(u) = f \text{ in a domain } \Omega \text{ and } u = 0 \text{ on } \partial\Omega.$$

The domain Ω is decomposed into two subdomains Ω_1 and Ω_2. We suppose that the problem is regular so that $u_i := u|_{\Omega_i}$, $i = 1,2$, is continuous and has continuous normal derivatives across the interface $\Gamma_i = \partial\Omega_i \cap \bar{\Omega}_j$, $i \neq j$. A modified Schwarz type method is considered.

$$\begin{array}{ll} \mathcal{L}u_1^{n+1} = f \quad \text{in } \Omega_1 & u_1^{n+1} = 0 \quad \text{on } \partial\Omega_1 \cap \partial\Omega \\ \mu_1 \nabla u_1^{n+1}.\mathbf{n}_1 + \mathcal{B}_1 u_1^{n+1} & = -\mu_1 \nabla u_2^n.\mathbf{n}_2 + \mathcal{B}_1 u_2^n \quad \text{on } \Gamma_1 \\ \mathcal{L}u_2^{n+1} = f \quad \text{in } \Omega_2 & u_2^{n+1} = 0 \quad \text{on } \partial\Omega_2 \cap \partial\Omega \\ \mu_2 \nabla u_2^{n+1}.\mathbf{n}_2 + \mathcal{B}_2 u_2^{n+1} & = -\mu_2 \nabla u_1^n.\mathbf{n}_1 + \mathcal{B}_2 u_1^n \quad \text{on } \Gamma_2 \end{array} \tag{3}$$

where μ_1 and μ_2 are real-valued functions and $\mathcal{B}_1$ and $\mathcal{B}_2$ are operators acting on the interfaces Γ_1 and Γ_2. For instance, $\mu_1 = \mu_2 = 0$ and $\mathcal{B}_1 = \mathcal{B}_2 = \text{Id}$ correspond to the algorithm (2); $\mu_1 = \mu_2 = 1$ and $\mathcal{B}_i = \alpha \in \mathbf{R}$, $i = 1,2$, has been proposed in [17] by P.L. Lions.

The question is:

Are there other possibilities in order to have convergence in a minimal number of steps?

In order to answer this question, we note that by linearity, the error e satisfies ($\mu_1 = \mu_2 = 1$)

$$\begin{aligned}
&\mathcal{L}(e_1^{n+1}) = 0 \quad \text{in } \Omega_1 \qquad e_1^{n+1} = 0 \quad \text{on } \partial\Omega_1 \cap \partial\Omega \\
&\nabla e_1^{n+1} \cdot \mathbf{n}_1 + \mathcal{B}_1(e_1^{n+1}) = -\nabla e_2^n \cdot \mathbf{n}_2 + \mathcal{B}_1(e_2^n) \quad \text{on } \Gamma_1 \\
&\mathcal{L}(e_2^{n+1}) = 0 \quad \text{in } \Omega_2 \qquad e_2^{n+1} = 0 \quad \text{on } \partial\Omega_2 \cap \partial\Omega \\
&\nabla e_2^{n+1} \cdot \mathbf{n}_2 + \mathcal{B}_2(e_2^{n+1}) = -\nabla e_1^n \cdot \mathbf{n}_1 + \mathcal{B}_2(e_1^n) \quad \text{on } \Gamma_2
\end{aligned}$$

The initial guess e_i^0 is arbitrary so that it is impossible to have convergence at step 1 of the algorithm. Convergence needs at least two iterations. Having $e_1^2 \equiv 0$ requires $-\nabla e_2^1.\mathbf{n}_2 + \mathcal{B}_1(e_2^1) \equiv 0$. The only meaningful information on e_2^1 is that $\mathcal{L}(e_2^1) = 0 \quad \text{in } \Omega_2$. In order to use this information, we introduce the DtN (Dirichlet to Neumann) map (a.k.a. Steklov-Poincaré): Let

$$\begin{aligned}
u_0 &: \Gamma_1 \to \mathbf{R} \\
\text{DtN}_2(u_0) &:= \nabla v.\mathbf{n}_{2|\partial\Omega_1 \cap \bar{\Omega}_2},
\end{aligned} \tag{4}$$

where $\mathbf{n}_2$ is the outward normal to $\Omega_2 \setminus \bar{\Omega}_1$, and v satisfies the following boundary value problem:

$$\begin{aligned}
\mathcal{L}(v) &= 0 \quad \text{in } \Omega_2 \setminus \bar{\Omega}_1 \\
v &= 0 \quad \text{on } \partial\Omega_2 \cap \partial\Omega \\
v &= u_0 \quad \text{on } \partial\Omega_1 \cap \bar{\Omega}_2.
\end{aligned}$$

Let $\mathcal{B}_1 := \text{DtN}_2$. This choice is optimal since we can check that $-\nabla e_2^1 \cdot \mathbf{n}_2 + \mathcal{B}_1(e_2^1) \equiv 0$. *The use of* $\mathcal{B}_i = \text{DtN}_j$ $(i \neq j)$ *as interface conditions in* (3) *is optimal: we have* (*exact*) *convergence in two iterations.*

The two-domain case for an operator with constant coefficients has been first treated in [13]. The multidomain case for a variable coefficient operator with both positive results [20] and negative conjectures [21] has been considered as well.

Remark 1. The main feature of this result is to be very general since it does not depend on the exact form of the operator $\mathcal{L}$ and can also be extended to more general systems or to coupled systems of equations as well with a proper care of the well posedness of the algorithm.

As an application, we take $\Omega = \mathbf{R}^2$ and $\Omega_1 = (-\infty, 0) \times \mathbf{R}$. Using a Fourier technique, it is possible to give the explicit form of the DtN operator for a constant coefficient operator. If $\mathcal{L} = \eta - \Delta$, the DtN map is a pseudo-differential operator whose symbol is

$$B_{i,\text{opt}}(k) = \sqrt{\eta + k^2},$$

i.e., $\mathcal{B}_{i,\text{opt}}(u)(0,y) = \int_{\mathbf{R}} B_{i,\text{opt}}(k)\hat{u}(0,k)e^{iky}\,dk$. This symbol is not polynomial in the Fourier variable k so that the operators and the above example shows, exact absorbing conditions are in general pseudo-differential. They correspond to exact absorbing conditions. These conditions are used on the artificial boundary resulting from the truncation of a computational domain. On this boundary, boundary conditions have to be imposed. The solution on the truncated domain depends on the choice of this artificial condition. We say that it is an exact absorbing boundary condition if the solution computed on the truncated domain is the restriction of the solution of the original problem. Surprisingly enough, the notions of exact absorbing conditions for domain truncation and that of optimal interface conditions in domain decomposition methods coincide.

As the above examples show, they are pseudodifferential. Therefore they are difficult to implement. Moreover, in the general case of variable coefficient operators and/or a curved boundary, the exact form of these operators is not known, although they can be approximated by partial differential operators which are easier to implement. The approximation of the DtN has been addressed by many authors since the seminal paper [8] by Engquist and Majda on this question.
It turns out that the approximations designed for domain truncation perform poorly when used in domain decomposition methods. There have been many research efforts in the last 15 years on how to tune approximate DtN maps so that they perform well in domain decomposition methods. The first works were based on Fourier techniques, see e.g. [9, 10, 14] and references therein. These approaches work fine for smooth coefficients operators. But when dealing with highly discontinuous coefficients, it is necessary to take a more algebraic approach, see [11, 18] in this direction. Results are promising but many issues are still open, see below.

3 Conclusion and Open Problems

Both approaches (Neumann-Neumann and optimized Schwarz methods) are robust thanks to Krylov methods. Neumann-Neumann, BDDC and FETI type methods are optimal but lack generality. Optimized Schwarz methods are general but are more difficult to tune. The main open problems are from a practical point of view

- the design of algebraic optimized interface conditions that are as efficient as the analytic ones
- the interplay between the optimized interface conditions and a coarse grid (see [15])

References

[1] Bjørstad, P.E., Espedal, M.S., Keyes, D.E., eds.: *Ninth International Conference on Domain Decomposition Methods*, 1997. Proceedings of the 9th International Conference on Domain Decomposition Methods in Bergen, Norway.

[2] Cai, X.-C., Sarkis, M.: A restricted additive Schwarz preconditioner for general sparse linear systems. *SIAM J. Sci. Comput.*, 21:239–247, 1999.

[3] Chan, T., Glowinski, R., Périaux, J., Widlund, O., eds.: *Domain Decomposition Methods*, Philadelphia, PA, 1989. SIAM. Proceedings of the Second International Symposium on Domain Decomposition Methods, Los Angeles, California, January 14–16, 1988.

[4] Chan, T.F., Mathew, T.P.: Domain decomposition algorithms. In *Acta Numerica, 1994*, pages 61–143. Cambridge University Press, 1994.
[5] Collino, F., Ghanemi, S., Joly, P.: Domain decomposition methods for harmonic wave propagation: a general presentation. *Comput. Meth. Appl. Mech. Eng.*, (2-4):171–211, 2000.
[6] Després, B.: Domain decomposition method and the Helmholtz problem. II. In *Second International Conference on Mathematical and Numerical Aspects of Wave Propagation (Newark, DE, 1993)*, pages 197–206, Philadelphia, PA, 1993. SIAM.
[7] Efstathiou, E., Gander, M.J.: Why Restricted Additive Schwarz converges faster than Additive Schwarz. *BIT*, 43(5):945–959, 2003.
[8] Engquist, B., Majda, A.: Absorbing boundary conditions for the numerical simulation of waves. *Math. Comp.*, 31(139):629–651, 1977.
[9] Gander, M.J.: Optimized Schwarz methods. *SIAM J. Numer. Anal.*, 44(2):699–731, 2006.
[10] Gander, M.J., Magoulès, F., Nataf, F.: Optimized Schwarz methods without overlap for the Helmholtz equation. *SIAM J. Sci. Comput.*, 24(1):38–60, 2002.
[11] Gerardo-Giorda, L., Nataf, F.: Optimized Schwarz methods for unsymmetric layered problems with strongly discontinuous and anisotropic coefficients. *J. Numer. Math.*, 13(4):265–294, 2005.
[12] Givoli, D.: *Numerical methods for problems in infinite domains*. Elsevier, 1992.
[13] Hagstrom, T., Tewarson, R.P., Jazcilevich, A.: Numerical experiments on a domain decomposition algorithm for nonlinear elliptic boundary value problems. *Appl. Math. Lett.*, 1(3), 1988.
[14] Japhet, C., Nataf, F., Rogier, F.: The optimized order 2 method. application to convection-diffusion problems. *Future Generation Computer Systems*,18(1):17–30, 2001.
[15] Japhet, C., Nataf, F., Roux, F.-X.: The Optimized Order 2 Method with a coarse grid preconditioner. application to convection-diffusion problems. In P. Bjørstad, M. Espedal, and D. Keyes, eds., *Ninth International Conference on Domain Decompositon Methods in Science and Engineering*, pages 382–389. Wiley, 1998.
[16] Lai, C.-H., Bjørstad, P.E., Cross, M., Widlund, O., eds.: *Eleventh International Conference on Domain Decomposition Methods*, 1998. Proceedings of the 11th International Conference on Domain Decomposition Methods in Greenwich, England, July 20–24, 1998.
[17] Lions, P.-L.: On the Schwarz alternating method. III: a variant for nonoverlapping subdomains. In T.F. Chan, R. Glowinski, J. Périaux, and O. Widlund, eds., *Third International Symposium on Domain Decomposition Methods for Partial Differential Equations , held in Houston, Texas, March 20–22, 1989*, Philadelphia, PA, 1990. SIAM.
[18] Magoulès, F., Roux, F.-X., Series, L.: Algebraic approach to absorbing boundary conditions for the Helmholtz equation. *Int. J. Comput. Math.*, 84(2): 231–240, 2007.

[19] Nataf, F.: Interface connections in domain decomposition methods. In *Modern methods in scientific computing and applications (Montréal, QC, 2001)*, vol. 75 of *NATO Sci. Ser. II Math. Phys. Chem.*, pages 323–364. Kluwer Academic, Dordrecht, 2002.
[20] Nataf, F., Rogier, F., de Sturler, E.: Optimal interface conditions for domain decomposition methods. Technical Report 301, CMAP (Ecole Polytechnique), 1994.
[21] Nier, F.: Remarques sur les algorithmes de décomposition de domaines. In *Seminaire: Équations aux Dérivées Partielles, 1998–1999*, Exp. No. IX, 26. École Polytech., 1999.
[22] Quarteroni, A., Valli, A.: *Domain Decomposition Methods for Partial Differential Equations*. Oxford Science, 1999.
[23] Smith, B.F., Bjørstad, P.E., Gropp, W.: *Domain Decomposition: Parallel Multilevel Methods for Elliptic Partial Differential Equations*. Cambridge University Press, 1996.
[24] Toselli, A., Widlund, O.: *Domain Decomposition Methods—Algorithms and Theory*, vol. 34 of *Springer Series in Computational Mathematics*. Springer, 2004.

The Development of Coarse Spaces for Domain Decomposition Algorithms

Olof B. Widlund

Courant Institute, 251 Mercer Street, New York, NY 10012, USA widlund@cims.nyu.edu

Summary. The importance of using coarse components, and thus at least one additional level, in the design of domain decomposition methods has been understood for at least twenty years. For many problems of interest, such a device, which provides at least a minimal amount of global transfer of information in each step, is necessary in order to obtain convergence rates which are independent of the number of subdomains. An historical overview, colored by the scientific history of its author, is given of the development of such coarse components of the domain decomposition algorithms. These algorithms are all preconditioned conjugate gradient methods or they are accelerated by using some alternative Krylov space method. The preconditioners are built from solvers of the given problem restricted to subdomains and a coarse approximation which often can be quite exotic.

1 Introduction

We will consider finite element approximations of, e.g., a self-adjoint scalar elliptic problem or the equations of linear elasticity. The domain Ω of the partial differential equation is subdivided into non-overlapping subdomains (substructures) Ω_i; there can be very many of them, in particular, when massively parallel computer systems are employed. Between the subdomains, we have the interface Γ; Γ_h is its set of finite element nodes. Each subdomain is the union of elements of the finite element triangulation.

There are two main families of domain decomposition algorithms: the *iterative substructuring algorithms*, using solvers of the finite element problems restricted to the Ω_i, each often with tens of thousands degrees of freedom, and the *overlapping Schwarz methods*, using solvers on a set of overlapping subdomains Ω_i', often obtained by adding layers of elements to the individual Ω_i's. Exact solvers are often used to solve these local problems as in much of traditional finite element practice.

The preconditioner of the finite element problem also include a coarse, global solver with a few degrees of freedom for each subdomain. A Krylov space method—conjugate gradients or GMRES—is always used to accelerate the convergence.

Early on, coarse spaces were not used and only continuous problems were considered; in fact it is unclear what a coarse problem then might be. Algorithms based

M. Bercovier et al. (eds.), *Domain Decomposition Methods in Science and Engineering XVIII*, Lecture Notes in Computational Science and Engineering,
DOI: 10.1007/978-3-642-02677-5,

on overlapping subdomains were considered by [2, 28, 31] and algorithms with non-overlapping subdomains, in a Poincaré-Steklov framework, by [1, 27].

In the 1980's, there were a number of studies for problems where the interfaces without cross points (or cross edges), i.e., all finite element nodes on the interface are common to the boundary of only two subdomains; the decompositions of the domains effectively were into strips.

2 Early Two-Level Domain Decomposition Methods

The successful introduction of a second, coarse level dates to the mid-eighties. In particular, the first and fourth paper in a series of four, [4, 5], were crucial for the development of the theory of iterative substructuring methods for more general decompositions; these papers existed at least in preprint form by the time of the first international conference on domain decomposition methods, DD1, held in Paris in January 1987. Already at that time, it was realized that a coarse component, to provide at least a minimal amount of global transfer of information across the entire domain, is required to obtain bounds which are independent of the number of subdomains.

In the first of these papers, on problems in two dimensions, the substructures are triangles and the coarse space is spanned by continuous, piece-wise linear functions on this coarse triangulation in a set-up resembling that of geometric multigrid. There is one local space for each of the edges of the interface. A $C(1+\log(H/h))^2$ bound is established for the condition number of the preconditioned operator; here H is a typical subdomain diameter and h that of the finite element triangles. These logarithmic factors arise when we partition the trace of finite element functions on the interface into a sum of functions with a nonzero trace only on one edge.

The main result is obtained in an analysis for one subdomain at a time. As a consequence, the constant C is independent of the number of subdomains and the result is valid uniformly for any scalar problem

$$-div(a(x)\,grad)u(x) = f(x),$$

where $a(x) = a_i, x \in \Omega_i$ with the a_i arbitrary, positive constants. An important tool, used in this work, is a finite element Sobolev inequality for plane domains:

$$\|u_h\|^2_{L^\infty(\Omega_i)} \leq C(1+\log(H/h))\|u_h\|^2_{H^1(\Omega_i)}.$$

This is a genuine finite element and best possible result; see [7].

Before discussing [5], which is regarded as the most important in the series, we consider the geometry of the decomposition of a domain in three dimensions. The interface Γ contains all the finite element nodes which belong to the closure of at least two subdomains. It is decomposed into *faces, edges,* and *vertices*: the nodes on a face F^{ij} belong to a pair of subdomains Ω_i and Ω_j, edges and vertices make up the boundary of faces with edges typically common to at least three subdomains, and vertices are end points of edges. Such decompositions can be defined even for quite irregular subdomains, such as those delivered by mesh partitioners. Each of these geometric objects can be defined in terms of an equivalence class of nodes with a common set of subdomain indices. For many iterative substructuring methods, as well as for some more recent methods based on overlapping decompositions, there are basis functions of coarse spaces directly associated with these geometric objects: $\theta_{\mathcal{F}ij}, \theta_{\mathcal{E}ik}$, and $\theta_{\mathcal{V}i\ell}$. They are defined by the value 1 on the set of nodes in question and vanish at all other nodes on Γ and they are discrete harmonic, i.e., the values inside the subdomains give a minimal energy extension. Therefore, they form a partition of unity for any subdomain which is interior to Ω.

The union of the edges and vertices of the interface in three dimensions is known as the *wire basket* and individual subdomains also have wire baskets. [5] concerns wire basket algorithms. Instead of working with a conventional coarse space, for which, to this day, no strong results, independent of the values of the a_i, have been derived for three dimensions, the coarse space functions are given in terms of their values on the wire basket of the subdomains. The values on a face are then given in terms of the values on its boundary; this establishes continuity across Γ. A corresponding interpolation operator, into the coarse space, will reproduce constants. Technically, this coarse space is of large dimension, but this is compensated for by using a particular inexact solver in terms of one variable per subdomain, namely the average values over the subdomain wirebaskets. The values at the subdomain nodes are then computed locally. We note that these first successful algorithms of this kind are also among the most complicated.

A version of these algorithms is developed and analyzed in the 1990 PhD thesis of Barry Smith. It was also implemented on parallel processors, see [29]. Smith then moved on to the development of PETSc. He also took the initiative to a joint project with Dryja and this author, which led to the development and analysis of a large number of *primal iterative substructuring* algorithms, see [13]. The analysis in that paper is carried out in an *abstract Schwarz* framework, which has its roots in a DD1 contribution of Lions [23]. All bounds are, with a few exceptions, of the form $C(1 + \log(H/h))^2$ and most of them are independent of coefficient jumps. Smith also later wrote a pioneering book, see [30].

Another important contribution at DD1, is a paper by [18]; their algorithms resembles one-level FETI methods. The importance of this work has been overlooked; see, however [32, Sec. 1.3.5].

By the time of DD2, the first two-level *additive Schwarz* methods had been developed and shown to be optimal and scalable, i.e., with convergence rates independent of the number of subdomains, for problems with moderately varying coefficients; cf. [12]. These preconditioners are built from solvers on the set of overlapping subdomains and a conventional coarse spaces just as that of [4]. At first, a generous overlap was assumed but the methods work most efficiently with modest overlap. This led to an analysis of the case of small overlap and the bound, with δ the overlap:

$$\kappa(T_{as}) \leq C(1 + H/\delta),$$

shown to be best possible by Brenner; see [6, 14].

Already at the time of DD3, it was realized that these and the iterative substructuring algorithms could be analyzed in a common *abstract Schwarz framework*; see [32, Chap. 2].

3 Additional Comments

It comes as no surprise to any student of multigrid that a global component of the preconditioner is very important. What makes the two families different, is that only two levels are required for a domain decomposition method even for very large problems. This limits the number of communication steps. The two-level overlapping Schwarz methods require two communication steps per iteration. One of them can be eliminated resulting in *restricted additive Schwarz* methods, invented by [8]. These algorithms have been studied extensively and they also typically require fewer iterations.

The great repertoire of coarse spaces has made it possible to develop fast methods with convergence rates independent of even large jumps in the material properties across the interface. It is also easy to extend the overlapping Schwarz methods to more than two levels and progress has also be made recently on introducing additional levels for the iterative substructuring methods of Sec. 4; see in particular [20, 33]. This work is increasingly relevant for very large problems and massively parallel computing systems for which the coarse space will be of very large dimension and presents a bottle neck.

The extension of any domain decomposition developed for scalar elliptic problems to the equations of linear elasticity requires a modification of the coarse spaces to accommodate the larger null space for these problems; in three dimensions, there are six rigid body modes of zero energy instead of a single constant. This *null space condition* for the coarse space was formalized in [24] and it is also explained well in [30]. In many cases, the extension is relatively routine, see [32, Chap. 8]. A successful approach begins by constructing a stable interpolation operator, which reproduces all rigid body modes, and with an energy that can be bounded uniformly or with a factor $C(1+\log(H/h))$.

4 Other Iterative Substructuring Methods

Other important domain decomposition algorithms date back at least to DD2, see [3]. This development led to *balancing Neumann-Neumann* methods with coarse space components; cf. [32, Sec. 6.2]. An important role in the description and analysis of the Neumann-Neumann algorithms is played by a family of weighted counting functions $\delta_i^\dagger$, associated with the individual $\partial\Omega_i$ and defined, for $\gamma \in [1/2, \infty)$, by a sum of contributions from the coefficient in Ω_i and its next neighbors;

$$\delta_i^\dagger(x) := \frac{a_i^\gamma}{\sum_{j \in N_x} a_j^\gamma}, \quad x \in \partial\Omega_{i,h} \cap \Gamma_h.$$

Here N_x is the set of indices j of the subregions such that $x \in \partial\Omega_{j,h}$. A subscript h denotes the set of nodes on the set in question. These functions provide a partition of unity:

$$\sum_i R_i^T \delta_i^{\dagger}(x) \equiv 1, \quad x \in \Gamma_h,$$

for any Ω_i such that $\partial\Omega_i \cap \partial\Omega_D = \emptyset$, and they span the coarse space of the algorithm. Here R_i^T provides an extension by zero to the nodes of $\Gamma \setminus \partial\Omega_i$. If the coefficients are constant in each subdomain, each of the $\delta_i^{\dagger}$ can be written as the linear combination of the face, edge, and vertex functions $\theta_{F^{ij}}, \theta_{E^{ik}}$, and $\theta_{V^{i\ell}}$, of the interface.

The local space V_i for the balancing methods has non-zero interface values only on $\partial\Omega_i$. A scaled Neumann problem, given by the bilinear form

$$\tilde{a}_i(u,v) = a_i \int_{\Omega_i} \nabla(\delta_i u) \cdot \nabla(\delta_i v) dx,$$

is used to define the local parts of a hybrid Schwarz method and $C(1+\log(H/h))^2$ bounds were established, with C independent of the number of substructures and of jumps in the coefficients across the interface, around 1995. These algorithms have proven very successful and have been used extensively, in a modified form, for problems of elasticity.

What is now called the *one-level FETI* methods were introduced in [17] and first analyzed in [25]. Instead of describing these methods, we will now consider the more recently developed FETI-DP and BDDC algorithms.

5 FETI-DP and BDDC

The FETI-DP methods were introduced in [15, 16] and the BDDC methods in [9]. These more recent methods only require the solution of positive definite problems. They are defined in terms of a set of *primal* continuity constraints which throughout the iteration; see Fig. 1. A pair of FETI-DP and BDDC preconditioned systems have essentially identical spectra if they employ the same primal constraints; see [26].

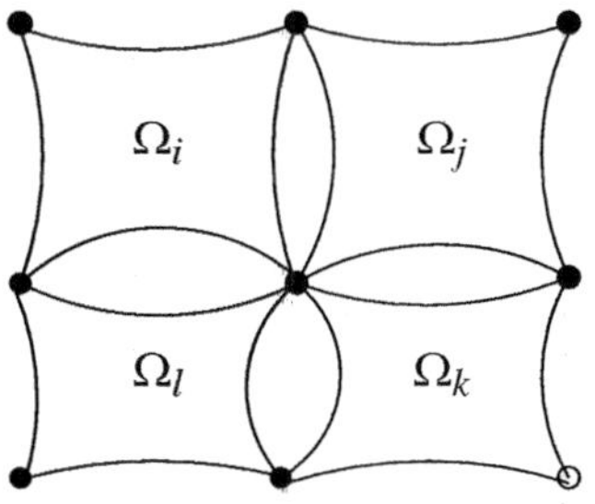

Fig. 1. Decomposition of subdomains for FETI-DP and BDDC methods.

The primal constraints in this case make the values at the subdomain vertices global, while we obtain multiple values at all other nodes on the interface. The partially subassembled stiffness matrix of this alternative finite element model is used to define the preconditioners. A linear system of equations of this kind has a positive definite matrix and it can be solved much less expensively than a system with the fully assembled matrix.

In a FETI-DP algorithm, the continuity at the edge nodes is enforced by using Lagrange multipliers and the rate of convergence is enhanced by solving Dirichlet problems on each subdomain in each iteration. The conjugate gradient algorithm is used to find the correct values of the Lagrange multipliers. The primal constraints provide a global component of these preconditioners.

In a BDDC algorithm, continuity is instead restored in each step by computing weighted averages across the interface. This leads to non-zero residuals at some nodes interior to the subdomains, and in each iteration, these residuals are eliminated by using subdomain Dirichlet solves.

For problems in three dimensions, primal variables associated with point constraints alone do not lead to competitive algorithms; this is technically closely related to the issues raised in early studies of primal iterative substructuring methods. Instead, or in addition, averages (and moments) over faces or, preferably edges, should have common values across the interface.

The selection a small and effective set of primal constraints for elasticity problems with large jumps in the Lamé parameters has been very challenging, see [21]. The resulting recipes have proven successful for very difficult problems, see [19]. In spite of the seemingly different coarse components of these algorithms, the tools of analysis are essentially the same as for the older iterative substructuring methods.

6 Additional Roles for Coarse Spaces

In work on incompressible Stokes, almost incompressible elasticity, and Maxwell's equations, the choice of coarse spaces requires additional care.

By the divergence theorem, a divergence-free extension of boundary data is only possible if there is a zero net flux across the boundary. If for a Schwarz method for almost incompressible elasticity a coarse component $\mathbf{u}_0$ of a given $\mathbf{u}$ can be chosen with same net fluxes across subdomain boundaries, then the interface values of the remainder, $\mathbf{w} := \mathbf{u} - \mathbf{u}_0$, will allow for a divergence free extension and a successful decomposition of $\mathbf{w}$ into local components. These ideas have been explored repeatedly for balancing Neumann-Neumann, FETI-DP, and BDDC algorithms; see e.g., [22] and more recently for overlapping Schwarz methods, see [10, 11], which use coarse spaces borrowed from primal substructuring methods. Taking account of the net flux across the subdomain boundaries is a necessity, for almost incompressible elasticity, since we have to make sure that a divergence free function can be partitioned into components in the same class; otherwise the energy of these local components would greatly exceed that of the given function. For Maxwell's equation, curl-free extension are desirable for very similar reasons.

References

[1] Agoshkov, V.I.: Poincaré-Steklov operators and domain decomposition methods in finite dimensional spaces. In R. Glowinski, G.H. Golub, G.A. Meurant, and J. Périaux, eds., *First International Symposium on Domain Decomposition Methods for Partial Differential Equations*, Philadelphia, PA, 1988. SIAM.
[2] Babuška, I.: Über Schwarzsche Algorithmen in partiellen Differentialgleichungen der mathematischen Physik. *ZAMM*, 37(7/8):243–245, 1957.

[3] Bourgat, J.-F., Glowinski, R., Le Tallec, P., Vidrascu, M.: Variational formulation and algorithm for trace operator in domain decomposition calculations. In T. Chan, R. Glowinski, J. Périaux, and O.Widlund, eds., *Domain Decomposition Methods. Second International Symposium on Domain Decomposition Methods*, pages 3–16, Philadelphia, PA, 1989. SIAM. Los Angeles, CA, Jan. 14–16, 1988.
[4] Bramble, J.H., Pasciak, J.E., Schatz, A.H.: The construction of preconditioners for elliptic problems by substructuring, I. *Math. Comp.*, 47(175):103–134, 1986.
[5] Bramble, J.H., Pasciak, J.E., Schatz, A.H.: The construction of preconditioners for elliptic problems by substructuring, IV. *Math. Comp.*, 53(187):1–24, 1989.
[6] Brenner, S.C.: Lower bounds of two-level additive Schwarz preconditioners with small overlap. *SIAM J. Sci. Comput.*, 21(5):1657–1669, 2000.
[7] Brenner, S.C., Sung, L.: Discrete Sobolev and Poincaré inequalities via Fourier series. *East-West J. of Numer. Math*, 8:83–92, 2000.
[8] Cai, X.-C., Sarkis, M.: A restricted additive Schwarz preconditioner for general sparse linear systems. *SIAM J. Sci. Comp.*, 21:239–247, 1999.
[9] Dohrmann, C.R.: A preconditioner for substructuring based on constrained energy minimization. *SIAM J. Sci Comput.*, 25(1):246–258, 2003.
[10] Dohrmann, C.R., Widlund, O.B.: A hybrid domain decomposition method for compressible and almost incompressible elasticity. Technical Report TR2008-919, Courant Institute, New York University, Dec. 2008.
[11] Dohrmann, C.R., Widlund, O.B.: An overlapping Schwarz algorithm for almost incompressible elasticity. Technical Report TR2008-912, Department of Computer Science, Courant Institute, New York University, May 2008.
[12] Dryja, M.: An additive Schwarz algorithm for two- and three-dimensional finite element elliptic problems. In T. Chan, R. Glowinski, J. Périaux, and O. Widlund, eds., *Domain Decomposition Methods. Second International Symposium on Domain Decomposition Methods*, 168–172, Philadelphia, PA, 1989. SIAM. Los Angeles, CA, Jan. 14–16, 1988.
[13] Dryja, M., Smith, B.F., Widlund, O.B.: Schwarz analysis of iterative substructuring algorithms for elliptic problems in three dimensions. *SIAM J. Numer. Anal.*, 31(6):1662–1694, Dec. 1994.
[14] Dryja, M., Widlund, O.B.: Domain decomposition algorithms with small overlap. *SIAM J. Sci.Comput.*, 15(3):604–620, May 1994.
[15] Farhat, C., Lesoinne, M., Le Tallec, P., Pierson, K., Rixen, D.: FETI-DP: A dual-primal unified FETI method- part I: A faster alternative to the two-level FETI method. *Internat. J. Numer. Methods Engrg.*, 50(7):1523–1544, 2001.
[16] Farhat, C., Lesoinne, M., Pierson, K.: A scalable dual-primal domain decomposition method. *Numer. Linear Algebra Appl.*, 7(7–8):687–714, 2000.
[17] Farhat, C., Roux, F.-X.: A Method of Finite Element Tearing and Interconnecting and its Parallel Solution Algorithm. *Internat. J. Numer. Meth. Engrg.*, 32: 1205–1227, 1991.
[18] Glowinski, R., Wheeler, M.F.: Domain decomposition and mixed finite element methods for elliptic problems. In R. Glowinski, G.H. Golub, G.A. Meurant,

and J. Périaux, eds., *First International Symposium on Domain Decomposition Methods for Partial Differential Equations*, 144–172, Philadelphia, PA, 1988. SIAM. Paris, Jan. 7–9, 1987.

[19] Klawonn, A., Rheinbach, O.: A parallel implementation of Dual-Primal FETI methods for three dimensional linear elasticity using a transformation of basis. *SIAM J. Sci. Comput.*, 28(5):1886–1906, 2006.

[20] Klawonn, A., Rheinbach, O.: Inexact FETI-DP methods. *Internat. J. Numer. Methods Engrg.*, 69:284–307, 2007.

[21] Klawonn, A., Widlund, O.B.: Dual-Primal FETI methods for linear elasticity. *Comm. Pure Appl. Math.*, 59(11):1523–1572, Nov. 2006.

[22] Li, J., Widlund, O.B.: BDDC algorithms for incompressible Stokes equations. *SIAM J. Numer. Anal.*, 44(6):2432–2455, 2006.

[23] Lions, P.-L.: On the Schwarz alternating method. I. In R. Glowinski, G.H. Golub, G.A. Meurant, and J. Périaux, eds., *First International Symposium on Domain Decomposition Methods for Partial Differential Equations*, pages 1–42, Philadelphia, PA, 1988. SIAM. Paris, Jan. 7–9, 1987.

[24] Mandel, J.: Iterative solvers by substructuring for the p-version finite element method. *Comp. Methods Appl. Mech. Engrg.*, 80:117–128, 1990.

[25] Mandel, J., Brezina, M.: Balancing domain decomposition for problems with large jumps in coefficients. *Math.Comp.*, 65(216):1387–1401, 1996.

[26] Mandel, J., Dohrmann, C.R., Tezaur, R.: An algebraic theory for primal and dual substructuring methods by constraints. *Appl. Numer. Math.*, 54:167–193, 2005.

[27] Quarteroni, A., Valli, A.: *Domain Decomposition Methods for Partial Differential Equations*. Oxford Science, 1999.

[28] Schwarz, H.A.: *Gesammelte Mathematische Abhandlungen*, vol. 2, 133–143. Springer, Berlin, 1890. First published in Vierteljahrsschrift der Naturforschenden Gesellschaft in Zürich, vol. 15, 1870, 272–286.

[29] Smith, B.F.: A parallel implementation of an iterative substructuring algorithm for problems in three dimensions. *SIAM J. Sci. Comput.*, 14(2):406–423, March 1993.

[30] Smith, B.F.: Bjørstad, P.E., Gropp, W.: *Domain Decomposition: Parallel Multilevel Methods for Elliptic Partial Differential Equations*. Cambridge University Press, 1996.

[31] Sobolev, S.L.: L'Algorithme de Schwarz dans la Théorie de l'Elasticité. *Comptes Rendus (Doklady) de l'Académie des Sciences de l'URSS*, IV((XIII) 6):243–246, 1936.

[32] Toselli, A., Widlund, O.: *Domain Decomposition Methods - Algorithms and Theory*, vol. 34 of *Springer Series in Computational Mathematics*. Springer, Berlin–Heidelberg–New York, 2005.

[33] Tu, X.: Three-level BDDC in two dimensions. *Internat. J. Numer. Methods Engrg.*, 69:33–59, 2007.

Part III

Contributed Presentations

Distributed Decomposition Over Hyperspherical Domains

Aron Ahmadia[1], David Keyes[1], David Melville[2], Alan Rosenbluth[2], and Kehan Tian[2]

[1] Department of Applied Physics and Applied Mathematics, Columbia University, New York, NY 10027, USA
[2] IBM T.J. Watson Research Center, Yorktown Heights, New York 10598, USA

Summary. We are motivated by an optimization problem arising in computational scaling for optical lithography that reduces to finding the point of minimum radius that lies outside of the union of a set of diamonds centered at the origin of Euclidean space of arbitrary dimension. A decomposition of the feasible region into convex regions suggests a heuristic sampling approach to finding the global minimum. We describe a technique for decomposing the surface of a hypersphere of arbitrary dimension, both exactly and approximately, into a specific number of regions of equal area and small diameter. The decomposition generalizes to any problem posed on a spherical domain where regularity of the decomposition is an important concern. We specifically consider a storage-optimized decomposition and analyze its performance. We also show how the decomposition can parallelize the sampling process by assigning each processor a subset of points on the hypersphere to sample. Finally, we describe a freely available C++ software package that implements the storage-optimized decomposition.

1 Global Optimization for Semiconductor Lithography Mask Design

In the newly heralded field of computational scaling [7], industrial scientists are now investigating a global minimization formulation of the problem of mask design for optimizing process windows in semiconductor lithography [5, 6]. We are considering the problem of forming an optimal mask design for the optical printing of a given 2-D target image, which is considered as a set of sampled target points that must be sufficiently illuminated for the image to correctly print. We attempt to minimize the total intensity of a set of d exposure modes, with each axis x_i corresponding to an intensity for mode i. In the space of the d exposure modes, each of the n samples of image features are represented as a d-dimensional diamond of infeasible space. Each sampled image feature is sufficiently illuminated by the set of exposure modes if their representative coordinate $\mathbf{x} \in \mathbf{R}^d$ lies outside the diamond. Sufficient illumination of all sampled target features in an exposure is achieved in all points outside of the union of the set of diamonds. Although a global minimum is ideal, the goal of the problem is to find good solutions in a reasonable amount of time. Additionally, global

M. Bercovier et al. (eds.), *Domain Decomposition Methods in Science and Engineering XVIII*, Lecture Notes in Computational Science and Engineering,
DOI: 10.1007/978-3-642-02677-5,

minima that satisfy all the constraints may still be rejected due to manufacturability considerations, so a good solution method will provide the global minima and may provide a set of the best available local minima within each orthant of the search space.

1.1 Convex Partitions of the Feasible Domain

The semiconductor lithography problem can be considered as the search for the global minima of a linear optimization problem subject to nonconvex constraints:

$$\text{minimize } \|\mathbf{x}\|_1 \text{ subject to } A_i(\mathbf{x}) \cdot \mathbf{x} \geq \mathbf{b}_i \; i = 1 \ldots n \tag{1}$$

We define the jth normalized principal axis of diamond A_i as $l_{i,j}\mathbf{c}_{i,j}$, with $\mathbf{c}_{i,j}$ representing the direction and $l_{i,j}$ the magnitude of the principal axis. We then note that each of the 2^d planes defining a half-space exterior of a diamond connects the d principal axes of the diamond. $A_i(\mathbf{x})$ can be considered as a function of the choices of signs for the vectors representing the principal axes of the diamond, the choice of connection to the 'positive' or 'negative' end of each principal axis uniquely determines one of the planes.

Since the diamond is an intersection of half-spaces, a point $\mathbf{x}$ is considered feasible with respect to the diamond if it lies inside any of the reflected (pointing outwards from the origin) half-spaces $A^+_{i,k}$ arising from the hyperplanes $A_{i,k}$. Each combination of the positive and negative principal $\mathbf{c}_j$ axes in a given diamond A_i is a set of d points that uniquely determine a constraint half-space $A^+_{i,k}$. For any given k and diamond A_i:

$$s_{i,k}(j) * \mathbf{c}_{i,j} \in A_{i,k} \;\; j = \{1 \ldots d\} \tag{2}$$

Since all of the constraint planes are linear and do not contain the origin:

$$\text{if } \mathbf{x}_1, \mathbf{x}_2, \ldots, \mathbf{x}_d \in A_k, \text{ and } \theta \in \mathbf{R}^d, \; \theta_j \geq 0, \; \sum_j^d \theta_j \geq 1$$

$$(\theta_1 \mathbf{x}_1 + \theta_2 \mathbf{x}_2 + \cdots + \theta_d \mathbf{x_d}) \in A_k^+$$

In particular:

$$\theta \in \mathbf{R}^d, \theta_j \geq 0, \; \sum_j^d \theta_j \geq 1 \tag{3}$$

$$(\theta_1 s_k(1)\mathbf{c}_1 + \theta_2 s_k(2)\mathbf{c}_2 + \cdots + \theta_d s_k(d)\mathbf{c}_d) \in A_k^+ \tag{4}$$

Because the set of principal axes $l_j\mathbf{c}_j$ for each diamond forms an orthogonal basis on $\mathbf{R}^d$, we can express:

$$\mathbf{x} = \sum_j^d l_j \theta_j \mathbf{c_j}, \;\; \theta_j = \mathbf{c}_j \cdot \mathbf{x} \tag{5}$$

Any combination of positive and negative $\theta_{i,j}$ can be converted to their absolute values with an appropriate choice of k, and we can say that $\mathbf{x}$ lies outside diamond i (by satisfying constraint half-space i,k) if:

$$\sum_{j=1}^{d} \frac{|\theta_{i,j}|}{l_{i,j}} \geq 1 \tag{6}$$

We may enumerate a single plane $A_{i,k}$ with a tuple, or ordered list of d signs $\mathbf{s_{i,k}}$, with the sign of the jth element $s_{i,k}(j)$ corresponding to the ends of the principal axes $\mathbf{c_{i,j}}$ it connects. We represent the concatenation of all d-tuples $s_{i,k}$ corresponding to a choice of plane for each of the n diamonds into a single n_l-tuple: $\mathbf{s}$, with $n_l = n * 2^d$. Similarly, we concatenate the list of all principal axes into a list $\{\mathbf{c_i}\}$, with $i = 1 \cdots n_l$.

We define a set of nonoverlapping regions $R_\mathbf{s}$ that collectively exhaust R^d, with a total of 2^{n_l} potential tuples $\mathbf{s}$ and corresponding regions $R_\mathbf{s}$.

$$R_\mathbf{s} : \{\mathbf{x} \in \mathbf{R}^d \mid \mathbf{sign}(\mathbf{c_i} \cdot \mathbf{x}) = s_i,\ i = 1, \ldots, n_l\} \tag{7}$$

Each $\mathbf{s}$ defines a convex region containing $R_\mathbf{s}$:

$$\sum_{j=1}^{d} \frac{s_{i,j}(\mathbf{c_{i,j}} \cdot \mathbf{x})}{l_{i,j}} \geq 1\ \forall i \tag{8}$$

If we enumerate the set of convex regions $R_\mathbf{s}$, the global minima is the best local minima from each of the regions.

2 Partitions of n-Space

We expect the true number of convex regions p in the space to be less than 2^{n_l}, as some intersections of diamond half-spaces will be empty. We seek an upper bound on p as a function of d, the dimensionality of the problem, and n, the number of diamonds. We consider each of the diamonds as a set of d cutting planes, with $\mathbf{c_{i,j}}$ represented as a cutting plane that intersects the origin. Upper bounds for the number of regions generated by origin-centered cutting planes in general position were established independently by [1], Perkins, Willis, and Whitmore (unpublished), and [8]. "Partitions of N-Space by Hyperplanes", [9] allows tighter bounds based on the degeneracy of the planes. We consider the diamonds to be non-degenerate, so the upper bounds for cutting planes in general position are sufficient:

$$p \leq B_d^n = 2 \sum_{i=0}^{d-1} \binom{n-1}{i} \tag{9}$$

3 Incomplete Search Heuristics

When applied to the semiconductor lithography problem, the coordinate axes represent linear combinations of the underlying physical variables that are chosen for their efficient average coupling to the features represented by the constraint diamonds. This selection process causes the modes to be strongly coupled to individual constraint features, which in turn causes the axes of the diamonds to be preferentially aligned with the coordinate axes. As a result the principal axes of the diamonds have a tendency to cluster together directionally, making the size distribution of the regions between intersecting ellipsoids very non-uniform, with the largest and deepest regions comprising a small fraction of the total. Empirically it is found that solutions which are adequately close to the global optimum can be found by searching only the largest regions. In [5], the authors suggest that directly sampling points on a hypersphere is a useful heuristic; see Fig. 1.

Fig. 1. Sampling reduces the number of regions to search by skipping small-angle regions, unsearched regions are darkly shaded.

The number of sampled regions n_s using a decomposition of the $(d-1)$-sphere embedded in R^d can be considered as a function of the search density per dimension ρ:

$$n_s(\rho) = \rho^d. \tag{10}$$

The number of sampled points is much more managable than the actual number of potential convex regions. We accept the currently unquantified risk of missing the global minimum in exchange for a more computationally tractable approximation to the problem.

4 Hypersphere Decomposition

4.1 Previous Work

We are now faced with the task of partitioning a $(d-1)$-sphere into regions of approximately equal area and small diameter. Fortunately, this topic has been well-studied in [3], based on a construction for 2-spheres introduced earlier in Zhou's

1995 PhD Thesis as well as unpublished work by E.B. Saff and I.H. Sloan. A domain decomposition method for particle methods on the 2-sphere was also introduced in [2]. A full introduction to the general algorithm for d-spheres into n_s partitions is impossible here, but we generalize the algorithm as a partitioning of m-spheres into regions which are then recursively partitioned as $(m-1)$-spheres.

4.2 A Memory-Efficient Tree Storage Scheme for Equal Area Hypersphere Regions

We are motivated by the large value of n_s to seek a compressed storage partition of n_s points. At each level j of the decomposition, starting at $j = d$, and ending at $j = 1$, we have some number of j-spheres embedded into $(j+1)$-space. We denote the number of j-spheres in the decomposition at level j by k_j. We propose using the procedure's recursive tree as a storage scheme for the points, avoiding the costs of storing full coordinate information for each point. This idea was originally proposed in [4] in section in 2.5 for spherical coding, though its potential for compressed storage was not fully explored. We are interested in forming a loose bound on the total number of nodes required for the storage of the tree. Since the number of nodes on the tree at level j is based on the number of spheres at level k_j, we seek an upper bound on k_j.

At each level j, we decompose each j-sphere into some finite number of $(j-1)$-spheres and 2 polar caps. The $(j-1)$-spheres correspond to "collars" of the j-spheres, and are assigned a number of regions proportional to their fractional area of the sphere. We impose an indexing on all the spheres for a given level j, such that if there are k_j spheres at level j, then the spheres are indexed from $i = 1,\ldots,k_j$, and let $k_{i,j}$ equal the number of $(j-1)$-spheres to decompose the i,j-th sphere into. Finally, we also affix $z_{i,j}$ to every sphere in the system, denoting the number of regions contained by sphere i,j.

If we fix $d = d-1$ and decompose the d-sphere, we have $k_d = 1$. We can now recursively build an estimate for k_{d-l-1} from k_{d-l} for $l = 0,\ldots,d-1$. We claim that $k_{d-l} = O(n^{l/d})$. This is true for the base case $l = 0$, and is true for all l if we can show that $k_{d-l-1} = O(n^{\frac{l+1}{d}})$. We use the fact that for a d-sphere being decomposed into s regions, the number k of $(d-1)$-spheres/collars it contains is:

$$k = \frac{\pi - 2\theta_{d,n}}{\frac{\sigma(S^d)}{s}^{\frac{1}{d}}} \tag{11}$$

Where $\theta_{d,n}$ is the polar cap angle for a d-sphere decomposed into n_s regions and $\sigma(S^d)$ is the measure of surface area of a d-sphere. At each level j, we can account for all regions contained by summing over the assigned regions for each j-sphere and the polar caps for all m-spheres, where $m > j$:

$$\sum_{i=1}^{k_j} z_{i,j} + 2\sum_{m=j+1}^{d+1} k_m = n \tag{12}$$

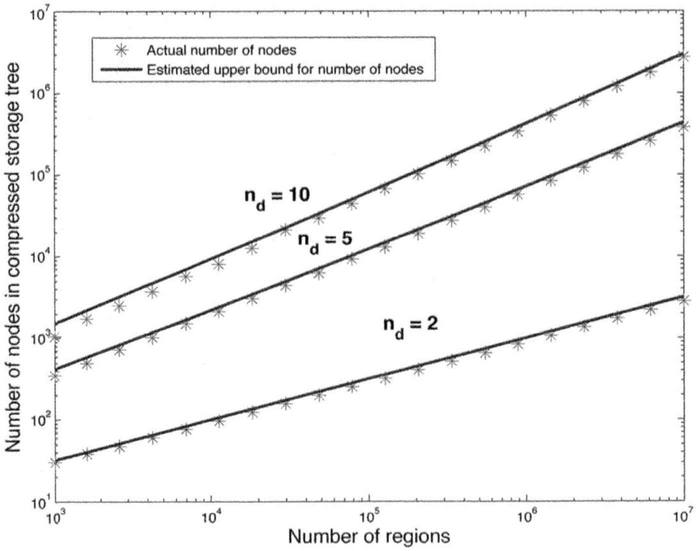

Fig. 2. Actual vs. Estimated Number of Nodes for $d = 2,5,10$ and between 1000 and 10,000,000 sample points

Trivially: $\sum_{i=1}^{k_j} z_{i,j} \leq n$. We now make the observation that the solution to max $\sum_{i=1}^{k}(p_i)^{1/j}$, subject to $\sum(p_i) \leq n$, is equal to $k(\frac{n}{k})^{1/j}$, and substitute:

$$k_{d-l-1} = O(n^{j/d})^{\frac{d-l-1}{d-l}} n^{\frac{1}{d-l}} \tag{13}$$

$$= O(n^{\frac{ld-l^2-l+d}{d(d-l)}}) \tag{14}$$

$$= O(n^{\frac{l+1}{d}}) \tag{15}$$

We substitute in equation (12) to obtain an estimate for the upper bound of the total nodes of the storage tree. We assume a constant $C = 1$, and:

$$K = \sum_{j=2}^{j=d} k_j = 2\sum_{j=2}^{j=d-1} s^{j/d} + s^{\frac{d-1}{d}} \tag{16}$$

The estimates were computed for $d = 2,5,10$ and $n_s = 10^3$ to $n_s = 10^7$, then compared against actual tree structures in Fig. 2.

4.3 Parallel Decomposition

Finally, we introduce an algorithm for two-level decomposition suitable for massively parallel distributed sampling of the n-sphere; see Fig. 3. This algorithm successfully distributed a parallel search over 2,048 Blue Gene nodes.

1. Apply Leopardi's algorithm to generate a decomposition along some subspace of the original space.

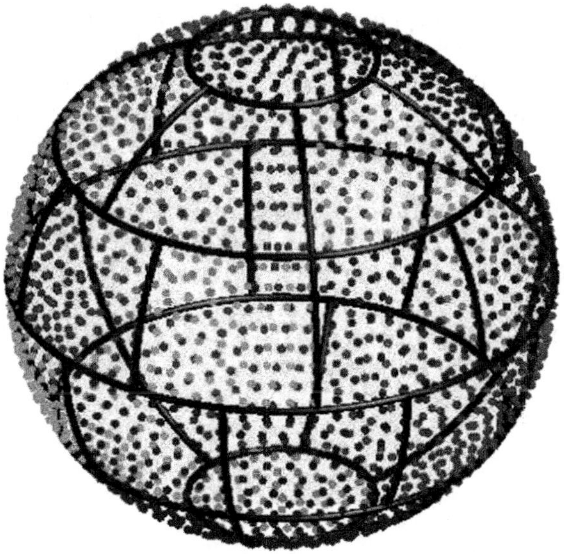

Fig. 3. Two-level distributed decomposition of a 2-sphere

2. For each decomposed region, apply Leopardi's algorithm again to sample points, along decomposed dimensions, the algorithm operates on region boundaries established in the first decomposition.

5 Ongoing Work

We wish to consider compressions of the tree structure by enforcing symmetry in the hypersphere decomposition. We are also interested in improving the performance of the tree structure code. A C++ implementation of the tree structure sampling code is available from http://aron.ahmadia.net/code.

Acknowledgement. The authors gratefully acknowledge Braxton Osting and Matias Courdurier for their insightful contributions to several sections of this paper.

References

[1] Cameron, S.H.: An estimate of the complexity requisite in a universal decision network. *Bionics Symposium, WADD Report*, pages 60–600, 1960.
[2] Egecioglu, Ö., Srinivasan, A.: Domain Decomposition for Particle Methods on the Sphere. In *Proceedings of the Third International Workshop on Parallel Algorithms for Irregularly Structured Problems*, pages 119–130. Lecture Notes in Computer Science, 117. Springer, London, 1996.
[3] Leopardi, P.: A partition of the unit sphere into regions of equal area and small diameter. *Electron. Trans. Numer. Anal.*, 25:309–327, 2006.

[4] Leopardi, P.: *Distributing points on the sphere: Partitions, separation, quadrature and energy*. PhD thesis, University of New South Wales, 2007.
[5] Rosenbluth, A.E., Bukofsky, S., Fonseca, C., Hibbs, M., Lai, K., Molless, A.F., Singh, R.N., Wong, A.K.K.: Optimum mask and source patterns to print a given shape. *J. Microlithogr. Microfabrication, Microsyst.*, 1:13-30, 2002.
[6] Rosenbluth, A.E., Melville, D., Tian, K., Lai, K., Seong, N., Pfeiffer, D., Colburn, M.: Global optimization of masks, including film stack design to restore TM contrast in high NA TCC's. *Proc. SPIE*, 6520:65200P, 2007.
[7] Singh, V.: Computational lithography: the new enabler of Moore's Law. *Proc. SPIE*, 682768271Q, 2007.
[8] Winder, R.O.:. Single stage threshold logic. *Switching Circuit Theory and Logical Design*, AIEE Special Publ. S-134, pages 321–322, 1960.
[9] Winder, R.O.: Partitions of N-Space by Hyperplanes. *SIAM J. Appl. Math.*, 811–818, 1966.

Domain Decomposition Preconditioning for Discontinuous Galerkin Approximations of Convection-Diffusion Problems

Paola F. Antonietti[1] and Endre Süli[2]

[1] MOX - Laboratory for Modeling and Scientific Computing, Dipartimento di Matematica, Politecnico di Milano, via Bonardi 9, 20133 Milano, ITALY. paola.antonietti@polimi.it

[2] Computing Laboratory, University of Oxford, Wolfson Building, Parks Road, Oxford OX1 3QD, UK. Endre.Suli@comlab.ox.ac.uk

1 Introduction

In the classical Schwarz framework for conforming approximations of nonsymmetric and indefinite problems [5, 6] the finite element space is optimally decomposed into the sum of a finite number of uniformly overlapped, two-level subspaces. In each iteration step, a coarse mesh problem and a number of smaller linear systems, which correspond to the restriction of the original problem to subregions, are solved instead of the large original system of equations. Based on this decomposition, domain decomposition methods of three basic type—additive, multiplicative and hybrid Schwarz methods—have been studied in the literature (cf. [4, 5, 6]). In [1, 2] it was shown that for discontinuous Galerkin (DG) approximations of purely elliptic problems optimal nonoverlapping Schwarz methods (which have no analogue in the conforming case) can be constructed. Moreover, it was proved that they exhibit spectral bounds analogous to the one obtained with conforming finite element approximations in the case of "small" overlap, making Schwarz methods particularly well-suited for DG preconditioning. Motivated by the above considerations, we study a class of nonoverlapping Schwarz preconditioners for DG approximations of convection-diffusion equations. The *generalized minimal residual* (GMRES) Krylov space-based iterative solver is accelerated with the proposed preconditioners. We discuss the issue of convergence of the resulting preconditioned iterative method, and demonstrate through numerical computations that the classical Schwarz convergence theory cannot be applied to explain theoretically the converge observed numerically.

2 Statement of the Problem and its DG Approximation

Given a bounded polyhedral domain $\Omega \subseteq \mathbb{R}^d$, $d = 2,3$, $f \in L^2(\Omega)$, and $g \in H^{1/2}(\partial\Omega)$, we consider the following elliptic convection-diffusion problem

M. Bercovier et al. (eds.), *Domain Decomposition Methods in Science and Engineering XVIII*, Lecture Notes in Computational Science and Engineering,
DOI: 10.1007/978-3-642-02677-5, © Springer-Verlag Berlin Heidelberg 2009

with constant coefficients:

$$-\varepsilon\Delta u + \beta\cdot\nabla u = f \text{ in } \Omega, \quad u = g \text{ on } \Gamma \equiv \partial\Omega, \tag{1}$$

where $\varepsilon > 0$ is the diffusion coefficient and $\beta \in \mathbb{R}^d$ is the velocity field.

We consider, for simplicity, shape-regular *quasi-uniform* partitions $\mathcal{T}_h$ of Ω with granularity $h > 0$, where each $K \in \mathcal{T}_h$ is the affine image of a fixed master element $\mathcal{K}$, *i.e.*, $K = F_K(\mathcal{K})$, where $\mathcal{K}$ is either the open unit d-simplex or the open unit d-hypercube in $\mathbb{R}^d$, $d = 2,3$. We denote by $\mathcal{F}_h$ the set of all faces of $\mathcal{T}_h$, and for $F \in \mathcal{F}_h$ we set $h_F = \mathrm{diam}(F)$. The symbol $\mathcal{F}_h^B$ will denote the set of all faces that lie on the boundary, Γ. For a given approximation order $\ell \geq 1$, we define the discontinuous Galerkin finite element space $V_h = \{v \in L^2(\Omega) : v|_K \circ F_K \in \mathcal{M}^\ell(\mathcal{K}) \quad \forall K \in \mathcal{T}_h\}$, where $\mathcal{M}^\ell(\mathcal{K})$ is either the space of polynomials of degree at most ℓ on $\mathcal{K}$, if $\mathcal{K}$ is the reference d-simplex, or the space of polynomials of degree at most ℓ in each variable on $\mathcal{K}$, if $\mathcal{K}$ is the reference d-hypercube.

We denote by ∇_h the elementwise application of the operator ∇, and, for $v \in V_h$ and $K \in \mathcal{T}_h$, v^+ (respectively, v^-) denotes the interior (respectively, exterior) trace of v defined on ∂K (respectively, $\partial K \setminus \Gamma$). Given $K \in \mathcal{T}_h$, the inflow and outflow parts of ∂K are defined

$$\partial_- K := \{x \in \partial K : \beta(x)\cdot\mathbf{n}_K(x) < 0\}, \quad \partial_+ K := \{x \in \partial K : \beta(x)\cdot\mathbf{n}_K(x) \geq 0\},$$

respectively, where $\mathbf{n}_K$ denotes the unit outward normal vector to ∂K.

For a parameter $\alpha \geq \alpha_{\min} > 0$ (at our disposal), and adopting the standard notation $\{\{\cdot\}\}$ for the face-average and $[\![\cdot]\!]$ for the jump operator [3], we define the bilinear form $B_h(\cdot,\cdot) : V_h \times V_h \to \mathbb{R}$ as

$$\begin{aligned} B_h(u,v) = {} & \int_\Omega \varepsilon\nabla_h u\cdot\nabla_h v\,\mathrm{d}x - \sum_{F\in\mathcal{F}_h}\int_F \{\{\varepsilon\nabla_h u\}\}\cdot[\![v]\!]\,\mathrm{d}s \\ & - \sum_{F\in\mathcal{F}_h}\int_F [\![u]\!]\cdot\{\{\varepsilon\nabla_h v\}\}\,\mathrm{d}s + \sum_{F\in\mathcal{F}_h}\int_F \alpha\varepsilon h_F^{-1}[\![u]\!]\cdot[\![v]\!] - \int_\Omega u\beta\cdot\nabla_h v\,\mathrm{d}x \\ & + \sum_{K\in\mathcal{T}_h}\int_{\partial_+ K}(\beta\cdot\mathbf{n}_K)\,u^+ v^+\,\mathrm{d}s + \sum_{K\in\mathcal{T}_h}\int_{\partial_- K\setminus\Gamma}(\beta\cdot\mathbf{n}_K)\,u^- v^+\,\mathrm{d}s. \end{aligned}$$

Then, the DG approximation of problem (1) reads as follows:

$$\text{Find } u_h \in V_h \text{ such that } B_h(u_h,v) = F_h(v) \ \forall v \in V_h, \tag{2}$$

where the functional $F_h(\cdot) : V_h \to \mathbb{R}$ is given by

$$\begin{aligned} F_h(v) := {} & \int_\Omega f v\,\mathrm{d}x + \sum_{F\in\mathcal{F}_h^B}\int_F \varepsilon g\nabla v^+\cdot\mathbf{n}_K\,\mathrm{d}s \\ & + \sum_{F\in\mathcal{F}_h^B}\int_F \varepsilon\alpha h_F^{-1} g v^+\,\mathrm{d}s + \sum_{K\in\mathcal{T}_h}\int_{\partial_- K\cap\Gamma}(\beta\cdot\mathbf{n}_K)\,g v^+\,\mathrm{d}s. \end{aligned}$$

Given a basis of V_h, any function $v \in V_h$ is uniquely determined by a set of degrees of freedom. Here and in the following, we use boldface notation to denote elements of the spaces of degrees of freedom (vectors in $\mathbb{R}^n$, and matrices in $\mathbb{R}^n \times \mathbb{R}^n$). If **B** is the *stiffness matrix* associated with the bilinear form $B_h(\cdot,\cdot)$ and the given basis, problem (2) can be rewritten as the system of linear equations $\mathbf{Bu} = \mathbf{F}$. In order to solve this system of linear equations efficiently by a Krylov space-based iterative solver (such as, for example, the GMRES method), suitable preconditioners have to be employed to accelerate the iterative scheme.

3 Nonoverlapping Schwarz Methods

We consider three levels of *nested* partitions of the domain Ω satisfying the previous assumptions: a subdomain partition $\mathcal{T}_N$ consisting of N nonoverlapping subdomains Ω_i, a coarse partition $\mathcal{T}_H$ (with mesh size H) and a fine partition $\mathcal{T}_h$ (with mesh size h). Next we introduce the key ingredients of the definition of the Schwarz preconditioners.

Local Solvers. For $i = 1,\dots,N$, we define the local DG spaces by

$$V_h^i := \{v \in V_h : v|_K = 0 \ \forall K \in \mathcal{T}_h,\ K \subset \Omega \setminus \Omega_i\}.$$

We note that a function in V_h^i is discontinuous and, as opposed to the case of conforming approximations, does not in general vanish on $\partial\Omega_i$. The classical extension (injection) operator from V_h^i to V_h is denoted by $R_i^{\mathrm{T}} : V_h^i \longrightarrow V_h$, $i = 1,\dots,N$. We define the local solvers $\mathcal{B}_i : V_h^i \times V_h^i \longrightarrow \mathbb{R}$ as

$$\mathcal{B}_i(u_i,v_i) := B_h(R_i^{\mathrm{T}} u_i, R_i^{\mathrm{T}} v_i) \ \ \forall u_i, v_i \in V_h^i, \quad i = 1,\dots,N.$$

Remark 1. Approximate local solvers, such as the ones proposed in [1, 2], could also be considered for the definition of the local components of the preconditioner.

Coarse Solver. For a given approximation order $0 \le p \le \ell$ we introduce the coarse space $V_H \equiv V_h^0 := \{v_0 \in L^2(\Omega) : v_0|_K \circ F_K \in \mathcal{M}^\ell(\mathcal{K}) \ \forall K \in \mathcal{T}_H\}$, and we define the *coarse solver* $\mathcal{B}_0 : V_h^0 \times V_h^0 \longrightarrow \mathbb{R}$ by

$$\mathcal{B}_0(u_0,v_0) := B_h(R_0^{\mathrm{T}} u_0, R_0^{\mathrm{T}} v_0) \quad \forall u_0, v_0 \in V_h^0,$$

where $R_0^{\mathrm{T}} : V_h^0 \longrightarrow V_h$ is the classical injection operator from V_h^0 to V_h.

For $0 \le i \le N$, let the projection operators $T_i : V_h \longrightarrow V_h^i \subset V_h$ be given by

$$\mathcal{B}_i(T_i u, v_i) := B_h(u, v_i) \quad \forall v_i \in V_h^i.$$

The additive and multiplicative Schwarz operator are defined by

$$T_{\mathrm{ad}} := \sum_{i=0}^{N} T_i, \quad T_{\mathrm{mu}} := I - (I - T_N)(I - T_{N-1}) \cdots (I - T_0),$$

respectively (cf. [5, 6]). The multiplicative Schwarz method is less amenable to parallelization than the additive method because the presence of the coarse solver T_0, which cannot be handled in parallel with the other local subproblem solvers, leads to a bottleneck for the whole algorithm. Motivated by the above observations, we also consider a *hybrid* operator in which the global operator T_0 is incorporated additively relative to the rest of the local solvers (see [4]):

$$T_{\text{hy}} := T_0 + I - (I - T_N)(I - T_{N-1}) \cdots (I - T_1).$$

The Schwarz operators can be written as products of suitable preconditioners, namely $\mathbf{M}_{\text{ad}}$, $\mathbf{M}_{\text{mu}}$ or $\mathbf{M}_{\text{hy}}$, and $\mathbf{B}$. Then, the Schwarz method consists of solving, by a suitable Krylov space-based iterative solver, the preconditioned system of equations $\mathbf{MBu} = \mathbf{MF}$, where $\mathbf{M}$ is either $\mathbf{M}_{\text{ad}}$, $\mathbf{M}_{\text{mu}}$ or $\mathbf{M}_{\text{hy}}$.

4 The Issue of Convergence

The abstract analysis of Schwarz methods for conforming approximations to non-symmetric elliptic problems, originally carried out by Cai and Widlund in [6], relies upon the GMRES convergence bounds of Eisenstat *et al.* [7]. According to [7], the GMRES method applied to the preconditioned system of equations does not stagnate (*i.e.*, the iterative method makes some progress in reducing the residual at each iteration step) provided that the symmetric part of T (where T is one of the Schwarz operators introduced in Sec. 3) is positive definite, and T is uniformly bounded. That is,

$$c_p(T) := \inf_{\substack{v \in V_h \\ v \neq 0}} \frac{\mathcal{S}_h(v, Tv)}{\mathcal{S}_h(v, v)} > 0, \quad C_p(T) := \sup_{\substack{v \in V_h \\ v \neq 0}} \frac{\|Tv\|_h}{\|v\|_h} \leq C, \tag{3}$$

where $\|\cdot\|_h$ is a suitable (mesh-dependent) norm on V_h in which the bilinear form $\mathcal{B}_h(\cdot,\cdot)$ is continuous and coercive, and where $\mathcal{S}_h(\cdot,\cdot)$ denotes the symmetric part of $\mathcal{B}_h(\cdot,\cdot)$. While the second condition can usually be shown to hold without difficulties, the first condition cannot, in general, be guaranteed. Indeed, as we demonstrate by numerical computations, $c_p(T)$ may be negative even in generic, non-pathological, cases. In Table 1 we show the computed values of $c_p(T_{\text{ad}})$ and $c_p(T_{\text{mu}})$ obtained with two choices of the global Péclet number $\text{Pe} := \|\beta\|_\infty |\Omega| / \varepsilon$ (that relates the rate of convection of a flow to its rate of diffusion) for the first test case considered in Section 5. Even though GMRES applied to the preconditioned systems does not stagnate and, in fact, converges in only a few iterations (cf. Section 5), $c_p(T) < 0$ once the spacing of the fine grid is sufficiently small.

Remark 2. Closer inspection reveals that, in the case of elliptic convection-dominated diffusion equations, the theory in [6] is far from satisfactory since, on the one hand, it relies upon the GMRES bounds from [7] that only provide *sufficient* conditions for non-stagnation of GMRES and, on the other hand, it requires the skew-symmetric part of the operator to be "small" relative to the symmetric part (typically a low-order compact perturbation). Clearly, such a requirement cannot be satisfied in the

Table 1. Estimate of $c_p(T)$: $\ell = p = 1$, $N = 16$, Cartesian grids.

(a) $c_p(T_{\mathrm{ad}})$: $\varepsilon = 10^{-1}$, $\beta = (1,1)^{\mathrm{T}}$

$H \downarrow h \rightarrow$	h_0	$h_0/2$	$h_0/4$	$h_0/8$
H_0	0.077	-0.008	-0.047	-0.067
$H_0/2$	-	0.101	0.037	0.005
$H_0/4$	-	-	0.117	0.050
$H_0/8$	-	-	-	0.119

(b) $c_p(T_{\mathrm{mu}})$: $\varepsilon = 10^{-3}$, $\beta = (1,1)^{\mathrm{T}}$

$H \downarrow h \rightarrow$	h_0	$h_0/2$	$h_0/4$	$h_0/8$
H_0	0.225	-0.553	-1.484	-2.795
$H_0/2$	-	0.114	-0.628	-1.554
$H_0/4$	-	-	0.114	-0.570
$H_0/8$	-	-	-	0.077

convection-dominated case. Similar conclusions have been drawn in [1, 2] in the case of nonoverlapping preconditioners for nonsymmetric DG approximations of the Laplace operator (where the skew-symmetric part of the operator happens to be of the same order as the symmetric part).

Remark 3. The comments above also apply in to the case of *generous overlapping* partitions (cf. [8]) under suitable additional assumptions on the size of the coarse mesh, *i.e.*, $H < H_0$. Closer inspection reveals that H_0 strongly depends on the size of the global Péclet number, making the analysis inapplicable in the convection-dominated case.

5 Numerical Experiments

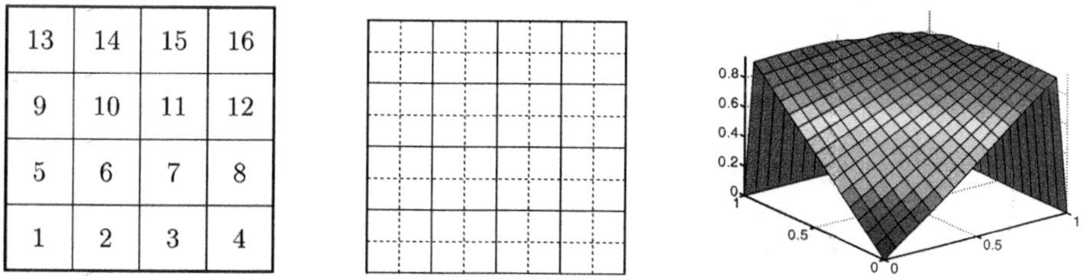

(a) (b) (c)

Fig. 1. (a) Subdomain ordering for $N = 16$; (b) initial coarse (solid line) and fine (dashed line) meshes; (c) the exact solution (4) for $\varepsilon = 10^{-2}$ (right).

We investigate the performance of our preconditioners while varying h, H and the Péclet number. We use a uniform subdomain partition of $\Omega = (0,1)^2$ consisting of 16 squares ordered as in Fig. 1(a). The initial coarse and fine refinements are depicted in Fig. 1(b). We denote by H_0 and h_0 the corresponding initial coarse and fine mesh sizes, respectively, and we consider $n = 1,2,3$ successive uniform refinements of the initial grids. The linear systemsof equations have been solved by GMRES with a (relative) tolerance set equal to 10^{-6} allowing a maximum of 100 (respectively, 600) iterations for the preconditioned (respectively, unpreconditioned) systems.

We set $\beta = (1,1)^{\mathrm{T}}$ and adjust the source term f and the boundary condition so that the exact solution is given by

$$u(x,y) = x + y - xy + \frac{1}{1 - \mathrm{e}^{-1/\varepsilon}} \left[\mathrm{e}^{-1/\varepsilon} - \mathrm{e}^{-(1-x)(1-y)/\varepsilon} \right]. \tag{4}$$

We note that for $0 < \varepsilon \ll 1$, *i.e.*, for $\mathsf{Pe} \gg 1$, the solution exhibits boundary layers along $x = 1$ and $y = 1$ (cf. Fig. 1(c) for $\varepsilon = 10^{-2}$).

Table 2. GMRES iteration counts: $\varepsilon = 1$.

	Additive				Multiplicative				Hybrid			
$H \downarrow h \rightarrow$	h_0	$h_0/2$	$h_0/4$	$h_0/8$	h_0	$h_0/2$	$h_0/4$	$h_0/8$	h_0	$h_0/2$	$h_0/4$	$h_0/8$
H_0	20	30	40	54	8	13	17	24	11	15	20	27
$H_0/2$	-	19	27	37	-	7	10	13	-	11	15	20
$H_0/4$	-	-	20	28	-	-	6	8	-	-	12	17
$H_0/8$	-	-	-	19	-	-	-	5	-	-	-	12
#iter(**B**)	58	109	204	371	58	109	204	371	58	109	204	371

Table 3. GMRES iteration counts: $\varepsilon = 10^{-1}$.

	Additive				Multiplicative				Hybrid			
$H \downarrow h \rightarrow$	h_0	$h_0/2$	$h_0/4$	$h_0/8$	h_0	$h_0/2$	$h_0/4$	$h_0/8$	h_0	$h_0/2$	$h_0/4$	$h_0/8$
H_0	23	34	48	62	11	15	21	29	12	17	24	30
$H_0/2$	-	20	30	41	-	8	11	16	-	12	16	20
$H_0/4$	-	-	21	29	-	-	7	10	-	-	12	17
$H_0/8$	-	-	-	19	-	-	-	6	-	-	-	11
#iter(**B**)	59	110	209	396	59	110	209	396	59	110	209	396

We compare the GMRES iteration counts for the additive, multiplicative and hybrid Schwarz preconditioners for different values of the Péclet number, working for the sake of simplicity with approximations with $\ell = p = 1$. The computed iteration counts obtained for $\varepsilon = 1, 10^{-1}, 10^{-3}, 10^{-4}$ are shown in Tables 2–5, respectively. Clearly, the multiplicative and the hybrid Schwarz preconditioners perform far better than the additive preconditioner. The results in Tables 2–5 show that for small Péclet numbers the iteration counts seem to increase with the Péclet number; whereas, whenever the problem becomes convection-dominated, *i.e.*, for $\mathsf{Pe} \gg 1$, the iteration counts needed for achieving the fixed tolerance decrease with the increase of the Péclet number. Moreover, in the convection-dominated regime the performance of the additive nonoverlapping preconditioner is comparable with the one in [8] in

Table 4. GMRES iteration counts: $\varepsilon = 10^{-3}$.

	Additive				Multiplicative				Hybrid			
$H \downarrow h \rightarrow$	h_0	$h_0/2$	$h_0/4$	$h_0/8$	h_0	$h_0/2$	$h_0/4$	$h_0/8$	h_0	$h_0/2$	$h_0/4$	$h_0/8$
H_0	15	21	26	33	6	8	10	14	8	10	12	16
$H_0/2$	-	17	24	32	-	5	8	13	-	8	11	15
$H_0/4$	-	-	18	27	-	-	6	10	-	-	9	14
$H_0/8$	-	-	-	20	-	-	-	5	-	-	-	10
#iter(**B**)	41	68	115	213	41	68	115	213	41	68	115	213

Table 5. GMRES iteration counts: $\varepsilon = 10^{-4}$.

	Additive				Multiplicative				Hybrid			
$H \downarrow h \rightarrow$	h_0	$h_0/2$	$h_0/4$	$h_0/8$	h_0	$h_0/2$	$h_0/4$	$h_0/8$	h_0	$h_0/2$	$h_0/4$	$h_0/8$
H_0	14	16	17	18	3	4	4	6	6	6	7	8
$H_0/2$	-	14	16	18	-	3	4	5	-	6	6	7
$H_0/4$	-	-	14	17	-	-	3	4	-	-	6	7
$H_0/8$	-	-	-	14	-	-	-	4	-	-	-	6
#iter(**B**)	40	67	119	215	40	67	119	215	40	67	119	215

the overlapping case, making the nonoverlapping version competitive in practical applications.

Table 6. GMRES iteration counts: multiplicative and hybrid (between parenthesis) Schwarz preconditioners.

	$\varepsilon = 10^{-1}$				$\varepsilon = 10^{-4}$			
$H \downarrow h \rightarrow$	h_0	$h_0/2$	$h_0/4$	$h_0/8$	h_0	$h_0/2$	$h_0/4$	$h_0/8$
H_0	13 (15)	20 (21)	27 (29)	36 (39)	5 (8)	6 (9)	8 (10)	10 (12)
$H_0/2$	-	11 (15)	16 (19)	21 (26)	-	5 (7)	6 (10)	9 (12)
$H_0/4$	-	-	10 (13)	14 (19)	-	-	5 (9)	7 (11)
$H_0/8$	-	-	-	8 (11)	-	-	-	6 (10)
#iter(**B**)	79	152	290	551	39	65	113	203

Finally, we investigate the effect of the subdomain ordering on the performance of the Schwarz preconditioner. We set $\beta = (-1,-1)^{\mathrm{T}}$, and we choose as exact solution one that is analogous to the exact solution considered so far but now such that u exhibits boundary layers along $x = 0$ and $y = 0$ for $0 < \varepsilon \ll 1$, so that the subdomains turn out to be ordered "downwind" (cf. Fig. 1(a)). In Table 6 we report the GMRES iteration counts obtained with the multiplicative and hybrid (in parenthesis) Schwarz method using $\ell = p = 1$.As expected, the subdomain ordering does affect the performance of the preconditioner and "downwind" ordering of subdomains can lead to an increase in the number of GMRES iterations.

Acknowledgement. This work was carried out while the first author was a visiting student at the Oxford University Computing Laboratory. She thanks OUCL for the kind hospitality.

References

[1] Antonietti, P.F.: Ayuso, B.: Schwarz domain decomposition preconditioners for discontinuous Galerkin approximations of elliptic problems: non-overlapping case. *M2AN Math. Model. Numer. Anal.*, 41(1):21–54, 2007.
[2] Antonietti, P.F.: Ayuso, B.: Multiplicative Schwarz methods for discontinuous Galerkin approximations of elliptic problems. *M2AN Math. Model. Numer. Anal.*, 42(3):443–469, 2008.
[3] Arnold, D.N., Brezzi, F., Cockburn, B., Marini, L.D.: Unified analysis of discontinuous Galerkin methods for elliptic problems. *SIAM J. Numer. Anal.*, 39(5):1749–1779 (electronic), 2001/02.
[4] Cai, X.-C.: An optimal two-level overlapping domain decomposition method for elliptic problems in two and three dimensions. *SIAM J. Sci. Comput.*, 14(1):239–247, 1993.
[5] Cai, X.-C., Widlund, O.B.: Domain decomposition algorithms for indefinite elliptic problems. *SIAM J. Sci. Statist. Comput.*, 13(1):243–258, 1992.
[6] Cai, X.-C., Widlund, O.B.: Multiplicative Schwarz algorithms for some nonsymmetric and indefinite problems. *SIAM J. Numer. Anal.*, 30(4):936–952, 1993.
[7] Eisenstat, S.C., Elman, H.C., Schultz, M.H.: Variational iterative methods for nonsymmetric systems of linear equations. *SIAM J. Numer. Anal.*, 20(2):345–357, 1983.
[8] Lasser, C., Toselli, A.: An overlapping domain decomposition preconditioner for a class of discontinuous Galerkin approximations of advection-diffusion problems. *Math. Comp.*, 72(243):1215–1238 (electronic), 2003.

Linearly Implicit Domain Decomposition Methods for Nonlinear Time-Dependent Reaction-Diffusion Problems

A. Arrarás, L. Portero and J.C. Jorge

Dpto. de Ingeniería Matemática e Informática, Universidad Pública de Navarra
Campus de Arrosadía, 31006 – Pamplona, Spain
{andres.arraras, laura.portero, jcjorge}@unavarra.es

Summary. A new family of linearly implicit fractional step methods is proposed for the efficient numerical solution of a class of nonlinear time-dependent reaction-diffusion equations. By using the method of lines, the original problem is first discretized in space via a mimetic finite difference technique. The resulting differential system of stiff nonlinear equations is locally decomposed by suitable Taylor expansions and a domain decomposition splitting for the linear terms. This splitting is then combined with a linearly implicit one-step scheme belonging to the class of so-called fractional step Runge-Kutta methods. In this way, the original problem is reduced to the solution of several linear systems per time step which can be trivially decomposed into a set of uncoupled subsystems. As compared to classical domain decomposition techniques, our proposal does not require any Schwarz iterative procedure. The convergence of the designed method is illustrated by numerical experiments.

1 Introduction

In this paper, we consider nonlinear parabolic initial-boundary value problems of the following form: Find $\psi : \Omega \times [0,T] \to \mathbb{R}$ such that

$$\begin{cases} \frac{\partial \psi(\underline{x},t)}{\partial t} = \mathrm{div}(K(\psi)\,\mathrm{grad}\,\psi) + g(\underline{x},t,\psi) + f(\underline{x},t), & (\underline{x},t) \in \Omega \times (0,T], \\ \psi(\underline{x},0) = \psi_0(\underline{x}), & \underline{x} \in \Omega, \\ \psi(\underline{x},t) = \psi_D(\underline{x},t), & (\underline{x},t) \in \partial\Omega \times [0,T], \end{cases} \tag{1}$$

where $\Omega \subseteq \mathbb{R}^2$, $K(\psi)$ is a 2×2 nonlinear symmetric positive-definite tensor, $g(\underline{x},t,\psi)$ is a nonlinear reaction term and $f(\underline{x},t)$ denotes the source/sink term. Initial and boundary data are given by $\psi_0(\underline{x})$ and $\psi_D(\underline{x},t)$, respectively. For the sake of simplicity, only Dirichlet boundary conditions are considered.

The numerical solution of problem (1) is carried out via the method of lines, thus combining a spatial discretization stage with the subsequent time integration process. For the first stage, we use a mimetic finite difference (MFD) method formulated on

M. Bercovier et al. (eds.), *Domain Decomposition Methods in Science and Engineering XVIII*, Lecture Notes in Computational Science and Engineering,
DOI: 10.1007/978-3-642-02677-5, © Springer-Verlag Berlin Heidelberg 2009

logically rectangular meshes. Our method extends the ideas discussed in [7] for linear parabolic problems to the nonlinear case (1) by introducing a quadratic bivariate interpolation approach in the discretization process. As for the time integration, the resulting system of nonlinear ordinary differential equations is locally decomposed by applying suitable Taylor expansions and a domain decomposition splitting for the linear terms. This kind of splitting was used in [2] for solving linear parabolic problems and has been recently surveyed by [5] in the context of regionally-additive schemes. Here, we combine such a technique with an extension of the class of linearly implicit fractional step methods designed and analyzed in [3, 4]. The totally discrete scheme is shown to be second-order convergent in both space and time under a mild stability restriction.

The remainder of the paper is divided into three sections. The first two briefly describe the spatial discretization and time integration processes. Also in Sec. 3, a linearly implicit splitting scheme due to Hundsdorfer and Verwer (cf. [1]) is introduced for comparison purposes. Finally, in the last section, some experiments illustrate the numerical behaviour of the proposed method.

2 Spatial Discretization

The spatial discretization of problem (1) is based on an MFD scheme derived from the support-operator method. This method, initially proposed in [6] and subsequently discussed in [7], provides a methodology for constructing discrete analogues of the invariant first-order differential operators appearing in the original problem (i.e., divergence and gradient).

Let us consider a discretization of Ω by means of a logically rectangular grid Ω_h, where h denotes the spatial mesh size. The first step in the MFD technique consists of choosing suitable degrees of freedom for semidiscrete scalar and vector functions: in this work, the former are defined at the cell centers of the mesh, while the latter are considered to be at the mesh nodes. We shall denote by V_h and $\tilde{V}_h$ the vector spaces of semidiscrete scalar and vector functions defined on the cell centers and nodes of Ω_h, respectively. As a second step, we equip these spaces with suitable scalar products, namely $[\cdot,\cdot]_{V_h}$ and $[\cdot,\cdot]_{\tilde{V}_h}$ (see [7] for details). The third step is to derive a discrete approximation to the divergence operator, $\mathrm{div}_h : \tilde{V}_h \to V_h$, which we shall refer to as the *prime operator*. Such an approximation is provided by the Gauss divergence theorem. Finally, the fourth step lies in defining a discrete gradient operator, $\widetilde{\mathrm{grad}}_h : V_h \to \tilde{V}_h$, as the adjoint to the discrete divergence div_h with respect to the previous scalar products, i.e.:

$$[\mathrm{div}_h \underline{\tilde{u}}_h, \varphi_h]_{V_h} \equiv [\underline{\tilde{u}}_h, -\widetilde{\mathrm{grad}}_h \varphi_h]_{\tilde{V}_h} \quad \forall \varphi_h \equiv \varphi_h(t) \in V_h,\ \forall \underline{\tilde{u}}_h \equiv \underline{\tilde{u}}_h(t) \in \tilde{V}_h. \tag{2}$$

Since $\widetilde{\mathrm{grad}}_h$ is somehow deduced from the so-called prime operator, we call it the *derived operator*. Within this framework, we shall denote by $\psi_h(t)$ and $g_h(t,\psi_h)$ the semidiscrete approximations to the scalar functions $\psi(\underline{x},t)$ and $g(\underline{x},t,\psi)$ at the cell centers of the mesh. Analogously, $f_h(t) \equiv r_h f(\underline{x},t)$, where r_h denotes the restriction operator to the cell centers of Ω_h.

The standard MFD method formulated in [7] is defined for linear problems in which $K \equiv K(\underline{x})$ does not depend on ψ. However, the more general case considered here requires an extension of this method to deal with the nonlinearity arising from $K(\psi)$. Let us briefly present the main features of such an extension. Recalling problem (1) and once we have defined the discrete operators div_h and $\widetilde{\text{grad}}_h$, we need to approximate the matrix-vector product $K(\psi)\,\text{grad}\,\psi$. For this product to be well-posed, since the components of $\widetilde{\text{grad}}_h\,\psi_h$ are given at the mesh nodes (as derived from (2)), the elements of $K(\psi)$ must also be evaluated at this location. Let us denote by $\tilde{\psi}_h$ the approximations to the unknown ψ at the nodes of Ω_h. Then, the discretization of tensor $K(\psi)$, given by $\tilde{K}_h(\tilde{\psi}_h)$, is obtained by suitably evaluating its elements at $\tilde{\psi}_h$. As a result, the second-order nonlinear term $\text{div}_h(\tilde{K}_h(\tilde{\psi}_h)\,\widetilde{\text{grad}}_h\,\psi_h)$ possess a local stencil involving nine cell-centered values ψ_h (as described in [7]) as well as four nodal values $\tilde{\psi}_h$ (due to the discrete tensor $\tilde{K}_h(\tilde{\psi}_h)$). In order to eliminate these values from the local stencil, we apply a quadratic bivariate interpolation method which permits to obtain $\tilde{\psi}_h$ as a linear combination of the corresponding nine values of ψ_h. Consequently, the discrete diffusion operator will be given by $A_h(\cdot) \equiv \text{div}_h(\tilde{K}_h(\cdot)\,\widetilde{\text{grad}}_h\,\cdot) : V_h \to V_h$ and the local stencil of $A_h(\psi_h)$ will thus have a compact nine-cell structure.

The discretization process described in this section gives rise to a stiff nonlinear differential system of the form:

$$\psi_h'(t) = F_h(t,\psi_h) \equiv A_h(\psi_h) + g_h(t,\psi_h) + f_h(t), \quad t \in (0,T], \tag{3}$$

with initial condition $\psi_h(0) = \psi_{h,0} \equiv r_h\psi_0(\underline{x})$. The MFD method has been theoretically proved to be second-order convergent when applied to linear elliptic problems with either Dirichlet or Neumann conditions discretized on smooth grids. Also if linear parabolic problems are considered, the numerical behaviour of this spatial discretization technique shows convergence of order 2 (cf. [7]).

3 Time Integration

In this section, we introduce a family of linearly implicit time integrators based on a splitting of the semidiscrete problem derived in (3). For that purpose, let us first consider a decomposition of the spatial domain Ω into s overlapping subdomains, i.e. $\Omega = \bigcup_{j=1}^{s}\Omega_j$, where $\Omega_j = \bigcup_{k=1}^{s_j}\Omega_{jk}$ such that $\Omega_{jk}\cap\Omega_{j\ell} = \emptyset$ if $k \neq \ell$. Associated to such a decomposition, we construct a sufficiently smooth partition of unity consisting of s functions $\rho_j : \Omega \to [0,1]$, for $j = 1,2,\ldots,s$, which satisfy the following properties:

$$\rho_j(\underline{x}) = \begin{cases} 0 & \text{if } \underline{x} \in \Omega\setminus\Omega_j, \\ h_j(\underline{x}) & \text{if } \underline{x} \in \bigcup_{\substack{k=1\\k\neq j}}^{s}(\Omega_j\cap\Omega_k), \\ 1 & \text{if } \underline{x} \in \Omega_j\setminus\bigcup_{\substack{k=1\\k\neq j}}^{s}(\Omega_j\cap\Omega_k), \end{cases} \tag{4}$$

where $0 \leq h_j(\underline{x}) \leq 1$ and $\sum_{j=1}^{s} h_j(\underline{x}) = 1$ for any $\underline{x}$ located in the overlapping regions. From these restrictions, it is obvious that $\Omega_j \equiv supp\,(\rho_j(\underline{x}))$, for $j = 1,2,\ldots,s$.

In order to introduce the time integration in a simple way, we divide the time interval $[0,T]$ into subintervals $[t_n, t_{n+1}]$ of the same length, where $t_n = n\tau$, for $n = 0,1,\ldots,N_T \equiv [T/\tau]$, and $\tau > 0$ is the constant time step. In the following, we shall denote by $\psi_{h,n}$ the numerical approximations to the semidiscrete solution values $\psi_h(t_n)$. Now, recalling the differential system (3), we consider the Taylor expansion of $A_h(\psi_h)$ around $\psi_{h,n}$:

$$A_h(\psi_h) = A_h(\psi_{h,n}) + J_h(\psi_{h,n})(\psi_h - \psi_{h,n}) + B_h(\psi_h, \psi_{h,n}), \tag{5}$$

where J_h denotes the Jacobian matrix $dA_h/d\psi_h$. If we consider $\breve{f}_h(\psi_{h,n}) \equiv A_h(\psi_{h,n}) - J_h(\psi_{h,n})\,\psi_{h,n}$ as an additional source/sink term, we can rewrite (5) as $A_h(\psi_h) = \breve{f}_h(\psi_{h,n}) + J_h(\psi_{h,n})\,\psi_h + B_h(\psi_h, \psi_{h,n})$. Note that the last term in this expression is nonlinear. Furthermore, using the partition of unity introduced in (4), we split the linear terms $J_h(\psi_{h,n})$, $\breve{f}_h(\psi_{h,n})$ and $f_h(t)$ as follows:

$$\begin{aligned}
2J_h(\psi_{h,n}) &= \sum_{j=1}^{s} J_h^j(\psi_{h,n}), &\quad& \text{where } J_h^j(\psi_{h,n}) = R_h(\rho_j(\underline{x}))J_h(\psi_{h,n}),\\
\breve{f}_h(\psi_{h,n}) &= \sum_{j=1}^{s} \breve{f}_h^j(\psi_{h,n}), && \text{where } \breve{f}_h^j(\psi_{h,n}) = R_h(\rho_j(\underline{x}))\breve{f}_h(\psi_{h,n}),\\
f_h(t) &= \sum_{j=1}^{s} f_h^j(t), && \text{where } f_h^j(t) = R_h(\rho_j(\underline{x}))f_h(t),
\end{aligned} \tag{6}$$

with $R_h(\rho_j(\underline{x}))$ being a diagonal matrix whose main diagonal is given by $r_h\rho_j(\underline{x})$. Finally, the right-hand side from (3) can be rewritten in the following form:

$$F_h(t, \psi_h) \equiv F_h^0(t, \psi_h) + F_h^1(t, \psi_h) + \cdots + F_h^s(t, \psi_h), \tag{7}$$

where $F_h^0(t, \psi_h) \equiv g_h(t, \psi_h) + B_h(\psi_h, \psi_{h,n})$ comprises the nonlinear part of $F_h(t, \psi_h)$, whereas $F_h^j(t, \psi_h) \equiv J_h^j(\psi_{h,n})\,\psi_h + \breve{f}_h^j(\psi_{h,n}) + f_h^j(t)$, for $j = 1,2,\ldots,s$, are linear non-homogeneous terms.

According to the ideas proposed in [3] for linear parabolic problems and subsequently adapted in [4] to the semilinear case, we can integrate (3) by using the splitting (7), together with the following fractional step method:

$$\begin{cases}
\psi_{h,n}^1 = \psi_{h,n},\\
\psi_{h,n}^2 = \psi_{h,n}^1 + \tau\sum_{k=1}^{2} \alpha_k F_h^{i_k}(t_n^k, \psi_{h,n}^k) + \frac{\tau}{2} F_h^0(t_n^1, \psi_{h,n}^1),\\
\psi_{h,n}^j = \psi_{h,n}^{j-1} + \tau\sum_{k=j-1}^{j} \alpha_k F_h^{i_k}(t_n^k, \psi_{h,n}^k), \quad j = 3,4,\ldots,2s-2,\\
\psi_{h,n}^{2s-1} = \psi_{h,n}^{2s-2} + \tau\sum_{k=2s-2}^{2s-1} \alpha_k F_h^{i_k}(t_n^k, \psi_{h,n}^k)\\
\qquad\qquad - \frac{\tau}{2} F_h^0(t_n^1, \psi_{h,n}^1) + \tau F_h^0(t_n^s, \psi_{h,n}^s),\\
\psi_{h,n+1} = \psi_{h,n}^{2s-1}, \qquad n = 0,1,\ldots,N_T-1,
\end{cases} \tag{8}$$

where $i_k = k$, for $k = 1,2,\ldots,s$, and $i_k = 2s-k$, for $k = s+1, s+2,\ldots,2s-1$. The intermediate times are $t_{n,1} = t_n$, $t_{n,k} = t_n + \tau/2$, for $k = 2,3,\ldots,2s-2$, and $t_{n,2s-1} =$

$t_n + \tau = t_{n+1}$, whereas the method coefficients are given by $\alpha_1 = \alpha_s = \alpha_{2s-1} = 1/2$ and $\alpha_k = 1/4\, \forall k \in \{2,3,\ldots,s-1\} \cup \{s+1,s+2,\ldots,2s-2\}$. Note that (8) is a linearly implicit one-step method with $(2s-1)$ internal stages belonging to the class of so-called fractional step Runge-Kutta (FSRK) methods (cf. [4]). It considers implicit contributions of the linear terms $\{F_h^j\}_{j=1}^s$, while explicitly handling the nonlinear term F_h^0. Recall that this term involves both the non-stiff reaction term $g_h(t,\psi_h)$ and the stiff remainder $B_h(\psi_h,\psi_{h,n})$. The former will not affect the stability of the scheme, provided it satisfies a Lipschitz condition (cf. [4]); by contrast, a mild stability restriction will arise due to the latter. A deeper insight on the stability properties of (8) will be provided in the last section.

Since (8) is an FSRK method, its internal stages consist of linear systems with the coefficient matrices $(I_h - \tau\alpha_j J_h^{i_j}(\psi_{h,n}))$, for $j = 2,3,\ldots,2s-1$. Owing to the domain decomposition splitting (6), each one of these linear systems involves the unknowns lying just in one of the subdomains $\{\Omega_j\}_{j=1}^s$. Moreover, since each subdomain Ω_j comprises s_j disjoint connected components, such a system can be easily decomposed into s_j uncoupled subsystems which allow a straightforward parallelization. As a difference with respect to classical domain decomposition methods, artificial boundary conditions are not required on each subdomain and, hence, no Schwarz iterative procedures are involved in the computations.

Following [4], the previous method can be proved to be of classical order 2. In fact, if we consider the case in which the number of levels $s = 2$ and apply the method to a linear parabolic problem, we recover the time integration process involved in the classical Peaceman-Rachford alternating direction implicit scheme. Therefore, (8) may be considered as a generalization of the Peaceman-Rachford scheme (cf. [3]).

As mentioned above, the conditional stability of (8) involves a mild stability restriction which makes it competitive with other existing linearly implicit splitting methods of order 2. For illustration, we shall compare our proposal with the so-called Hundsdorfer and Verwer scheme analyzed in [1]. This scheme is based on the technique of stabilizing corrections and, when applied to problem (3) with splitting (7), it leads to:

$$\left\{\begin{array}{lll} \psi_{h,n}^0 &= \psi_{h,n} + \tau F_h(t_n,\psi_{h,n}), & \\ \psi_{h,n}^j &= \psi_{h,n}^{j-1} + \theta\tau\big(F_h^j(t_{n+1},\psi_{h,n}^j) - F_h^j(t_n,\psi_{h,n})\big), & j = 1,2,\ldots,s, \\ \widehat{\psi}_{h,n}^0 &= \psi_{h,n}^0 + \sigma\tau\big(F_h(t_{n+1},\psi_{h,n}^s) - F_h(t_n,\psi_{h,n})\big), & \\ \widehat{\psi}_{h,n}^j &= \widehat{\psi}_{h,n}^{j-1} + \theta\tau\big(F_h^j(t_{n+1},\widehat{\psi}_{h,n}^j) - F_h^j(t_{n+1},\psi_{h,n}^s)\big), & j = 1,2,\ldots,s, \\ \psi_{h,n+1} &= \widehat{\psi}_{h,n}^s, & n = 0,1,\ldots,N_T-1. \end{array}\right. \tag{9}$$

For any given θ, the scheme (9) is conditionally convergent of classical order 2, whenever $\sigma = \frac{1}{2}$, and of order 1 otherwise. Although the stability restriction of this method is similar to that of (8), it requires two more implicit stages in order to achieve the same accuracy.

4 Numerical Results

This section shows the numerical behaviour of methods (8) and (9) in the solution of nonlinear parabolic problems of type (1). In particular, let us consider (1) posed on the unit square $\Omega \equiv \{\underline{x} = (x,y) \in \mathbb{R}^2 : 0 < x < 1,\ 0 < y < 1\}$. Tensor $K(\psi)$ is a symmetric positive-definite nonlinear matrix defined as $K(\psi) = Q(\theta)D(\psi)Q(\theta)^T$, where $Q(\theta)$ is a 2×2 rotation matrix with angle $\theta = \pi/4$ and $D(\psi)$ is a 2×2 diagonal matrix whose diagonal entries are $1+\psi^2$ and $1+8\psi^2$. The nonlinear reaction term is chosen to be $g(\psi) = -(1+\psi^2)e^{-\psi}$, whereas the source/sink term $f(\underline{x},t)$ and both initial and Dirichlet boundary conditions are defined in such a way that $\psi(x,y,t) = e^{11-4t}x^4(1-x)^4y^4(1-y)^4$ is the exact solution of the problem.

The discretization of the spatial domain Ω is based on the construction of a smooth curvilinear grid $\Omega_h \equiv \{(\tilde{x}_{i,j}, \tilde{y}_{i,j})\}_{i,j=1}^N$ with coordinates:

$$\tilde{x}_{i,j} = \xi_{i,j} + 10\,\xi_{i,j}\,(1-\xi_{i,j})\,(\tfrac{1}{2}-\xi_{i,j})\,\eta_{i,j}\,(1-\eta_{i,j}),$$
$$\tilde{y}_{i,j} = \eta_{i,j} + 10\,\eta_{i,j}\,(1-\eta_{i,j})\,(\tfrac{1}{2}-\eta_{i,j})\,\xi_{i,j}\,(1-\xi_{i,j}),$$

where $\xi_{i,j} = (i-1)h$, $\eta_{i,j} = (j-1)h$ and $h = 1/(N-1)$. This grid is obtained from a uniform grid, by using an analytical transformation. Fig. 1 shows an example of such a grid for $N = 17$.

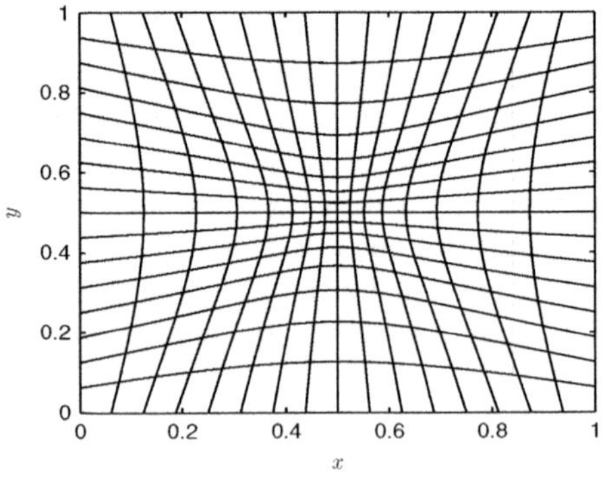

Fig. 1. Logically rectangular grid for $N = 17$.

Afterwards, we consider a decomposition of Ω into $s = 4$ overlapping subdomains $\{\Omega_j\}_{j=1}^s$, each of which involves $s_j = 4$ disjoint connected components, for $j = 1,2,3,4$. Related to such a decomposition, we define a smooth partition of unity consisting of a sequence of functions $\{\rho_j(\underline{x})\}_{j=1}^s$ based on (4). This partition of unity is displayed on Fig. 2, where the overlapping subdomains are given by $\Omega_j \equiv supp\,(\rho_j(\underline{x}))$.

For the time integration of this test problem, we consider the linearly implicit FSRK method (8) as well as the Hundsdorfer and Verwer scheme (9), with $\theta = 1$ and $\sigma = 1/2$. Let us introduce the global error at time $t = t_n$ as $E_{h,\tau} = r_h\psi(\underline{x},t_n) - \psi_{h,n}$, for $n = 1,2,\ldots,N_T$. Under certain discrete norm $\|\cdot\|_h$ and suitable stability restrictions between h and τ, it holds that $\|E_{h,\tau}\|_h \le C(h^2+\tau^2)$, being C a positive constant

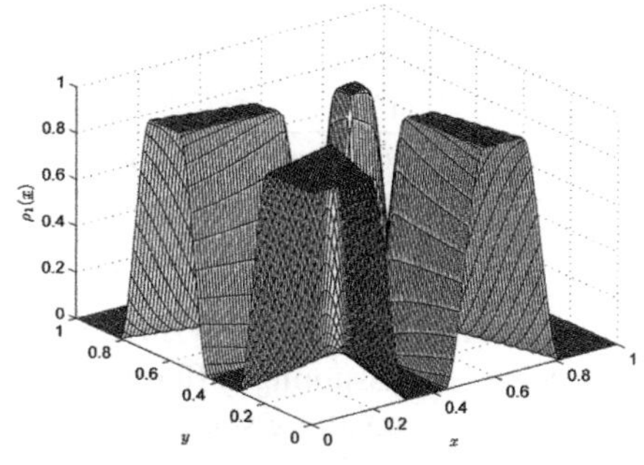

Function $\rho_1(\underline{x})$

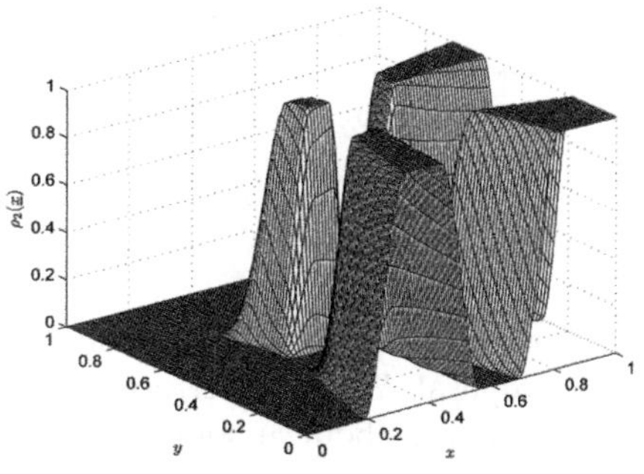

Function $\rho_2(\underline{x})$

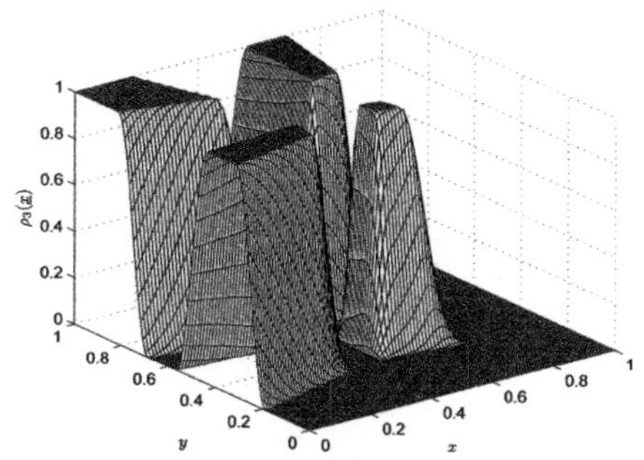

Function $\rho_3(\underline{x})$

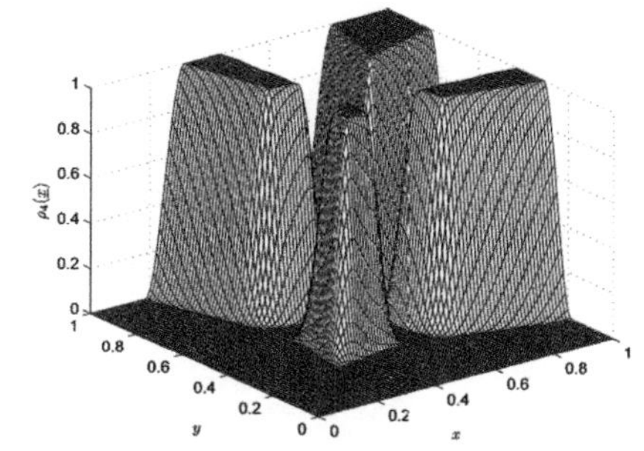

Function $\rho_4(\underline{x})$

Fig. 2. Smooth partition of unity $\{\rho_j(\underline{x})\}_{j=1}^{s}$ related to $\{\Omega_j\}_{j=1}^{s}$, for $s = 4$.

independent of h and τ. In our convergence study, we shall measure these errors by using the discrete L^2-norm in space and the discrete maximum norm in time, denoted by $\|E_{h,\tau}\|_2$. Tables 1 and 2 present the asymptotic behaviour of the global errors when the scheme (8) is used for different values of h and τ. As expected, it is shown to be conditionally convergent of order 2 in both space (see Table 1) and time (see Table 2).

Table 1. Global errors obtained in method (8) for $\tau = 5 \cdot 10^{-8}$.

h	$h_0 = 2^{-4}$	$h_0/2$	$h_0/4$	$h_0/8$	$h_0/16$	$h_0/32$
$\|E_{h,\tau}\|_2$	4.530E-2	3.305E-3	5.028E-4	1.120E-4	2.639E-5	6.574E-6

Finally, Table 3 compares the stability restrictions arising between h and τ when both methods are applied to this example. Here, we compute the maximum time steps τ_h^{PR} and τ_h^{HV} which make (8) and (9) respectively stable for different mesh sizes h. In view of the numerical results, we can conclude that both schemes converge

Table 2. Global errors obtained in method (8) for $h = 2^{-7}$.

τ	$\tau_0 = 10^{-4}$	$\tau_0/2$	$\tau_0/4$	$\tau_0/8$	$\tau_0/16$	$\tau_0/32$
$\|E_{h,\tau}\|_2$	1.424E-5	3.562E-6	8.908E-7	2.227E-7	5.568E-8	1.392E-8

under a non-severe stability limitation which is revealed to be slightly milder for our proposal. We have performed additional experiments assuming different types of solutions on both smooth and non-smooth grids and the resulting stability restrictions preserve a similar behaviour. Therefore, the generalization of the Peaceman-Rachford method given by (8) may be considered as a remarkable alternative to other existing linearly implicit splitting methods of order 2.

Table 3. Maximum time steps τ_h^{PR} and τ_h^{HV} permitted for different mesh sizes h.

h	$h_0 = 2^{-4}$	$h_0/2$	$h_0/4$	$h_0/8$	$h_0/16$	$h_0/32$
τ_h^{PR}	2.30E-3	1.90E-3	7.30E-4	3.00E-4	1.25E-4	5.10E-5
τ_h^{HV}	2.20E-4	7.05E-5	2.12E-5	6.43E-6	1.90E-6	5.80E-7

Acknowledgement. This research was partially supported by the Spanish Ministry of Science and Education under Research Project MTM2007-63772 and by Government of Navarre under Research Project CTP-07/R-10.

References

[1] in 't Hout, K.J., Welfert, B.D.: Stability of ADI schemes applied to convection-diffusion equations with mixed derivative terms. *Appl. Numer. Math.*, 57:19–35, 2007.

[2] Mathew, T.P., Polyakov, P.L., Russo, G., Wang, J.: Domain decomposition operator splittings for the solution of parabolic equations. *SIAM J. Sci. Comput.*, 19:912–932, 1998.

[3] Portero, L., Jorge, J.C.: A generalization of Peaceman-Rachford fractional step method. *J. Comput. Appl. Math.*, 189:676–688, 2006.

[4] Portero, L., Jorge, J.C.: A new class of second order linearly implicit fractional step methods. *J. Comput. Appl. Math.*, 218:603–615, 2008.

[5] Samarskiĭ, A.A., Matus, P.P., Vabishchevich, P.N.: *Difference schemes with operator factors.* Mathematics and its Applications, 546. Kluwer, Dordrecht, 2002.

[6] Samarskiĭ, A.A., Tishkin, V., Favorskiĭ, A., Shashkov, M.Y.: Operational finite-difference schemes. *Differ. Equ.*, 17:854–862, 1981.

[7] Shashkov, M. *Conservative finite-difference methods on general grids.* Symbolic and Numeric Computation Series. CRC, Boca Raton, FL, 1996.

NKS for Fully Coupled Fluid-Structure Interaction with Application

Andrew T. Barker[1] and Xiao-Chuan Cai[2]

[1] Department of Applied Mathematics, University of Colorado, Boulder, CO 80309-0526, andrew.barker@colorado.edu
[2] Department of Computer Science, University of Colorado, Boulder, CO 80309-0430, cai@cs.colorado.edu

1 Introduction

Newton-Krylov-Schwarz algorithms have been used in many areas and are often quite scalable and robust. In this paper we explore the application of Schwarz type domain decomposition preconditioners to some fully coupled systems for fluid-structure interaction. In particular, we are interested in developing a scalable parallel framework for the simulation of blood flow in human arteries [11]. In [2, 3], coupled fluid-structure problems are solved in 3D for patient-specific artery models, with emphasis on accurately representing vessel geometry, on constitutive model for the artery walls, and other physical concerns. In this paper we focus on a class of parallel domain decomposition algorithms for solving the coupled systems and report on the robustness and parallel scalability of the algorithms.

Very often in the simulation of fluid-structure interaction, fluid and structure are iteratively coupled, as in [4, 5, 7]. That is, fluid and structure subproblems are solved alternately (or in parallel), passing boundary conditions between them, until the solutions are compatible at the fluid-structure interface, and then the simulation proceeds to the next time step. However, this approach often requires small timesteps, can become unstable, and can reduce the order of accuracy of the solution [8]. In contrast, we use fully monolithic coupling, where the fluid and the structure are solved together as one system.

2 Governing Equations

We use a linear elastic model for the structure. The primary variable in the structure equations is the displacement vector $\mathbf{x}_s$. Define σ_s as the stress-strain relation or Cauchy stress tensor

$$\sigma_s = \lambda_s(\nabla \cdot \mathbf{x}_s)I + 2\mu_s(\nabla \mathbf{x}_s + \nabla \mathbf{x}_s^T)$$

M. Bercovier et al. (eds.), *Domain Decomposition Methods in Science and Engineering XVIII*, Lecture Notes in Computational Science and Engineering,
DOI: 10.1007/978-3-642-02677-5,

where λ_s and μ_s are the Lamé constants. The equilibrium equation for linear elasticity is

$$\rho_s \frac{\partial^2 \mathbf{x}_s}{\partial t^2} = \nabla \cdot \sigma_s + \mathbf{f}_s. \tag{1}$$

We fix the structure displacement $\mathbf{x}_s = 0$ on the dry, non-interaction boundary Γ_S; the boundary conditions on the fluid-structure interaction boundary Γ_w will be presented when we discuss the fluid-structure coupling.

The mesh points of our fluid domain move, and the displacements of the mesh nodes from their original reference configuration define a separate field that we need to represent. For the grid displacements $\mathbf{x}_f$, we simply use the Laplace equation

$$\Delta \mathbf{x}_f = 0 \tag{2}$$

in the interior of the domain, following [9]. In our numerical simulations this simple relation gives a smooth grid as the boundaries of the domain move, rarely causing problems with ill-conditioned elements. The boundary conditions for this field are either fixed zero Dirichlet conditions (at the inlet and outlet of the fluid domain) or are prescribed to follow the movement of the structure.

We model blood as a viscous incompressible Newtonian fluid, using the Navier-Stokes equations written in the ALE frame

$$\left.\frac{\partial \mathbf{u}_f}{\partial t}\right|_Y + [(\mathbf{u}_f - \omega_g) \cdot \nabla]\mathbf{u}_f + \frac{1}{\rho_f}\nabla p = \nu_f \Delta \mathbf{u}_f, \tag{3}$$

$$\nabla \cdot \mathbf{u}_f = 0. \tag{4}$$

Here $\mathbf{u}_f$ is the fluid velocity vector and p is the pressure. The given data include the fluid density ρ_f, and $\nu_f = \mu_f / \rho_f$, the kinematic viscosity. External body forces are ignored. Also, $\omega_g = \partial \mathbf{x}_f / \partial t$ is the velocity of the moving mesh in the ALE frame and the Y indicates that the time derivative is to be taken with respect to the ALE coordinates, not the Eulerian coordinates [9].

Boundary conditions for the fluid equations consist of a no-slip condition $\mathbf{u}_f = 0$ at rigid walls Γ_f, a Dirichlet condition where $\mathbf{u}_f$ takes a given profile at the inlet Γ_i, and a zero traction condition

$$\sigma_f \cdot \mathbf{n} = \mu_f (\nabla \mathbf{u}_f \cdot \mathbf{n}) - p\mathbf{n} = 0 \tag{5}$$

on the outlet Γ_o, where σ_f is the Cauchy stress tensor for the fluid and $\mathbf{n}$ is the unit outward normal.

At the fluid-structure interface we require that the structure velocity match the fluid velocity $\mathbf{u}_f = \partial \mathbf{x}_s / \partial t$ and we also enforce that the moving mesh must follow the solid movement $\mathbf{x}_f = \mathbf{x}_s$, so that the solid can maintain a Lagrangian description. The coupling of traction forces at the boundary can be written $\sigma_s \cdot \mathbf{n} = \sigma_f \cdot \mathbf{n}$ where $\mathbf{n}$ is the unit normal vector at the fluid-solid interface.

3 Spatial Discretization

Because of space constraints, we omit the full derivation of the weak form of the governing equations. We note two interesting points here. First, because of our moving grid, the variational spaces in which we seek a solution to the fluid subproblem are time-dependent. Second, the variational spaces associated with the fluid subproblem and the mesh subproblem depend implicitly on the current solution to the structure subproblem, as this solution provides essential boundary conditions for the fluid and mesh subproblems.

The spatial discretization is done with quadrilateral finite elements, with a conforming discretization at the fluid-structure interface, so that no special interpolation scheme is necessary to move information between fluid and structure.

We write the structure displacement vector $\mathbf{x}_s$ as $\mathbf{x}_s \approx \sum_j \varphi_j(x)x_j(t)$ and denote the vector of coefficients x_j as x_s. Using this approximation, we arrive at the semi-discrete system

$$M_s \frac{\partial^2 x_s}{\partial t^2} + C_s \frac{\partial x_s}{\partial t} + K_s x_s = F \tag{6}$$

where $C_s = \alpha M_s + \beta K_s$ is an added Rayleigh damping matrix where α and β are small parameters; typically $\alpha \approx 0.1$ and $\beta \approx 0.01$ [6].

We use biquadratic quadrilateral finite elements in our ALE discretization of the moving mesh. We approximate $\mathbf{x}_f \approx \sum_j \xi_j(x)x_j(t)$. This is a standard finite-element discretization of the Laplace equation resulting in $K_m x_f = 0$ with boundary conditions that depend on the structure subproblem.

The fluid is discretized with the LBB-stable $Q_2 - Q_1$ finite elements. Using finite-dimensional approximations $\mathbf{u}_f \approx \sum_j \varphi_j(x,t)u_j(t)$ and $p \approx \sum_j \psi_j(x,t)p_j(t)$ we can write the semi-discrete Navier-Stokes equations in the ALE frame as

$$M_f \frac{\partial u}{\partial t} + B(u)u + K_f u - Q^T p = M_f f, \tag{7}$$

$$Qu = 0 \tag{8}$$

where M_f is a mass matrix, $B(u)$ represents the nonlinear convective operator, K_f is the discrete Laplacian, and Q is the discrete divergence operator.

The mesh displacement continuity and velocity continuity conditions are enforced directly at each timestep; we replace rows of the matrix corresponding to these degrees of freedom with rows representing the equations $x_s = x_f$, and similarly for the velocity. We also need to discretize the traction force that the fluid exerts on the solid boundary, namely $\sigma_f \cdot \mathbf{n} = \mu_f(\nabla \mathbf{u}_f \cdot \mathbf{n}) - p\mathbf{n}$. The result has block matrix form

$$\sigma_f \cdot \mathbf{n} = \begin{pmatrix} A_{uu} \; A_{uv} \; A_{up} \\ A_{vu} \; A_{vv} \; A_{vp} \end{pmatrix} \begin{pmatrix} u_f \\ v_f \\ p \end{pmatrix} = (A_u \; A_p) \begin{pmatrix} u \\ p \end{pmatrix}. \tag{9}$$

This will be inserted as a force in the discrete form of the structure equations to enforce the traction matching condition at the fluid-structure interface.

4 Temporal Discretization

We use the trapezoid rule $y^{n+1} = y^n + (\Delta t/2)\left(f^{n+1} + f^n\right)$ which is a second-order accurate implicit scheme for all our time discretization.

For the structure time-stepping, we follow [6] in implementing the trapezoid rule by reducing the order of (6) from second order to first order. Our new vector of unknowns includes both solid displacement and velocity, $y = (x_s, \partial x_s/\partial t)^T$. Then

$$\frac{\partial y}{\partial t} = f(y,t) = \begin{pmatrix} \dfrac{\partial x_s}{\partial t} \\ M^{-1}(F(t) - K_s x_s - C_s \dfrac{\partial x_s}{\partial t}) \end{pmatrix}.$$

The trapezoid rule for this differential algebraic equation can be written

$$My^{n+1} = My^n + \frac{\Delta t}{2}\left[Ky^{n+1} + Ky^n + F^{n+1} + F^n\right]$$

where

$$M = \begin{pmatrix} I & \\ & M_s \end{pmatrix}, \quad K = \begin{pmatrix} & I \\ -K_s & -C_s \end{pmatrix}.$$

The moving mesh, like the continuity equation for the fluid, is enforced independent of time. So we simply require

$$K_m x_f^{n+1} = 0$$

at each time step.

Rescaling pressure by the timestep Δt, we apply a slightly modified version of the trapezoid rule to (7) to get

$$Mu^{n+1} = Mu^n + \frac{1}{2}\left[(S + \Delta t R^{n+1})u^{n+1} + (S + \Delta t R^n)u^n\right]$$

where

$$M = \begin{pmatrix} M_f & 0 \\ 0 & 0 \end{pmatrix}, \quad R^n = \begin{pmatrix} -B(u^n) - K_f & 0 \\ 0 & 0 \end{pmatrix}, \quad S = \begin{pmatrix} 0 & -Q^T \\ Q & 0 \end{pmatrix}.$$

We use the same time-stepping scheme for fluid and structure, so we can simply put the discretized fluid and structure problems together in one system with coupling enforced implicitly. In summary, we have

$$(M + W)y^{n+1} - My^n - \frac{\Delta t}{2}(Ky^{n+1} + Ky^n) - \frac{\Delta t}{2}(F^{n+1} + F^n) = 0 \qquad (10)$$

where

$$y^n = \begin{pmatrix} u^n \\ \Delta t p^n \\ x_f^n \\ x_s^n \\ \dot{x}_s{}^n \end{pmatrix}, \quad M = \begin{pmatrix} M_f & & & & \\ & & & & \\ & & & & \\ & & & I & \\ & & & & M_s \end{pmatrix},$$

$$W = \begin{pmatrix} & & & \\ & & & \\ & & K_m & \\ A_u & A_p & & \\ & & & \end{pmatrix}, \quad K = \begin{pmatrix} -B-K_f & -(1/\Delta t)Q^T & & & \\ (1/\Delta t)Q & & & & \\ & & & & \\ & & & & I \\ & & & -K_s & -C_s \end{pmatrix}.$$

Though written in matrix form, many of the operators above are nonlinear. In particular the B term depends on u_f, and the K_f, M_f and Q terms depend on the moving mesh x_f. This implies that we have a Jacobian of the form

$$J = \begin{pmatrix} J_f & -Q^T & Z_m & & \\ Q & & Z_c & & \\ & & K_m & & \\ A_u & A_p & & I & -(\Delta t/2)I \\ & & & (\Delta t/2)K_s & M_s + (\Delta t/2)C_s \end{pmatrix} \tag{11}$$

where J_f is the Jacobian of the nonlinear term in the momentum equation and Z_m and Z_c are the nonlinear contributions of the moving mesh to the momentum and continuity equations. The form of Z_m and Z_c are unknown, and our implementation of the Jacobian simply ignores them, which is a reasonable approximation as long as the mesh movement is slow, i.e., the timestep is sufficiently small.

5 Solving the Nonlinear System

At each timestep, we solve the nonlinear system (10) with an inexact Newton method with line search. At each Newton step we solve a preconditioned linear system of the form $J(y)M^{-1}(Ms) = z$ for the Newton correction s, where M^{-1} is a one-level additive Schwarz preconditioner [10, 12, 13]. In this domain decomposition preconditioner, the formation of subdomains does not consider the fluid-structure boundary, so that a subdomain may contain fluid elements, structure elements, or both. Subdomain solves are done by LU factorization with homogeneous Dirichlet boundary conditions on the boundaries for all solution variables, including the fluid pressure.

In practice, we order the unknowns for the Jacobian system not by field ordering as in (11), but by element ordering. The choice of ordering can have significant effect on the convergence properties of the solver. By this choice, the nonzero-block structure is banded. That is, within each element the unknowns are ordered as in (11), but globally the matrix looks like the nine-point stencil for a Poisson equation.

6 Numerical Results

Our solver is implemented using PETSc [1]. All computations are performed on an IBM BlueGene/L supercomputer at the National Center for Atmospheric Research with 1024 compute nodes.

We begin all our simulations with zero initial conditions for structure displacement and fluid velocity, therefore compatibility between fluid and structure is easily satisfied in the initial conditions. In all the numerical results in this paper, we use a timestep $\Delta t = 0.01$, a Young's modulus $E = 1.0 \cdot 10^5$, we stop the linear solver when the preconditioned residual has decreased by a factor of 10^{-4} and we stop the Newton iteration when the nonlinear residual has decreased by a factor of 10^{-6}. We set GMRES to restart every 40 iterations, and have the structural damping parameters $\alpha = 0.1, \beta = 0.01$. Simulations begin with zero initial conditions and proceed 10 timesteps, reporting average walltime and nonlinear iteration count per timestep, and average GMRES iterations per Newton step.

Our fluid-structure interaction simulations can deal with large deformations of the computational grid without the quality of the mesh degrading and without affecting convergence, and we maintain sufficient spatial resolution to resolve vortices and other interesting flow features.

The scalability of our algorithm is presented in Table 1. Our method scales well with respect to number of processors and scales fairly well with respect to problem size. It is also worth noting the large grid sizes and processor counts that we have used with success. The growth in GMRES iterations for large processor counts suggests that the less than perfect speedup could probably be improved by use of a multilevel preconditioner.

unknowns	np	GMRES	Newton	time
	64	9.3	5.0	123.44
$1.0 \cdot 10^6$	128	13.4	5.0	57.11
	256	18.2	5.0	36.41
	512	24.0	5.0	22.08
	128	17.5	4.8	125.31
	256	21.5	4.8	66.11
$2.1 \cdot 10^6$	512	29.7	4.8	39.97
	1024	35.9	4.8	22.90
	2048	40.0	4.8	17.25
	128	15.1	4.7	198.34
	256	20.1	4.7	100.23
$2.6 \cdot 10^6$	512	29.3	4.7	46.50
	1024	40.0	4.7	28.11
	2048	48.6	4.7	21.10

Table 1. Speedup and scalability. In this table ASM overlap $\delta = 2$, Reynolds number = 132.02, $\nu_s = 0.30$

Our simulation is also robust with respect to physical parameters. In Table 2a, we show numerical results for various Reynolds numbers. Many blood flow simulations, for example [2], use Reynolds numbers in the range 30–100, but we can exceed that without much difficulty. As the Poisson ratio ν_s approaches $1/2$ the structure becomes incompressible and the structure problem becomes more numerically challenging; our solver is fairly robust also in this respect (results not shown). In some FSI methods, the case where fluid and structure densities are nearly equal is particularly difficult. Our monolithic coupling avoids this difficulty, see Table 2b.

ρ_s	ρ_f	GMRES	Newton
10^{-6}	10^{-6}	57.8	2.5
10^{-6}	10^{-3}	41.1	3.8
10^{-6}	1.0	7.3	5.6
10^{-6}	10.0	5.8	7.9
1.0	10^{-6}	59.0	2.3
1.0	10^{-3}	40.5	3.7
1.0	1.0	7.5	5.4
1.0	10.0	5.8	7.9
10^6	10^{-6}	60.4	2.4
10^6	10^{-3}	64.7	2.3
10^6	1.0	24.6	4.0
10^6	10.0	11.0	4.2

unknowns	Re	GMRES	Newton
	33.00	12.0	4.8
	66.01	12.1	4.8
$2.1 \cdot 10^6$	132.02	12.2	4.8
	264.03	12.5	4.8
	1056.12	12.7	10.0
	33.00	12.9	4.7
	66.01	13.1	4.7
$2.6 \cdot 10^6$	132.02	13.5	4.7
	264.03	13.5	4.7
	1056.12	13.7	9.9

Table 2. (a) Sensitivity of algorithm to various fluid densities (ρ_f) and solid densities (ρ_s); these problems have $6.5 \cdot 10^5$ unknowns and tests are done with 128 processors. (b) Sensitivity to Reynolds number with 256 processors. In both (a) and (b) ASM overlap $\delta = 8$ and $\nu_s = 0.30$

7 Conclusion

Accurate modeling of blood flow in compliant arteries is a computational challenge. In order to meet this challenge, we need not only to model the physics accurately but also to develop scalable algorithms for parallel computing. In this paper we develop a Newton-Krylov-Schwarz solver that scales well in parallel and is effective for solving the implicitly coupled fluid-structure interaction problem. Our method is quite robust with respect to different vessel geometries, Reynolds numbers, Poisson ratios, densities, spatial mesh sizes and time step sizes.

Acknowledgement. We thank K. Hunter, C. Lanning, R. Shandas, A. Quarteroni, F. Hwang, J. Wilson, and X. Yue for their help and suggestions in the project. The research was supported in part by DOE under DE-FC02-01ER25479 and DE-FC02-04ER25595, and in part by NSF under grants ACI-0305666, CNS-0420873, CCF-0634894, and CNS-0722023.

References

[1] Balay, S., Buschelman, K., Eijkhout, V., Gropp, W., Kaushik, D., Knepley, M., McInnes, L., Smith, B., Zhang, H.: PETSc users manual. Technical Report ANL-95/11 - Revision 2.1.5, Argonne National Laboratory, 2004.

[2] Bazilevs, Y., Calo, V.M., Zhang, Y., Hughes, T.J.R.: Isogeometric fluid-structure interaction analysis with applications to arterial blood flow. *Comput. Mech.*, 38:310–322, 2006.

[3] Figueroa, C., Vignon-Clementel, I., Jansen, K., Hughes, T., Taylor, C.: A coupled momentum method for modeling blood flow in three-dimensional deformable arteries. *Comput. Methods Appl. Mech. Engrg.*, 195:5685–5706, 2006.

[4] Formaggia, L., Gerbeau, J.F., Nobile, F., Quarteroni, A.: On the coupling of 3D and 1D Navier-Stokes equations for flow problems in compliant vessels. *Comput. Methods Appl. Mech. Engrg.*, 191:561–582, 2001.

[5] Formaggia, L., Gerbeau, J.F., Nobile, F., Quarteroni, A.: Numerical treatment of defective boundary conditions for the Navier-Stokes equations. *SIAM J. Numer. Anal.*, 40:376–501, 2002.

[6] Hughes, T.: The Finite Element Method: Linear Static and Dynamic Finite Element Analysis. Dover, 2000.

[7] Hunter, K., Lanning, C., Chen, S., Zhang, Y., Garg, R., Ivy, D., Shandas, R.: Simulations of congenital septal defect closure and reactivity testing in patient-specific models of the pediatric pulmonary vasculature: a 3D numerical study with fluid-structure interaction. *J. Biomech. Eng.*, 128(4):564–572, 2006.

[8] E. Kuhl, S. Hulshoff, and R. de Borst. An arbitrary Lagrangian Eulerian finite-element approach for fluid-structure interaction phenomena. *Internat. J. Numer. Methods Engrg.*, 57:117–142, 2003.

[9] Nobile, F.: Numerical Approximation of Fluid-Structure Interaction Problems with Application to Haemodynamics. PhD thesis, École Polytechnique Fédérale de Lausanne, 2001.

[10] Ovtchinnikov, S., Dobrian, F., Cai, X.-C., Keyes, D.E.: Additive Schwarz-based fully coupled implicit methods for resistive Hall magnetohydrodynamic problems. *J. Comput. Phys.*, 225(2):1919–1936, 2007. ISSN 0021-9991.

[11] Quarteroni, A., Tuveri, M., Veneziani, A.: Computational vascular fluid dynamics: problems, models and methods. *Comput. Vis. Sci.*, 2:163–197, 2000.

[12] Quarteroni, A., Valli, A.: Domain Decomposition Methods for Partial Differential Equations. Oxford University Press, Oxford, 1999.

[13] Toselli, A., Widlund, O.: Domain Decomposition Methods—Algorithms and Theory. Springer, Berlin, 2005.

Weak Information Transfer between Non-Matching Warped Interfaces

Thomas Dickopf and Rolf Krause

University of Bonn, Institute for Numerical Simulation
Wegelerstraße 6, 53115 Bonn, Germany
{dickopf,krause}@ins.uni-bonn.de

Summary. We consider the information transfer between non-matching finite element meshes arising from domain decomposition. Dealing with complex three-dimensional geometries, especially in the case of computational mechanics and nonlinear contact problems, one can usually not achieve a decomposition of the global domain with mere planar interfaces in a sensible way. Thus, subdomains with warped interfaces emerge which, after an independent discretization, yield a geometrically non-conforming decomposition with small gaps and overlaps. In this paper, we employ a mortar approach and develop a method for the assembly of a discrete coupling operator providing a stable information transfer across such geometrically distinct warped interfaces.

1 Introduction

The efficient realization of an exchange of discrete information between geometrically non-conforming interfaces in three-dimensional space is of high interest in many applications. In case a domain is decomposed, even if the real boundaries of the subdomains coincide, the discrete interior interfaces formed by independently generated meshes will not. Besides, in computational mechanics often the use of an a priori decomposition into structural parts with different mechanical properties is advisable. This specifically holds true for contact problems, where the actual interface is unknown in advance.

In this paper, the discretization of the coupling constraints is done in a weak sense by a mortar approach [2], proposing the use of an L^2-projection between non-matching meshes to allow for optimal error estimates. Here, motivated by [4, 8, 9], we develop an efficient method to compute the emerging discrete transfer operator for meshes on warped interfaces exhibiting gaps and overlaps. The whole extent of its applicability becomes clear when we use our operator to simulate the transmission of forces between colliding bodies.

M. Bercovier et al. (eds.), *Domain Decomposition Methods in Science and Engineering XVIII*, Lecture Notes in Computational Science and Engineering,
DOI: 10.1007/978-3-642-02677-5, © Springer-Verlag Berlin Heidelberg 2009

2 Discrete Information Transfer

Let Γ^k, $k \in \{m,s\}$, be a two-dimensional connected submanifold of $\mathbb{R}^3$ with boundary. In applications each of these surfaces naturally appears as a subset of the boundary of a three-dimensional domain Ω^k, $k \in \{m,s\}$. In particular, it is assumed that all the surface information exchange between the domains Ω^m and Ω^s takes place across the segments Γ^m and Γ^s. For simplicity we do not consider crosspoints of more than two interfaces.

In order to prescribe matching conditions expressing the mutual information transfer, we assume a sufficiently smooth, bijective mapping $\boldsymbol{\Phi} : \Gamma^s \to \Gamma^m$ to be given, which relates the opposite interfaces. Although, in general, such a mapping is part of the overall solution and not achievable a priori, a reasonable discrete version reflecting coupling in normal direction can be found by a linearization, which we discuss later on. Then, for a Sobolev function $u = (u^m, u^s) \in \prod_{k\in\{m,s\}} H^{\frac{1}{2}}(\Gamma^k)$, emerging as the respective traces of H^1-functions defined on Ω^k, $k \in \{m,s\}$, the transmission conditions which for second order partial differential equations commonly have to be realized are

$$[u] = 0, \quad \frac{\partial u^m}{\partial \boldsymbol{n}} \circ \boldsymbol{\Phi} = \frac{\partial u^s}{\partial \boldsymbol{n}}, \quad \text{a. e. on } \Gamma^s. \tag{1}$$

Here, $[u] := u^s - u^m \circ \boldsymbol{\Phi}$ is the jump of u across the interface $\Gamma := \Gamma^m \cup \Gamma^s$ and $\frac{\partial}{\partial \boldsymbol{n}}$ denotes the appropriate normal derivative on Γ.

The enforcement of discrete constraints corresponding to (1), which we present now, is motivated by the understanding that, in a very general sense, with non-matching meshes a pointwise coupling yields indeed a conforming approximation but does not provide optimal discretization error estimates. So, we employ a mortar approach, see [2], and impose a weak matching condition by the introduction of suitable Lagrange multipliers on the interface.

Let $\mathcal{T}^k$ be a shape regular surface mesh of Γ^k, $k \in \{m,s\}$, made up of triangles and quadrilaterals. In applications these meshes are inherited from unstructured volume meshes of the domains Ω^m and Ω^s, consisting of tetrahedrons, hexahedrons, pyramids, and prisms. We denote the nodes of $\mathcal{T}^k$ by $\mathcal{N}^k$. On both surface meshes, we use the space of Lagrangian conforming finite elements of first order $X_h(\Gamma^k)$ and denote its nodal basis functions as $(\lambda_p^k)_{p\in\mathcal{N}^k}$ with $\lambda_p^k(q) = \delta_{pq}$, $p,q \in \mathcal{N}^k$, $k \in \{m,s\}$. Then, the unconstrained product finite element space is given as $X_h := \prod_{k\in\{m,s\}} X_h(\Gamma^k)$. All finite element functions will be marked by the subscript $_h$.

We define a discrete multiplier space $M_h \subseteq X_h(\Gamma^s)$ and fix a basis $(\psi_p)_{p\in\mathcal{N}^s}$. In fact, M_h turns out to be an approximation space for the normal derivative on Γ and has to be chosen compatibly. Because the multiplier space is associated with the mesh $\mathcal{T}^s$ and because the values on Γ^m will constrain the values on Γ^s, we call entities with superscript ^{s} slave (or non-mortar), whereas entities with superscript ^{m} are referred to as master (or mortar). Then, the well-known weak "zero jump condition" of the mortar method from [2] is

$$\int_{\Gamma^s} \psi_h \cdot [u_h]\, d\boldsymbol{a} = 0 \quad \forall\, \psi_h \in M_h.$$

Inspired by these weak coupling constraints, we use the representations of u_h and ψ_h in the chosen bases of X_h and M_h, respectively, and define the discrete mortar transfer operator via its algebraic representation, $\boldsymbol{T} := \boldsymbol{D}^{-1}\boldsymbol{B}$, with the entries

$$\begin{aligned} d_{pq} &:= \int_{\Gamma^s} \psi_p \lambda_q^s \, d\boldsymbol{a}, & p,q &\in \mathcal{N}^s, \\ b_{pq} &:= \int_{\Gamma^s} \psi_p (\lambda_q^m \circ \boldsymbol{\Phi}) \, d\boldsymbol{a}, & p &\in \mathcal{N}^s,\ q \in \mathcal{N}^m. \end{aligned} \tag{2}$$

The transfer operator $\boldsymbol{T}$ maps discrete values on the master side via the multiplier space M_h to the slave side. More precisely, for $v^m \in X_h(\Gamma^m)$ the function $\boldsymbol{T}v^m$ is the L^2-projection of $v^m \circ \boldsymbol{\Phi}$ onto $X_h(\Gamma^s)$. Now, two possible algebraic forms of the discrete matching conditions are

$$(\boldsymbol{D}u_h^s)_p - (\boldsymbol{B}u_h^m)_p = 0 \quad \text{or} \quad (u_h^s)_p - (\boldsymbol{T}u_h^m)_p = 0 \quad \forall\, p \in \mathcal{N}^s. \tag{3}$$

The left variant $(3)_1$ can be used as a constraint in the saddle point formulation of a coupled problem. The right one $(3)_2$ allows for either the elimination of the degrees of freedom on the slave side or the application of a Dirichlet–Neumann type algorithm as in [4]. Note that the approach is indeed non-conforming, i.e. the weak coupling constraints (3) do generally not guarantee that the stronger condition (1) is satisfied.

The constraints involving the mortar transfer operator $\boldsymbol{T}$ can easily be adjusted for the approximation of a variational inequality, e.g., stemming from a free boundary value problem. If $\boldsymbol{D}$ is a diagonal matrix with positive entries, which can be achieved by using dual Lagrange multipliers as in [6, 7] or mass lumping, this results in ordinary inequality constraints for all $p \in \mathcal{N}^s$. Naturally, our coupling approach is not limited to scalar valued problems.

Finally, to prove optimal discretization error estimates for the global approximation of the considered problem, all discrete function spaces have to be chosen appropriately. In particular, a uniform inf-sup condition between the finite element spaces on Ω^m and Ω^s and the multiplier space M_h needs to hold. Then, a proof can be carried out following [2] in case of a linear problem and following [7] in case of a free boundary value problem.

3 The Discrete Coupling Operator

Even in the case of matching interfaces (but possibly non-matching meshes) the assembly of the master-slave coupling is intricate because intersections of arbitrary element faces have to be computed. But dealing with geometrically non-conforming decompositions exhibiting gaps and overlaps between warped interfaces, we are obliged to meet further challenges, since the unknown mapping $\boldsymbol{\Phi}$ directly enters the definition of the transfer operator in (2). A first idea for the handling of this more sophisticated case by projecting the meshes of the opposite interfaces onto an explicitly given two-dimensional submanifold of $\mathbb{R}^3$ can be found in [4]. In [9] the interfaces are projected onto a plane varying with the slave side, instead. Then, the coupling terms are computed by numerical integration on intersections of projected faces in this plane. A further possibility is the automatic construction of an approximate identifying mapping $\boldsymbol{\Phi}_h$ as in [8].

Here, we derive an algorithm, which assembles the discrete coupling operator from local information exclusively. We compute triangulated intersections of two respective faces in a locally adjusted projection plane but, unlike before, carry out the quadrature on the possibly warped slave side Γ^s directly. Going beyond [9], we give a sound derivation of our element-based approach, which does not use any parametrizations of the two-dimensional faces and is also suitable for isoparametric elements.

Let $\mathfrak{F}^m$ and $\mathfrak{F}^s$ be the sets of master and slave faces, respectively. Only to ease the derivation of the algorithm, we assume a bounded set $U \subset \mathbb{R}^2$ and global parametrizations $\boldsymbol{\varphi}_k : U \to \Gamma^k$ of Γ^k, $k \in \{m,s\}$, to be given so that $\boldsymbol{\Phi} = \boldsymbol{\varphi}_m \circ \boldsymbol{\varphi}_s^{-1}$. This means that points on the interfaces which have the same preimages in the parameter domain U are identified. We point out that the parametrizations will shortly be replaced by suitable discrete and local versions, which are immediately computable from the geometric information already available in finite element programs.

In the following, we denote the three-dimensional element belonging to the face F^k from the mesh $\mathfrak{T}^k$ by E^k. We generically denote the respective reference element of E^m and E^s by $\hat{E}$. The coordinate transformation from $\hat{E}$ to E^k is called $\boldsymbol{G}_{\hat{E}\to E^k}$ and the affine transformation between two triangles T_1 and T_2 is $\boldsymbol{G}_{T_1 \to T_2}$.

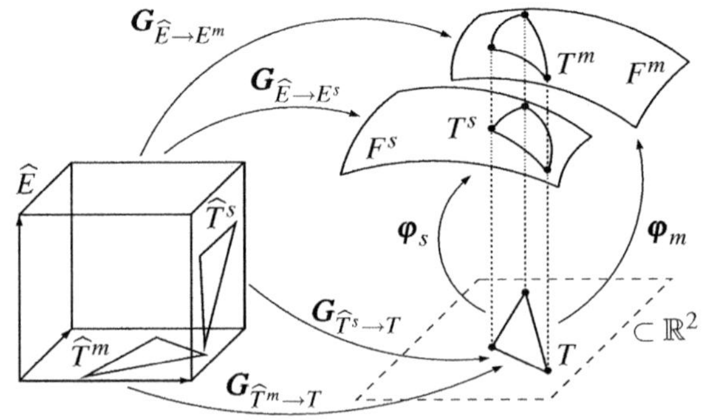

Fig. 1. Derivation of the assembly algorithm for the discrete coupling operator.

We use the decompositions of Γ^m and Γ^s induced by the meshes $\mathfrak{T}^m$ and $\mathfrak{T}^s$ and observe the fact that the set $\{\boldsymbol{\Phi}^{-1}(F^m)|\ F^m \in \mathfrak{F}^m\}$ is a partition of the slave side Γ^s. In particular, for each slave face $F^s \in \mathfrak{F}^s$ we have $\cup_{F^m \in \mathfrak{F}^m} \left(F^s \cap (\boldsymbol{\varphi}_s \circ \boldsymbol{\varphi}_m^{-1})(F^m)\right) = F^s$. Then, integrating by substitution, we write formally,

$$
\begin{aligned}
b_{pq} &= \sum_{\substack{F^s \in \mathfrak{F}^s \\ F^m \in \mathfrak{F}^m}} \left(\int_{F^s \cap (\boldsymbol{\varphi}_s \circ \boldsymbol{\varphi}_m^{-1})(F^m)} \psi_p \cdot (\lambda_q^m \circ \boldsymbol{\varphi}_m \circ \boldsymbol{\varphi}_s^{-1}) \, d\boldsymbol{a} \right) \\
&= \sum_{\substack{F^s \in \mathfrak{F}^s \\ F^m \in \mathfrak{F}^m}} \left(\int_{\boldsymbol{\varphi}_s^{-1}(F^s) \cap \boldsymbol{\varphi}_m^{-1}(F^m)} (\psi_p \circ \boldsymbol{\varphi}_s) \cdot (\lambda_q^m \circ \boldsymbol{\varphi}_m) \cdot |\det \nabla \boldsymbol{\varphi}_s| \, d\boldsymbol{a} \right).
\end{aligned}
$$

We now assume that each intersection $\boldsymbol{\varphi}_s^{-1}(F^s) \cap \boldsymbol{\varphi}_m^{-1}(F^m) \subset \mathbb{R}^2$ can be divided into finitely many triangles, and for each triangle T we denote the corresponding triangles on the interfaces by $T^k := \boldsymbol{\varphi}_k(T)$, $k \in \{m,s\}$, see Fig. 1 in case E^m and E^s are hexahedrons. Then, we transfer the triangles to the reference element with the inverses of the three-dimensional transformations, namely $\hat{T}^k := \boldsymbol{G}_{\hat{E} \to E^k}^{-1}(T^k)$. Now the two-dimensional affine transformations $\boldsymbol{G}_{\hat{T}^k \to T}$ from these triangles $\hat{T}^k$ to the triangle T can easily be computed. Hence, we have $\boldsymbol{\varphi}_k|_T \equiv \boldsymbol{G}_{\hat{E} \to E^k} \circ \boldsymbol{G}_{\hat{T}^k \to T}^{-1}$ and are in a position to continue the above formal calculation for the contribution of each triangle T separately,

$$
\begin{aligned}
&\int_T (\psi_p \circ \boldsymbol{\varphi}_s) \cdot (\lambda_q^m \circ \boldsymbol{\varphi}_m) \cdot |\det \nabla \boldsymbol{\varphi}_s| \, d\boldsymbol{a} = \\
&|\det \nabla \boldsymbol{G}_{\hat{T}^s \to T}^{-1}| \int_T (\hat{\psi}_p \circ \boldsymbol{G}_{\hat{T}^s \to T}^{-1}) \cdot (\hat{\lambda}_q \circ \boldsymbol{G}_{\hat{T}^m \to T}^{-1}) \cdot |\det \nabla \boldsymbol{G}_{\hat{E} \to E^s}(\boldsymbol{G}_{\hat{T}^s \to T}^{-1})| \, d\boldsymbol{a}. \quad (4)
\end{aligned}
$$

At this we use the representations via the shape functions on the reference element, $\psi_p \circ \boldsymbol{G}_{\hat{E} \to E^s} = \hat{\psi}_p$ and $\lambda_q^m \circ \boldsymbol{G}_{\hat{E} \to E^m} = \hat{\lambda}_q$. By abuse of notation $\nabla \boldsymbol{G}_{\hat{E} \to E^s}$ stands for its restriction to the corresponding faces in the domain $\hat{E}$ and the codomain E^s, respectively. Thus, the exclusive use of the three-dimensional finite element transfomations supersedes the additional introduction of two-dimensional parametrizations of warped faces. Besides, we note that $|\det \nabla \boldsymbol{G}_{\hat{T}^s \to T}^{-1}| = \frac{|\hat{T}^s|}{|T|}$ is constant because $\boldsymbol{G}_{\hat{T}^s \to T}$ is an affine mapping.

These considerations lead to the understanding that the entries b_{pq} of the coupling matrix $\boldsymbol{B}$ can be computed as a sum of integrals of the form (4) over triangles. This requires that suitable approximations of the parametrizations $\boldsymbol{\varphi}_k$ are known. In fact there is no need to establish any parametrizations explicitly. We only have to replace all triangles T, T^m, and T^s by approximating ones to allow for the evaluation of the right hand side of (4). For this purpose, we introduce the following algorithm. Firstly, the intersections and their triangulations are computed in a projection plane locally adjusted to the slave side. Secondly, the triangles T^m and T^s on the respective interfaces Γ^m and Γ^s are created by an inverse projection. Then, an appropriate quadrature formula can be applied on the respective reference elements directly.

Algorithm

(A1) Build an octree data structure to determine which master and slave faces are "close" to each other.

(A2) Loop over all slave faces $F^s \in \mathfrak{F}^s$.

(B1) Loop over all master faces $F^m \in \mathcal{F}^m$.
- (C1) Only continue if F^m is "close" to F^s.
- (C2) Apply a Householder reflection H so that $H(\boldsymbol{n}^s) = \boldsymbol{e}_3$, where $\boldsymbol{n}^s$ is a suitably chosen outer normal of the current slave face.
- (C3) Compute $\tilde{F}^k$ as the convex hull of the corners of F^k projected onto the $\boldsymbol{e}_1\boldsymbol{e}_2$-plane, $k \in \{m,s\}$.
- (C4) Compute the intersection $\tilde{F}^m \cap \tilde{F}^s$ and a triangulation $\cup T_i$.
- (C5) Loop over all triangles T_i.
  - (D1) Perform an inverse projection of the corners of T_i to get corresponding triangles T_i^m and T_i^s on the original faces F^m and F^s, respectively.
  - (D2) Use the transformation $\boldsymbol{G}^{-1}_{\hat{E}\to E^k}$ to compute the triangle $\hat{T}_i^k$ on the reference element, $k \in \{m,s\}$.
  - (D3) Use a two-dimensional quadrature formula to create weights ω_l and integration points $\boldsymbol{x}_l^m$ and $\boldsymbol{x}_l^s$ on the triangles $\hat{T}_i^m$ and $\hat{T}_i^s$, respectively.
  - (D4) Set $\omega_l' := \omega_l \, |\det \nabla \boldsymbol{G}_{\hat{E}\to E^s}(\boldsymbol{x}_l^s)| \, |\hat{T}_i^s|$.
  - (D5) Add the contribution of triangle T_i,

$$b_{pq} \mapsto b_{pq} + \sum_l \omega_l' \hat{\psi}_p(\boldsymbol{x}_l^s)\hat{\lambda}_q(\boldsymbol{x}_l^m), \quad p \in \mathcal{N}^s,\ q \in \mathcal{N}^m,$$

$$d_{pq} \mapsto d_{pq} + \sum_l \omega_l' \hat{\psi}_p(\boldsymbol{x}_l^s)\hat{\lambda}_q(\boldsymbol{x}_l^s), \quad p,q \in \mathcal{N}^s.$$

- (C6) End of loop over triangles T_i.

(B2) End of loop over master faces F^m.

(A3) End of loop over slave faces F^s.

The one-time creation of the octree in step (A1), which is of complexity $\mathcal{O}(|\mathcal{N}| \cdot \log(|\mathcal{N}|))$, guarantees that the remaining steps of the algorithm have optimal complexity $\mathcal{O}(|\mathcal{N}^s|)$. For the efficient computation of the intersection $\tilde{F}^m \cap \tilde{F}^s$ and a Delaunay triangulation in step (C4) we use the quickhull algorithm QHULL [1]. The additionally needed input, a point which is a priori known to lie within the intersection, is computed by a modified simplex algorithm particularly detecting whether an intersection is empty.

We carry out an extensive analysis of our method in context of the numerical simulation of multi-body contact problems in [3]. In particular, we show that our algorithm can be interpreted as an a priori approximation $\boldsymbol{\Phi}_h$ of the actual (contact) mapping by a composition of local projections and inverse projections. Although, there, $\boldsymbol{\Phi}_h$ is piecewisely defined and possibly discontinuous at the edges of the slave faces, those considerations close the gap to [8].

Regarding the algorithm as an elaborate construction of an approximate mapping $\boldsymbol{\Phi}_h$ and subsequent numerical integration in step (D5), we note that the integral is not necessarily evaluated exactly since, in general, the integrand is not a polynomial. An analysis of the additional consistency error due to inexact constraints has not been achieved yet. But a similar problem arises for the mortar method in case of geometrically matching interfaces if a quadrature rule is used only based on either $\mathcal{T}^m$ or $\mathcal{T}^s$, see [5] and the references therein.

4 Numerical Results

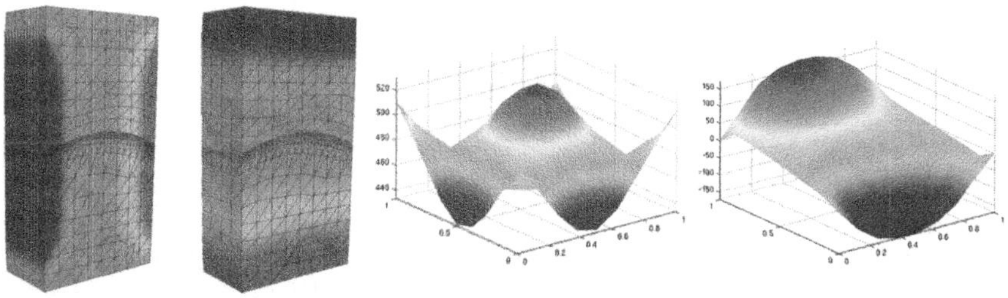

Fig. 2. Cut through deformed body with u_1 and u_3; normal stresses and one tangential stress component at the non-matching warped interfaces (from left to right).

Our numerical studies, not all presented here, show that the coupling by means of the developed discrete mortar transfer operator results in a discretization with optimal convergence for diverse problem classes and performs very well in various geometrical situations. Here, we consider two vector valued examples from computational mechanics assuming linear elastic material behavior. We are not concerned with complex overall geometries but rather focus on the information transfer across complicated interfaces.

The first example virtually reflects the ideal case where two bodies have to be glued at interfaces coinciding in the continuous setting. In case dual multipliers on warped interior interfaces are used and $\mathcal{T}^s$ is considerably coarser than $\mathcal{T}^m$, the authors in [6] observe artificial oscillations of the deformations as well as the stresses. In contrast, even if $h_s/h_m \approx 8/1$, our method yields a smooth solution and does not require any stabilization of the dual multipliers, see Fig. 2. Moreover, one can see that the occurring interface stresses are very well resolved although there are only 81 nodes on the slave side.

As second example we present the numerical solution of a variational inequality arising from a contact problem. Figure 3 shows two separate bodies with bulging interfaces being pressed by non-symmetric Dirichlet boundary conditions at the top and the bottom. Here, for the coupling of the entirely independent hexahedral meshes, which only happens in normal direction, we use standard nodal multipliers and lumping of the matrix $\boldsymbol{D}$. Finally, we note that the computed discrete contact stresses are quite smooth despite the large variations in the local shape of the colliding interfaces.

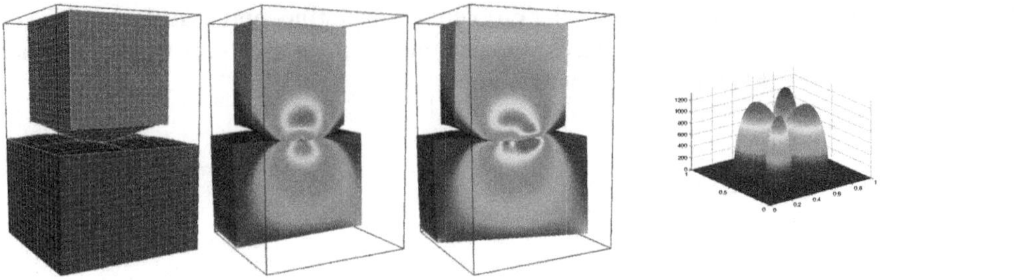

Fig. 3. Initial geometry with bulges (left); different cuts through deformed bodies with von Mises stresses (center); normal stresses at contact boundary (right).

References

[1] Barber, C.M., Dobkin, D.P., Huhdanpaa, H.: The quickhull algorithm for convex hulls. *ACM Trans. Math. Software*, 22(4):469–483, 1996.

[2] Bernardi, C., Maday, Y., Patera, A.T.: A new nonconforming approach to domain decomposition: the mortar element method. In H. Brezis and J. L. Lions, eds., *Nonlinear partial differential equations and their applications*, vol. 299 of *Pitman Res. Notes Math.*, pages 13–51. Harlow: Longman Scientific & Technical, New York, 1994.

[3] Dickopf, T., Krause, R.: Efficient simulation of multi-body contact problems on complex geometries: a flexible decomposition approach using constrained minimization. *Internat. J. Numer. Methods Engrg.*, 2009.

[4] Eck, C., Wohlmuth, B.: Convergence of a contact-Neumann iteration for the solution of two-body contact problems. *Math. Models Methods Appl. Sci.*, 13(8):1103–1118, 2003.

[5] Falletta, S.: The approximate integration in the mortar method constraint. In O. Widlund and D. Keyes, editors, *Domain Decomposition Methods in Science and Engineering XVI*, volume 55 of *Lect. Notes Comput. Sci. Eng.*, pages 555–563. Springer, 2007.

[6] Flemisch, B., Wohlmuth, B.: Stable Lagrange multipliers for quadrilateral meshes of curved interfaces in 3d. *Comput. Methods Appl. Mech. Engrg.*, 196(8):1589–1602, 2007.

[7] Hüeber, S., Wohlmuth, B.: An optimal a priori error estimate for non-linear multibody contact problems. *SIAM J. Numer. Anal.*, 43(1):157–173, 2005.

[8] Krause, R., Sander, O.: Fast solving of contact problems on complicated geometries. In R. Kornhuber et al., eds., *Domain Decomposition Methods in Science and Engineering*, vol. 40 of *Lect. Notes Comput. Sci. Eng.*, pages 495–502. Springer, 2005.

[9] Puso, M.A.: A 3d mortar method for solid mechanics. *Internat. J. Numer. Methods Engrg.*, 59(3):315–336, 2004.

Computational Tool for a Mini-Windmill Study with SOFT

M. Garbey, M. Smaoui, N. De Brye, and C. Picard

Department of Computer Science, University of Houston, Houston, TX 77204 USA
garbey@cs.uh.edu

1 Introduction and Motivation

In this paper, we present a parallel computational framework for the completely automated design of a Vertical Axis Fluid Turbine (VAFT) . Simulation, Optimum design, Fabrication and Testing (SOFT) of the VAFT is integrated into a hardware/software environment that can fit into a small office space.

The components of the four steps design loop are as follows

1. **Simulation**: We use a parallel CFD algorithm to run a direct simulation of the fluid structure interaction problem. We derive from that computation the torque and the average rotation speed for a given friction coefficient on the rotor shaft and an average flow speed. Our objective is to get the most power out of the windmill, consequently the highest rotation speed possible.
2. **Optimization**: We optimize the shape of the blade section with a genetic algorithm and/or a surface response. The evaluation of the objective function (average rotation speed) corresponds to the direct simulation of the Navier Stokes flow interacting with the rotating turbine, until reaching a stationary regime. Because this simulation is compute-intensive, we distribute the evaluation of the objective function for the different shapes (gene or parameter combinations) on a network of computers using an embarrassingly parallel algorithm.
3. **Fabrication**: The optimization procedure results in a supposedly optimum shape in the chosen design space. This shape is sent to a 3-D printer that fabricates the real turbine. This turbine is set up such that it can be easily mounted on a standard base equipped with an electric alternator/generator.
4. **Testing**: The windmill is tested in a mini wind tunnel. The electric output is measured and a video camera can directly monitor the windmill rotation through the transparent wall of the wind tunnel. This information can be analyzed by the computer system and comparison with the simulation is assessed. Figure 1 gives a graphical overview of the SOFT concept.

This four-steps loop can be repeated as many times as needed. Eventually, artificial intelligence tools such as Bayesian networks can be added to close the design

M. Bercovier et al. (eds.), *Domain Decomposition Methods in Science and Engineering XVIII*, Lecture Notes in Computational Science and Engineering,
DOI: 10.1007/978-3-642-02677-5,

loop efficiently. This component would decide when to test other classes of design characterized by the number of blades, the number of stages in the turbine, the use of baffles to channel the flow etc.; see Fig. 2

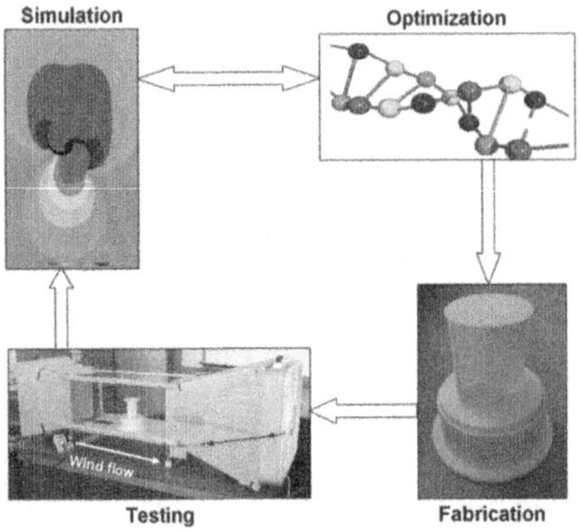

Fig. 1. SOFT concept

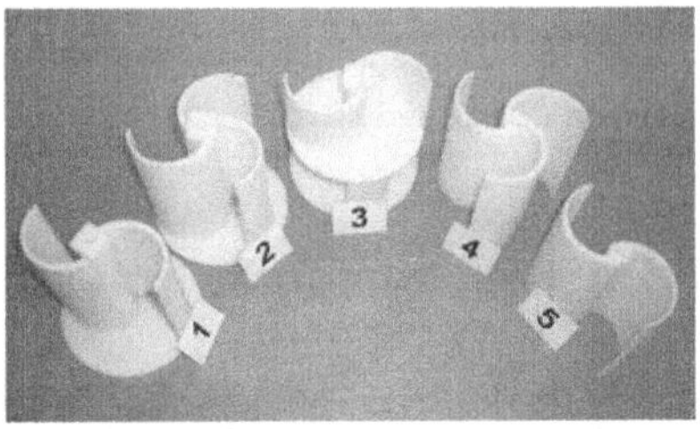

Fig. 2. Collection of VAFT Shapes

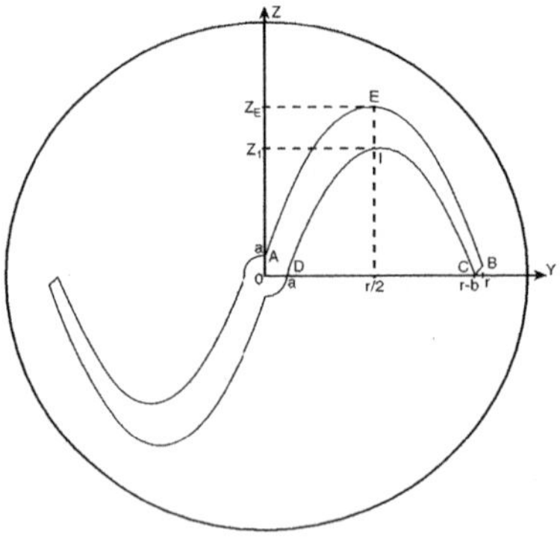

Fig. 3. Design of the turbine

We will concentrate here on two dimensional computation with a simplified two scoop blade that is symmetric with respect to its shaft as in Fig. 3.

We have chosen to optimize the VAFT in low speed flow condition, with Reynolds number in the range (100–2000). We do not need a priori to deal with complex turbulent flow neither stability issues in the fluid structure interaction. One of the possible applications is to power remote sensors with VAFT when other energy sources are more difficult to manage. We are also interested in low Reynolds number flows that are characteristic of micro air vehicle [8].

This project has some obvious pedagogic components that can motivate undergraduate students to do science! However, in reality, a critical step in the process is obviously the CFD method to test the VAFT performance: the numerical simulator should be robust, extremely fast but accurate enough to discriminate bad design from good design. We will discuss an immersed boundary method and domain decomposition solver that we have tentatively developed to satisfy this ambitious program. We first describe the incompressible Navier Stokes Solver.

2 Flow Solver

We use the penalty method introduced by Caltagirone and his co-workers [3] that is simpler to implement than our previous boundary fitted methods [4] and applies naturally to flow in a domain with moving walls [7].

The flow of incompressible fluid in a rectangular domain $\Omega = (0, L_x) \times (0, L_y)$ with prescribed values of the velocity on $\partial\Omega$ obeys the NS equations:

$$\begin{aligned} \partial_t U + (U \cdot \nabla)U + \nabla p - \nu \nabla \cdot (\nabla U) = f \quad & \text{in } \Omega \\ div(U) = 0 \quad & \text{in } \Omega \\ U = \mathbf{g} \quad & \text{on } \partial\Omega. \end{aligned}$$

We denote by $U(x,y,t)$ the velocity with components (u_1, u_2) and by $p(x,y,t)$ the normalized pressure of the fluid. ν is a kinematic viscosity.

With an immersed boundary approach the domain Ω is decomposed into a fluid subdomain Ω_f and a moving rigid body subdomain corresponding to the blade Ω_b. In the L_2 penalty method the right hand side f is a forcing term that contains a mask function Λ_{Ω_b}

$$\begin{aligned} \Lambda_{\Omega_b}(x,y) = 1 \quad & \text{if } (x,y) \in \Omega_b, \\ 0 \quad & \text{elsewhere,} \end{aligned}$$

and is defined as

$$f = -\frac{1}{\eta} \Lambda_{\Omega_b} \{U - U_b(t)\}. \tag{1}$$

U_b is the velocity of the moving blade and η is a small positive parameter that tends to 0.

A formal asymptotic analysis helps us to understand how the penalty method matches the no slip boundary condition on the interface $S_b^f = \bar{\Omega}_f \cap \bar{\Omega}_b$ as $\eta \to 0$. Let us define the following expansion:

$$U = U_0 + \eta\, U_1, \quad p = p_0 + \eta\, p_1.$$

Formally we obtained at leading order,

$$\frac{1}{\eta} \Lambda_{\Omega_b} \{U_0 - U_b(t)\} = 0,$$

that is

$$U_0 = U_b \quad \text{for } (x,y) \in \Omega_b.$$

The leading order terms U_0 and p_0 in the fluid domain Ω_f satisfy the standard set of NS equations:

$$\begin{aligned} \partial_t U_0 + (U_0 \cdot \nabla)U_0 + \nabla p_0 - \nu \nabla \cdot (\nabla U_0) = 0 \quad & \text{in } \Omega_f \\ div(U_0) = 0 \quad & \text{in } \Omega. \end{aligned}$$

At next order we have in Ω_b,

$$\nabla p_0 + U_1 + Q_b = 0, \tag{2}$$

where

$$Q_b = \partial_t U_b + (U_b \cdot \nabla) U_b - \nu \nabla \cdot (\nabla U_b).$$

Further the wall motion U_b must be divergence free which is the case for a rigid body. In conclusion, the flow evolution is dominated by the NS equations in the flow domain, and by the Darcy law with very small permeability inside the rotor.

In this framework, the efficiency of the NS code relies essentially on the design of robust and efficient parallel solvers for linear operators of the following two types

$$-\varepsilon\Delta + \delta u \cdot \nabla + Id \quad \text{and} \quad -\Delta.$$

To be more specific, time stepping uses a multi-step projection scheme. Space discretization is done with a staggered grid. The convection is processed with the method of characteristic. Since the penalty term is linear it is trivial to make that term implicit in time stepping. Finally we use a combination of Aitken-Schwarz as the domain decomposition solver with block LU decomposition per subdomain. We refer to [5] for an extensive report on the performance of that parallel solver on multiple computer architecture and a comparison with other solvers such as multigrid or Krylov methods. We are going now to describe a key aspect that is the Fluid Structure Interaction (FSI) approach we have followed here.

3 FSI

First let us discuss the computation of the torque. The main difficulty with the penalty method is that the flow field is not differentiable at the fluid structure interface. The computation of the drag forces exerted on the blade cannot be done directly with the standard formula

$$F = \int_{\partial\Omega_b} \sigma(U,p) n \, d\gamma,$$

where $\sigma(U,p) = \frac{1}{2}\nu(\nabla U + (\nabla U)^t) - p\,I$.

Using the observation of [2], we can compute this force with an integral on the gradient of pressure *inside* the blade:

$$F = \lim_{\eta \to 0} \int_{\Omega_b} \nabla p \, dx,$$

which ends up with the simple formula using the momentum equation:

$$F = \lim_{\eta \to 0} \frac{-1}{\eta} \int_{\Omega_b} U - U_b \, dx. \tag{3}$$

The computation of the torque is done by summing up the contribution of (3) to the torque, cell-wise and inside the blade. We take into account only the interior cells to avoid the singularity at the wall and leave out from the calculation all the cells that intersect the boundary of the blade $\partial\Omega_b$. The verification on the computational efficiency of this technic has been done with static torque calculation. We checked that when the penalty parameter goes to zero, $\eta \longrightarrow 0$, the numerical error is rapidly dominated by the grid accuracy. As $h \longrightarrow 0$, we observe first order convergence. Finally we did compare our torque computation with Adina's computation. Adina is a commercial finite element code that uses an arbitrary Lagrangian-Eulerian formulation with displacements compatibility and traction equilibrium at the blade interface.

We did some fine mesh calculation with Adina of the static toque, i.e for fixed orientation of the blade, and use that numerical solution as reference. We found that it was easy to maintain a 10% accuracy compare to the reference solution computed with Adina with moderated grid size and Reynolds number of order a few hundreds, provided that the tip of the blade had a thickness of at least 3 to 4 mesh points.

Let us discuss now the FSI algorithm based on this torque calculation. It is classic to apply the second Newton law and advance the rotor accordingly: we alternate then the flow solver and solid rotation. Unfortunately, while the penalty method is very robust, this solution is ill-conditioned, due to the stiffness of the coupling and sensitivity to the noisy calculation of the Torque. As a matter of fact, we can expect small high frequency oscillation in time of the torque calculation with rotating blades as Cartesian cell enter/leaves the domain of computation Ω_b.

"Thinking parallel" leads to a completely different new solution to solve this FSI. Based on extensive FSI simulations with Adina of various blade designs, we have observed that the velocity of the rotor can be represented accurately with few Fourier modes:

$$\frac{\partial \Phi}{\partial t} = \Phi_0 + \Phi_1 \sin(\Theta) + \Phi_2 \cos(\Theta) + \cdots .$$

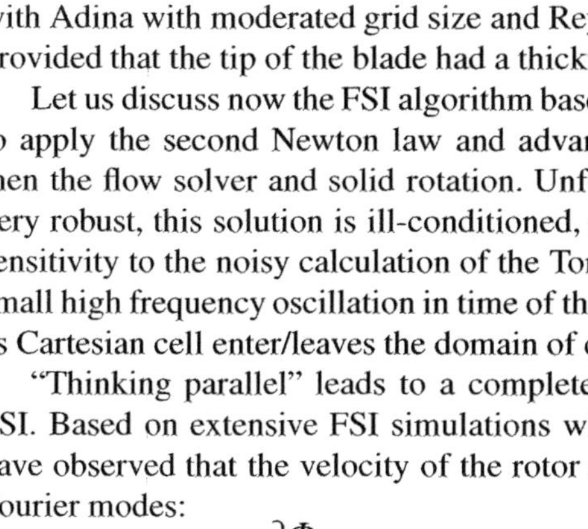

Fig. 4. Comparison of various Blades at different Reynolds number

Figure 4 gives a representative example of such an Adina calculation of the rotating velocity speed of the blade. Since we are interested in comparing design to take decision, it does not take a lot of accuracy to compare blade performances [6].

Our solution is then to apply a forcing speed to the blade and compute the torque exerted on the rotor. We first generate a surface response that approximates the torque with a family of a few coefficient in the Fourier expansion. The idea is second to optimize the periodic rotating speed $\frac{\partial \Phi}{\partial t}$ function on the surface response that satisfies at best the second Newton law:

$$\min_{(\phi_0,\phi_1,\phi_2,\cdots)} \left\| I\frac{\partial^2 \Phi}{\partial t^2} - T \right\|_{(0,P)}. \tag{4}$$

Solving this minimization problem requires a regularization. We can indeed post-process the noisy results of the torque calculation with various angular speed obtained with the penalty method and modest grid size; see Fig. 5. Figure 6 is the result of Fourier filtering on the data of Fig. 5 with a second order filter. This regularization makes the minimization of (4) easy to process and robust with respect to noisy torque calculation. More importantly once the surface response for $T(\phi_0, \phi_1, \phi_2, \ldots)$ is generated, we can solve the optimum design with different load on the wind mill, by changing our objective function (4) only.

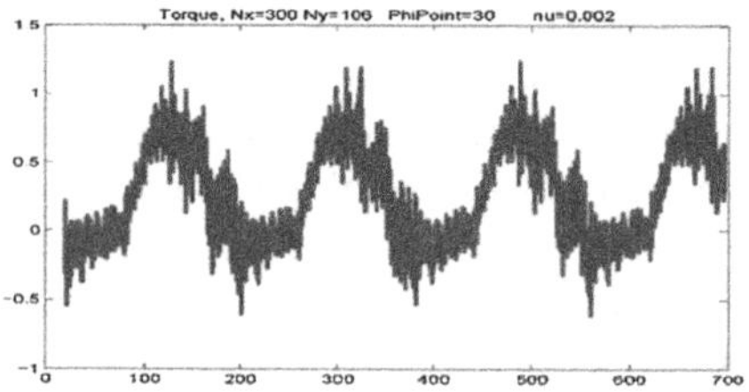

Fig. 5. Non-filtered torque (Reynolds = 250)

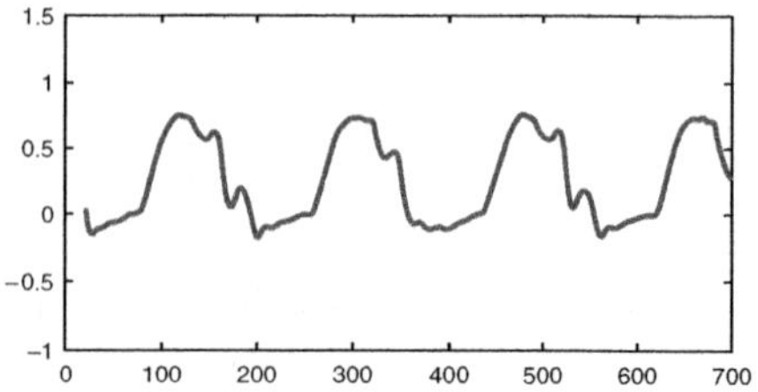

Fig. 6. Filtered torque (Reynolds = 250)

We shall now discuss the potential of this method for parallel processing.

4 Parallel Computing Scenario and Conclusion

In the short history of parallel computing, the tendency has been to solve larger and larger problems to get performance rather than reducing the exectution time for fixed (modest) size problems. The second is needed in optimum design while the first is for grand challenge problems only. Nowadays computers have hundreds of processors, and most standard algorithms with modest grid size problems cannot take advantage of this potential. The Sicortex system for example offers a very cost effective 72 cores parallel system in a standard desktop PC box format, that uses 200 Watts only. This sounds as a good motivation to come up with algorithms for small problem such as the two dimension NS FSI problem considered in this paper and which can take advantage of such a resource.

We observe that sampling the space for low order speed approximation (4) generates $O(100)$ independent tasks with embarrassing parallelism. Comparing design between various blade shapes can be done either by surface response and/or stochastic algorithms such as genetic algorithm or alternatively particle swarm algorithm. This adds a second level of large scale parallelism. We speculate that this approach that relies heavily on the robustness of the (parallel) domain decomposition CFD solver can run with volunteer computing effort such as offered by BOINC [1].

To conclude this paper, we have presented the SOFT concept to design VAFT *automatically* and a domain decomposition algorithm that can be a robust numerical engine for the FSI simulation of the VAFT. We found in our recent experience that this project had a positive impact to motivate our students in science and possibly improve our student enrollment. It is somewhat fascinating to our students to build real turbine with a numerical algorithm. We are currently running simulations to test the limits of our FSI/Immersed Boundary approach with Reynolds numbers much larger than in the present paper. This is, indeed, a very critical issue for the applicability of our method. However we believe that in principle one can reuse any existing NS codes into our FSI and optimization design framework to tackle larger windmill designs, that have to run in the turbulent boundary atmospheric layer where urban VAFTs should operate.

References

[1] Anderson, D.P., Christensen, C., Allen, B.: Designing a runtime system for volunteer computing. In *SC '06. Proceedings of the ACM/IEEE SuperComputing 2006 Conference*, pages 33–43, November 2006.

[2] Angot, P., Bruneau, C.-H., Fabrie, P.: A penalization method to take into account obstacles in incompressible viscous flows. *Numer. Math.*, 81(4):497–520, February 1999.

[3] Arquis, E., Caltagirone, J.P.: Sur les conditions hydrodynamiques au voisinage d'une interface milieu fluide-milieux poreux: Application a la convection naturelle. *C. R. Acad. Sci. Paris Sér. II*, 299:1–4, 1984.

[4] Garbey, M., Vassilevski, Y.V.: A parallel solver for unsteady incompressible 3d navier-stokes equations. *Parallel Comput.*, 27(4):363–389, March 2001.

[5] Hadri, B., Garbey, M.: A fast Navier-Stokes flow simulator tool for image based CFD. *J. Algorithms Comput. Technol.*, 2(4): 527–556, 2008.

[6] Queipo, N.V., Haftka, R.T., Shyy, W., Goel, T., Vaidyanathan, R., Kevin, T.P.: Surrogate-based analysis and optimization. *Progr. Aerospace Sc.*, 41(1):1–28, January 2005.

[7] Schneider, K., Farge, M.: Numerical simulation of the transient flow behaviour in tube bundles using a volume penalization method. *J. Fluids Struct.*, 20(4): 555–566, May 2005.

[8] Shyy, W., Lian, Y., Tang, J., Viieru, D., Liu, H.: *Aerodynamics of Low Reynolds Number Flyers*. Cambridge Aerospace Series. Cambridge University Press, Cambridge, MA, 2007.

On Preconditioners for Generalized Saddle Point Problems with an Indefinite Block

Piotr Krzyżanowski

University of Warsaw, ul. Banacha 2, 02-097 Warszawa, Poland;
piotr.krzyzanowski@mimuw.edu.pl

1 Introduction

In many applications one needs to solve a discrete system of linear equations with a symmetric block matrix

$$\mathcal{M}\begin{pmatrix} u \\ p \end{pmatrix} \equiv \begin{pmatrix} A & B^T \\ B & 0 \end{pmatrix} \begin{pmatrix} u \\ p \end{pmatrix} = \begin{pmatrix} f \\ g, \end{pmatrix} \tag{1}$$

where the block $A = A^T$ is not necessarily positive definite and may even be singular. Such situation occurs, for example, after suitable finite element discretization of the generalized Stokes problem, cf. [5],

$$-\Delta u - \omega u + \nabla p = f, \tag{2}$$

$$\operatorname{div} u = 0, \tag{3}$$

when for large enough ω one cannot preserve the ellipticity of $-\Delta - \omega$. Another example is the time-harmonic Maxwell equation, see [7],

$$\nabla \times \nabla \times u - \omega u + \nabla p = f, \tag{4}$$

$$\operatorname{div} u = 0, \tag{5}$$

where large enough ω again results in an indefinite A. Although the whole system matrix (1) remains invertible when $\omega = 0$, the matrix A which then corresponds to the discrete curl-curl operator, has a large kernel.

In these examples, the discrete problem matrix (1) is ill conditioned with respect to the mesh parameter h. Our aim in this paper is to analyze block preconditioners for such systems, for which the preconditioned conjugate residuals (PCR) method, see [8], converges independently of h.

Block preconditioning allows for an efficient reuse of existing methods of preconditioning problems of simpler structure, such as symmetric positive definite systems. Actually, block diagonal or triangular preconditioners decompose in a natural

M. Bercovier et al. (eds.), *Domain Decomposition Methods in Science and Engineering XVIII*, Lecture Notes in Computational Science and Engineering,
DOI: 10.1007/978-3-642-02677-5,

way the large system (1) into several smaller and simpler problems. Since domain decomposition based preconditioners are very well developed for symmetric and positive definite problems, and high quality software, such as PETSc, see [1] contains implementations of very robust methods, the use of block preconditioners may be a reasonable solution method instead of more involved methods.

We present an analysis of some block preconditioning algorithms within a general framework, essentially assuming only that equation (1) is well posed and that $\mathcal{M}$ is symmetric. Our analysis is valid for inexact block solvers and shows that a successful preconditioner can be based on preconditioners for symmetric positive definite sub-problems. Let us note that probably the first observation that (diagonal) preconditioners based on positive definite blocks are applicable even in the case when A is not necessarily positive definite, was made in [9]. Here we generalize this observation to various kinds of block preconditioners.

2 General Assumptions

Let $\bar{V}, \bar{W}$ be real Hilbert spaces with scalar products denoted by $((\cdot,\cdot))$ and $(\cdot,\cdot)$, respectively. The norms in these spaces, induced by the inner products, will be denoted by $\|\cdot\|$ and $|\cdot|$. We consider a family of finite dimensional subspaces indexed by the parameter $h \in (0,1)$: $V_h \subset \bar{V}, \quad W_h \subset \bar{W}$. If V_h, W_h come from finite element approximations, the dimension of these subspaces increases for decreasing h.

Following [4], let us introduce three continuous bilinear forms: $a : \bar{V} \times \bar{V} \to R$, $b : \bar{V} \times \bar{W} \to R$, $c : \bar{W} \times \bar{W} \to R$. We assume that $a(\cdot,\cdot)$ is symmetric and there exists a constant α, independent of h, such that

$$\exists \alpha > 0 \quad \forall h \in (0,1) \qquad \inf_{v \in V_h^0, v \neq 0} \; \sup_{u \in V_h^0, u \neq 0} \frac{a(u,v)}{\|u\| \, \|v\|} \geq \alpha, \tag{6}$$

where $V_h^0 = \{v \in V_h : \forall q \in W_h \quad b(v,q) = 0\}$. We shall also assume that the finite dimensional spaces V_h and W_h satisfy the uniform LBB condition,

$$\exists \beta > 0 \quad \forall h \in (0,1) \quad \forall p \in W_h \qquad \sup_{v \in V_h, v \neq 0} \frac{b(v,p)}{\|v\|} \geq \beta \, |p|. \tag{7}$$

Remark 1. Condition (6), when related to our motivating problems, generalized Stokes (2)–(3) or time-harmonic Maxwell equations (4)–(5), imposes some conditions on the values of ω, e.g., in the latter case, $\sqrt{\omega}$ has to be distinct from any Maxwell eigenvalue of the discrete problem.

From now on, we drop the subscript h to simplify the notation. In what follows we consider preconditioners for a family of finite dimensional problems:

Problem 1. Find $(u,p) \in V \times W$ such that

$$\mathcal{M} \begin{pmatrix} u \\ p \end{pmatrix} \equiv \begin{pmatrix} A & B^* \\ B & 0 \end{pmatrix} \begin{pmatrix} u \\ p \end{pmatrix} = \begin{pmatrix} F \\ G \end{pmatrix}. \tag{8}$$

The operators in (8) are defined by:

$$A : V \to V, \quad ((Au,v)) = a(u,v) \quad \forall u,v \in V,$$
$$B : V \to W, \quad (Bu,p) = b(u,p) \quad \forall u \in V,\, p \in W,$$

while the right hand side components $F \in V$, $G \in W$ satisfy $((F,v)) \equiv \langle\langle f,v \rangle\rangle$ and $(G,w) \equiv \langle g,w \rangle$, where f,g are given linear continuous functionals on $\bar{V}$, $\bar{W}$, and $\langle\langle \cdot,\cdot \rangle\rangle$, $\langle \cdot,\cdot \rangle$ denote the duality pairing in $\bar{V}$, $\bar{W}$, respectively. B^* denotes the formal adjoint operator to B, i.e. $(Bu,p) = ((u,B^*p))$ for all $u \in V$, $p \in W$. Let us recall the key theorem which we shall use throughout the paper. This is the classical result on the stability of (8):

Lemma 1. [4] *Under the above assumptions, there exits a unique pair* $(u,p) \in V \times W$ *which solves* (8). *Moreover,*

$$\|u\| + |p| \lesssim \|F\| + |G|. \tag{9}$$

Here, and it what follows, $x \lesssim y$ means that there exists a positive constant C, independent of x, y and h, such that $x \le Cy$. Similarly, $x \simeq y$ will denote that both $x \lesssim y$ and $y \lesssim x$ hold.

We introduce two more operators, $A_0 : V \to V$ and $J_0 : W \to W$. We assume that they are self-adjoint, their inverses are easy to apply, and that they define inner products spectrally equivalent to $((\cdot,\cdot))$ and $(\cdot,\cdot)$, respectively:

$$((A_0u,u)) \simeq ((u,u)) \quad \forall u \in V, \tag{10}$$
$$(J_0p,p) \simeq (p,p) \quad \forall p \in W. \tag{11}$$

In other words, we shall always assume that A_0 and J_0 define good preconditioners for the Grammian matrices for the chosen bases in V and W, respectively. For example, the A_0 preconditioner may be constructed using very efficient domain decomposition or multigrid techniques; for J_0, in some cases such as the generalized Stokes problem, one can also apply a very cheap diagonal scaling instead of domain decomposition.

With any X-elliptic, selfadjoint operator G we may associate an energy norm of $x \in X$, $\|f\|_G = ((Gf,f))^{1/2}$. From (10)–(11) it directly follows that the energy norms defined by A_0, J_0 and their inverses are equivalent, with constants independent of h, to the original norms in appropriate spaces:

Lemma 2. *For any* $f \in V$ *and* $g \in W$,

$$\|f\|_{A_0} \simeq \|f\| \simeq \|f\|_{A_0^{-1}}, \tag{12}$$
$$|g|_{J_0} \simeq |g| \simeq |g|_{J_0^{-1}}. \tag{13}$$

Lemma 3. *The norms of* A, B, A_0, J_0, $\mathcal{M}$ *in appropriate spaces are bounded independently of* h,

$$\|A\|_{V \to V}, \quad \|B\|_{V \to W}, \quad \|A_0\|_{V \to V}, \quad \|J_0\|_{W \to W}, \quad \|\mathcal{M}\|_{V \times W \to V \times W} \lesssim 1.$$

Moreover,

$$\|A_0^{-1}\|_{V \to V}, \quad \|J_0^{-1}\|_{W \to W}, \quad \|\mathcal{M}^{-1}\|_{V \times W \to V \times W} \lesssim 1.$$

In the rest of the paper, we shall analyze preconditioners for the Preconditioned Conjugate Residual method (PCR), which is known to be applicable to indefinite symmetric systems, provided the preconditioner is a symmetric, positive definite operator. Other methods may also be applicable, such as QMR, BiCGStab, GMRES, etc. When applied to $\mathcal{M}$ with a preconditioner $\mathcal{P}$, its convergence rate, according to [8], depends on the quantity $\kappa(\mathcal{P}^{-1}\mathcal{M}) = \rho(\mathcal{P}^{-1}\mathcal{M})\rho(\mathcal{M}^{-1}\mathcal{P})$, where ρ denotes the spectral radius of a matrix.

3 Block Diagonal Preconditioner

In the section, we recall a result regarding the block diagonal preconditioner,

$$\mathcal{M}_D = \begin{pmatrix} A_0 & 0 \\ 0 & J_0 \end{pmatrix}.$$

This preconditioner has been thoroughly analyzed for symmetric saddle point problems, assuming either V-ellipticity, see [11], or only V^0-ellipticity of A, see e.g. [9, Sec. 3.2]. These results directly apply to the more general case, when A only satisfies (6). Actually, the only non-trivial property of $\mathcal{M}$ which is required in the proof is the stability result of Lemma 1.

Lemma 4 ([9]). *The preconditioned operator* $\mathcal{P}_D = \mathcal{M}_D^{-1}\mathcal{M}$ *satisfies*

$$\kappa(\mathcal{P}_D) \lesssim 1.$$

4 Block Upper Triangular Preconditioner

Another preconditioner for the operator $\mathcal{M}$ is based on a block upper triangular matrix

$$\mathcal{M}_U = \begin{pmatrix} A_0 & B^* \\ 0 & J_0 \end{pmatrix}. \tag{14}$$

The preconditioned operator $\mathcal{P}_U = \mathcal{M}_U^{-1} M$ is equal to

$$\mathcal{P}_U = \begin{pmatrix} A_0 & \\ & J_0 \end{pmatrix}^{-1} \begin{pmatrix} A - B^* J_0^{-1} B & B^* \\ B & 0 \end{pmatrix}, \tag{15}$$

so the triangular preconditioner acts as the diagonal preconditioner $\mathcal{M}_D$ applied to an *augmented* matrix

$$\tilde{\mathcal{M}}_U = \begin{pmatrix} A - B^* J_0^{-1} B & B^* \\ B & 0 \end{pmatrix}.$$

Remark 2. Usually, block systems (1) are augmented by *adding* a non-negative matrix to A, see e.g. [2] or [6]. Klawonn's preconditioner also results in a positively augmented matrix, cf. [10]. Here, we end up with a negatively augmented matrix, that is, we *subtract* a non-negative definite matrix from A. Numerical results provided in the final section, as well as some theoretical considerations, indicate that this approach improves the overall convergence of the iterative solver.

Due to the decomposition (15), it is still possible to use a PCR method to solve the preconditioned problem. The analysis of the upper triangular preconditioner reduces to the previous case of block diagonal preconditioning.

Lemma 5. *Lemma* 3 *holds for the augmented matrix* $\tilde{\mathcal{M}}_U$.

Applying the estimates from the block-diagonal case and using this lemma, we conclude that

Theorem 1. $\kappa(\mathcal{P}_U) \lesssim 1$.

5 Lower Block Triangular Preconditioner

It is also possible, with some additional assumptions, to analyze, in the same framework, the lower triangular block preconditioner

$$\mathcal{M}_L = \begin{pmatrix} A_0 & 0 \\ B & J_0 \end{pmatrix}. \tag{16}$$

The preconditioned operator $\mathcal{P}_L = \mathcal{M}_L^{-1} M$ then equals

$$\mathcal{P}_L = \begin{pmatrix} A_0 - A & \\ & J_0 \end{pmatrix}^{-1} \begin{pmatrix} A - AA_0^{-1}A & (A_0 - A)A_0^{-1}B^* \\ BA_0^{-1}(A_0 - A) & -BA_0^{-1}B^* \end{pmatrix}. \tag{17}$$

so the upper triangular preconditioner acts as a diagonal preconditioner

$$\mathcal{M}_{DL} = \begin{pmatrix} A_0 - A & \\ & J_0 \end{pmatrix}$$

applied to some symmetric matrix

$$\tilde{\mathcal{M}}_L = \begin{pmatrix} A - AA_0^{-1}A & (A_0 - A)A_0^{-1}B^* \\ BA_0^{-1}(A_0 - A) & -BA_0^{-1}B^* \end{pmatrix}.$$

See [3] for an analysis of the case when A is positive definite. In order to use the PCR framework, which requires the preconditioner to be positive definite, we have to assume some scaling of A_0; see [3].

Theorem 2. *If there exists a constant $m > 0$, independent of h, such that*

$$(((A_0 - A)u, u)) > m((u,u)) \quad \forall u \in V, \tag{18}$$

then

$$\kappa(\mathcal{P}_L) \lesssim 1.$$

6 Numerical Experiments

The numerical experiments were performed using a MATLAB implementation of a Taylor-Hood finite element discretization of the generalized Stokes problem (2)–(3) on a unit square, with homogeneous boundary condition for the velocity. The discretization resulted in a matrix $A = D - \omega M$, where D is the discrete Laplacian and M corresponds to the velocity mass matrix. We conducted two kinds of tests. First, we experimented with $A_0 = D + M$ and $J_0 = M$ (the pressure mass matrix), calling this preconditioner as the "exact" preconditioner. Then, in order to show a more realistic application, we used "inexact" preconditioners with A_0^{-1} defined as the incomplete Cholesky solve of $D+M$, with drop tolerance 10^{-3}.

In both cases, we investigated the convergence rate of the block diagonal, upper triangular and lower triangular preconditioners discussed above, for several values of ω and varying mesh size h. The stopping criterion was the reduction of the residual norm by a factor of 10^6. To provide sufficient scaling for the A_0 block in the lower triangular preconditioner, we have set $A_0 = 2D + M$ in $\mathcal{M}_L$ in the "exact" case. For comparison with the upper triangular solver, we also included a diagonally preconditioned positively augmented system, see Remark 2,

$$\mathcal{P}_{aug} = \mathcal{P}_D^{-1} \cdot \begin{pmatrix} A + B^T J_0^{-1} B & B^T \\ B & 0 \end{pmatrix}.$$

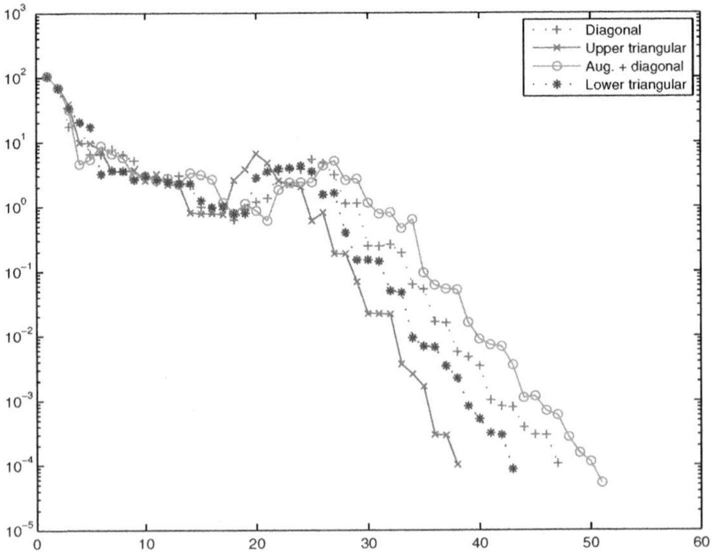

Fig. 1. A comparison of convergence histories of the PCR using four preconditioners for discretized generalized Stokes problem with $\omega = 10$. Exact A_0 solver (see details in the text).

Table 1. Iteration counts for various parameters and preconditioners; left panel: "exact" case, right panel: "inexact" case.

ω	h	$\mathcal{P}_D$	$\mathcal{P}_U$	$\mathcal{P}_{aug}$	$\mathcal{P}_L$	$\mathcal{P}_D$	$\mathcal{P}_U$	$\mathcal{P}_{aug}$	$\mathcal{P}_L$
10	1/4	39	32	39	33	56	44	57	47
10	1/64	45	38	50	43	116	113	130	101
100	1/4	73	66	86	84	103	92	126	109
100	1/64	133	114	150	135	144	128	171	119

As expected, good preconditioners such as those used in the "exact" case, led to iteration counts virtually independent of h. On the other hand, the number of iterations seems to grow sublinearly with the increase of ω, cf. Table 1.

7 Conclusions

Block preconditioning using optimal preconditioners for simple symmetric positive definite operators leads to optimal results with respect to the mesh size h under very mild assumptions on the A block in (1). There is a connection between the (left-) upper triangular preconditioning and the augmented Lagrangian method, with a prospective advantage of the former over the latter.

A general drawback of these preconditioners is that, in some situations, for example, when $A = D - \omega M$ with both D and M positive semidefinite (the case of time-harmonic Maxwell's equations), our bounds also depend on ω. Clearly, in such a case, if ω is very large, one should rather, instead of A, treat M as the dominant term in this block. How to choose the inexact preconditioning blocks in a robust way so that the block preconditioners would perform well independently of ω remains an open problem.

References

[1] Balay, S., Buschelman, K., Gropp, W.D., Kaushik, D., Knepley, M.G., McInnes, L.C., Smith, B.F., Zhang, H.: PETSc Web page, 2001. http://www.mcs.anl.gov/petsc.

[2] Benzi, M., Liu, J.: Block preconditioning for saddle point systems with indefinite $(1,1)$ block. *Internat. J. Comput. Math.*, 84(8):1117–1129, 2007.

[3] Bramble, J.H., Pasciak, J.E.: A preconditioning technique for indefinite systems resulting from mixed approximations of elliptic problems. *Math. Comp.*, 50 (181):1–17, 1988.

[4] Brezzi, F., Fortin, M.: *Mixed and hybrid finite element methods*, vol. 15 of *Springer Series in Computational Mathematics*. Springer, New York, 1991. ISBN 0-387-97582-9.

[5] Cliffe, K.A., Garratt, T.J., Spence, A.: Eigenvalues of block matrices arising from problems in fluid mechanics. *SIAM J. Matrix Anal. Appl.*, 15(4):1310–1318, 1994.

[6] Golub, G.H., Greif, C., Varah, J.M.: An algebraic analysis of block diagonal preconditioner for saddle point systems. *SIAM J. Matrix Anal. Appl.*, 27(3): 779–792 (electronic), 2005.

[7] Greif, C., Schötzau, D.: Preconditioners for the discretized time-harmonic Maxwell equations in mixed form. *Numer. Linear Algebra Appl.*, 14(4):281–297, 2007.

[8] Hackbusch, W.: *Iterative solution of large sparse systems of equations*, vol. 95 of *Applied Mathematical Sciences*. Springer, New York, 1994. Translated and revised from the 1991 German original.

[9] Klawonn, A.: *Preconditioners for Indefinite Problems*. PhD thesis, Universität Münster, Germany, 1996.

[10] Klawonn, A.: Block-triangular preconditioners for saddle point problems with a penalty term. *SIAM J. Sci. Comput.*, 19(1):172–184 (electronic), 1998. Special issue on iterative methods (Copper Mountain, CO, 1996).

[11] Klawonn, A.: An optimal preconditioner for a class of saddle point problems with a penalty term. *SIAM J. Sci. Comput.*, 19(2):540–552 (electronic), 1998.

Lower Bounds for Eigenvalues of Elliptic Operators by Overlapping Domain Decomposition

Yuri A. Kuznetsov

Department of Mathematics, University of Houston, 651 Philip G. Hoffman Hall, Houston, TX 77204-3008, USA kuz@math.uh.edu

Summary. In this paper, we consider a new approach to estimation from below of the lowest eigenvalues of symmetric positive definite elliptic operators. The approach is based on the overlapping domain decomposition procedure and on the replacement of subdomain operators by special low rank perturbed scalar operators. The algorithm is illustrated by applications to model problems with mixed boundary conditions and strongly discontinuous coefficients.

1 Introduction

In this paper, we propose a new approach for estimations from below of the lowest eigenvalues of a symmetric elliptic operator

$$\mathcal{L} = -\sum_{i,j=1}^{d} \frac{\partial}{\partial x_i} a_{ij} \frac{\partial}{\partial x_j} + c. \tag{1}$$

Here, $a = (a_{ij})$ is a symmetric uniformly positive definite $d \times d$ matrix with piecewise smooth bounded entries a_{ij}, $i, j = \overline{1, d}$, c is a nonnegative piecewise smooth bounded function, and $d = 2, 3$. Without loss of generality, we assume that the matrix $a = a(x)$ and the coefficient $c = c(x)$, $x \in \mathbb{R}^d$ are piecewise constant.

We consider the eigenvalue problem

$$\mathcal{L}w = \lambda w \tag{2}$$

in a bounded domain $\Omega \subset \mathbb{R}^d$ with the boundary $\partial\Omega$ subject to the boundary conditions

$$\begin{aligned} w &= 0 \quad \text{on } \Gamma_D, \\ \mathbf{u} \cdot \mathbf{n} - \sigma w &= 0 \quad \text{on } \Gamma_R, \end{aligned} \tag{3}$$

where $\mathbf{u} = -a\nabla w$ is the flux vector-function, $\Gamma_D = \overline{\Gamma}_D$ is a Dirichlet part of $\partial\Omega$, Γ_R is a Robin part of $\partial\Omega$, $\sigma = \sigma(x)$, $x \in \Gamma_R$, is a nonnegative piecewise constant function,

M. Bercovier et al. (eds.), *Domain Decomposition Methods in Science and Engineering XVIII*, Lecture Notes in Computational Science and Engineering,
DOI: 10.1007/978-3-642-02677-5, © Springer-Verlag Berlin Heidelberg 2009

and $\mathbf{n}$ is the outward unit normal to Γ_R. In the case $\sigma \equiv 0$ the Robin boundary condition becomes the Neumann boundary condition. We assume that $\Gamma_D \cup \overline{\Gamma}_R = \partial\Omega$. For the sake of simplicity, we assume that Ω is either a polygon ($d = 2$), or a polyhedron ($d = 3$).

It is well known that all the eigenvalues λ in (2), (3) are real, nonnegative, and the lowest eigenvalue λ_1 is the solution of the minimization problem

$$\lambda_1 = \inf_{v \in V, \|v\|_2 = 1} \Phi(v). \tag{4}$$

Here,

$$\Phi(v) = \int_\Omega \Big[\sum_{i,j=1}^{d} a_{ij} \frac{\partial v}{\partial x_i} \frac{\partial v}{\partial x_j} + cv^2 \Big] \, \mathrm{dx} + \int_{\Gamma_R} \sigma v^2 \, \mathrm{ds}. \tag{5}$$

and

$$V = \{ v : v \in H^1, \ v = 0 \text{ on } \Gamma_D \}, \tag{6}$$

where $H^1 \equiv H^1(\Omega)$ is the Sobolev space and $\|v\|_2$ is the $L_2(\Omega)$ norm of v.

In the only case $\sigma \equiv 0$ on Γ_R, $\overline{\Gamma}_R = \partial\Omega$, and $c \equiv 0$ in Ω (the Neumann problem) the minimal eigenvalue λ_1 is equal to zero. Otherwise, λ_1 is positive. In any case, λ_1 is a single eigenvalue, and an eigenfunction $w_1 = w_1(x)$ does not change its sign in Ω (for instance, $w_1(x) > 0$ for all $x \in \Omega$). For the Neumann problem, we denote the minimal nonzero eigenvalue by λ_2. This eigenvalue may be multiple.

Estimations from above for the minimal (or the minimal nonzero) eigenvalue in (2), (3) can be obtained by the Ritz method, in particular, by using the P_1 finite element method. In many practical applications, estimations from below are much more important. In particular situations (see [1]), the estimates from below can be obtained by using the finite difference discretization of (2), (3). Another method is described in [4]. The latter method is rather limited and very complicated for implementation.

In this paper, we propose a new method to derive estimations from below for the minimal eigenvalue λ_1 (minimal nonzero eigenvalue λ_2) in (2), (3). The method is based on a partitioning of the domain Ω into simpler shaped subdomains. We assume that we are able to derive explicit estimates from below of the lowest eigenvalues of the eigenvalue problems in subdomains. The accuracy of the estimates depends on a partitioning into subdomains. Thus, the method does not always provide sufficiently reliable (or practically acceptable) estimates from below of the lowest eigenvalues.

The paper is organized as follows. In Section 2, we describe the new method on the functional level. The finite element justification of the method is given in Section 3.

2 Description of the Method

Let Ω be partitioned into $m \geq 1$ polygonal, $d = 2$ (polyhedral, $d = 3$), open overlapping subdomains Ω_k, $k = \overline{1,m}$, i.e. $\Omega = \bigcup_{k=1}^m \Omega_k$. We define m quadratic functionals

$$\Phi_k(v) = \int_\Omega \sum_{i,j=1}^{d} a_{ij}^{(k)} \frac{\partial v}{\partial x_i}\frac{\partial v}{\partial x_j}\,\mathrm{d}x + \int_{\Gamma_R} \sigma^{(k)} v^2\,\mathrm{d}s \tag{7}$$

where $a^{(k)} = (a_{ij}^{(k)})$ are symmetric $d \times d$ matrices with piecewise constant entries $a_{ij}^{(k)}$, $i, j = \overline{1, d}$, $\sigma^{(k)}$ are nonnegative piecewise constant functions defined on Γ_R, and $v \in V$. We assume that the matrices $a^{(k)}$ are positive definite in Ω_k and $a^{(k)} = 0$ in $\Omega \setminus \overline{\Omega}_k$, and that the functions $\sigma^{(k)}$ are zero on $\Gamma_R \setminus \partial\Omega_k$, $k = \overline{1, m}$. To this end, the formulae

$$\Phi_k(v) = \int_{\Omega_k} \sum_{i,j=1}^{d} a_{ij}^{(k)} \frac{\partial v}{\partial x_i}\frac{\partial v}{\partial x_j}\,\mathrm{d}x + \int_{\Gamma_{R,k}} \sigma^{(k)} v^2\,\mathrm{d}s, \tag{8}$$

where $\Gamma_{R,k} = \Gamma_R \cap \partial\Omega_k$, gives an alternative definition for $\Phi_k(v)$, $k = \overline{1, m}$.

We assume that

$$a = \sum_{k=1}^{m} a^{(k)} \quad \text{in } \Omega \tag{9}$$

$$\sigma = \sum_{k=1}^{m} \sigma_k \quad \text{on } \Gamma_R. \tag{10}$$

Then, under the latter assumptions we get

$$\Phi(v) = \sum_{k=1}^{m} \Phi_k(v) + \int_\Omega c v^2\,\mathrm{d}x. \tag{11}$$

Let us consider the eigenvalue problems

$$\begin{aligned} \mathcal{L}_k w &= \mu w && \text{in } \Omega_k,\\ w &= 0 && \text{on } \Gamma_D \cap \partial\Omega_k,\\ \mathbf{u}^{(k)} \cdot \mathbf{n}_k &= 0 && \text{on } \partial\Omega_k \setminus \partial\Omega,\\ \mathbf{u}^{(k)} \cdot \mathbf{n}_k - \sigma^{(k)} w &= 0 && \text{on } \Gamma_{R,k}, \end{aligned} \tag{12}$$

where

$$\begin{aligned} \mathcal{L}_k &= -\sum_{i,j=1}^{d} \frac{\partial}{\partial x_i} a_{ij}^{(k)} \frac{\partial}{\partial x_j}\\ \mathbf{u}^{(k)} &= -a^{(k)} \nabla w, \end{aligned} \tag{13}$$

and $\mathbf{n}_k$ is the outward unit normal to $\partial\Omega_k$, $k = \overline{1, m}$.

We partition the subdomains Ω_k, $k = \overline{1, m}$, into two groups. For the subdomains Ω_k in the first group we assume that $\Gamma_D \cap \partial\Omega_k = \emptyset$ and $\sigma^{(k)} = 0$ on $\Gamma_{R,k}$ (or $\Gamma_{R,k} = \emptyset$), $1 \le k \le m$. For the subdomains Ω_k in the first group, the minimal eigenvalue $\mu_0^{(k)}$ in (12) is equal to zero and

$$w_0 = \frac{1}{|\Omega_k|^{1/2}} \quad \text{in } \Omega_k, \tag{14}$$

where $|\Omega_k|$ is the area of Ω_k, $d = 2$ (volume of Ω_k, $d = 3$), is the corresponding L_2-normalized positive eigenfunction. We denote the minimal (lowest) nonzero eigenvalue in (12) by $\mu_1^{(k)}$, $1 \le k \le m$.

All other subdomains Ω_k, $1 \le k \le m$, we put into the second group. For a subdomain Ω_k in the second group the minimal eigenvalue in (12) is positive. We denote this eigenvalue also by $\mu_1^{(k)}$, $1 \le k \le m$.

It is obvious (see [3]) that for any subdomain Ω_k in the first group the inequality

$$\Phi_k(v) \ge \mu_1^{(k)} (P_k v,\, v) \equiv \mu_1^{(k)} \int_\Omega (P_k v)\, v \,\mathrm{dx} \tag{15}$$

holds for any $v \in V$ where the operator P_k is defined by

$$(P_k v)(x) = \begin{cases} 0, & x \in \Omega \setminus \overline{\Omega}_k, \\ v(x) - \frac{1}{|\Omega_k|} \int_{\Omega_k} v(x')\,\mathrm{dx}', & x \in \Omega_k. \end{cases} \tag{16}$$

For the subdomains Ω_k in the second group the inequality (15) also holds with the operator P_k defined by

$$(P_k v)(x) = \begin{cases} 0, & x \in \Omega \setminus \overline{\Omega}_k, \\ v(x), & x \in \Omega_k. \end{cases} \tag{17}$$

In both cases, the operator P_k is an orthogonal L_2-projector, i.e. $P_k = P_k^*$ and $P_k^2 = P_k$.

Let us assume that we have a set of positive numbers μ_k which estimate from below the eigenvalues $\mu_1^{(k)}$ in (15), i.e. $\mu_1^{(k)} \ge \mu_k > 0$, $k = \overline{1,\, m}$, and define the operator

$$P = \sum_{k=1}^{m} \mu_k P_k. \tag{18}$$

By the definition,

$$\Phi(v) \ge (Pv,\, v) + (cv,\, v) \quad \text{for all } v \in V. \tag{19}$$

Thus, in the case of a positive definite operator $\mathfrak{L}$ we obtain

$$\lambda_1 = \min_{v \in V, ||v||_2 = 1} \Phi(v) \ge \nu_1 = \min_{v \in L_2, ||v||_2 = 1} [(Pv,\, v) + (cv,\, v)] \tag{20}$$

where ν_1 is the minimal eigenvalue of the eigenvalue problem

$$Pv + cv = \nu v \quad \text{in } \Omega. \tag{21}$$

In the case of the Neumann problem, the minimal nonzero eigenvalue λ_2 in (2), (3) is estimated from below by the minimal nonzero eigenvalue ν_2 in (21), i.e.

$$\lambda_2 \geq \nu_2 = \min_{v \in L_2, \|v\|_2=1, (v,1)=0} (Pv, v). \tag{22}$$

The set $\widehat{\Gamma} = \bigcup_{k=1}^{m} \partial\Omega_k$ partitions Ω into $\hat{n}$ polygonal, $d = 2$ (polyhedral, $d = 3$), subdomains $\widehat{G}_s$, $s = \overline{1, \hat{n}}$. We impose an additional partitioning of Ω into n, $n \geq \hat{n}$, nonoverlapping subdomains G_s with boundaries ∂G_s, $s = \overline{1, n}$, such that the set $\widehat{\Gamma}$ belongs to the set $\Gamma = \bigcup_{s=1}^{n} \partial G_s$. We assume that each of the subdomains G_s, $s = \overline{1, n}$, is simply connected and does not coincide with any of the subdomains Ω_k, $k = \overline{1, m}$. We also assume that the coefficient $c = c(x)$ is constant in each of the subdomains G_s, $s = \overline{1, n}$.

Define the set of orthogonal projectors Q_k by

$$(Q_k v)(x) = \begin{cases} 0, & x \in \Omega \setminus \overline{G}_k, \\ \frac{1}{|G_k|} \int_{G_k} v(x')\,\mathrm{d}x', & x \in G_k, \end{cases} \tag{23}$$

where $v \in L_2(\Omega)$. Then, the mean values v_s in G_s of a function $v \in L_2(\Omega)$ are defined by

$$v_s = Q_s v, \quad s = \overline{1, n}. \tag{24}$$

Assume that a subdomain Ω_k, $1 \leq k \leq m$, belongs to the first group and

$$\overline{\Omega}_k = \bigcup_{s=1}^{t} \overline{G}_s. \tag{25}$$

Then, it is obvious that

$$P_k v = v - \frac{1}{|\Omega_k|} \sum_{s=1}^{t} |G_s| v_s, \tag{26}$$

where $v_s = Q_s v$, $s = \overline{1, t}$.

Let ν be an eigenvalue in (21) and W be the set of all the eigenfunctions corresponding to this eigenvalue. A simple analysis of equations (21) in subdomains G_s, $s = \overline{1, n}$, shows that W always contains a function which is a constant in each of the subdomains G_s, $s = \overline{1, n}$.

It follows that in (20) and (21) we can replace the space L_2 by the space V_h of functions which are constant in each of the subdomains G_s, $s = \overline{1, n}$, i.e. the definitions of ν_1 in (20) and ν_2 in (22) can be replaced by

$$\nu_1 = \min_{v \in V_h, \ \|v\|_2=1} [(Pv, v) + (cv, v)], \tag{27}$$

$$\nu_2 = \min_{v \in V_h, \ \|v\|_2=1, \ (v, 1)=0} (Pv, v), \tag{28}$$

respectively.

The variational problems (27) and (28) result in the algebraic eigenvalue problems

$$K\bar{w} = \nu M\bar{w}, \quad \bar{w} \in \mathbb{R}^n, \tag{29}$$

with the diagonal $n \times n$ matrix

$$M = \text{diag}\{|G_1|,\ \dots\ ,\ |G_n|\}\ .$$

The matrix K for problem (27) is symmetric and positive definite. The matrix K for problem (28) is symmetric and positive semidefinite with the explicitly known one-dimensional null-space.

Remark 1. The replacement of the space V by the space $L_2(\Omega)$ in (20) and (22) can be justified by using the convergence results for the P_1 finite element method for eigenvalue problem (4)–(6) on quasiuniform regular shaped triangular mesh/tetrahedral meshes. To prove the latter statement, we have to apply the proposed method to the P_1 discretization of (2)–(3) on the meshes which are conforming with respect to the partitioning of Ω into subdomains G_s, $s = \overline{1,\ n}$.

Remark 2. The requirement $\Omega = \bigcup_{k=1}^{m} \Omega_k$ in the beginning of Section 2 can be replaced by the following weaker requirement. Namely, we may require that each two points in Ω should be connected by a curve γ in $\bigcup_{k=1}^{m} \Omega_k$. For instance, the partitioning of the unit square $\Omega = (0;\ 1) \times (0;\ 1)$ into rectangles $\Omega_1 = (0;\ 0.5) \times (0;\ 1)$, $\Omega_2 = (0.5;\ 1) \times (0;\ 1)$, and $\Omega_3 = (0;\ 1) \times (0;\ 0.5)$ is admissible (see Example 2 in the next section).

3 Two Simple Examples

Example 1. Let Ω be the unit square and ω be a simply connected subdomain in Ω. We denote by δ the area of ω and assume that $\mathcal{L} = -\triangle + c$ where $\triangle$ denotes the Laplace operator, $\partial\Omega = \overline{\Gamma}_N$ and the coefficient c in (1) equals to a positive constant c_ω in ω and zero in $\Omega \setminus \overline{\omega}$. We choose $m = 1$, i.e. $\Omega_1 = \Omega$, and partition Ω into subdomains $G_1 = \omega$ and $G_2 = \Omega \setminus \overline{\omega}$, i.e. $n = 2$. Applying the algorithm described in the previous section with $\mu^{(1)} = \pi^2$ we get $K = M\widehat{K}$ where

$$\widehat{K} = \begin{pmatrix} \pi^2(1-\delta) + c_\omega & -1+\delta \\ -\delta & \delta \end{pmatrix} \tag{30}$$

and $M = \text{diag}\{\delta;\ 1-\delta\}$. Computing the minimal eigenvalue ν_1 of (20), we get the estimate

$$\lambda_1 \geq \nu_1 > \frac{c_\omega \delta}{2(\pi^2 + c_\omega)}\ . \tag{31}$$

Example 2. Let Ω be the unit square partitioned into three subdomains $\Omega_1 = (0;\ 1) \times (0;\ \delta)$, $\Omega_2 = (\delta;\ 1) \times (0;\ 1)$, and $\Omega_3 = (0;\ 1) \times (\delta;\ 1)$ as shown in Fig. 1 where $\delta \in (0;\ 1)$. We assume that $\mathcal{L} = -\triangle$ and $\Gamma_D = \{(x_1, x_2)\ :\ x_1 = 0,\ x_2 \in (0;\ \delta)\}$. In Figure 1, we show the partitioning of Ω into rectangles G_i, $i = \overline{1,\ 4}$.

We define the operators $\mathcal{L}_k$ by setting $a^{(k)} = a_k I_2$, $k = 1,2,3$. Here, I_2 denotes the identity 2×2 matrix and the functions a_k, $k = 1,2,3$, are defined as follows:

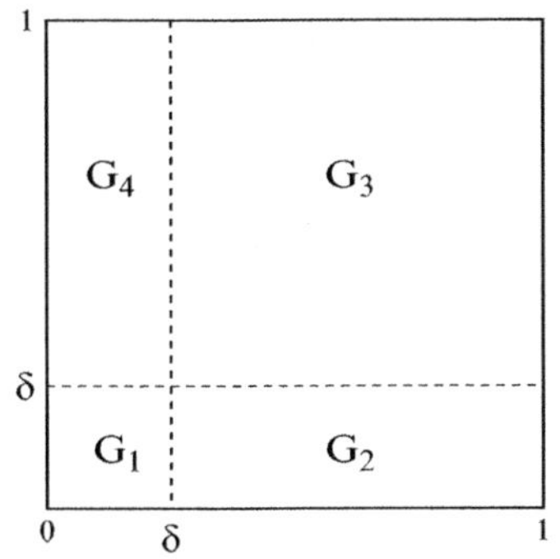

Fig. 1. Partitioning of Ω into rectangles G_i, $i = \overline{1,\ 4}$

$$a_1 = \begin{cases} 1 & \text{in } G_1, \\ 0.5 & \text{in } G_2, \end{cases}$$
$$a_2 = 0.5 \quad \text{in } G_2 \bigcup G_3, \tag{32}$$
$$a_3 = \begin{cases} 1 & \text{in } G_3, \\ 0.5 & \text{in } G_4. \end{cases}$$

Applying the algorithm described in the previous section with $\mu^{(1)} = \mu^{(2)} = \mu^{(3)} = \pi^2/2$, we get $K = M\widehat{K}$ where

$$\widehat{K} = \frac{\pi^2}{2} \begin{pmatrix} 0.25 & 0 & 0 & 0 \\ 0 & 2-\delta & -1+\delta & 0 \\ 0 & -\delta & 2\delta & -\delta \\ 0 & 0 & -1+\delta & 1-\delta \end{pmatrix}. \tag{33}$$

By using the straightforward calculations, we derive the estimate

$$\lambda_1 \geq \nu_1 \geq \|\widehat{K}\|_\infty^{-1} = \frac{\pi^2}{2} \cdot \frac{\delta(1-\delta)(2-\delta)}{(1+\delta)(3-\delta)}. \tag{34}$$

Thus, in the case $\delta \ll 1$ we get the asymptotic estimate

$$\lambda_1 \geq \frac{\pi^2}{3}\delta. \tag{35}$$

Acknowledgement. The author is grateful to O. Boiarkine for preparation of this paper and to the referee for the useful remarks.

References

[1] Forsythe, G.E., Wasow, W.R.: *Finite-Difference Methods for Partial Differential Equations.* Applied Mathematics Series. Wiley, New York, 1960.

[2] Gould, S.H.: *Variational methods for eigenvalue problems. An introduction to the Weinstein method of intermediate problems.* Oxford University Press, London, 1966.
[3] Kuznetsov, Y.A.: Two-level preconditioners with projectors for unstructured grids. *Russian J. Numer. Anal. Math. Modelling*, 15(3-4):247–255, 2000.
[4] Weinstein, A., Stenger, W.: *Methods of intermediate problems for eigenvalues. Theory and ramifications*, vol. 89 of *Mathematics in Science and Engineering*. Academic, New York–London, 1972.

From the Boundary Element Domain Decomposition Methods to Local Trefftz Finite Element Methods on Polyhedral Meshes

Dylan Copeland[1], Ulrich Langer[2], and David Pusch[3]

[1] Institute of Computational Mathematics, Johannes Kepler University Linz, dylan.copeland@jku.at

[2] Institute of Computational Mathematics, Johannes Kepler University Linz, ulanger@numa.uni-linz.ac.at

[3] RICAM, Austrian Academy of Sciences, david.pusch@oeaw.ac.at

Summary. We derive and analyze new boundary element (BE) based finite element discretizations of potential-type, Helmholtz and Maxwell equations on arbitrary polygonal and polyhedral meshes. The starting point of this discretization technique is the symmetric BE Domain Decomposition Method (DDM), where the subdomains are the finite elements. This can be interpreted as a local Trefftz method that uses PDE-harmonic basis functions. This discretization technique leads to large-scale sparse linear systems of algebraic equations which can efficiently be solved by Algebraic Multigrid (AMG) methods or AMG preconditioned conjugate gradient methods in the case of the potential equation and by Krylov subspace iterative methods in general.

1 Introduction

We introduce new finite element methods based on the symmetric boundary element domain decomposition method presented in [5], which can be applied with general polygonal or polyhedral meshes. That is, each element of the mesh may be any polygon or polyhedron, since we treat the elements as subdomains. There are many important practical applications where one wants to discretize PDEs on such kinds of meshes without further decomposition of the polyhedra, see e.g. [6] and [1]. Boundary integral operators are utilized to obtain a method which solves for traces of the solution on the element boundaries, from which the solution may be obtained via a representation formula. Simple, low-order boundary element spaces are used to approximate traces on the element surfaces, yielding a finite element method with PDE-harmonic basis functions.

Since boundary integral operators are used only locally, piecewise constant coefficients are admissible, and the coupling of boundary element functions is local. Consequently, sparse linear systems are obtained, which can be solved by Krylov iterative methods. For the potential equation, the resulting system is symmetric and

M. Bercovier et al. (eds.), *Domain Decomposition Methods in Science and Engineering XVIII*, Lecture Notes in Computational Science and Engineering,
DOI: 10.1007/978-3-642-02677-5,

positive definite, and algebraic multigrid (see [9]) is a very effective preconditioner in the conjugate gradient solver.

2 The Potential Equation

Let $\Omega \subset \mathbb{R}^d$ be a bounded domain with a polygonal ($d = 2$) or polyhedral ($d = 3$) Lipschitz boundary $\Gamma = \partial\Omega$, and let $d \in \{2,3\}$ be the dimension of the computational domain Ω. In this section, we assume for simplicity that $d = 2$. As a model problem, we consider the Dirichlet boundary value problem (BVP) for the potential equation

$$-\mathrm{div}(a(x)\nabla u(x)) = f(x) \quad \text{for } x \in \Omega, \quad u(x) = g(x) \quad \text{for } x \in \Gamma. \tag{1}$$

We assume that the coefficient a is piecewise constant, $f \in L_2(\Omega)$, and $g \in H^{1/2}(\Gamma)$. Further, we suppose that there is a non–overlapping decomposition of our domain Ω into e_h shape-regular polygonal elements Ω_i such that

$$\overline{\Omega} = \bigcup_{i=1}^{e_h} \overline{\Omega}_i, \quad \Omega_i \cap \Omega_j = \emptyset \quad \text{for } i \neq j, \quad \Gamma_i = \partial\Omega_i, \quad \overline{\Gamma}_{ij} = \overline{\Gamma}_i \cap \overline{\Gamma}_j \tag{2}$$

and that $a(x) = a_i > 0$ for $x \in \Omega_i$, $i = 1,\ldots,e_h$. The domain Ω is assumed to be scaled in such a way that $\mathrm{diam}(\Omega_i) = O(h) \leq h_0 < 1/2$ for all $i = 1,\ldots,e_h$. Under the assumptions made above, there obviously exists a unique weak solution $u \in H^1(\Omega)$ of the BVP (1).

Using the local Dirichlet-to-Neumann map

$$a_i \partial u/\partial \nu_i = a_i S_i u_{|\Gamma_i} - N_i f \quad \text{on } \Gamma_i, \tag{3}$$

we observe that the variational formulation of (1) is equivalent to the associated variational formulation on the skeleton $\Gamma_S = \Gamma_{S,h} = \cup_{i=1}^{e_h} \Gamma_i$ (see, e.g., [7]): find $u \in H^{1/2}(\Gamma_S)$ with $u = g$ on Γ such that

$$\sum_{i=1}^{e_h} \int_{\Gamma_i} a_i (S_i u_i)(x)\, v_i(x) ds_x = \sum_{i=1}^{e_h} \int_{\Gamma_i} (N_i f(x)) v_i(x) ds_x \tag{4}$$

for all $v \in H_0^{1/2}(\Gamma_S)$, where $u_i = u_{|\Gamma_i}$ and $v_i = v_{|\Gamma_i}$ denote the traces of u and v on Γ_i, respectively. The Steklov–Poincaré operator S_i and the Newton potential operator N_i have different representations (see again [7]). Here we are using the symmetric representation

$$S_i = D_i + (\frac{1}{2}I + K_i')V_i^{-1}(\frac{1}{2}I + K_i) : H^{1/2}(\Gamma_i) \to H^{-1/2}(\Gamma_i) \tag{5}$$

of the local Steklov–Poincaré operator S_i via the local single layer potential integral operator $V_i : H^{-1/2}(\Gamma_i) \to H^{1/2}(\Gamma_i)$, the local double layer potential operator

$K_i : H^{1/2}(\Gamma_i) \to H^{1/2}(\Gamma_i)$, its adjoint $K_i' : H^{-1/2}(\Gamma_i) \to H^{-1/2}(\Gamma_i)$, and the local hypersingular boundary integral operator $D_i : H^{1/2}(\Gamma_i) \to H^{-1/2}(\Gamma_i)$, see, e.g., [11] for the definition and properties of these boundary integral operators. The operator N_i is defined by the equation

$$N_i = V_i^{-1}\widetilde{N}_{i,0} : \widetilde{H}^{-1}(\Omega_i) \to H^{-1/2}(\Gamma_i), \tag{6}$$

where the Newton potential operator $\widetilde{N}_{i,0}$ is given by the relation

$$(\widetilde{N}_{i,0}f)(x) = \int_{\Omega_i} U^*(x-y)f(y)dy, \ \ x \in \Gamma_i. \tag{7}$$

Here $U^*(x) = -(1/2\pi)\log|x|$ and $U^*(x) = 1/(4\pi|x|)$ denotes the fundamental solution of the negative Laplace operator $-\Delta$ for $d = 2$ and $d = 3$, respectively.

For simplicity (higher-order versions can be constructed in the same way), we use continuous piecewise linear boundary element functions for approximating the potential u on the skeleton Γ_S and piecewise constant boundary element functions for approximating the normal derivatives $t_i = \partial u/\partial \nu_i$ on the boundary Γ_i of the polygonal element Ω_i. This yields the element stiffness matrices

$$\mathbf{S}_{i,h} = a_i\mathbf{D}_{i,h} + a_i\big(0.5\,\mathbf{I}_{i,h}^\top + \mathbf{K}_{i,h}^\top\big)\big(\mathbf{V}_{i,h}\big)^{-1}\big(0.5\,\mathbf{I}_{i,h} + \mathbf{K}_{i,h}\big) \tag{8}$$

and the element load vectors

$$\mathbf{f}_{i,h} = \mathbf{I}_{i,h}^\top\big(\mathbf{V}_{i,h}\big)^{-1}\mathbf{f}_{i,h}^N, \tag{9}$$

where the matrices $\mathbf{V}_{i,h}$, $\mathbf{K}_{i,h}$, $\mathbf{D}_{i,h}$ and $\mathbf{I}_{i,h}$ arise from the BE Galerkin approximation to the single layer potential operator V_i, double layer potential operator K_i, hypersingular integral operator D_i and the identity operator I_i living on Γ_i, respectively. $\mathbf{I}_{i,h}$ is nothing but the mass matrix. The vector $\mathbf{f}_{i,h}^N$ is defined by the Newton potential identity

$$(\mathbf{f}_{i,h}^N, \mathbf{t}_{i,h}) = \int_{\Gamma_i}\int_{\Omega_i} U^*(x-y)f(y)dy\, t_{h,i}(x)ds_x \tag{10}$$

for all vectors $\mathbf{t}_{i,h}$ corresponding to the piecewise constant functions $t_{h,i}$ on Γ_i. Now, we obtain the BE-based FE system

$$\mathbf{S}_h\mathbf{u}_h = \mathbf{f}_h \tag{11}$$

by assembling the stiffness matrix $\mathbf{S}_h$ and the load vector $\mathbf{f}_h$ from the element stiffness matrices (8) and the element load vectors (9), respectively, and by incorporating the Dirichlet boundary condition as usual.

The solution of (11) provides an approximation to the Dirichlet trace of the solution to (1) on the boundary $\partial\Omega_i$ of all elements Ω_i, $i = 1,\ldots,e_h$. Applying the Dirichlet-to-Neumann map locally (i.e. element-wise), we may obtain an approximate solution $\tilde{u}_h$to u in each element Ω_i via the representation formula (see, e.g., [7] or [11]).

Following [5] we immediately obtain the discretization error estimate

$$\|u - u_h\|_h \leq c(u)h^{3/2} \tag{12}$$

in the mesh-dependent norm $\|v\|_h^2 := \sum_{i=1}^{e_h} \|v|_{\Gamma_i}\|_{H^{1/2}(\Gamma_i)}^2$ for a sufficiently (piecewise) smooth solution u, where u_h is the continuous piecewise linear function on the skeleton $\Gamma_{S,h}$ corresponding to the Dirichlet nodal values and to the nodal values from the solution vector $\mathbf{u}_h$ of (11). The discretization error estimate (12) yields the usual $O(h)$ estimate of the discretization error $u - \tilde{u}_h$ in the $H^1(\Omega)$-norm.

In our first numerical experiments we solve the Laplace equation in $\Omega = (0,1) \times (0,1)$ with prescribed Dirichlet conditions on the boundary $\Gamma = \partial\Omega$. The Dirichlet datum g is given as the trace of the function $g(x) = \log\|x - x^*\|$ on Γ, where the singularity $x^* = (1.1, 1.1)^\top$ is located outside the computational domain Ω.

Figure 1 shows a close-up view of a polygonal mesh that was generated with the help of a software tool from the group of Olaf Steinbach at the TU Graz. Table 1

Fig. 1. Close-up view of a polygonal mesh

provides numerical results for 3 meshes (with N_h nodes) which were separately generated by the software tool mentioned above. The coarsest, the intermediate, and the finest mesh contain polygonal elements with a maximum of 13, 7 and 8 nodes, respectively. The systems of algebraic equations were solved by the Preconditioned Conjugate Gradient (PCG) method. The preconditioner is defined by a standard Algebraic MultiGrid (AMG) method implemented in the AMG package PEBBLES developed by [9]. More precisely, the AMG preconditioning step consists of one symmetric V-cycle with one pre-smoothing step and one post-smoothing step. The AMG level denotes the number of levels used in the algebraic multigrid process. The auxiliary coarse grid matrices are constructed by Galerkin projection. We observe that the times for constructing the stiffness matrices and for setting up the AMG also depend on the number of nodes of the polygons. The termination condition for the PCG iterations is defined as the reduction of the initial error by the factor $\varepsilon_{it} = 10^{-12}$

with respect to the $S_h C_h^{-1} S_h$-energy norm. We remark that the $S_h C_h^{-1} S_h$-energy norm is close to the S_h-energy norm if C_h is a good preconditioner for S_h. The nearly constant iteration numbers demonstrate the excellent preconditioning properties of the AMG preconditioner. The last two columns provide discretization errors in the $L_2(\Gamma_S)$- and $L_2(\Omega)$-norms. Note that the $L_2(\Gamma_S)$-norm is a mesh-dependent norm.

	AMG			PCG			
N_h	Level	S_h	Setup	Cycle	Iter.	$\|\|u-u_h\|\|_{0,\Gamma_S}$	$\|\|u-u_h\|\|_{0,\Omega}$
44249	4	4.3	17.2	0.35	15	1.29 E-4	4.88 E-6
71735	4	4.3	9.1	0.42	17	1.42 E-4	3.85 E-6
247250	5	17.0	82.0	1.80	17	6.73 E-5	1.51 E-6

Table 1. Numerical results for the polygonal mesh (CPU time in seconds).

3 The Helmholtz Equation

Let $\Omega \subset \mathbb{R}^3$ be a bounded domain with a polyhedral Lipschitz boundary $\Gamma = \partial\Omega$. As a model problem, we consider the interior Dirichlet BVP for the Helmholtz equation

$$-\Delta u(x) - \kappa^2 u(x) = 0 \quad \text{for } x \in \Omega, \quad u(x) = g(x) \quad \text{for } x \in \Gamma. \tag{13}$$

We assume that the wavenumber $\kappa > 0$ is piecewise constant and not an interior eigenvalue, and $g \in H^{1/2}(\Gamma)$. The case of a non-zero source function can be treated via the Newton potential operator as in the previous section, but we omit this for simplicity.

Given a domain decomposition satisfying (2), the BE-based FE method for the Helmholtz equation is formally identical to the method presented in the previous section for the potential equation (with $a = 1$ and $f = 0$). That is, the variational method is simply equation (4), with the Steklov–Poincaré operator S_i still given by (5) but with different operators D_i, K_i, and V_i. The appropriate boundary integral operators are given, e.g., in [8], [10], or [11], and the representation formula holds with the Helmholtz fundamental solution $U^*(x) = e^{i\kappa|x|}/(4\pi|x|)$.

4 The Maxwell Equations

Under the same assumptions on Ω and κ, we consider the interior Dirichlet BVP for the time-harmonic Maxwell equation

$$\mathbf{curl}\,\mathbf{curl}\,\mathbf{u} - \kappa^2 \mathbf{u} = \mathbf{0} \qquad \text{in } \Omega, \quad \gamma_t \mathbf{u} := \mathbf{u} \times \mathbf{n} = \mathbf{g} \quad \text{on } \Gamma, \tag{14}$$

where $\mathbf{n}$ is the outward unit normal vector. Developing a method of the form (4) for (14) involves quite technical trace spaces and boundary integral operators, so we only outline the main results here.

[2] defined the operator div_Γ and the appropriate function space $\mathbf{X} := \mathbf{H}_{\|}^{-1/2}(\mathrm{div}_\Gamma, \Gamma)$ for the tangential trace γ_t on Lipschitz polyhedral domains, for which $\gamma_\tau : \mathbf{H}(\mathbf{curl}, \Omega) \to \mathbf{X}$ is linear and continuous. In [3], potential operators $\boldsymbol{\Psi}_E, \boldsymbol{\Psi}_M : \mathbf{X} \to \mathbf{H}(\mathbf{curl}\,\mathbf{curl}, \Omega)$ are defined such that the representation formula

$$\mathbf{u} = \boldsymbol{\Psi}_M(\gamma_t \mathbf{u}) + \boldsymbol{\Psi}_E(\gamma_N \mathbf{u}) \tag{15}$$

holds, where $\gamma_N \mathbf{u} := \kappa^{-1} \gamma_t \mathbf{curl}\,\mathbf{u}$ is the Neumann trace.

Defining the boundary integral operators $C, M : \mathbf{X} \to \mathbf{X}$ by

$$\begin{aligned} C &:= \{\gamma_t\}_\Gamma \circ \boldsymbol{\Psi}_E = \{\gamma_N\}_\Gamma \circ \boldsymbol{\Psi}_M, \\ M &:= \{\gamma_t\}_\Gamma \circ \boldsymbol{\Psi}_M = \{\gamma_N\}_\Gamma \circ \boldsymbol{\Psi}_E, \end{aligned}$$

where $\{\}_\Gamma$ denotes the average across Γ, the Dirichlet-to-Neumann map can be expressed as

$$S : \mathbf{X} \to \mathbf{X}, \qquad S := C + (\frac{1}{2}I + M)C^{-1}(\frac{1}{2}I - M).$$

The inverse $C^{-1} : \mathbf{X} \to \mathbf{X}$ is given by [3, Corollary 5.5], and this representation of S is symmetric with respect to the bilinear form $< \mathbf{v}, \mathbf{w} >_{\tau,\Gamma} := \int_\Gamma (\mathbf{w} \times \mathbf{n}) \cdot \mathbf{v}$.

Since the trace operator γ_t is oriented with respect to the normal vector, in order to define a trace operator on the mesh skeleton we arbitrarily choose a global normal vector field $\mathbf{n}_S$ on the skeleton Γ_S. Then $\gamma_t^S := (\mathbf{n}_S \cdot \mathbf{n}_i)\gamma_{t,i}$, on each Γ_i^h, uniquely defines a tangential trace on Γ_S^h. Now the space $\gamma_t^S(\mathbf{H}(\mathbf{curl}, \Omega))$ is denoted $\mathbf{X}_S := \mathbf{H}_{\|}^{-1/2}(\mathrm{div}_\Gamma, \Gamma)$, with the mesh-dependent norm $\|\mathbf{v}\|^2_{\mathbf{X}_s} := \sum_{i=1}^{e_h} \|(\mathbf{n}_S \cdot \mathbf{n}_i)\mathbf{v}|_{\Gamma_i^h}\|^2_{\mathbf{X}_i}$.

The Maxwell skeleton variational formulation is to find $\mathbf{v} \in \mathbf{X}_S$ satisfying $\mathbf{v} = \mathbf{g}$ on Γ and

$$\sum_{i=1}^{e_h} < S_i \mathbf{v}_i, \mathbf{w}_i >_{\tau,\Gamma_i^h} = 0, \qquad \text{for all } \mathbf{w} \in \mathbf{X}_S \text{ such that } \mathbf{w}_{|\Gamma} = 0.$$

The space $\mathbf{X}_S$ can be approximated by the lowest-order Raviart-Thomas space defined on a triangular mesh of Γ_S. Galerkin discretization of the operators C_i, M_i, and I_i then yields matrices $\mathbf{C}_{i,h}$, $\mathbf{M}_{i,h}$, and $\mathbf{I}_{i,h}$, respectively, whereby we define the approximation

$$\mathbf{S}_{i,h} := \mathbf{C}_{i,h} + (\frac{1}{2}\mathbf{I}_{i,h} + \mathbf{M}_{i,h})\mathbf{C}_{i,h}^{-1}(\frac{1}{2}\mathbf{I}_{i,h} - \mathbf{M}_{i,h})$$

to S_i. Assembling the local element matrices $\mathbf{S}_{i,h}$ and incorporating the boundary conditions (where $\mathbf{u}_h \times \mathbf{n}$ approximates $\mathbf{g}$) results in a system of linear algebraic equations similar to the system (11).

In Table 2 we report the results of some numerical experiments in solving the linear system (11) in the Maxwell case by the GMRES iterative solver without a preconditioner. A preconditioner remains to be derived, so the iteration counts grow in these computations. The error relative to the exact solution $\nabla \times (U^*(x - (1.5, 0, 0))x)$ in the mesh-dependent norm of $\mathbf{L}_2(\Gamma_S)$ is of order $O(h^{1/2})$. Since the area of the mesh skeleton Γ_S grows in proportion to h^{-1}, one may consider the $\mathbf{L}_2(\Gamma_S)$-error of the trace to be of order $O(h)$. Also, the $\mathbf{L}_2(\Omega)$-norm of the error is of order $O(h)$, comparable to the standard finite element method with the lowest-order Nedelec elements. We refer the reader to [4] for more numerical results.

h	Edges	Iter.	$\|u-u_h\|_{0,\Gamma_S}$	$\|u\|_{0,\Gamma_S}$	$\|u-u_h\|_{0,\Omega}$
1/8	2156	243	5.09 E-2	0.490	9.19 E-3
1/16	16024	556	3.70 E-2	0.677	4.81 E-3
1/32	123440	1219	2.73 E-2	0.948	2.52 E-3
1/64	968800	2987	2.13 E-2	1.33	1.39 E-3

Table 2. Tetrahedral mesh of the unit cube $\Omega = (0,1)^3$, $\kappa = 1$.

5 Conclusions

Our technique can obviously be generalized to potential equations with piecewise smooth coefficients $a(.)$ and, therefore, to nonlinear potential equations arising, e.g., in electromagnetics. On each element Ω_i, $a(.)$ can be approximated by its value at the center of gravity of Ω_i. Moreover, we can easily construct polygonal and polyhedral elements with special geometric features like small holes and inclusions. In particular, periodic structures allow a fast generation of the finite element equations or a fast matrix-vector multiplication. The generalization to problems for which the fundamental solution is locally known (for frozen coefficients) is obviously feasible. The methods presented here can be applied to acoustic and electromagnetic scattering problems by coupling with BEM in the unbounded exterior domain. We mention that one and the same technique is used for generating the finite and the boundary element equations. The latter issues is addressed in detail in a paper by [4].

Acknowledgement. The authors gratefully acknowledge the financial support by the Austrian Science Fund (FWF) under the grant P19255.

References

[1] Brezzi, F., Lipnikov, K., Shashkov, M.: Convergence of mimetic finite difference method for diffusion problems on polyhedral meshes. *SIAM J. Numer. Anal.*, 43(3):1872–1896, 2005.

[2] Buffa, A., Ciarlet, Jr., P.: On traces for functional spaces related to Maxwell's equations. I. An integration by parts formula in Lipschitz polyhedra. *Math. Methods Appl. Sci.*, 24:9–30, 2001.

[3] Buffa, A., Hiptmair, R., von Petersdorff, T., Schwab, C.: Boundary element methods for Maxwell transmission problems in Lipschitz domains. *Numer. Math.*, 95:459–485, 2003.

[4] Copeland, D.: Boundary-element-based finite element methods for Helmholtz and Maxwell equations on general polyhedral meshes. *Int. J. Appl. Math. Comput. Sci.*, 5(1):60–73, 2009.

[5] Hsiao, G.C., Wendland, W.L.: Domain decomposition in boundary element methods. In R. Glowinski, Y.A. Kuznetsov, G. Meurant, J. Périaux and O.B. Widlund, eds., *Proceedings of the Fourth International Symposium on Domain Decomposition Methods for Partial Differential Equations, Moscow, May 21-25, 1990*, pages 41–49, Philadelphia, 1991. SIAM.

[6] Kuznetsov, Y., Lipnikov, K., Shashkov, M.: Mimetic finite difference method on polygonal meshes for diffusion-type problems. *Comput. Geosci.*, 8(4):301–324, December 2004.

[7] Langer, U., Steinbach, O.: Coupled finite and boundary element domain decomposition methods. In M. Schanz and O. Steinbach, eds., *Boundary Element Analysis: Mathematical Aspects and Application*, vol. 29 of *Lecture Notes in Applied and Computational Mechanics*, pages 29–59, Berlin, 2007. Springer.

[8] McLean, W.: *Strongly elliptic systems and boundary integral equations*. Cambridge University Press, Cambridge, 2000.

[9] Reitzinger, S.: *Algebraic Multigrid Methods for Large Scale Finite Element Equations*. Reihe C - Technik und Naturwissenschaften. Universitätsverlag Rudolf Trauner, Linz, 2001.

[10] Sauter, S., Schwab, C.: *Randelementemethoden: Analyse, Numerik und Implementierung schneller Algorithmen*. Teubner, Stuttgart, Leipzig, Wiesbaden, 2004.

[11] Steinbach, O.: *Numerical Approximation Methods for Elliptic Boundary Value Problems. Finite and Boundary Elements*. Springer, New York, 2008.

An Additive Neumann-Neumann Method for Mortar Finite Element for 4th Order Problems

Leszek Marcinkowski

Department of Mathematics, University of Warsaw, Banacha 2, 02-097 Warszawa, Poland,
Leszek.Marcinkowski@mimuw.edu.pl

Summary. In this paper, we present an additive Neumann-Neumann type parallel method for solving the system of algebraic equations arising from the mortar finite element discretization of a plate problem on a nonconforming mesh. Locally, we use a conforming Hsieh-Clough-Tocher macro element in the subdomains. The proposed method is almost optimal i.e. the condition number of the preconditioned problem grows poly-logarithmically with respect to the parametes of the local triangulations.

1 Introduction

Many real life phenomenas and technical problems are modelled by partial differential equations. A way of constructing an effective approximation of the differential problem is to introduce one global conforming mesh and then to set an approximate discrete problem. However often it is required to use different approximation methods or independent local meshes in some subregions of the original domains. This may allow us to make an adaptive changes of the local mesh in a substructure without modifying meshes in other subdomains. A mortar method is an effective method of constructing approximation on nonconforming triangulations, cf. [1, 13].

There are many works for iterative solvers for mortar method for second order problem, see e.g. [2, 3, 6, 7] and references therein. But there is only a limited number of papers investigating fast solvers for mortar discretizations of fourth order elliptic problems, cf. [8, 10, 14].

In this paper, we focus on a Neumann-Neumann type of algorithm for solving a discrete problem arising from a mortar type discretization of a fourth order model elliptic problem with discontinuous coefficients in 2D. We consider a mortar discretization which use Hsieh-Clough-Tocher (HCT) elements locally in subdomains. Our method of solving system of equations is a Neumann-Neumann type of algorithm constructed with the help of Additive Schwarz Method (ASM) abstract framework. The obtained results are almost optimal i.e. it shown that the number of CG iteration applied to the preconditioned system grows only logarithmically with the ratio $H/\underline{h}$ and is independent of the jumps of the coefficients.

M. Bercovier et al. (eds.), *Domain Decomposition Methods in Science and Engineering XVIII*, Lecture Notes in Computational Science and Engineering,
DOI: 10.1007/978-3-642-02677-5,

2 Discrete Problem

In this section, we introduce a model problem and discuss its mortar discretization.

We consider a polygonal domain Ω in the plane which is partitioned into disjoint polygonal subdomains Ω_k such that $\overline{\Omega} = \bigcup_{k=1}^N \overline{\Omega}_k$ with $\overline{\Omega}_k \cap \overline{\Omega}_l$ being an empty set, an edge or a vertex (crosspoint). We assume that these subdomains form a coarse triangulation of the domain which is shape regular the sense of [5].

The model differential problem is to find $u^* \in H_0^2(\Omega)$ such that

$$a(u^*, v) = \int_\Omega fv\,dx \quad \forall v \in H_0^2(\Omega), \tag{1}$$

where $f \in L^2(\Omega)$,

$$H_0^2(\Omega) = \{u \in H^2(\Omega) : u = \partial_n u = 0 \text{ on } \partial\Omega\}$$

and

$$a(u,v) = \sum_{k=1}^N \int_{\Omega_k} \rho_k [u_{x_1x_1} v_{x_1x_1} + 2\, u_{x_1x_2} v_{x_1x_2} + u_{x_2x_2} v_{x_2x_2}]\, dx.$$

Here ρ_k are any positive constant and ∂_n is a normal unit normal derivative.

A quasiuniform triangulation $T_h(\Omega_k)$ made of triangles is introduced in each subdomain Ω_k, and let $h_k = \max_{\tau \in T_h(\Omega_k)} \operatorname{diam}(\tau)$ be the parameter of this triangulation, cf. e.g. [4].

Let Γ_{ij} denote the interface between two subdomains Ω_i and Ω_j i.e. the open edge that is common to these subdomains, i.e. $\overline{\Gamma}_{ij} = \overline{\Omega}_i \cap \overline{\Omega}_j$. We also introduce a global interface $\Gamma = \bigcup_i \overline{\partial\Omega_i \setminus \partial\Omega}$.

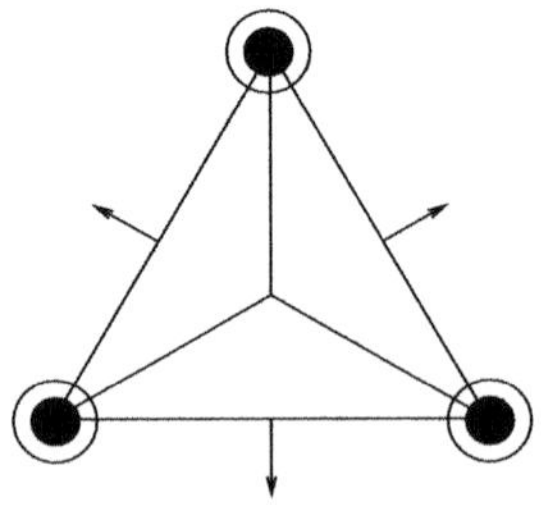

Fig. 1. HCT element.

We can now introduce local finite element spaces. Let $X_h(\Omega_k)$, be the finite element space defined as follows, cf. Fig. 1:

$$\begin{aligned} X_h(\Omega_k) = \{u \in C^1(\Omega_k) : \; & u \in P_3(\tau_i),\ \tau_i \in T_h(\Omega_k), \text{ for triangles } \tau_i, \\ & i = 1,2,3, \text{ formed by connecting the vertices of} \\ & \text{any } \tau \in T_h(\Omega_k) \text{ to its centroid, and} \\ & u = \partial_n u = 0 \text{ on } \partial\Omega_k \cap \partial\Omega\}, \end{aligned}$$

where $P_3(\tau_i)$ is the function space of cubic polynomials defined over τ_i.

Next a global space $X_h(\Omega)$ is defined as $X_h(\Omega) = \prod_{i=1}^N X_h(\Omega_k)$.

Each edge Γ_{ij} inherits two 1D triangulations made of segments that are edges of elements of the triangulations of Ω_i and Ω_j, respectively. In this way, each Γ_{ij} is provided with two independent and different 1D meshes which are denoted by $T_{h,i}(\Gamma_{ij})$ and $T_{h,j}(\Gamma_{ij})$, cf. Fig. 2.

One of the sides of Γ_{ij} is defined as a mortar (master) one, denoted by γ_{ij} and the other as a nonmortar (slave) one denoted by δ_{ji}. Let the mortar side of Γ_{ij} be chosen by the condition: $\rho_j \leq \rho_i$, (i.e. here, the mortar side is the i-th one).

For each interface Γ_{ij} two test spaces are defined: $M_t^h(\delta_{ji})$ the space formed by C^1 smooth piecewise cubic functions on the slave h_j triangulation of δ_{ji}, i.e $T_{h,j}(\Gamma_{ji})$, which are piecewise linear in the two end elements, and $M_n^h(\delta_{ji})$ the space of continuous piecewise quadratic functions on the elements of triangulation of $T_{h,j}(\Gamma_{ji})$, which are piecewise linear in the two end elements of this triangulation.

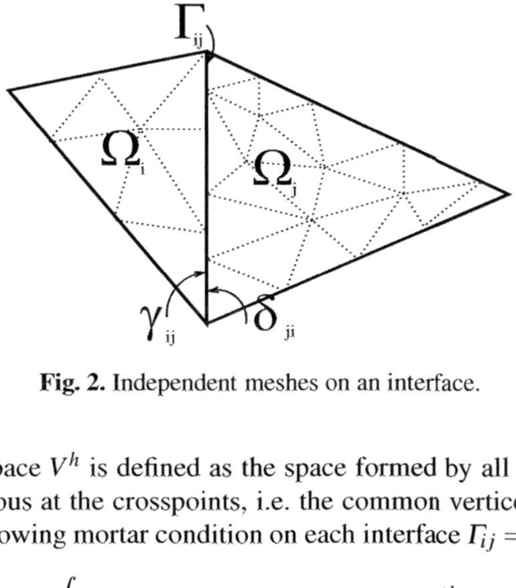

Fig. 2. Independent meshes on an interface.

The discrete space V^h is defined as the space formed by all function in $X_h(\Omega)$, which are continuous at the crosspoints, i.e. the common vertices of substructures, and satisfy the following mortar condition on each interface $\Gamma_{ij} = \delta_{ji} = \gamma_{ij} \subset \Gamma$:

$$\int_{\delta_{ji}} (u_i - u_j)\phi \, ds = 0 \qquad \forall \phi \in M_t^h(\delta_{ji}), \tag{2}$$

$$\int_{\delta_{ji}} (\partial_n u_i - \partial_n u_j)\psi \, ds = 0 \qquad \forall \psi \in M_n^h(\delta_{ji}).$$

It is worth mentioning that $u \in V^h$ has discontinuous ∇u at a crosspoint c_r, i.e. ∇u has as many values as the number of substructures with this crosspoint c_r.

Our discrete problem is to find $u_h^* \in V^h$ such that

$$a_h(u_h^*, v) = \int_\Omega fv \, dx \qquad \forall v \in V^h, \tag{3}$$

where $a_h(u,v) = \sum_{k=1}^N a_k(u,v)$ for

$$a_k(u,v) = \int_{\Omega_k} \rho_k [u_{x_1x_1} v_{x_1x_1} + 2\, u_{x_1x_2} v_{x_1x_2} + u_{x_2x_2} v_{x_2x_2}]\, dx.$$

This problem has a unique solution and for error estimates, we refer to [9].

3 Neumann-Neumann Method

In this section, we introduce our Neumann-Neumann method.

For the simplicity of presentation, we assume that our subdomains Ω_k are triangles which form a coarse triangulation of Ω.

We introduce a splitting of $u \in X_k(\Omega_k)$ into two $a_k(\cdot,\cdot)$ orthogonal parts: $u = P_k u$ and discrete biharmonic part $H_h = u - P_k u$, where $P_k u \in X_{h,0}(\Omega_k)$ is defined by

$$a_k(P_k u, v) = a_k(u,v) \qquad \forall v \in X_{h,0}(\Omega_k)$$

with $X_{h,0}(\Omega_k) = X_h(\Omega_k) \cap H_0^2(\Omega_k)$. The discrete biharmonic part of u: $H_k u = u - P_k u \in X_h(\Omega_k)$ satisfies

$$\begin{cases} a_k(H_k u, v) = 0 & \forall v \in X_{h,0}(\Omega_k), \\ Tr\, H_k u = Tr\, u & \text{on } \partial\Omega_k, \end{cases} \tag{4}$$

where $Tr\, u = (u, \nabla u)$. Let $Hu = (H_1 u, \ldots, H_N u)$ denotes the part of $u \in X_h(\Omega)$ which is discrete biharmonic in all subdomains. We also set

$$\tilde{V}^h = HV^h = \{u \in V^h : u \text{ is discrete biharmonic in all } \Omega_k\} \tag{5}$$

Each function in $\tilde{V}^h$ is uniquely defined by the values of all degree of freedoms associated with all HCT nodal points i.e. the vertices and the midpoints, which are on masters and at crosspoints since the values of the degrees of freedom corresponding to the HCT nodes in the interior of a nonmortar (slave) are defined by the mortar conditions (2) and that the values of the degrees of freedom of the nodes interior to the subdomains are defined by (4).

3.1 Local Subspaces

For each subdomain Ω_i, we introduce an extension operator $E_i : X_h(\Omega_i) \to V^h$ as $E_i u = \tilde{E}_i u + \hat{P}_i u$, where $\hat{P}_i u = (0,\ldots,0,P_i u,0,\ldots,0)$ and $\tilde{E}_i : X_h(\Omega_i) \to \tilde{V}^h$ is defined as follows:

- $\tilde{E}_i u(x) \in \tilde{V}^h$, i.e. it is discrete biharmonic in all subdomains,
- $\tilde{E}_i u(x) = u(x)$ and $\nabla \tilde{E}_i u(x) = \nabla u(x)$ and $\partial_n \tilde{E}_i u(m) = \partial_n u(m)$ for an x a nodal point (vertex) and a midpoint m of an element of $T_{h,i}(\Gamma_{ij})$ for any mortar $\gamma_{ij} \subset \partial\Omega_i$,
- $\nabla \tilde{E}_i u(v) = \nabla u(v)$ for any vertex v of substructure Ω_i,
- $\tilde{E}_i u(c_r) = \frac{1}{N(c_r)} u(c_r)$ for any crosspoint c_r which is a vertex of $\partial\Omega_i$. Here, $N(c_r)$ is the number of domains which have c_r as a vertex.

- $Tr\,\tilde{E}_i u = 0$ on remaining masters and at the crosspoints which are not on $\partial\Omega_i$.

The values of the degrees of freedom of $E_i u$ on a slave are defined by the mortar conditions (2) and in subdomains Ω_j, $j \neq i$ by (4).

We also have

$$\sum_{k=1}^{N} E_k u_k = u$$

for any $u = (u_1, \ldots, u_N) \in V^h$.

We next define local spaces $V_k = E_k V_c^h(\Omega_k)$, where $V_c^h(\Omega_k)$ is the subspace of $X_h(\Omega_k)$ of functions that have zero values at all vertices of Ω_k. Local bilinear forms are defined over $V_c^h(\Omega_k)$ as $b_k(u,v) = a_k(u,v)$. Note that $u \in V_k$ can be nonzero only in Ω_i and Ω_j for a j such that Γ_{ij} is a common edge of Ω_i and Ω_j and its master side is associated with Ω_i.

3.2 Coarse Space

For any $u = (u_1, \ldots, u_N) \in X_h(\Omega)$, we introduce $I_0 u \in V^h$ which is defined solely by the values of u at crosspoints, i.e. the common vertices of substructures in Ω as

$$I_0 u = \sum_{k=1}^{N} E_k (I_{H,k} u_k), \tag{6}$$

where $I_{H,k} u_k \in X_h(\Omega_k)$ is a linear interpolant of u_k at the three vertices of a triangular substructure Ω_k.

Next let us define a coarse space as

$$V_0 = I_0 V^h,$$

and a coarse bilinear form

$$b_0(u,v) = \left(1 + \log\left(\frac{H}{\underline{h}}\right)\right)^{-1} a_h(u,v),$$

where $H = \max_k H_k$ for $H_k = \mathrm{diam}(\Omega_k)$, and $\underline{h} = \min_k h_k$. Note that the dimension of V_0 equals to the number of crosspoints.

We see that $V^h = V_0 + \sum_{k=1}^{N} V_k$.

Next following the Additive Schwarz Method (ASM) abstract scheme, special projection-like operators are introduced: $T_k : V_k \to V^h$ for $k = 0, \ldots, N$ by

$$b_0(T_0 u, v) = a_h(u,v) \qquad \forall v \in V_0 \tag{7}$$

and let $T_k u = E_k \hat{T}_k u$ for $k = 1, \ldots, N$, where $\hat{T}_k u \in V_c^h(\Omega_k)$ is defined by

$$b_k(\hat{T}_k u, v) = a_h(u, E_k v) \qquad \forall v \in V_c^h(\Omega_k). \tag{8}$$

The operator T_k is symmetric and nonnegative definite over V^h in the terms of the form $a_h(u,v)$.

Finally, an ASM operator $T : V_h \to V_h$ is defined by

$$T = T_0 + \sum_{k=1}^{N} T_k.$$

We then replace problem (3) by a new equivalent one:

$$Tu_h^* = g, \tag{9}$$

where $g = \sum_{k=0}^{N} g_i$ and $g_i = T_i u_h^*$ for u_h^* the solution of (3).

The main result of this paper is the following theorem:

Theorem 1. *For any $u \in V^h$, it holds that*

$$c\, a_h(u,u) \le a_h(Tu,u) \le C\left(1+\log\left(\frac{H}{\underline{h}}\right)\right)^2 a_h(u,u),$$

where $H = \max_k H_k$ with $H_k = \operatorname{diam}(\Omega_k)$, $\underline{h} = \min_k h_k$, and c, C are positive constants independent of all mesh parameters h_k, H_k and the coefficients ρ_k.

Sketch of the Proof

We present here only a sketch of the proof which is based on the abstract ASM scheme, cf. e.g. [12].

We have to check three key assumptions, cf. [12]. For our method the assumption II (Strengthened Cauchy-Schwarz Inequalities), is satisfied with a constant independent of the number of subdomains by a coloring argument.

Note that T_0 is the orthogonal projection onto V_0 (in terms of the bilinear form $a_h(\cdot,\cdot)$) which is scaled by $(1+\log(H/\underline{h}))^{-1}$, i.e., we have

$$a_h(u,u) = (1+\log(H/\underline{h}))\, b_0(u,u) \qquad \forall u \in V_0.$$

It can also be shown following the lines of proof of [11] that

$$a_h(E_k u, E_k u) \le C_1\, (1+\log(H/\underline{h}))^2 b_k(u,u) \qquad \forall u \in V_k,$$

where C_1 is a constant independent of mesh parameters and subdomain coefficients. Thus these two estimates yields that the constant ω in the assumption III (Local Stability), is bounded by $C_1\,(1+\log(H/\underline{h})^2$.

It remains to prove assumption I (Stable Decomposition), i.e., we have to prove that there exists a positive constant C_0^2 such that for any $u \in V_h$ there are $w_0 \in V_0$ and $w_k \in V_k$, $k = 1,\dots,N$ such that $u = w_0 + \sum_{k=1}^{N} E_k w_k$ and

$$b_0(w_0,w_0) + \sum_{k=1}^{N} b_k(w_k,w_k) \le C_0^2 a_h(u,u). \tag{10}$$

We first define decomposition for $u = (u_1,\dots,u_N) \in V^h$. Let $w_0 = I_0 u$ and $w_k = u_k - I_{h,k} u_k \in V_c^h(\Omega_k)$. Note that

$$w_0 + \sum_{k=1}^{N} E_k w_k = I_0 u + \sum_{k=1}^{N} E_k (u_k - I_{h,k} u_k) = \sum_{k=1}^{N} E_k u_k = u.$$

Next, we see that

$$\begin{aligned}\sum_{k=1}^{N} b_k(w_k, w_k) &= \sum_{k=1}^{N} \rho_k |u_k - I_{H,k} u_k|^2_{H^2(\Omega_k)} \\ &= \sum_{k=1}^{N} \rho_k |u_k|^2_{H^2(\Omega_k)} = a_h(u,u).\end{aligned} \tag{11}$$

Again following the lines of proof of [11], we can show that

$$a_h(I_0 u, I_0 u) \leq C_0^2 (1 + \log(H/\underline{h})) a_h(u,u),$$

where C_0^2 is a constant independent of mesh parameters and subdomain coefficients, thus

$$b_0(w_0, w_0) = (1 + \log(H/\underline{h}))^{-1} a_h(I_0 u, I_0 u) \leq C_0^2 \, a_h(u,u).$$

The last estimate and (11) yield us the bound in (10) and this concludes the sketch of the proof.

Acknowledgement. This work was partially supported by Polish Scientific Grant N/N201/ 0069/33.

References

[1] Bernardi, C., Maday, Y., Patera, A.T.: A new nonconforming approach to domain decomposition: the mortar element method. In *Nonlinear partial differential equations and their applications. Collège de France Seminar, vol. XI (Paris, 1989–1991)*, vol. 299 of *Pitman Res. Notes Math. Ser.*, pages 13–51. Longman Sci. Tech., Harlow, 1994.

[2] Bjørstad, P.E., Dryja, M., Rahman, T.: Additive Schwarz methods for elliptic mortar finite element problems. *Numer. Math.*, 95(3):427–457, 2003.

[3] Braess, D., Dahmen, W., Wieners, C.: A multigrid algorithm for the mortar finite element method. *SIAM J. Numer. Anal.*, 37(1):48–69, 1999.

[4] Brenner, S.C., Scott, L.R.: *The mathematical theory of finite element methods*, vol. 15 of *Texts in Applied Mathematics*. Springer, New York, 2nd ed., 2002.

[5] Brenner, S.C., Sung, L.-Y.: Balancing domain decomposition for nonconforming plate elements. *Numer. Math.*, 83(1):25–52, 1999.

[6] Dryja, M.: A Neumann-Neumann algorithm for a mortar discetization of elliptic problems with discontinuous coefficients. *Numer. Math.*, 99:645–656, 2005.

[7] Kim, H.H., Widlund, O.B.: Two-level Schwarz algorithms with overlapping subregions for mortar finite elements. *SIAM J. Numer. Anal.*, 44(4):1514–1534, 2006.

[8] Marcinkowski, L.: Domain decomposition methods for mortar finite element discretizations of plate problems. *SIAM J. Numer. Anal.*, 39(4):1097–1114, 2001.
[9] Marcinkowski, L.: A mortar element method for some discretizations of a plate problem. *Numer. Math.*, 93(2):361–386, 2002.
[10] Marcinkowski, L.: An Additive Schwarz Method for mortar Morley finite element discretizations of 4th order elliptic problem in 2d. *Electron. Trans. Numer. Anal.*, 26:34–54, 2007.
[11] Marcinkowski, L.: A Neumann-Neumann algorithm for a mortar finite element discretization of 4th order elliptic problems in 2d. Tech. Report 173, Institute of Applied Mathematics and Mechanics, Warsaw University, June 2007.
[12] Toselli, A., Widlund, O. *Domain decomposition methods—algorithms and theory*, vol. 34 of *Springer Series in Computational Mathematics*. Springer, Berlin, 2005.
[13] Wohlmuth, B.I.: *Discretization Methods and Iterative Solvers Based on Domain Decomposition*, vol. 17 of *Lectures Notes in Computational Science and Engineering*. Springer, Berlin, 2001.
[14] Xu, X., Li, L., Chen, W.: A multigrid method for the mortar-type Morley element approximation of a plate bending problem. *SIAM J. Numer. Anal.*, 39(5): 1712–1731, 2001/02.

A Numerically Efficient Scheme for Elastic Immersed Boundaries

F. Pacull[1] and M. Garbey[2]

[1] Fluorem - Ecully 69134 France - fpacull@fluorem.com
[2] Dept. of Computer Science, University of Houston, Houston, TX 77204, USA

1 Introduction

The main approaches to simulate fluid flows in complex moving geometries, use either moving-grid or immersed boundary techniques [5, 6, 7]. This former type of methods imply re-meshing, which are expensive computationally in the fluid/elastic-structure interaction cases that involve large structure deformations. In contrast, in the immersed boundary techniques, the effect of the boundary is applied remotely to the fluid by a constraint/penalty on the governing equations or a locally modified discretization/stencil: the fluid mesh is then globally independent of the moving interface, described by Lagrangian coordinates, and the effect of the interaction is introduced into the fluid variables at the Eulerian grid points next to the interface.

Many applications of fluid/flexible-body interaction simulations with large deformation are in bio-engineering. The accuracy of the input data in such a problem is not very high and one may prefer to emphasis the robustness of the numerical method over high accuracy of the solution process. A major advantage of the Immersed Boundary Method (IBM), pioneered by C.S. Peskin [10], is the high level of uniformity of mesh and stencil, avoiding the critical interpolation processes of the cut-cell/direct methods. Based on the standard finite-difference method, the IBM allows highly efficient domain decomposition techniques to be implemented. In other words, the difficulty of simulating dynamical interaction phenomena with complex geometries can be overcome by implementing, in a fast and easy way, large fine grid parallel computations that takes full advantage of a uniform stencil on an extended regular domain, as described in [3, 4], for blood flow applications. We are first going to recall the IBM formulation.

2 Discretization of the IBM

A complete and accurate introduction to the IBM can be found in [10]. Here is a brief description of the fluid/elastic interface model unified into a set of coupled PDEs. The incompressible Navier-Stokes system is written as:

M. Bercovier et al. (eds.), *Domain Decomposition Methods in Science and Engineering XVIII*, Lecture Notes in Computational Science and Engineering,
DOI: 10.1007/978-3-642-02677-5,

$$\rho\left[\frac{\partial V}{\partial t}+(V\cdot\nabla)V\right] = -\nabla P+\mu\Delta V+F \tag{1}$$

$$\nabla\cdot V = 0 \tag{2}$$

The IBM requires the extrapolation of the Lagrangian vector f into the Eulerian vector field F from the RHS of (1). In the IBM of Peskin, we use a distribution of Dirac delta functions δ for that purpose:

$$F(x,t)=\int_{\Gamma} f(s,t)\delta\left(x-X(s,t)\right)ds=\begin{cases} f(s,t) & \text{if } x=X(s,t)\\ 0 & \text{otherwise}\end{cases} \tag{3}$$

The motion of the immersed boundary should match the motion of the neighboring fluid particles because of a no-slip boundary condition. Eq.(4) approximates this no-slip boundary condition using the Dirac delta function as an interpolating tool for V, from Ω to Γ:

$$\frac{\partial X(s,t)}{\partial t}=\int_{\Omega} V(x,t)\delta(x-X(s,t))dx=\begin{cases} V(X(s,t),t) & \text{if } x=X(s,t)\\ 0 & \text{otherwise}\end{cases} \tag{4}$$

The immersed boundary obeys a linear elastic model. We use Hooke's law of elasticity, i.e. the tension $\mathcal{T}$ of the immersed boundary is a linear function of the strain. For a one-dimensional boundary, we have:

$$\mathcal{T}(s,t)=\sigma\left|\frac{\partial X(s,t)}{\partial s}\right|, \tag{5}$$

where σ is the boundary elasticity coefficient. The local elastic force density f is defined as:

$$f(s,t)=\frac{\partial(\mathcal{T}(s,t)\tau(s,t))}{\partial s}, \quad \tau(s,t)=\frac{\partial X(s,t)/\partial s}{|\partial X(s,t)/\partial s|}. \tag{6}$$

τ is the unit tangent vector to Γ. Finally, by plugging (5) into the set of equations in (6), we get:

$$f(s,t)=\sigma\frac{\partial^2 X(s,t)}{\partial s^2} \tag{7}$$

The practical implementation of the IBM of Peskin offers dozens of different possibilities regarding the choice of the temporal scheme, the space discretization, the discrete approximation of the Dirac function and so on. There is clearly a compromise between the stability of the scheme that suffers from sharp numerical interfaces for the pressure, that should be discontinuous, and accuracy that needs this numerical feature. We refer to the thesis of the first author [9] and its bibliography for an extensive comparison of possible implementations for standard benchmark problems such as the oscillation relaxation of stretched bubble toward its equilibrium or the motion of the bubble in a cavity flow test case. These benchmark problems show clearly that the IBM method does not preserve the volume of the bubble. The "numerical porosity" of the IBM is a drawback of this immersed boundary technique. We are going to present a volume conservation method that fixes this problem.

3 Volume Conservation Method Based on Constrained Optimization

As described in [8], most of the existing methods to improve volume conservation use a local change of stencil on the Eulerian grid, or require the computation of the normal to the boundary at each of its points to build a constrained local interpolation operator for the boundary velocity. The following method is global and uses the initial volume enclosed by the immersed boundary as a control objective. We represent the position vector of the immersed boundary with a global polynomial. In the 2D bubble test case, it is natural to use a Fourier expansion. Initially the discretization points on the immersed boundary are equally spaced in the curvilinear space. The motion of these discrete points follows the fluid flow because of the no-slip boundary condition. At all time, therefore, these discretization points are a regular transformation of the original distribution and can be used in the Fourier representation. For convenience, we are going to define the global volume conservation problem in the case of a closed immersed boundary like in our benchmark problems above.

After discretization, the immersed boundary is represented by a finite set of grid points:

$$\{X_i\}_{0\le i\le M-1} = \{X_{1,i}, X_{2,i}\}_{0\le i\le M-1}. \tag{8}$$

$\{X_{1,i}\}_{0\le i\le M-1}$ is the vector of the horizontal components of the moving points and $\{X_{2,i}\}_{0\le i\le M-1}$ that of their vertical components. We assume that M is even and then define $K \equiv \frac{M}{2}$. The Fourier expansion of X is, for $j = 1,2$ and $0 \le i \le M-1$:

$$\hat{X}_{j,i}(\tilde{\alpha}) = \frac{1}{M}\left[\alpha^A_{j,0} + 2\sum_{k=1}^{K-1}\left(\alpha^A_{j,k}\cos\left(2\pi k\frac{i}{M}\right) + \alpha^B_{j,k}\sin\left(2\pi k\frac{i}{M}\right)\right) + \alpha^A_{j,K}(-1)^i\right] \tag{9}$$

where

$$\alpha^A_{j,k} = \sum_{i=0}^{M-1} X_{j,i}\cos\left(2\pi k\frac{i}{M}\right), \quad j = 1,2,\ 0 \le k \le K,$$

$$\alpha^B_{j,k} = \sum_{i=0}^{M-1} X_{j,i}\sin\left(2\pi k\frac{i}{M}\right), \quad j = 1,2,\ 1 \le k \le K-1,$$

$$\alpha^B_{j,0} = \alpha^B_{j,K} = 0, \qquad j = 1,2.$$

Let us introduce the notation:

$$\tilde{\alpha} = (\alpha^A_{1,0}..\alpha^A_{1,K}\alpha^B_{1,1}..\alpha^B_{1,K-1}\alpha^A_{2,0}..\alpha^A_{2,K}\alpha^B_{2,1}..\alpha^B_{2,K-1}) \tag{10}$$

and

$$\alpha = (\alpha^A_{1,1}..\alpha^A_{1,K-1}\alpha^B_{1,1}..\alpha^B_{1,K-1}\alpha^A_{2,1}..\alpha^A_{2,K-1}\alpha^B_{2,1}..\alpha^B_{2,K-1}). \tag{11}$$

It is easy to compute analytically the area of the bubble using (9) and Green's theorem:

$$\text{Area}(\alpha) = \frac{4\pi}{M^2} \sum_{k=1}^{K-1} k(\alpha_{1,k}^A \alpha_{2,k}^B - \alpha_{2,k}^A \alpha_{1,k}^B) \tag{12}$$

If $\text{Area}(\alpha) = V_0$ at time zero, our constraints writes $c(\alpha) = \text{Area}(\alpha) - V_0 = 0$.

We perform a least square minimization of the position change, constrained with the area preservation. The function to minimize is:

$$\begin{aligned} F(\tilde{\alpha}) &= \| \{X_i - \hat{X}_i(\tilde{\alpha})\}_{0 \le i \le M-1} \|_2^2 \\ &= \sum_{i=0}^{M-1} \left((X_{1,i} - \hat{X}_{1,i}(\tilde{\alpha}))^2 + (X_{2,i} - \hat{X}_{2,i}(\tilde{\alpha}))^2 \right) \end{aligned} \tag{13}$$

Since the input variable of the constraint $c(\alpha) = 0$ is α and not $\tilde{\alpha}$, the search space of the minimization is $\mathbb{R}^{4(K-1)}$ and not $\mathbb{R}^{4K}$. The coefficients

$$(\alpha_{1,0}^A, \alpha_{1,K}^A, \alpha_{2,0}^A, \alpha_{2,K}^A)$$

are fixed. These coefficients are not related to the area, but control the global position of the immersed boundary in the domain.

We can express the constrained minimization problem as follows:

$$\min_{\alpha / c(\alpha) = 0} F(\alpha) \quad \text{with } F(\alpha) = F(\tilde{\alpha})|_{(\alpha_{1,0}^A, \alpha_{1,K}^A, \alpha_{2,0}^A, \alpha_{2,K}^A)}. \tag{14}$$

We are going to show that this optimization problem has a unique solution.

We define the Lagrangian $L(\alpha, \lambda)$ with $\alpha \in \mathbb{R}^{4(K-1)}$, $\lambda \in \mathbb{R}$:

$$L(\alpha, \lambda) = F(\alpha) + \lambda c(\alpha) \tag{15}$$

F and c are both twice continuously differentiable with respect to α. A nice property of the Hessian $\nabla_{\alpha\alpha} F$ is that it is independent of α and diagonal:

$$\nabla_{\alpha\alpha} F = \frac{4}{M} I_{4(K-1)} \tag{16}$$

This is also true for $\nabla_{\alpha\alpha} c$ in this particular test-case with area formula (12):

$$\nabla_{\alpha\alpha} c = \frac{4\pi}{M^2} \begin{bmatrix} 0 & & & +B_K \\ & & -B_K & \\ & -B_K & & \\ +B_K & & & 0 \end{bmatrix}, \tag{17}$$

$$B_K = \begin{bmatrix} 1 & & & 0 \\ & 2 & & \\ & & \ddots & \\ 0 & & & K-1 \end{bmatrix}. \tag{18}$$

So, for a given pair (α^*, λ^*) and for all α in the neighborhood of α^*, we have:

$$\alpha^T \nabla_{\alpha\alpha} L(\alpha^*, \lambda^*) \alpha = \alpha^T \nabla_{\alpha\alpha} F \alpha + \lambda^* \alpha^T \nabla_{\alpha\alpha} c \alpha \tag{19}$$

$$= \frac{4}{M} \alpha^T \alpha + \frac{8\pi\lambda^*}{M^2} \sum_{k=1}^{K-1} k(\alpha^A_{1,k}\alpha^B_{2,k} - \alpha^A_{2,k}\alpha^B_{1,k}) \tag{20}$$

$$= \frac{4}{M} \alpha^2 + 2\lambda^* \text{Area}(\alpha) \tag{21}$$

Hence, $\lambda^* > 0$ implies:

$$\alpha^T \nabla_{\alpha\alpha} L(\alpha^*, \lambda^*) \alpha > 0 \tag{22}$$

Whenever we find a pair (α^*, λ^*) such that $\nabla_\alpha F(\alpha^*) = 0$, $c(\alpha^*) = 0$ and $\lambda^* > 0$, the first-order Lagrangian sufficiency condition is satisfied. This implies that α^* is a strict local minimum.

We note here that $\nabla_{\alpha\alpha} F$ has the same form for any geometry of Γ in any dimension, since it simply represents the norm of the correction on X. We also have here an elegant analytical formulation for the constraint in the "Bubble" test case. The functions $F(\alpha)$ and $c(\alpha)$ are easy to compute as well as $\nabla_\alpha F(\alpha)$, $\nabla_\alpha c(\alpha)$ and the Hessians of F and c, respectively $\nabla_{\alpha\alpha} F$ and $\nabla_{\alpha\alpha} c$, are constant. In the general situation, one may have to use a more complicated numerical method to get the elements of the minimization problem (14) regarding the constraint. In the numerical implementation for the "Bubble" test case, we solve:

$$\begin{pmatrix} \nabla_\alpha L(\alpha, \lambda) \\ \nabla_\lambda L(\alpha, \lambda) \end{pmatrix} = \begin{pmatrix} 0 \\ 0 \end{pmatrix} \tag{23}$$

using the classical Newton-Raphson algorithm. A good initial solution for the iterative Newton algorithm is the α coefficients corresponding to the actual position of the boundary X before the correction. In most cases, the volume of the immersed boundary is found to evolve relatively slowly with respect to time. Then it is not necessary to perform this minimization at every time step.

Compared to the traditional method, the Fourier expansion allows a fairly compact representation of the interface, without any loss of accuracy.

To reduce the computational load of the minimization process, one may want to restrict the search space of the α coefficients to a finite space of dimension smaller than M the number of discrete points that support the immersed interface. Because of the high order accuracy of the Fourier expansion, we work with the first $\frac{K}{4}$ of the Fourier expansion coefficients $(\alpha_{j,k}, \beta_{j,k})$, $j = 1, 2$. This corresponds roughly to the idea that one needs at least 4 mesh points to represent a wave period. This compact representation of the immersed closed boundary has several advantages. First, it filters out the high wave frequency components of the position vector, and removes most of the noise in the force term. Second, it can also drastically speed up an implicit IBM scheme, by reducing the search space for the Newton algorithm. In [9], we implemented the Inexact Newton Backtracking Method of M. Pernice and H.F. Walker in an implicit IBM scheme, in which the unknown is the position of the moving boundary at the next time step. We used a Krylov method to solve the inexact Newton condition and an associated Jacobian matrix approached by finite-differences. Without a Fourier representation of the boundary position, the size of the Newton search space in two space dimensions is $2M$, while here it becomes $\frac{M}{2}$.

For the "bubble" test cases discussed earlier we observe a perfect global volume preservation. Most of the minimization work is done in the very first step and we observe then a regular and fast evolution of $\nabla_\alpha L(\alpha,\lambda)$ while $c(\alpha) = \nabla_\lambda L(\alpha,\lambda)$ requires only two or three steps to be negligible. The high Fourier modes should be cut off so that the volume correction does not make the system numerically unstable: the small corrections can originate high frequencies oscillations that will be self-exciting in the interaction.

We will now to present an application of our IBM implementation that is non trivial and can take advantage of the Fourier representation of the immersed interface.

4 Application of the IBM and Conclusion

Let us consider a single bubble in a long rectangular cavity at rest. The flow velocity at the initial time is null and the bubble is a circle. We equip the moving bubble with a membrane that can contract or dilate periodically by forcing the boundary elasticity coefficient in the Hooke law. We set:

$$\sigma(\theta) = \sigma_0\left(1 + \sigma_1\left(1 + \sin\left(2\pi\frac{t}{P}\right)\right)(\cos(\theta) + 1)\right), \tag{24}$$

where $\theta \in (0,2\pi)$ is the angle in the polar moving coordinate system attached to the bubble. To be more specific if X is a point attached to the membrane, the corresponding θ stays invariant with respect to X, no matter the X motion. $\theta \in (\frac{\pi}{2}, 3\frac{\pi}{2})$ corresponds to the anterior side of the bubble, and the posterior side is the opposite side of the bubble.

The elasticity coefficient increases and decreases periodically in time with period P. The variation of the elasticity coefficient in time is most pronounced around $\theta = 0$, and has less variation around $\theta = \pi$. The largest contraction and relaxation move of the membrane happens around $\theta = 0$.

During the contraction phase, the posterior side of the membrane projects the liquid inside the bubble toward the forward side. While this mass of liquid travels forward, the posterior side of the membrane relaxes. It results overall in a motion of the bubble forward. We observe from the numerical simulation that this motion is maintained by two nice symmetric vortices that companion the bubble motion. Fig. 1 shows an established forward motion of the bubble with the bubble starting at rest near the right side of the cavity. In this simulation, we have $(\sigma_0, \sigma_1) = (200, 5)$, and the elasticity coefficient varies in the interval $200 \leq \sigma(\theta) \leq 4200$. The center of the bubble is centered on a symmetry axis of the rectangular cavity at $t = 0$. The solution conserves this symmetry as time goes on. However, if the position of the bubble is slightly shifted away of this axis at time $t = 0$, the no-slip boundary flow condition rapidly breaks the symmetry. The direction of motion becomes unstable and the bubble starts a (slow) chaotic motion. In practice a tail attached at the bubble at the $\theta = 0$

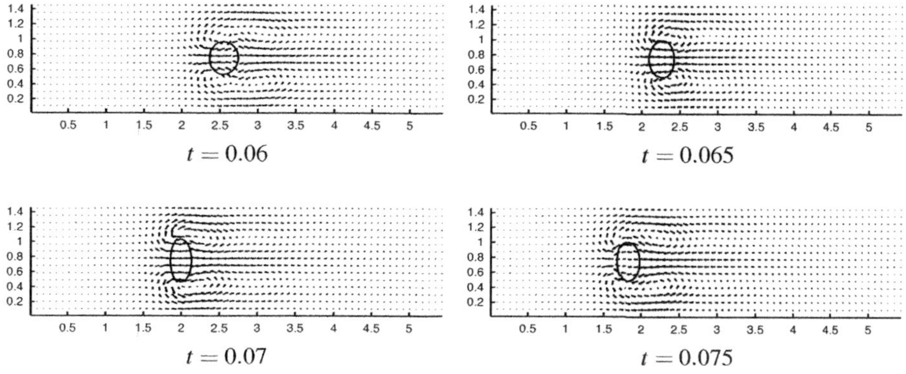

Fig. 1. Active bubble motion with time period 0.02

angle position should stabilize this motion because of viscosity forces. Alternatively, a tandem of two "active" bubbles can pilot a larger body. We are currently studying the optimum design of this setup using our simulation tool.

This motion of the "active bubble" with no flapping fins nor flagella, or helicoidal motion [1] is an amazing example of fluid dynamic. We found particularly fascinating that the IBM technique, that is relatively simple to implement, gives access to such complex fluid flow problems.

References

[1] Cortez, R., Cowen, N., Dillon, R., Fauci, L.: Simulation of Swimming Organisms; Couling Internal Mechanics with External Fluid Dynamics, *Comput. Sci. Engrg.*, 6(3): 38–45, 2004.

[2] Francois, M., Shyy, W.: Computations of Drop Dynamics with the Immersed Boundary Method, part 1: Numerical Algorithm and Buoyancy-Induced Effect, *Numer. Heat Transfer B*, 44:101–118, 2003.

[3] Garbey, M., Pacull, F.: Toward A Matlab MPI Parallelized Immersed Boundary Method. In *Parallel Computational Fluid Dynamics, Theory and Applications*, pages 397–404. A. Deane et al., eds. Elsevier, 2006.

[4] Garbey, M., Pacull, F.: A Versatile Incompressible Navier-Stokes Solver for Blood Flow Application, *Internat. J. Numer. Methods Fluids* 54(5):473–496, 2007.

[5] Glowinski, R.: A Fictitious Domain Approach to the Direct Numerical Simulation of Incompressible Flow Past Moving Rigid Bodies: Application to Particle Flow, *J. Comput. Phys.*, 162:363–426, 2001.

[6] Lee, L., LeVeque, R.: Immersed Interface Methods for Incompressible Navier-Stokes Equations, *SIAM J. Sci. Comput.*, 25:832–856, 2003.

[7] Mittal, R., Iaccarino, G.: Immersed Boundary Methods, *Ann. Rev. Fluid. Mech.*, 37:239–261, 2005.

[8] Newren, E.P.: *Enhancing The Immersed Boundary Method Stability, Volume Conservation, And Implicit Solvers*, PhD thesis, Department of Mathematics, University of Utah, 2007.
[9] Pacull, F.: *A Numerical Study of The Immersed Boundary Method and Application to Blood Flow*, PhD thesis, Department of Mathematics, University of Houston, 2006.
[10] Peskin, C.S.: The Immersed Boundary Method, *Acta Numer.*, 11:479–517, 2002.

A Domain Decomposition Method Based on Augmented Lagrangian with a Penalty Term

Chang-Ock Lee[1] and Eun-Hee Park[2]

[1] Department of Mathematical Sciences, KAIST, Daejeon, 305-701, South Korea
colee@kaist.edu

[2] Department of Mathematical Sciences, KAIST, Daejeon, 305-701, South Korea
mfield@kaist.ac.kr

Summary. An iterative substructuring method with Lagrange multipliers is considered for the second order elliptic problem, which is a variant of the FETI-DP method. The standard FETI-DP formulation is associated with a saddle-point problem which is induced from the minimization problem with a constraint for imposing the continuity across the interface. Starting from the slightly changed saddle-point problem by addition of a penalty term with a positive penalization parameter η, we propose a dual substructuring method which is implemented iteratively by the conjugate gradient method. In spite of the absence of any preconditioners, it is shown that the proposed method is numerically scalable in the sense that for a large value of η, the condition number of the resultant dual problem is bounded by a constant independent of both the subdomain size H and the mesh size h. We discuss computational issues and present numerical results.

1 Introduction

Let us consider the following Poisson model problem

$$\begin{aligned} -\Delta u &= f \quad \text{in } \Omega, \\ u &= 0 \quad \text{on } \partial\Omega, \end{aligned} \tag{1}$$

where Ω is a bounded polygonal domain in $\mathbb{R}^2$ and f is a given function in $L^2(\Omega)$. For simplicity, we assume that Ω is partitioned into two nonoverlapping subdomains $\{\Omega_i\}_{i=1}^2$ such that $\overline{\Omega} = \bigcup_{i=1}^2 \overline{\Omega_i}$. It is well-known that problem (1) is equivalent to the constrained minimization

$$\min_{\substack{v_i \in H^1(\Omega_i) \\ v_i = 0 \text{ on } \partial\Omega \cap \partial\Omega_i \\ v_1 = v_2 \text{ on } \partial\Omega_1 \cap \partial\Omega_2}} \sum_{i=1}^{2} \left(\frac{1}{2} \int_{\Omega_i} |\nabla v_i|^2 \, dx - \int_{\Omega_i} f v_i \, dx \right). \tag{2}$$

In a domain-decomposition approach, a key point is how to convert the constrained minimization problem (2) into an unconstrained one. The most popular methods,

M. Bercovier et al. (eds.), *Domain Decomposition Methods in Science and Engineering XVIII*, Lecture Notes in Computational Science and Engineering,
DOI: 10.1007/978-3-642-02677-5,

developed for different purposes are the Lagrangian method, the method of penalty functions, and the augmented Lagrangian method. Such various ideas have been introduced for handling constraints as the continuity across the interface in (2) (see [4, 6, 8]). The FETI-DP method is one of the most advanced dual substructuring methods, which introduces Lagrange multipliers to enforce the continuity constraint by following the Lagrangian method. In this paper, we propose a dual iterative substructuring algorithm which deals with the continuity constraint across the interface using the augmented Lagrangian method. Many studies of the augmented Lagrangian method have been done in the frame of domain-decomposition techniques which belong to families of nonoverlapping Schwarz alternating methods, variants of FETI method, etc. (cf. [1, 3, 8, 11])

This paper is organized as follows. In Section 2, we introduce a saddle-point formulation for an augmented Lagrangian with a penalty term. Section 3 provides a dual iterative substructuring method and presents algebraic condition number estimates. In Section 4, we mainly deal with computational issues in view of implementation of the proposed method and show the numerical results. For details omitted here due to space restrictions, we refer the reader to [9].

2 Saddle-Point Formulation

Let $\mathcal{T}_h$ denote a quasi-uniform triangulation on Ω. We consider the discretized variational problem for (1): find $u_h \in X_h$ such that

$$a(u_h, v_h) = (f, v_h) \quad \forall v_h \in X_h, \tag{3}$$

where $a(u_h, v_h) = \int_\Omega \nabla u_h \cdot \nabla v_h \, dx$ and $(f, v_h) = \int_\Omega f v_h \, dx$. Here, X_h is the standard $\mathcal{P}_1$-conforming finite element space.

Before proposing a constrained minimization problem whose minimizer has a connection with the solution of (3), we introduce some commonly-used notations. We assume that Ω is decomposed into N non-overlapping subdomains $\{\Omega_k\}_{k=1}^N$ such that

(i) Ω_k is a polygonally shaped open subset of Ω.
(ii) the decomposition $\{\Omega_k\}_{k=1}^N$ of Ω is geometrically conforming.
(iii) Γ_{kl} denotes the common interface of two adjacent subdomains Ω_k and Ω_l.

Let us use $\mathcal{T}_{h_k}$ to denote a quasi-uniform triangulation of Ω_k, where we have matching grids on the boundaries of neighboring subdomains across the interfaces. On each Ω_k, we set a finite-dimensional subspace X_h^k of $H^1(\Omega_k)$:

$$X_h^k = \{v_h^k \in \mathcal{C}^0(\overline{\Omega}_k) \mid \forall \tau \in \mathcal{T}_{h_k},\ v_h^k|_\tau \in \mathcal{P}_1(\tau),\ v_h^k|_{\partial\Omega\cap\partial\Omega_k} = 0\}.$$

Next, we define a bilinear form on $X_h^c \times X_h^c$:

$$a_h(u, v) = \sum_{k=1}^N \int_{\Omega_k} \nabla u \cdot \nabla v \, dx.$$

where $X_h^c = \{v = (v_h^k)_k \in \prod_{k=1}^N X_h^k \mid v \text{ is continuous at each corner}\}$.

It is well-known that solving the finite element problem (3) is equivalent to solving the saddle-point formulation: find a saddle point $(u_h, \lambda_h) \in X_h^c \times \mathbb{R}^E$ such that

$$\mathcal{L}(u_h, \lambda_h) = \max_{\mu_h \in \mathbb{R}^E} \min_{v_h \in X_h^c} \mathcal{L}(v_h, \mu_h) = \min_{v_h \in X_h^c} \max_{\mu_h \in \mathbb{R}^E} \mathcal{L}(v_h, \mu_h), \tag{4}$$

where the Lagrangian $\mathcal{L} : X_h^c \times \mathbb{R}^E \to \mathbb{R}$ is defined by

$$\mathcal{L}(v, \mu) = \mathcal{J}(v) + \langle Bv, \mu \rangle = \frac{1}{2} a_h(v, v) - (f, v) + \langle Bv, \mu \rangle.$$

Here, B is a signed Boolean matrix such that for any $v \in X_h^c$, $Bv = 0$ which enforces the continuity of v across the interface.

Now, we shall slightly change the saddle-point formulation (4) by addition of a penalty term to the Lagrangian $\mathcal{L}$. Let J_η be a bilinear form on $X_h^c \times X_h^c$ defined as

$$J_\eta(u, v) = \sum_{k<l} \frac{\eta}{h} \int_{\Gamma_{kl}} (u^k - u^l)(v^k - v^l)\, ds, \quad \eta > 0,$$

where $h = \max_{k=1,\cdots,N} h_k$. Given the augmented Lagrangian $\mathcal{L}_\eta$ defined by

$$\mathcal{L}_\eta(v, \mu) = \mathcal{L}(v, \mu) + \frac{1}{2} J_\eta(v, v),$$

we consider the following saddle-point problem:

$$\mathcal{L}_\eta(u_h, \lambda_h) = \max_{\mu_h \in \mathbb{R}^E} \min_{v_h \in X_h^c} \mathcal{L}_\eta(v_h, \mu_h) = \min_{v_h \in X_h^c} \max_{\mu_h \in \mathbb{R}^E} \mathcal{L}_\eta(v_h, \mu_h). \tag{5}$$

Based on the characterization of a saddle-point formulation like problem (5) by a variational problem in [7], it can be shown that the saddle-point of (5) is equivalent to the solution of the following variational problem: find $(u_h, \lambda_h) \in X_h^c \times \mathbb{R}^E$ such that

$$\begin{aligned} a_\eta(u_h, v_h) + \langle v_h, B^T \lambda_h \rangle &= (f, v_h) \quad && \forall v_h \in X_h^c, \\ \langle Bu_h, \mu_h \rangle &= 0 && \forall \mu_h \in \mathbb{R}^E. \end{aligned} \tag{6}$$

Moreover, the primal solution u_h of (6) is exactly equal to the solution of the variational problem (3).

3 Iterative Substructuring Method

The saddle-point formulation (6) is expressed in the following algebraic form

$$\begin{bmatrix} A_{\Pi\Pi} & A_{\Pi e} & 0 \\ A_{\Pi e}^T & A_{ee}^\eta & B_e^T \\ 0 & B_e & 0 \end{bmatrix} \begin{bmatrix} u_\Pi \\ u_e \\ \lambda \end{bmatrix} = \begin{bmatrix} f_\Pi \\ f_e \\ 0 \end{bmatrix}, \tag{7}$$

where u_Π denotes the degrees of freedom (dof) at the interior nodes and the corners, u_e those on the edge nodes on the interface except at the corners. After eliminating u_Π and u_e in (7), we have the following system for the Lagrange multipliers:

$$F_\eta \lambda = d_\eta \tag{8}$$

where

$$F_\eta = B_e S_\eta^{-1} B_e^T, \quad d_\eta = B_e S_\eta^{-1}(f_e - A_{\Pi e}^T A_{\Pi\Pi}^{-1} f_\Pi).$$

Here, $S_\eta = S + \eta J = (A_{ee} - A_{\Pi e}^T A_{\Pi\Pi}^{-1} A_{\Pi e}) + \eta J$. Noting that F_η is symmetric positive definite, we solve the resultant dual system (8) iteratively by the conjugate gradient method (CGM). Hence, the key issue is to provide a sharp estimate for the condition number of F_η.

Note that J in S_η is represented as $J = B_e^T D(J_B) B_e$ where $D(J_B)$ is a block diagonal matrix such that the diagonal block J_B is a positive definite matrix induced from

$$\frac{1}{h}\int_{\Gamma_{ij}} \varphi\psi \, ds \quad \forall \varphi, \psi \in X_h^c|_{\Gamma_{ij}}.$$

Let us denote by Λ the space of vectors of dof associated with the Lagrange multipliers where the norm $\|\cdot\|_\Lambda$ and the dual norm $\|\cdot\|_{\Lambda'}$ are defined by

$$\|\mu\|_\Lambda^2 = \mu^T D(J_B)\mu \;\; \forall \mu \in \Lambda \quad \text{and} \quad \|\lambda\|_{\Lambda'} = \max_{\mu\in\Lambda} \frac{|\langle \lambda, \mu\rangle|}{\|\mu\|_\Lambda} \;\; \forall \lambda \in \Lambda.$$

In order to derive bounds on the extreme eigenvalues of F_η, we first mention some useful properties.

Lemma 1. *For $S = A_{ee} - A_{\Pi e}^T A_{\Pi\Pi}^{-1} A_{\Pi e}$, there exists a constant $C > 0$ such that*

$$v_e^T S v_e \le C v_e^T J v_e \quad \forall v_e \perp \mathrm{Ker} B_e.$$

Proposition 1. *Let $\|\cdot\|_{S_\eta}$ be the norm induced by the symmetric positive definite matrix S_η. For any $\lambda \in \mathbb{R}^E$,*

$$\lambda^T F_\eta \lambda = \max_{v_e \neq 0} \frac{|v_e^T B_e^T \lambda|^2}{\|v_e\|_{S_\eta}^2}.$$

From Lemma 1 and Proposition 1, we have

Theorem 1. *For any $\lambda \in \Lambda$, we have that*

$$\frac{1}{C+\eta}\|\lambda\|_{\Lambda'}^2 \le \lambda^T F_\eta \lambda \le \frac{1}{\eta}\|\lambda\|_{\Lambda'}^2.$$

where C is the constant estimated in Lemma 1.

Using Theorem 1 based on Lemma 3.1 in [10], we now give the estimate of the condition number $\kappa(F_\eta)$.

Corollary 1. *We have the condition number estimate of the dual system (8)*

$$\kappa(F_\eta) \le \left(\frac{C}{\eta}+1\right)\kappa(J_B), \quad C = \frac{\lambda_{\max}^S}{2\lambda_{\min}^{J_B}},$$

where $\lambda_{\max}^S$ and $\lambda_{\min}^{J_B}$ are the maximum eigenvalue of S and the minimum eigenvalue of J_B, respectively. Furthermore, the constant C is independent of the subdomain size H and the mesh size h.

Corollary 2. *For a sufficiently large η, there exists a constant C^* independent of h and H such that*

$$\kappa(F_\eta) \le C^*.$$

In particular, assuming that each triangulation $\mathcal{T}_{h_k}$ on Ω_k is uniform, $C^* = 3$.

Remark 1. To the best of our knowledge, the algorithm with such a constant bound of the condition number is unprecedented in the field of domain decomposition. Adding the penalization term J_η to the FETI-DP formulation results in a strongly scalable algorithm without any domain-decomposition-based preconditioners even if it is redundant in view of equivalence relations among the concerned minimization problems.

4 Computational Issues and Numerical Results

4.1 Computational Issues

In focusing on the implementation of the proposed algorithm, the saddle-point formulation in form of (7) is rewritten as follows

$$\begin{bmatrix} K_{rr}^\eta & K_{rc} & B_r^T \\ K_{rc}^T & K_{cc} & 0 \\ B_r & 0 & 0 \end{bmatrix} \begin{bmatrix} u_r \\ u_c \\ \lambda \end{bmatrix} = \begin{bmatrix} f_r \\ f_c \\ 0 \end{bmatrix}, \tag{9}$$

where u_c denotes the dof at the corners and u_r the remaining of dof. Eliminating u_r and u_c in (9) yields

$$F_\eta \lambda = d_\eta \tag{10}$$

where

$$F_\eta = F_{rr} + F_{rc}F_{cc}^{-1}F_{rc}^T, \quad d_\eta = d_r - F_{rc}F_{cc}^{-1}d_c.$$

In view of implementation, the difference with the FETI-DP method ([4]) is that we invert K_{rr}^η that contains the penalization parameter η. To compare our algorithm with the FETI-DP method, we need to make a more careful observation of behavior of $(K_{rr}^\eta)^{-1}$. Note that

$$K_{rr}^\eta = K_{rr} + \eta \tilde{J} = \begin{bmatrix} A_{ii} & A_{ie} \\ A_{ie}^T & A_{ee} \end{bmatrix} + \begin{bmatrix} 0 & 0 \\ 0 & \eta J \end{bmatrix}$$

where $J = B_e^T D(J_B) B_e$. Thanks to the specific type of discrete Sobolev inequality in Lemma 3.4 of [2], we get the following estimate.

Theorem 2. *For each $\eta > 0$, we have that*

$$\kappa(K_{rr}^{\eta}) \lesssim \left(\frac{H}{h}\right)^2 \left(1 + \log\frac{H}{h}\right)(1+\eta).$$

Theorem 2 shows how severely η damages the property of K_{rr}^{η} as η is increased. Since K_{rr}^{η} is solved iteratively, it might be expected that the large condition number of K_{rr}^{η} shown above may cause the computational cost relevant to K_{rr}^{η} to be more expensive. We shall establish a good preconditioner for K_{rr}^{η} in order to remove a bad effect of η. We introduce the preconditioner M as follows

$$M = \overline{K}_{rr} + \eta\tilde{J} = \begin{bmatrix} A_{ii} & 0 \\ 0 & A_{ee} \end{bmatrix} + \begin{bmatrix} 0 & 0 \\ 0 & \eta J \end{bmatrix}.$$

Theorem 3. *The condition number of the preconditioned problem grows asymptotically as*

$$\kappa(M^{-1}K_{rr}^{\eta}) := \frac{\lambda_{\max}(M^{-1}K_{rr}^{\eta})}{\lambda_{\min}(M^{-1}K_{rr}^{\eta})} \lesssim \frac{H}{h}\left(1 + \log\frac{H}{h}\right).$$

4.2 Numerical Results

Let Ω be $[0,1]^2 \subset \mathbb{R}^2$. We consider the Poisson problem with the exact solution

$$u(x,y) = y(1-y)\sin(\pi x).$$

The reduced dual problem (10) is solved iteratively by CGM. We monitor the convergence of CGM with the stopping criterion $\frac{\|r_k\|}{\|r_0\|} \leq \text{TOL}$, where r_k is the dual residual error on the k-th CG iteration and TOL$=10^{-8}$. We decompose Ω into N_s square subdomains with $N_s = 1/H \times 1/H$, where each subdomain is partitioned into $2 \times H/h \times H/h$ uniform triangular elements.

First, we make a comparison between our proposed method and the FETI-DP method from the viewpoint of the conditioning of the related matrices F_{η} and F. Table 1 shows that the condition number $\kappa(F_{\eta})$ and the CG iteration number remain almost constant when the mesh is refined and the number N_s of subdomains is increased while keeping the ratio H/h constant. Moreover, we observe numerically that the condition number of F_{η} is bounded by the constant 3 independently of h and H, while the condition number in the FETI-DP method grows with increasing H/h (cf. [4, 5]). In addition, it is shown in Table 1 that the proposed method is superior to the FETI-DP method in the number of CG iterations for convergence. In Table 2, the condition number of K_{rr}^{η} and $M^{-1}K_{rr}^{\eta}$ are listed to show how well the designed preconditioner M for $(K_{rr}^{\eta})^{-1}$ performs. It confirms that the influence of η on $\kappa(K_{rr}^{\eta})$ is completely removed after adopting M.

Table 1. Comparison between the proposed method ($\eta = 10^6$) and the FETI-DP method ($\eta = 0$)

N_s	$\frac{H}{h}$	$\eta = 10^6$		$\eta = 0$	
		iter. no	$\kappa(F_\eta)$	iter. no	$\kappa(F)$
4×4	4	3	2.0938	14	7.2033
	8	7	2.7170	23	2.2901e+1
	16	13	2.9243	33	5.9553e+1
	32	14	2.9771	48	1.4707e+2
8×8	4	3	2.0938	18	7.9241
	8	7	2.7170	32	2.5668e+1
	16	12	2.9245	48	6.7409e+1
16×16	4	3	2.0938	19	7.9461
	8	7	2.7170	34	2.6324e+1

Table 2. Performance of preconditioner M for $(K_{rr}^{\eta})^{-1}$ where $N_s = 4\times4$

η	$\frac{H}{h}=4$		$\frac{H}{h}=8$		$\frac{H}{h}=16$	
	$\kappa(K_{rr}^{\eta})$	$\kappa(M^{-1}K_{rr}^{\eta})$	$\kappa(K_{rr}^{\eta})$	$\kappa(M^{-1}K_{rr}^{\eta})$	$\kappa(K_{rr}^{\eta})$	$\kappa(M^{-1}K_{rr}^{\eta})$
0	43.2794	14.8532	228.0254	40.0332	1.1070e+3	104.3459
1	34.5773	11.8232	161.1716	28.7437	7.0562e+2	68.3468
10^1	91.3072	11.4010	420.1058	28.1835	1.8390e+3	67.6093
10^2	8.5119e+2	11.3525	3.9824e+3	28.1232	1.7513e+4	67.5325
10^3	8.4538e+3	11.3475	3.9616e+4	28.1170	1.7430e+5	67.5247
10^4	8.4480e+4	11.3470	3.9596e+5	28.1164	1.7421e+6	67.5240
10^5	8.4474e+5	11.3469	3.9593e+6	28.1164	1.7420e+7	67.5239
10^6	8.4473e+6	11.3469	3.9593e+7	28.1164	1.7420e+8	67.5239
10^7	8.4473e+7	11.3469	3.9593e+8	28.1164	1.7420e+9	67.5238

5 Conclusions

In this paper, we have proposed a dual substructuring method based on an augmented Lagrangian with a penalty term. Unlike other substructuring methods, it is shown that without any preconditioners, the designed method is scalable in the sense that for a large penalty parameter η, the condition number of the relevant dual system has a constant bound independent of H and h. In addition, we dealt with an implementational issue. An optimal preconditioner with respect to η is established in order to increase the ease of use and the practical efficiency of the presented method.

Acknowledgement. This work was partially supported by the SRC/ERC program of MOST/ KOSEF(R11-2002-103).

References

[1] Bavestrello, H., Avery, P., Farhat, C.: Incorporation of linear multipoint constraints in domain-decomposition-based iterative solvers. Part II: Blending FETI-DP and mortar methods and assembling floating substructures. *Comput. Methods Appl. Mech. Engrg.*, 196:1347–1368, 2007.

[2] Bramble, J.H., Pasciak, J.E., Schatz, A.H.: The construction of preconditioners for elliptic problems by substructuring. I. *Math. Comp.*, 47:103–134, 1986.

[3] Farhat, C., Lacour, C., Rixen, D.: Incorporation of linear multipoint constraints in substructure based iterative solvers. Part I: A numerically scalable algorithm. *Internat. J. Numer. Methods Engrg.*, 43:997–1016, 1998.

[4] Farhat, C., Lesoinne, M., Pierson, K. A scalable dual-primal domain decomposition method. *Numer. Linear Algebra Appl.*, 7:687–714, 2000.

[5] Farhat, C., Mandel, J., Roux, F.-X.: Optimal convergence properties of the FETI domain decomposition method. *Comput. Methods Appl. Mech. Engrg.*, 115: 365–385, 1994.

[6] Farhat, C., Roux, F.-X.: A method of finite element tearing and interconnecting and its parallel solution algorithm. *Internat. J. Numer. Methods Engrg.*, 32: 1205–1227, 1991.

[7] Girault, V., Raviart,P.-A.: *Finite element methods for Navier-Stokes equations.* Springer, Berlin, 1986.

[8] Glowinski, R., LeTallec, P.: Augmented Lagrangian interpretation of the nonoverlapping Schwarz alternating method. In *Third International Symposium on Domain Decomposition Methods for Partial Differential Equations (Houston, TX, 1989)*, pages 224–231. SIAM, Philadelphia, PA, 1990.

[9] Lee, C.-O., Park, E.-H.: A dual iterative substructuring method with a penalty term. In *KAIST DMS Applied Mathematics Research Report Series 07-2.* KAIST, 2007.

[10] Mandel, J., Tezaur, R.: Convergence of a substructuring method with Lagrange multipliers. *Numer. Math.*, 73:473–487, 1996.

[11] Le Tallec, P., Sassim, T.: Domain decomposition with nonmatching grids: augmented Lagrangian approach. *Math. Comp.*, 64:1367–1396, 1995.

Parallelization of a Constrained Three-Dimensional Maxwell Solver

F. Assous[1], J. Segré[2], and E. Sonnendrücker[3]

[1] Department of Mathematics and Statistics, Bar-Ilan University, Israel
franck.assous@netscape.net
[2] CEA-Saclay, DEN/DM2S/SFME, France jacques.segre@cea.fr
[3] Institut de Recherche Mathématique Avancée (IRMA), CNRS-Université Louis Pasteur, Strasbourg, France sonnen@math.u-strasbg.fr

Summary. The numerical solution of very large 3D electromagnetic field problems are challenging for various applications in the industry. In this paper, we propose a nonoverlapping domain decomposition approach for solving the 3D Maxwell equations on MIMD computers, based on a mixed variational formulation. It is especially well adapted for the solution of the Vlasov-Maxwell equations, widely used to simulate complex devices like particle injectors or accelerators. This approach in particular leads to reuse without modification most of an existing sequential code.

1 Introduction

In order to simulate complex devices like particle injectors and accelerators, we need in some cases a full three-dimensional code for the solution of the Vlasov-Maxwell equations. A three-dimensional code [7] has been written for this purpose and has already been used for many applications (see [9]). This code solves the instationary Maxwell equations with continuous approximations of the electromagnetic field. The time-stepping numerical scheme is explicit thanks to a mass lumping procedure and leads to an efficient algorithm. Moreover, in order to handle precisely the conditions on the divergence of the fields, these are considered as constraints. They are dualized, using a Lagrange multiplier, which yields a saddle-point variational formulation. In this paper, we propose a domain decomposition approach for the parallelization of this constrained 3D Maxwell solver. This choice allows us to reuse a large part of the sequential code for the solution on each subdomain. We first recall the constrained wave equation formulation of Maxwell's equations. Then we introduce a adapted variational formulation, the continuity at the interfaces being imposed by duality using Lagrange multipliers. Next, we describe the discretization and derive a linear system suitable for multiprocessor solution. The preconditioned Uzawa algorithm used for the solution of this system is then described. And finally we present an exemple of numerical application.

M. Bercovier et al. (eds.), *Domain Decomposition Methods in Science and Engineering XVIII*, Lecture Notes in Computational Science and Engineering,
DOI: 10.1007/978-3-642-02677-5, © Springer-Verlag Berlin Heidelberg 2009

2 Constrained Wave Equation Formulation

Let Ω be a bounded, open subset of $\mathbb{R}^3$, and Γ its boundary. We denote by $\mathbf{n}$ the unit outward normal to Γ. Let c, ε_0 and μ_0 be respectively the light velocity, the dielectric permittivity and the magnetic permeability, the Maxwell equations in vacuum read:

$$\frac{\partial \vec{E}}{\partial t} - c^2\nabla\times\vec{B} = -\frac{1}{\varepsilon_0}\vec{J}, \quad \nabla\cdot\vec{E} = \frac{\rho}{\varepsilon_0}, \tag{1}$$

$$\frac{\partial \vec{B}}{\partial t} + \nabla\times\vec{E} = 0, \qquad \nabla\cdot\vec{B} = 0, \tag{2}$$

where $\vec{E}$ and $\vec{B}$ are the electric and magnetic fields respectively. the charge and current densities ρ and $\vec{J}$ satisfy the charge conservation equation:

$$\frac{\partial \rho}{\partial t} + \nabla\cdot\vec{J} = 0\,. \tag{3}$$

These quantities depend on the space variable $\vec{x}$ and the time variable t. It is well known that when Maxwell's equations are used in a Particle in Cell code, as the continuity equation (3) is not generally satisfied numerically, special care needs to be taken so that the Poisson equation $\nabla\cdot\vec{E} = \frac{\rho}{\varepsilon_0}$ remains satisfied throughout the length of the computation [4]. The same problem occurs for the $\nabla\cdot\vec{B} = 0$ condition on some unstructured meshes when the divergence of a curl is not close enough to zero. If these constraints were not satisfied then spurious modes could polute the numerical solution. This problem was dealt with in [1] by using a constrained wave equation formulation of Maxwell's equations that we recall in the case of perfectly conducting boundary conditions. These are the only ones that we shall consider here, as the case of any artificial boundary is not an issue for the parallelization. The electric field is then computed using the following equations:

$$\frac{\partial^2 \vec{E}}{\partial t^2} + c^2\nabla\times\nabla\times\vec{E} - \nabla p = -\frac{1}{\varepsilon_0}\frac{\partial \vec{J}}{\partial t}, \quad \nabla\cdot\vec{E} = \frac{\rho}{\varepsilon_0}, \tag{4}$$

together with the perfectly conducting condition $\vec{E}\times\vec{n} = 0$ on the boundary Γ, and the initial condition $\vec{E}(t=0) = \vec{E}_0$. Moreover, dealing with a second-order problem, we add an initial condition for $\partial_t\vec{E}$, directly obtained from (1) as $t = 0$. To enforce the divergence constraint on the electric field we have introduced the Lagrange multipliers p to dualize the constraint in (1). The treatment on the magnetic field is performed in the same way.

3 Variational Formulations

Let us first introduce a few notations. The bounded domain Ω is subdivided into N disjoint subdomains that we denote by Ω_i, $1 \le i \le N$. The boundary between subdomains i and j, if not empty, will be denoted by Σ_{ij} and the whole internal boundary of subdomain i will be denoted by $\Sigma_i = \cup_j \Sigma_{ij}$. Moreover for a distribution $\mathbf{T} \in H^{-1/2}(\Sigma_{ij})^3$ and a function $\mathbf{f} \in H^{1/2}(\Sigma_{ij})^3$, $\langle T, f\rangle_{\Sigma_{ij}}$ denotes the corresponding duality product. Let us also recall the definitions of the functional spaces:

$$\begin{aligned} H(curl,\Omega) &= \{\vec{E} \in L^2(\Omega)^3, \nabla \times \vec{E} \in L^2(\Omega)^3\},\\ H(div,\Omega) &= \{\vec{E} \in L^2(\Omega)^3, \nabla \cdot \vec{E} \in L^2(\Omega)\},\\ H_0(curl,\Omega) &= \{\vec{E} \in H(curl,\Omega),\ \vec{E} \times \vec{n} = 0 \text{ on } \Gamma\}. \end{aligned}$$

In fact we are dealing with a time-dependant problem and we should include this dependency in the definition of the functionnal spaces. For sake of simplicity we will only assume that every formulations in the sequel hold for almost any t in the time interval $[0, \mathcal{T}]$. The variational formulation for the constrained equation of the electric field on the whole domain, is obtained first by multiplying the wave equation in (4) by $\vec{F} \in H_0(curl,\Omega) \cap H(div,\Omega)$ (denoted $H_0(curl,div,\Omega)$). Then integrating by parts over Ω, we get a first mixed variational formulation which is well posed since the well known inf-sup condition [3, 5] is fulfilled. Adding $c^2 \int_\Omega \nabla \cdot \vec{E}\, \nabla \cdot \vec{F}\, d\vec{x}$ to its LHS and $c^2/\varepsilon_0 \int_\Omega \rho \nabla \cdot \vec{F}\, d\vec{x}$ to its RHS, we get an augmented variational formulation which reads:

Find $(\vec{E}, p) \in H_0(curl,div,\Omega) \times L^2(\Omega)$ *such that :*

$$\frac{d^2}{dt^2}\int_\Omega \vec{E}\cdot\vec{F}\, d\vec{x} + c^2\Big(\int_\Omega \nabla\times\vec{E}\cdot\nabla\times\vec{F}\, d\vec{x} + \int_\Omega \nabla\cdot\vec{E}\,\nabla\cdot\vec{F}\, d\vec{x}\Big) + \int_\Omega p\nabla\cdot\vec{F}\, d\vec{x}$$

$$= -\frac{1}{\varepsilon_0}\frac{d}{dt}\int_\Omega \vec{J}\cdot\vec{F}\, d\vec{x} + c^2/\varepsilon_0 \int_\Omega \rho\nabla\cdot\vec{F}\, d\vec{x} \quad \forall \vec{F} \in H_0(curl,div,\Omega), \tag{5}$$

$$\int_\Omega \nabla\cdot\vec{E}\, q\, d\vec{x} = \frac{1}{\varepsilon_0}\int_\Omega \rho q\, d\vec{x} \quad \forall q \in L^2(\Omega). \tag{6}$$

This formulation is well posed as well. In order to get a Maxwell solver suitable for multiprocessor computation, we introduce a variational formulation, which allows to treat each subdomain Ω_i separately. The continuity conditions are expressed on the tangential and the normal part separately. The continuity conditions across the interfaces Σ_{ij}, i.e. between the different subdomains, are written as $[\vec{E} \times \vec{n}_i]_{\Sigma_{ij}} = 0$ and $[\vec{E} \cdot \vec{n}_i]_{\Sigma_{ij}} = 0$ where $[.]_{\Sigma_{ij}}$ is the jump across Σ_{ij} and $\vec{n}_i$ the unit normal outward vector to Ω_i. Now, to handle these conditions, we enforce the continuity of the electric field by duality, introducing Lagrange multipliers on the subdomain interfaces. This method is similar in spirit to the dual Schur complement method as in [6]. A dualization procedure was also used in [2] to deal with continuity at material interfaces. We thus introduce the new unknowns $\vec{\lambda}_{ij}$, which are the Lagrange multipliers of the above constraints. We define the functional space associated to the broken domain with no continuity requirement at the interfaces:

$$X_0 = \{\vec{E} \in L^2(\Omega)^3, \vec{E}_{|\Omega_i} \in H(curl,\Omega_i) \cap H(div,\Omega_i), \vec{E} \times \vec{n} = 0 \text{ on } \Gamma\}.$$

Next we define the trace space on the internal boundaries

$$M = \left\{ \vec{\mu} \in \prod_{ij} (H^{1/2}(\Sigma_{ij}))^3;\ \exists \vec{F} \in H_0(curl, div, \Omega) \text{ with } \vec{F}_{|\Sigma_{ij}} = \vec{\mu}_{|\Sigma_{ij}} = \vec{\mu}_{ij} \right\}.$$

We shall decompose at any point of an internal boundary which is not shared by more than two subdomains any trace vector $\vec{\mu}$ in μ_n its normal component and $\vec{\mu}_T$ its tangential component.The orientation of $\vec{n}$ is chosen so that the normal be outward for the subdomain with the smallest index. This decomposition is well defined almost everywhere on the internal boundary. We recall that the scalar components of the tangential traces of fields in $H(curl, \Omega_i)$ along Σ_{ij}, as well as the normal traces of fields in $H(div, \Omega_i)$ are defined in $H^{-1/2}(\Sigma_{ij})$. Then, the spaces $H^{1/2}(\Sigma_{ij})$ will be the natural functional spaces for these Lagrange multipliers $\vec{\lambda}_{ij}$. Then, from the augmented formulation (5)-(6) the new variational formulation defined on the whole broken domain Ω reads:

Find $(\vec{E}, p, \vec{\lambda}) \in X_0 \times L^2(\Omega) \times M$ *such that* :

$$\frac{d^2}{dt^2} \sum_i \int_{\Omega_i} \vec{E} \cdot \vec{F} \, d\vec{x} + c^2 \Big(\sum_i \int_{\Omega_i} \nabla \times \vec{E} \cdot \nabla \times \vec{F} \, d\vec{x} + \sum_i \int_{\Omega_i} \nabla \cdot \vec{E} \, \nabla \cdot \vec{F} \, d\vec{x} \Big)$$
$$+ \sum_i \int_{\Omega_i} p \nabla \cdot \vec{F} \, d\vec{x} + \sum_{ij} (\langle \lambda_n, [\vec{F} \cdot \vec{n}] \rangle_{\Sigma_{ij}} + \langle \vec{\lambda}_T, [\vec{F} \times \vec{n}] \rangle_{\Sigma_{ij}})$$
$$= -\frac{1}{\varepsilon_0} \frac{d}{dt} \sum_i \int_{\Omega_i} \vec{J} \cdot \vec{F} \, d\vec{x} + c^2/\varepsilon_0 \sum_i \int_{\Omega_i} \rho \nabla \cdot \vec{F} \, d\vec{x} \quad \forall \vec{F} \in X_0, \tag{7}$$

$$\sum_i \int_{\Omega_i} \nabla \cdot \vec{E} \, q \, d\vec{x} = \frac{1}{\varepsilon_0} \sum_i \int_{\Omega_i} \rho q \, d\vec{x} \quad \forall q \in L^2(\Omega), \tag{8}$$

$$\sum_{ij} (\langle \mu_n, [\vec{E} \cdot \vec{n}] \rangle_{\Sigma_{ij}} + \langle \vec{\mu}_T, [\vec{E} \times \vec{n}] \rangle_{\Sigma_{ij}}) = 0 \quad \forall \vec{\mu} \in M, \tag{9}$$

Following the strategy by Raviart and Thomas [8] it has been proven that this problem has a unique solution $(\vec{E}, p, \vec{\lambda})$ of which $(\vec{E}, p)$ is the solution to the problem posed in the whole domain:

Theorem 1. *Assuming that Ω is a convex polyhedron, problem* (7)–(9) *has a unique solution* $(\vec{E}, p, \vec{\lambda}) \in X_0 \times L^2(\Omega) \times M$. *Moreover,* $(\vec{E}, p) \in H_0(curl, \Omega) \cap H(div, \Omega) \times L^2(\Omega)$ *is the solution to the problem* (5)–(6) *and we have* $\lambda_n = (\frac{c^2}{\varepsilon_0} \rho - c^2 \nabla \cdot \vec{E} - p)_{\Sigma_{ij}}$, $\vec{\lambda}_T = c^2 (\nabla \times \vec{E})_{T|\Sigma_{ij}}$ *on* Σ_{ij}.

4 Space and Time Discretization

We assume that the domain Ω is first meshed with tetrahedra and then a mesh partitioner is used to subdivide the mesh into disjoint sub-meshes which correspond to the subdomains Ω_i, so that the intersection of the subdomains consists of faces of tetrahedra which coincide on each side. Following the method described in [1, 2], Taylor-Hood elements are used. For this purpose the coarse mesh of tetrahedra $\mathcal{T}_{2h}$

is subdivided, each tetrahedron being subdivided into eight sub-tetrahedra to give the finer mesh $\mathcal{T}_h$. We shall denote by $(\varphi_k)_k$ the P^1 basis functions on the finer mesh and by $(\psi_l)_l$ the P^1 basis functions associated to the coarser mesh. Let us also denote by $P_h^1(\Omega_i)$ the P^1 space defined on the fine mesh of Ω_i and $P_{2h}^1(\Omega_i)$ the P^1 space defined on the coarse mesh of Ω_i. We define $V_{hi} \subset P_h^1(\Omega_i)^3$ the finite dimensional space associated to $H_0(curl,\Omega_i) \cap H(div,\Omega_i)$ and $L_{2h} \subset P_{2h}^1(\Omega_i)$ the finite dimensional space associated to $L^2(\Omega_i)$ in the conforming finite element approximation, see [1] for more precisions. We can now introduce the finite dimensional space $T_{hij} = \{\overrightarrow{\tau} \in P_h^1(\Sigma_{ij})^3, \overrightarrow{\tau}(x) \cdot \overrightarrow{n}_i = 0\}$ for the discretization of the interfaces.

Then, we introduce the matrices associated to the different terms in the variational formulation. For the domain Ω_i, we denote by M_i the lumped mass matrix for vectors on the fine mesh and M_{2i} the lumped mass matrix corresponding to scalars on the coarse mesh. We denote by K_i the matrix corresponding to $c^2 \int_{\Omega_i} \nabla \times \overrightarrow{E}_i \cdot \nabla \times \overrightarrow{F}_i d\overrightarrow{x} + c^2 \int_{\Omega_i} \nabla \cdot \overrightarrow{E}_i \nabla \cdot \overrightarrow{F}_i d\overrightarrow{x}$, L_i the matrix corresponding to $\int_{\Omega_i} \nabla \cdot \overrightarrow{E}_i q_i d\overrightarrow{x}$ and R_{ij} the matrix corresponding to $\langle \overrightarrow{E}_i \cdot \overrightarrow{\mu}_{ij} \rangle$. Moreover for any matrix A, A^T denotes the transpose of A. In order to verify the discrete inf-sup condition, the electric field is approximated on the finer mesh $\mathcal{T}_h$ (with the subscript h), whereas the Lagrange multiplier p is approximated on the coarser mesh $\mathcal{T}_{2h}$ (with the subscript $2h$). With this notation problem (7)–(9) discretized in space becomes

$$\frac{d^2}{dt^2} M_i \overrightarrow{E}_{hi}(t) + K_i \overrightarrow{E}_{hi} + L_i^T p_{2hi} + \sum_j \varepsilon_{ij} R_{ij}^T \overrightarrow{\lambda}_{hij} = -\frac{1}{\varepsilon_0} \frac{d}{dt} M_i \overrightarrow{J}_{hi}(t) \tag{10}$$

$$L_i \overrightarrow{E}_{hi}(t) = \frac{1}{\varepsilon_0} M_{2i} \rho_{2hi} \tag{11}$$

$$R_{ij}(\overrightarrow{E}_{hi} - \overrightarrow{E}_{hj}) = 0, \tag{12}$$

with ε_{ij} defined by $\varepsilon_{ij} = 1$ if $i < j$ and $\varepsilon_{ij} = -1$ if $i > j$. For time differentiation we choose an explicit centered scheme of order two (the leap-frog scheme), where Δt is the time-step and $t_n = n\Delta t$ are the discrete times. In order to enforce the constraints numerically the Lagrange multipliers are defined at the most advanced time steps. This yields, in each of the subdomain Ω_i, the following matrix problem which needs to be solved at each time step:

$$M_i \overrightarrow{E}_{hi}^{n+1} + L_i^T p_{2hi}^{n+1} + \sum_j \varepsilon_{ij} R_{ij}^T \overrightarrow{\lambda}_{hij}^{n+1} = \overrightarrow{F}_i^n \tag{13}$$

$$L_i \overrightarrow{E}_{hi}^{n+1} = \frac{1}{\varepsilon_0} M_{2i} \rho_{2hi}^{n+1} \tag{14}$$

$$R_{ij}(\overrightarrow{E}_{hi}^{n+1} - \overrightarrow{E}_{hj}^{n+1}) = 0, \tag{15}$$

where $\overrightarrow{F}_i^n$ contains all the terms being known at time t_{n+1}.

Let us now give an expression of the full linear system, involving all the subdomains. We denote $\overrightarrow{u} = (\overrightarrow{E}_1, p_1, \ldots, \overrightarrow{E}_N, p_N)^T$ and $\overrightarrow{\lambda} = (\overrightarrow{\lambda}_{12}, \ldots)^T$. Then the linear system to be solved has the form:

$$\begin{pmatrix} A & R^T \\ R & 0 \end{pmatrix} \begin{pmatrix} \overrightarrow{u} \\ \overrightarrow{\lambda} \end{pmatrix} = \begin{pmatrix} \overrightarrow{G} \\ 0 \end{pmatrix} \tag{16}$$

where

$$A = \begin{bmatrix} A_1 & 0 & 0 \\ 0 & \ddots & 0 \\ 0 & 0 & A_N \end{bmatrix}, \quad A_i = \begin{pmatrix} M_i & L_i^T \\ L_i & 0 \end{pmatrix}$$

and $\overrightarrow{G}$ is the vector built up with the right-hand sides of (13) and (14). We chose to solve this system with an iterative algorithm, similar to the Uzawa algorithm. Noticing that we can eliminate the unknowns $(\overrightarrow{E}_1, p_1, \ldots, \overrightarrow{E}_N, p_N)$ in the system to get $RA^{-1}R^T\overrightarrow{\lambda} = RA^{-1}\overrightarrow{G}$, the Uzawa algorithm amounts to using a conjugate gradient algorithm on this latter system. The solution of this system involves the inversion of A which amounts to the local solution on each subdomain of the original constrained problem which was solved in the sequential code.

5 Solution of the Doubly Constrained System

In order to solve this doubly constrained system we shall use two embedded preconditioned Uzawa algorithms. The preconditioner of the outer Uzawa problem must be an approximate inverse of $RA^{-1}R^T$. We first remark that the columns of R corresponding to the degrees of freedom p_i are identically null. Therefore we have to find an approximate of $\tilde{R}\tilde{A}^{-1}\tilde{R}^T$ where $\tilde{R}$ (resp. $\tilde{A}$) is the $N \times m$ (resp. $N \times N$) block submatrix extracted from R (resp. A) by eliminating the blocks related to the p_i's. The analysis of the inner system yields that on each sub-domain i:

$$\tilde{A}_i = M_i^{-1} - M_i^{-1}L_i^T(L_iM_i^{-1}L_i^T)^{-1}L_iM_i^{-1}.$$

The simplest preconditionner of the outer Uzawa problem is therefore defined as $P_{out} = \tilde{R}D\tilde{R}_T$ where D is a block diagonal matrix, each block D_i being a diagonal approximation of $\tilde{A}_i$. Noticing that $L_iM_i^{-1}L_i^T$ is the inner Uzawa operator we chose $D_i = diag(M_i^{-1} - M_i^{-1}L_i^TP_{in,i}^{-1}L_iM_i^{-1})$ with $P_{in,i}$ the preconditionner of the inner Uzawa problem. At every iteration of the outer Uzawa algorithm, we have to solve on each subdomain Ω_i the linear system:

$$\begin{cases} M_i\overrightarrow{E}_i + L_i^T p_i = \overrightarrow{b}_i \\ L_i\overrightarrow{E}_i = c_i \end{cases}$$

with the inner Uzawa algorithm. Note that, thanks to the chosen distribution of the matrix R, the outer Uzawa algorithm involves only local matrix vector products and reductions in addition to the local inner Uzawa solves.

6 Numerical Application

We present a classical test case related to the time evolution of a cavity resonant mode. We consider a cubic cavity enclosed in a perfect conductor in a cube of side equal to one. At time $t = 0$ we initialize the field components in the whole domain with the analytical expressions calculated at the initial time. Then the field values obtained at the final computational time $t = T_f$ can be compared with the exact solution. The cube is discretized by cutting each side into 16 pieces and then each resulting smaller cube into 6 tetrahedra. This gives us the coarse mesh. The associated fine mesh then consists of 196608 elements. We performed the domain decomposition by hand using the specificity of our mesh. The fields depicted on Fig. 1 enable us to verify visually that the results are correct, which is confirmed by comparison to the analytical results. The results are identical for the runs on different numbers of processors. We have also verified that even with an irregular partitionning such as those obtained with Metis the results are correct as well. In order to verify the

Fig. 1. Components E_x and E_z computed on 4 processors

efficiency of the parallelization, we ran this test case on 1, 2, 4, 8 and 16 processors. Except, when going from 1 to 2 processors which does not give any improvement due to the overhead linked to the outer Uzawa the speed-up is proportional to the number of processors which corresponds to the optimal performance one can expect from the parallelization. For instance, for 300 time steps without diagnostics on an Origin 2000 with R10000 processors, the computation times are 6 min 27 s for one processor, 3 min 33 s for 4 processors and 53 s for 16 processors. However, these results about the efficiency of this parallelization algorithm must be assessed with regard to the accuracy achieved on the continuity of the solution at the interfaces and moreover to the error between the result on one processor and the results on several ones.

7 Conclusion

In this paper, we presented a nonoverlapping domain decomposition approach for solving the three-dimensional time-dependent Maxwell equations. It is constructed from a mixed variational formulation, as a constraint on the divergence is taken into account explicitely. For this purpose, it is especially well adapted for the solution of the Vlasov-Maxwell equations, widely used in the framework of plasma physics or hyperfrequency devices simulations. The domain decomposition methodology we chose to implement has the important asset, which led us to choose it, that it enables to reuse without modification most of the existing sequential code. It requires only to add an external Uzawa algorithm in order to enforce the continuity of the fields at the subdomain interfaces.

References

[1] Assous, F., Degond, P., Heintzé, E., Raviart, P.A., Segré, J.: On a finite element method for solving the three-dimensional Maxwell equations. *J. Comput. Phys.*, 109:222–237, 1993.

[2] Assous, F., Degond, P., Segré, J.: Numerical approximation of the maxwell equations in inhomogeneous media by a p^1 conforming finite element method. *J. Comput. Phys.*, 128:363–380, 1996.

[3] Babuška, I.: The finite element method with Lagrange multipliers. *Numer. Math.*, 20:179–192, 1973.

[4] Birdsall, C.K., Langdon, A.B.: *Plasmas physics via computer simulation.* McGraw-Hill, New-York, 1985.

[5] Brezzi, F.: On the existence, uniqueness and approximation of saddle point problems arising from Lagrange multipliers. *Rev. Française Automat. Informat. Rechereche Opérationnelle Sér. Rouge*, 8(R2):129–151, 1974.

[6] Farhat, C., Roux, F.-X.: An unconventional domain decomposition method for an efficient parallel solution of large-scale finite element systems. *SIAM J. Sci. Statist. Comput.*, 13(1):379–396, 1992.

[7] Heintzé, E.: *Résolution des équations de Maxwell tridimensionnelles instationnaires par une méthode d'éléments finis*. PhD thesis, University Paris 6, France, 1992.

[8] Raviart, P.-A., Thomas, J.-M.: Primal hybrid finite element methods for 2nd order elliptic equations. *Math. Comp.*, 31(138):391–413, 1977.

[9] Secroun, A., Mens, A., Segré, J., Assous, F., Piault, E., Rebuffie, J.: Modeling of a microchannel plate working in pulsed mode. In *Proc. SPIE, 22nd International Congress on High-Speed Photography and Photonics*, 2869: pages 132–138, 1997.

A Discovery Algorithm for the Algebraic Construction of Optimized Schwarz Preconditioners

Amik St-Cyr[1] and Martin J. Gander[2]

[1] National Center for Atmospheric Research, 1850 Table Mesa Drive, Boulder CO 80305. amik@ucar.edu

[2] University of Geneva, 2-4 rue du Lièvre, CP 64 CH-1211 Genève Martin.Gander@math.unige.ch

Summary. Optimized Schwarz methods have been developed at the continuous level; in order to obtain optimized transmission conditions, the underlying partial differential equation (PDE) needs to be known. Classical Schwarz methods on the other hand can be used in purely algebraic form, which have made them popular. Their performance can however be inferior compared to that of optimized Schwarz methods. We present in this paper a discovery algorithm, which, based purely on algebraic information, allows us to obtain an optimized Schwarz preconditioner for a large class of numerically discretized elliptic PDEs. The algorithm detects the nature of the elliptic PDE, and then modifies a classical algebraic Schwarz preconditioner at the algebraic level, using existing optimization results from the literature on optimized Schwarz methods. Numerical experiments using elliptic problems discretized by Q_1-FEM, P_1-FEM, and FDM demonstrate the algebraic nature and the effectiveness of the discovery algorithm.

1 Introduction

Optimized Schwarz methods are based on transmission conditions between subdomains which are different from the classical Dirichlet conditions. The transmission conditions are adapted to the partial differential equation in order to lead to faster convergence of the method. Optimized transmission conditions are currently available for many types of scalar PDEs: for Poisson problems including a diagonal weight, see [3], for indefinite Helmholtz problems, see [4] and for advection reaction diffusion problems, see [2, 5]. More recently, it was shown that one can easily transform a classical algebraic Schwarz preconditioner such as restricted additive Schwarz (RAS) methods, into an optimized one, by simply changing some matrix entries in the local subdomain matrices, see [6]. However, in order to know what changes to make, one needs to know what the underlying PDE is, and thus the optimized RAS method has so far not become a black box solver, in contrast to classical RAS. We propose in this paper a discovery algorithm which is able to extract all the required information from the given matrix, and thus to make optimized RAS into a black box solver, for discretizations of the elliptic partial differential equation

M. Bercovier et al. (eds.), *Domain Decomposition Methods in Science and Engineering XVIII*, Lecture Notes in Computational Science and Engineering,
DOI: 10.1007/978-3-642-02677-5,

$$\nu\Delta u + \mathbf{a}\cdot\nabla u - \eta u = f \quad \text{in } \Omega, \tag{1}$$

with suitable boundary conditions. Here, ν, $\mathbf{a}$ and η are all functions of $\mathbf{x}$, and by a suitable choice, we can handle all elliptic PDEs for which currently optimized transmission conditions are known. In this paper, we focus on Robin transmission conditions, which are of the form

$$(\partial_n + p)u_i = (\partial_n + p)u_j \tag{2}$$

at the interface between subdomain i and j. In general, the optimized scalar parameter p depends on the local mesh size h, the overlap width Ch, the interface diameter L, and the coefficients of the underlying PDE, i.e. $p = p(h, Ch, L, \nu, \mathbf{a}, \eta)$.

A discretization of (1) leads to a linear system of equations of the form

$$A\mathbf{u} = (K + S + M)\mathbf{u} = \mathbf{f}, \tag{3}$$

where K is the stiffness matrix, $K\mathbb{I} = 0$ with $\mathbb{I}$ the vector of all ones, S is skew-symmetric from the advection term of the PDE, $S\mathbb{I} = 0$, and M is a mass matrix from the η term in the PDE. For a black box preconditioner, only the matrix A is given, and the decomposition (3) needs to be extracted automatically, in addition to the mesh size and the diameter of the interface, in order to use the existing formulas for the optimized parameter p in a purely algebraic fashion. We also need to extract a normal derivative for (2) algebraically.

In what follows, we assume that we are given the restriction operators R_j and $\tilde{R}_j$ of a restricted additive Schwarz method for the linear system (3), and the associated classical subdomain matrices $A_j := R_j A R_j^T$. The restricted additive Schwarz preconditioner for (3) would then be $\sum_j \tilde{R}_j^T A_j^{-1} R_j$, and the optimized restricted additive Schwarz method is obtained by slightly modifying the local subdomain matrices A_j at interface nodes, in order to obtain $\tilde{A}_j$, which represent discretizations with Robin, instead of Dirichlet boundary conditions, see [6]. In order for this replacement to lead to an optimized Schwarz method, an algebraic condition needs to be satisfied, which requires a minimal overlap and a certain condition at cross-points; for details, see [6].

2 Discovery Algorithm

We now describe how appropriate matrices $\tilde{A}_j$ for an optimized Schwarz preconditioner can be generated algebraically for discretizations of the PDE (1) given in matrix form (3). There are three steps in the algorithm to generate the modified $\tilde{A}_j$:

1. Interface detection.
2. Extraction of physical and discretization parameters.
3. Construction of the optimized transmission condition.

2.1 Interface Detection

For a matrix $A \in \mathbb{R}^{N\times N}$, let $\mathcal{S}(A)$ be its **canonical index** set, i.e. the set of integers going from 1 to N, and let $\mathbf{c} \in \mathbb{N}^N$ be its **multiplicity**, i.e. c_i contains the total number of non-zero entries in the corresponding row of A. For a subdomain decomposition given by restriction matrices R_j, let the matrix $A_j = R_j A R_j^T$ have the canonical index $\mathcal{S}(A_j)$ with multiplicity $\mathbf{c}_j$. Then the set of indices $\mathcal{B}(A_j)$ representing the interfaces of subdomain j corresponds to the non-zero entries of $\mathbf{c}_j - R_j\mathbf{c}$, where $\mathbf{c}$ is the multiplicity of A. The set $\mathcal{B}(A_j)$ indicates which rows of the matrix A_j need to be modified in order to obtain $\tilde{A}_j$ for an optimized preconditioner.

2.2 Extraction of Physical and Discretization Parameters

We start by guessing the decomposition (3) of A by computing

$$S = \frac{1}{2}(A - A^T), \quad M = \mathrm{diag}(A\mathbb{I}), \quad K = \frac{1}{2}(A + A^T) - M. \tag{4}$$

This approach does not necessarily find the same parts one would obtain by knowing the discretization: for example we can only guess a lumped mass matrix and not discover an upwind scheme. The parts we obtain however correspond to a decomposition relevant for the problem.

Definition 1. *Using (4) for each interface node i, we define the*

- **local viscosity indicator:** $\nu_i := \sum_j |K_{ij}|/(2(c_i - 1))$,
- **local advection indicator:** $\alpha_i := \max_j |S_{ij}|$,
- **local zeroth order term indicator:** $\eta_i := sign(K_{ii})M_{ii}$.

These three indicators are enough to reveal the PDE-like properties of the matrix at the interface:

1. $\nu_i > 0$, $\eta_i = 0$ and $\alpha_i = 0$: Poisson equation.
2. $\nu_i > 0$, $\eta_i > 0$ and $\alpha_i = 0$: Poisson equation with weight, or implicit heat equation, see [3].
3. $\nu_i > 0$, $\eta_i < 0$ and $\alpha_i = 0$: indefinite Helmholtz equation, see [4].
4. $\nu_i > 0$, $\eta_i = 0$ and $\alpha_i \neq 0$: advection-diffusion equation, see [5].
5. $\nu_i > 0$, $\eta_i > 0$ and $\alpha_i \neq 0$: implicit advection-diffusion equation, see [2].

In the other cases (except if (1) has been multiplied by minus one, which can also be treated similarly), optimized transmission conditions have not yet been analyzed, and we thus simply apply RAS for that particular row. We next have to estimate the local mesh size h_i. The indicators above contain in general this information, for example for a standard five point finite difference discretization of $\eta - \nu\Delta$, we would obtain $\nu_i = \frac{\nu}{h^2}$, but we cannot detect the mesh size h separately without further information. In addition, the algebraic equations could have been scaled by h, or h^2, or any other algebraically useful diagonal scaling. However, in general, the optimized parameter p is also scaled accordingly: the analytical formulas for p all contain the size of the interface L and the mesh spacing h in a certain relation. Since the latter are both interface related quantities, we use the trace of the discovered Laplacian in order to estimate them.

Definition 2. *The relevant local mesh size at point $i \in \mathcal{B}(A_k)$ is*

$$h_i^k \approx \Big(\sum_j |(K_k)_{ij}|/(2c_i - 1)\Big)^{-1/2}, \tag{5}$$

where K_k is the trace of the discovered Laplacian and c_i is its associated multiplicity.

We finally need to estimate the interface diameter L of each interface. To this end, we need to discover the dimension of the problem is. This can be achieved using the ratio of interior nodes versus interface nodes in each subdomain. Solving the equation (# denotes the cardinality of the set)

$$\#\mathcal{B}(A_k) = (\#\mathcal{S}(A_k))^{\frac{d-1}{d}} \tag{6}$$

for d in each subdomain, we obtain an estimate for the dimension denoted by $\bar{d}_k$. We accept a fractional dimension because it is not uncommon for example for three dimensional domains to represent thin shells.

Definition 3. *The diameter of each interface $L = L_{jk}$ between subdomain j and k is estimated for 2d and 3d problems by*

$$L_{jk} := (\#\mathcal{B}_k(A_j))^{-\frac{\bar{d}_k - 2}{\bar{d}_k - 1}} \sum_{i \in \mathcal{B}_k(A_j)} h_i^j, \tag{7}$$

where $\mathcal{B}_k(A_j)$ denotes the interface nodes of subdomain j with subdomain k.

2.3 Construction of the Optimized Transmission Condition

In order to construct an algebraic approximation to the Robin transmission condition (2), we need a normal derivative approximation. Suppose that row i was identified as an interface node. For this row, we can partition the indices denoting the position in the row with non-zero elements into three sets:

1. the diagonal entry denoted by set $\{i\}$,
2. the off-diagonal entries that are not involved in the interface denoted by $\mathcal{I}_i$ for interior,
3. the off-diagonal entries that are on the interface, denoted by $\mathcal{F}_i$.

These indices take values in the set of integers indexing the full matrix A, but in order to simplify what follows, we re-label these indices from 1 to J. Let $\{\mathbf{x}_j\}_{j=1}^J$ be a set of arbitrary spatial points with associated scalar weights $\{w_j\}_{j=1}^J$, and let $\delta\mathbf{x}_{ji} = \mathbf{x}_j - \mathbf{x}_i$. In order to define a normal derivative at the point $\mathbf{x}_i$, we assume that

$$\|\delta \mathbf{x}_{ji}\| \le h, \quad \text{and} \quad \sum_{j\in\mathfrak{F}_i} w_j \delta \mathbf{x}_{ji} = O(h^2), \tag{8}$$

and we define an approximate unit outward normal vector $\mathbf{n}$ at $\mathbf{x}_i$ by

$$\mathbf{n} = -\sum_{j\in\mathfrak{I}_i} w_j \delta \mathbf{x}_{ji} / \|\sum_{j\in\mathfrak{I}_i} w_j \delta \mathbf{x}_{ji}\|. \tag{9}$$

A situation might arise were the set $\mathfrak{I}_i$ is empty. In this case the connectivity of the matrix must be exploited in order to find a second set of points connected to the points in $\mathfrak{F}_i$. By removing the points lying on any boundary a new set $\mathfrak{I}_i$ can be generated. This procedure can be repeated until the set is non-empty. Let the vectors τ_k, $k = 1,\ldots,d-1$ be an orthonormal basis spanning the tangent plane implied by $\mathbf{n}$ at $\mathbf{x}_i$, i.e. $\mathbf{n}\cdot\tau_k = 0$.

Proposition 1. *If conditions (8) are satisfied, and in addition $w_i = -\sum_{j\neq i} w_j$, then for a sufficiently differentiable function u around $\mathbf{x}_i$, we have*

$$-\frac{\sum_{j=1}^{J} w_j u(\mathbf{x}_j)}{\|\sum_{j\in\mathfrak{I}_i} w_j \delta \mathbf{x}_{ji}\|} = \mathbf{n}\cdot\nabla u(\mathbf{x}_i) + O(h). \tag{10}$$

Proof. Using a Taylor expansion, and the sum condition on w_j, we obtain

$$\begin{aligned}
\sum_{j=1}^{J} w_j u(\mathbf{x}_j) &= \sum_{j=1}^{J} w_j (u(\mathbf{x}_i) + \delta \mathbf{x}_{ji}\cdot\nabla u(\mathbf{x}_i) + O(h^2)) \\
&= (w_i + \sum_{j\neq i} w_j) u(\mathbf{x}_i) + \sum_{j\neq i} w_j \delta \mathbf{x}_{ji}\cdot\nabla u(\mathbf{x}_i) + O(h^2) \\
&= \nabla u(\mathbf{x}_i)\cdot \sum_{j\neq i} w_j \delta \mathbf{x}_{ji} + O(h^2).
\end{aligned}$$

Now using the second condition in (8), and the decomposition of the gradient into normal and tangential components, $\nabla u(\mathbf{x}_i) = u_0\mathbf{n} + \sum_{k=1}^{d-1} u_k \tau_k$, we get

$$\begin{aligned}
\sum_{j=1}^{J} w_j u(\mathbf{x}_j) &= \nabla u(\mathbf{x}_i)\cdot \sum_{j\in\mathfrak{I}_i} w_j \delta \mathbf{x}_{ji} + O(h^2), \\
&= u_0 \mathbf{n}\cdot \sum_{j\in\mathfrak{I}_i} w_j \delta \mathbf{x}_{ji} + \sum_{k=1}^{d-1} u_k \tau_k \cdot \sum_{j\in\mathfrak{I}_i} w_j \delta \mathbf{x}_{ji} + O(h^2).
\end{aligned}$$

The double sum vanishes, since the sum on j equals $\mathbf{n}$ up to a multiplicative constant and $\mathbf{n}\cdot\tau_k = 0$. Now using the definition of the approximate normal $\mathbf{n}$, and using that $u_0 = \mathbf{n}\cdot\nabla u(\mathbf{x}_i)$, leads to the desired result.

Note that the formula for the approximation of the normal derivative (10) does not need the explicit computation of a normal or tangential vector at the interface.

Definition 4. *An approximation $A_i^{\mathfrak{I}}$ to the normal derivative is generated from matrix A at a line i having a non-empty set $\mathfrak{I}_i$ by performing (in this order): $a_{ii} = 0$, $a_{ij} = 0$ for $j \in \mathfrak{F}_i$, $a_{ii} = -\sum_{j\neq i} a_{ij}$.*

There are also optimized Schwarz methods with higher order transmission conditions, which use tangential derivatives at the interfaces. Such methods involve for the Poisson equation the Laplace-Beltrami operator at the interface, see [3], or more generally the remaining part of the partial differential operator, see for example [2] or [1]. If we want to use higher order transmission conditions also at the algebraic level, we need to extract the corresponding discretization stencil at the interface as well. This stencil has the same dimensions as $A_i^{\mathcal{J}}$ and contains all the coefficients lying in $\mathcal{F}$.

Definition 5. *The complement of $A_i^{\mathcal{J}}$ is the matrix $A_i^{\mathcal{F}}$ generated from matrix A at a line i having a non-empty set $\mathcal{J}_i$ by performing (in this order): $a_{ii} = 0$, $a_{ij} = 0$ for $j \in \mathcal{J}_i$, $a_{ii} = -\sum_{j \neq i} a_{ij}$.*

The matrices used to detect the nature of the PDE cannot be employed in the construction of the optimized transmission operator, because they might be rank deficient. The detected mass matrix could be employed, if one is present. However, for more generality, we choose the diagonal mass matrix for an interface node i as $D_i = \mathbf{h}_i^2 A_{ii}$: its sign is correct for the elliptic operator for the definite case; for the indefinite case, it needs to be multiplied by -1. From Definitions 4 and 5, and the first of the assumptions (8), we see that the normal and complement matrices are both $O(1)$ (the complement is a difference of 2 normal derivatives at the interface divided by h). However, the entries in the matrix are proportional to $1/h^2$. Thus the normal derivative needs a scaling factor of $1/h$. Consequently, both the mass and complement matrices are divided by h.

The algebraic representation of the transmission condition for domain k in the matrix is then given by

$$T_k \equiv \operatorname{diag}\left(\frac{\mathbf{p}_k}{\mathbf{h}_k}\mathbf{D}_k\right) + A_k^{\mathcal{J}} + \operatorname{diag}\left(\frac{\mathbf{q}_k}{\mathbf{h}_k}\right)A_k^{\mathcal{F}}, \tag{11}$$

where the division of a vector by a vector is component wise.

3 Numerical Experiments

We consider three different discretizations, FDM, Q_1-FEM, and P_1-FEM applied to an *a priori unknown* positive definite Helmholtz operator. In all cases the solution is $u(x,y) = \sin(\pi x)\sin(\pi y)$ on the domain $(0,1) \times (0,1)$ with Dirichlet boundary conditions. We present results for the iterative form of the algorithm and its acceleration by GMRES. For all cases a starting vector containing noise in $(0,1)$ was employed.

In the first set of experiments, see Table 1, a square corner $(0,1/2) \times (0,1/2)$ is considered as one of the two subdomains, and the L-shaped rest is the other subdomain. These domains are uniformly discretized for the first experiment by a finite difference method, and for the second experiment by a Q1 finite element method. In each experiment, an overlap of two mesh sizes is added. We can see from Table 1 that the optimized Schwarz methods generated purely algebraically from the global

h		1/8	1/16	1/32	1/64	1/128	1/256
Q_1-FEM:							
iterative:	RAS	6	15	32	67	136	275
iterative:	O0	8	14	23	33	48	65
iterative:	O2	8	13	19	24	30	36
GMRES:	RAS	4	6	10	13	19	26
GMRES:	O0	5	7	9	12	15	19
GMRES:	O2	5	7	9	11	13	16
FDM:							
iterative:	RAS	7	16	32	67	136	275
iterative:	O0	8	16	27	42	63	90
iterative:	O2	8	15	24	33	43	53
GMRES:	RAS	4	8	11	17	24	35
GMRES:	O0	4	6	8	10	14	17
GMRES:	O2	5	6	9	10	12	14

Table 1. Structured corner domain: the same algebraic algorithm was employed

matrix perform significantly better than the classical Schwarz method, both for the iterative and the GMRES accelerated versions.

We next show an example of a triangularly shaped decomposition of the square into two subdomains, as shown in Fig. 1. The discretization is now performed using

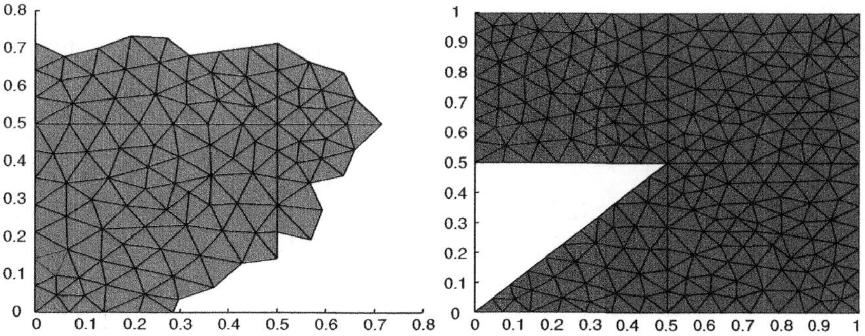

Fig. 1. Left panel: triangularly shaped domain Ω_1 extended by an overlap of size $3h$. Right panel: non-convex domain Ω_2.

an unstructured triangular mesh and a P1 finite element discretization. We show in Table 2 again a comparison of the iteration counts for the classical and various optimized Schwarz methods, obtained purely at the algebraic level with the discovery algorithm. We observe again that substantial gains are possible.

Triangles		534	2080	8278
P_1-FEM:				
iterative:	RAS	16	35	71
iterative:	O0	12	22	32
iterative:	O2	11	18	24
GMRES:	RAS	11	14	20
GMRES:	O0	8	11	15
GMRES:	O2	8	11	14

Table 2. Unstructured corner domain: the same algebraic algorithm was employed

Acknowledgement. The National Center for Atmospheric Research is sponsored by the National Science Foundation. The first author's travel to Geneva, and the second author were partly supported by the Swiss National Science Foundation Grant 200020-1 17577/1.

References

[1] Bennequin, D., Gander, M.J., Halpern, L.: A homographic best approximation problem with application to optimized Schwarz waveform relaxation. *Math. Comp.*, 78:185–223, 2009.

[2] Dubois, O.: *Optimized Schwarz Methods for the Advection-Diffusion Equation and for Problems with Discontinuous Coefficients.* PhD thesis, McGill University, June 2007.

[3] Gander, M.J.: Optimized Schwarz methods. *SIAM J. Numer. Anal.*, 44(2):699–731, 2006.

[4] Gander, M.J., Magoulès, F., Nataf, F.: Optimized Schwarz methods without overlap for the Helmholtz equation. *SIAM J. Sci. Comput.*, 24(1):38–60, 2002.

[5] Japhet, C., Nataf, F., Rogier, F.: The optimized order 2 method. Application to convection-diffusion problems. *Future Generation Computer Systems FUTURE*, 18(1):17–30, 2001.

[6] St-Cyr, A., Gander, M.J., Thomas, S.J.: Optimized multiplicative, additive and restricted additive Schwarz preconditioning. *SIAM J. Sci. Comput.*, 29(6):2402–2425, 2007.

On the Convergence of Optimized Schwarz Methods by way of Matrix Analysis

Sébastien Loisel and Daniel B. Szyld

Temple University {szyld|loisel}@temple.edu

Summary. Domain decomposition methods are widely used to solve in parallel large linear systems of equations arising in the discretization of partial differential equations. Optimized Schwarz Methods (OSM) have been the subject of intense research because they lead to algorithms that converge very quickly. The analysis of OSM has been a very challenging research area and there are currently no general proofs of convergence for the optimized choices of the Robin parameter in the case of overlap. In this article, we apply a proof technique developed for the analysis of Schwarz-type algorithms using matrix analysis techniques and specifically using properties of matrix splittings, to the Optimized Schwarz algorithms. We thus obtain new general convergence results, but they apply only to large Robin parameters, which may not be the optimal ones.

1 Introduction

Schwarz iterative methods for the solution of boundary value problems have been extensively studied; see, e.g., [12, 14, 16]. When some of these methods are used as parallel preconditioners, they have been shown to scale perfectly in many thousands of processors; see, e.g., [6].

The main idea is to split the overall domain Ω into multiple subdomains $\Omega = \Omega_1 \cup \Omega_2 \cup \cdots \cup \Omega_p$, then solve Dirichlet problems on each subdomain, and iterate. This is usually refered to as multiplicative Schwarz methods. There are many variants, which we name after the type of boundary condition used on the artificial interfaces, for instance Schwarz-Neumann or Schwarz-Robin. While the classical Schwarz methods (with Dirichlet boundary conditions on the artificial interface) are very well understood, [12, 14, 16], less progress has been made in the theory of Schwarz-Robin methods. In this paper, we present a new convergence result for large Robin parameters, which may be useful in certain situations.

One difficulty with Schwarz-Robin algorithms is the choice of the real parameter α for the differential operator $\alpha u + D_\nu u$ (D_ν is a normal derivative) on the artificial interface. The first analysis of a Schwarz-Robin method was performed with some generality by Lions [9] for the case of zero overlap. It is well-known that overlap usually improves the convergence rate of Schwarz algorithms. Detailed studies for the

M. Bercovier et al. (eds.), *Domain Decomposition Methods in Science and Engineering XVIII*, Lecture Notes in Computational Science and Engineering,
DOI: 10.1007/978-3-642-02677-5,

overlap case do exist for the case of simple domains, such as rectangles, half planes, or hemispheres, and the main analytical tool is the use of Fourier transforms. For a history, review, analysis and extensive bibliography of such methods, see [5]. For general domains, and two overlapping subdomains, Kimn [7, 8], proved convergence of the method under certain conditions; see also [10].

The main contribution of this paper is to show convergence of the (multiplicative) Schwarz-Robin iteration for elliptic problems on p general subdomains of a general two-dimensional domain, when α is sufficiently large.

For classical Schwarz methods, algebraic representations have been proposed; see [1, 4, 11] and references therein. Our approach is inspired by some of these papers, and by [7, 8, 15], and it was prompted by some general matrix analytic convergence results in the very recent paper [3] based on the theory of splittings. We are able to apply these new results using an inverse trace inequality, which we prove in Lemma 1 in Sec. 2.

1.1 Model Problem and Notation

Let Ω be an open region in the plane. For simplicity, we consider the problem

$$\Delta u = f \ \text{ in } \Omega, \qquad \text{with } u = 0 \ \text{ on } \partial\Omega; \tag{1}$$

although our proof holds, mutatis mutandis, for any symmetric and coercive elliptic operator.

We use a piecewise linear finite element discretization of (1). We denote by $T = T_h$ the triangulation of the finite element method (which depends on the mesh parameter h), and by v_i its vertices.

For the description of the discretized problem, we abuse the notation, and call u the vector of nodal values which approximate the function u at the nodes v_i. The discretized problem is then

$$Au = f,$$

where the $n \times n$ matrix A has entries

$$A_{ij} = \int_\Omega \nabla\phi_i \cdot \nabla\phi_j,$$

with ϕ_i and ϕ_j being piecewise linear basis functions corresponding to vertices v_i and v_j of the finite element discretization. The finite element space is denoted $V_{h,0}(\Omega) \subset H_0^1(\Omega)$.

We now define Schwarz and Optimized Schwarz algorithms. First, we introduce the notion of *restriction matrices*. This is a matrix R of the form

$$R = \begin{bmatrix} 0\,0\,1\,0\,0\,0 \\ 0\,1\,0\,0\,0\,0 \\ 0\,0\,0\,0\,1\,0 \end{bmatrix},$$

i.e, the $m \times n$ matrix R is formed by taking m rows from an $n \times n$ permutation matrix. The transpose R^T is known as a prolongation matrix. Note that $RR^T = I_m$, the identity in the m-dimensional space.

Given a *domain decomposition* $\Omega = \Omega_1 \cup \dots \cup \Omega_p$, such that each Ω_i is a union of triangles in T_h, we can form restriction matrices $R_1, \dots, R_p$ which restrict to those vertices in the *interior* of Ω_i. These matrices are uniquely determined up to permutation of the rows.

Let $u_0 \in V_0(\Omega)$. The (multiplicative) Schwarz iteration can be phrased algebraically as

$$u_{k+\frac{j}{p}} = u_{k+\frac{j-1}{p}} + M_j(f - Au_{k+\frac{j-1}{p}}) \quad \text{for } k = 0,1,2,\dots \text{ and } j = 1,2,\dots,p.$$

This recurrence relation can be concisely stated in terms of the error terms $u_k - u$:

$$u_{k+1} - u = (I - M_pA)\dots(I - M_1A)(u_k - u) \quad \text{for } k = 0,1,2,\dots, \tag{2}$$

where $M_i = R_i^T(R_iAR_i^T)^{-1}R_i$, $i = 1,\dots,p$. If we define $A_i = R_iAR_i^T$, we can write instead $M_i = R_i^T A_i^{-1} R_i$. If the subdomain Ω_i has n_i vertices in its interior, then A_i is an $n_i \times n_i$ matrix with entries

$$A_{i,jk} = \int_{\Omega_i} \nabla\phi_j \cdot \nabla\phi_k,$$

where $\phi_1, \dots, \phi_{n_i}$ are the piecewise linear basis functions of $V_0(\Omega_i)$. Note that A_i is the finite element discretization of a Dirichlet problem in Ω_i. The main idea of Optimized Schwarz Methods is to use Robin problems on the subdomains, instead of Dirichlet problems. This means that we must replace A_i by a matrix $\widetilde{A}_i$ of a Robin problem. The FEM discretization of a Dirichlet problem, (e.g., A_i) does not have any degrees of freedom along the boundary Ω_i (which is why our restriction matrices correspond to the vertices in the interior of Ω_i). However, the FEM discretization of a Robin problem contains degrees of freedom along the boundary. We want to keep the same matrices R_i for both algorithms, which means that the domain decomposition $\Omega_1^\circ, \dots, \Omega_p^\circ$ for the Robin version is not the same as the domain decomposition $\Omega_1, \dots, \Omega_p$ of the Dirichlet version.

For $i = 1,\dots,p$, define Ω_i° to be the set of triangles in Ω_i which do not have a vertex on $\overline{\partial\Omega_i \setminus \partial\Omega}$. The matrix $\widetilde{A}_i$ of the Neumann problem for Ω_i° is

$$\widetilde{A}_{i,jk} = \int_{\Omega_i^\circ} \nabla\phi_j \cdot \nabla\phi_k,$$

where $\phi_1, \dots, \phi_{n_i}$ are the piecewise linear basis functions of $V_0(\Omega_i)$. The matrix of the Robin problem with real parameter $\alpha > 0$ is then $\widetilde{A}_i + \alpha B_i$, where

$$B_{i,jk} = \int_{\partial\Omega_i^\circ} \phi_j \phi_k.$$

Optimized Schwarz algorithms use the matrices

$$\widetilde{M}_i = R_i^T(\widetilde{A}_i + \alpha B_i)^{-1}R_i \tag{3}$$

instead of $M_i = R_i^T A_i^{-1} R_i$, and the iteration is then

$$u_{k+1} - u = (I - \widetilde{M}_p A) \dots (I - \widetilde{M}_1 A)(u_k - u) \quad \text{for } k = 0, 1, 2, \dots \tag{4}$$

A related algorithm uses a coarse grid correction. This is achieved by choosing R_0 to be an averaging operator of the form

$$R_0 = \begin{bmatrix} 0.3 & 0.7 & 0 & 0 & 0 & 0 & 0 & 0 \\ 0 & 0 & 0.25 & 0.4 & 1 & 0.15 & 0 & 0 \\ 0 & 0 & 0 & 0 & 0 & 0 & 0.28 & 1 \end{bmatrix};$$

i.e., an $n_0 \times n$ matrix with non-negative entries. Usually, we choose R_0 so that R_0^T is the matrix of an interpolation operator from a coarse grid T_H to the fine grid T_h, where $H \gg h$. We then let $M_0 = R_0^T A_0^{-1} R_0$, where $A_0 = R_0 A R_0^T$ and we use $\widetilde{M}_i$ for $i = 1, \dots, p$. The coarse grid corrected OSM then fits our framework and the iteration is

$$u_{k+1} - u = (I - \widetilde{M}_p A) \dots (I - \widetilde{M}_1 A)(I - M_0 A)(u_k - u) \quad \text{for } k = 0, 1, 2, \dots \tag{5}$$

As we refine the mesh, the triangulation T_h changes. We may also want to change the domain decomposition $\Omega_1, \dots, \Omega_{p(h)}$, increasing the number of subdomains $p(h)$ as h goes to zero, in such a way that the amount of work per subdomain remains constant.

2 Convergence of OSM

In this section, we prove that the OSM (4) converges. We further assume that the triangulation T_h is quasi-uniform; see, e.g., [2].

Lemma 1 (Inverse Trace Inequality). *Let Ω be a domain with a quasi-uniform triangulation $T = T_h$. Let U be a set of triangles in T and let Γ be a set of edges in T. Let $V(\Omega, \Gamma, U)$ denote the space of piecewise linear functions on T which are zero at every vertex of U outside of Γ. Then, there is a constant $C < \infty$ which depends on the regularity parameters of T_h, but not on h, Γ, U or u, such that*

$$\int_U (\nabla u)^2 \leq C h^{-1} \int_\Gamma u^2, \tag{6}$$

for every $u \in V(\Omega, \Gamma, U)$.

Proof. What makes this inverse trace inequality possible is that the quadratic form $\int_U (\nabla u)^2$ depends only on the function values of u at the vertices of Γ. Indeed, let A_U be the matrix of $\int_U (\nabla u)^2$ on the space $V(\Gamma, U)$, with entries $A_{U,ij} = \int_U \nabla \phi_i \cdot \nabla \phi_j$ when both v_i and v_j are on Γ, and with $A_{U,ij} = 0$ otherwise. We introduce the "vertexwise" norm $u^T u = \sum_{v_i \in \Gamma} u^2(v_i)$. The matrix A_U has the same sparsity pattern, or is sparser than the matrix A of the elliptic problem. Hence, the bandwidth of A_U is independent of Γ and U, but does depend on the regularity parameters of T_h. By the regularity of T_h as $h \to 0$, A (and hence A_U) has a bandwidth $N(h) < N < \infty$, uniformly in h. By [13, §6.3.2], the entries of A_U are bounded by some constant C_1 which is independent of h. Hence, $u^T H u \leq C_2 u^T u$, where C_2 depends only on the regularity parameters of T_h. Likewise, by a variant of [2, §6.2], there is a C_3 such that $u^T u \leq h^{-1} C_3 \int_\Gamma u^2$. Putting these together, we obtain (6).

To prove our main result, we must first cite a Theorem from [3], where the notation M^H stands for the conjugate transpose of M.

Theorem 1. [3, Theorem 3.15] *Let $A \in \mathbb{C}^{n \times n}$ be Hermitian and positive semidefinite. Let $\widetilde{M}_i \in \mathbb{C}^{n \times n}$, $i = 1, \ldots, p$, be such that*

(i) $\ker A\widetilde{M}_i A = \ker \widetilde{M}_i A$.
(ii) *There exists a number $\gamma > 1/2$ such that*

$$\widetilde{M}_i + \widetilde{M}_i^H - 2\gamma \widetilde{M}_i^H A \widetilde{M}_i \tag{7}$$

is positive semidefinite on the range of A, for $i = 1, \ldots, p$.
(iii) $\bigcap_{i=1}^{p} \ker \widetilde{M}_i A = \ker A$.

Then the iteration (4) *converges for any initial vector u_0.*

We note that in our case, the matrices are real and symmetric and therefore $\widetilde{M}_i^H = \widetilde{M}_i$. We also note that Theorem 1 is general and does not require that the matrices $\widetilde{M}_i$ have the particular form discussed in this paper.

Theorem 2. *There is a constant C, which depends only on the regularity parameters of T_h, such that for all $\alpha \geq Ch^{-1}$, the OSM iteration defined by* (4) *converges for any initial vector u_0.*

Proof. We verify the hypotheses of Theorem 1. Since A is injective, part (i) is automatically satisfied. Part (iii) is also easily checked: it suffices to show that $\bigcap_{i=1}^{p} \ker \widetilde{M}_i = \{0\}$. Let f be such that $\widetilde{M}_i f = 0$ for $i = 1, \ldots, p$. Then, $R_i f = (\widetilde{A}_i + \alpha B_i) 0 = 0$ for $i = 1, \ldots, p$. Hence, $f|_{\Omega_i} = 0$ for $i = 1, \ldots, p$. Since $\Omega = \Omega_1 \cup \cdots \cup \Omega_p$, we have that $f = 0$.

For part (ii), we multiply (7) on the left by $(\widetilde{A}_i + \alpha B_i) R_i$ and on the right by its transpose $R_i^T (\widetilde{A}_i + \alpha B_i)$, using the expression of the symmetric matrix $\widetilde{M}_i$, and that $R_i R_i^T = I_{n_i}$, then it follows that all we need to show is that $u^T (\widetilde{A}_i + \alpha B_i - \gamma A_i) u \geq 0$ for every $u \in V_0(\Omega_i)$. We show this now for $\gamma = 1$. We calculate

$$\begin{aligned} u^T (\widetilde{A}_i + \alpha B_i - A_i) u &= \int_{\Omega_i^\circ} (\nabla u)^2 + \alpha \int_{\partial \Omega_i^\circ} u^2 - \int_{\Omega_i} (\nabla u)^2 \\ &= \alpha \int_{\partial \Omega_i^\circ} u^2 - \int_{U_i} (\nabla u)^2, \end{aligned}$$

where $U_i = \Omega_i \setminus \Omega_i^\circ$. Since the function u is in $V_0(\Omega_i) \subset V(\Omega_i, \partial \Omega_i^\circ, U_i)$, the result follows from Lemma 1. □

Remark 1. A choice of parameter $\alpha = O(h^{-1})$ is not an optimal value (see, e.g., [5]) and will typically yield a convergence rate which is not much better than the classical Schwarz algorithm (2). In fact, in [15] a similar two-parameter variant of OSM is given, and for specific choices of these parameters which closely correspond to our choice of $\alpha = O(h^{-1})$, the method coincides with classical Schwarz; see equation (4.3) of [15].

3 Numerical Experiments

We present a numerical experiment for problem (1) on an L-shaped region split into twelve subdomains, and we have $h = 1/32$. We use piecewise linear elements. The domain, the subdomains and the mesh are depicted in Figure 1. In Figure 2 we show

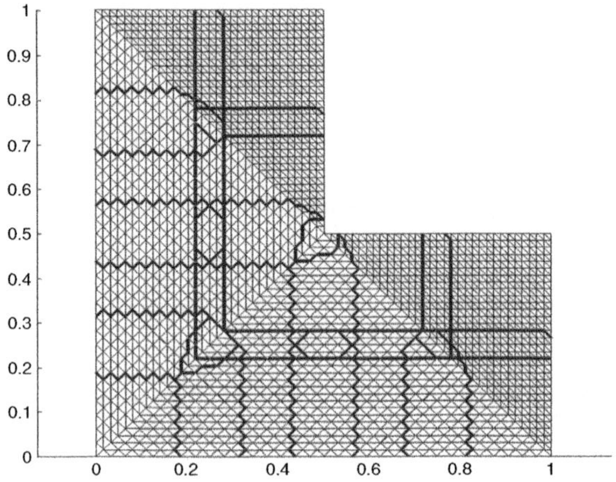

Fig. 1. L-shaped domain subdivided into twelve overlapping subdomains

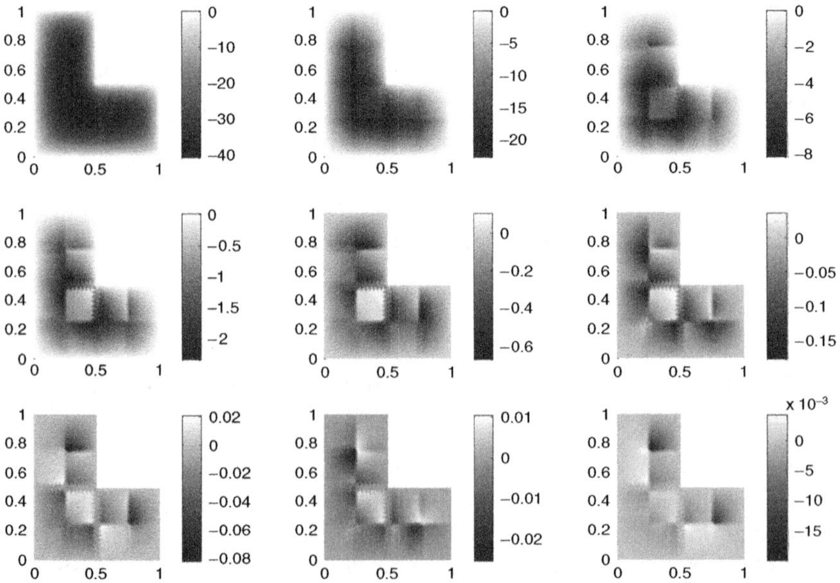

Fig. 2. Error terms of the same nine iterates with twelve subdomains.

the error terms of the first 9 iterations of the OSM algorithm using $\alpha = 5$ for these 12 subdomains. Note the scale in each of the error iterates.

References

[1] Benzi, M., Frommer, A., Nabben, R., Szyld, D.B.: Algebraic theory of multiplicative Schwarz methods. *Numer. Math.*, 89:605-639, 2001.

[2] Brenner, S.C., Scott, L.R.: *The Mathematical Theory of Finite Element Methods*, 2nd ed. Springer, New York, 2002.

[3] Frommer, A., Nabben, R., Szyld, D.B.: Convergence of stationary iterative methods for Hermitian semidefinite linear systems and applications to Schwarz methods. *SIAM J. Matrix Anal. Appl.*, 30:925–938, 2008.

[4] Frommer, A., Szyld, D.B. Weighted max norms, splittings, and overlapping additive Schwarz iterations. *Numer. Math.*, 83:259–278, 1999.

[5] Gander, M.J.: Optimized Schwarz Methods. *SIAM J. Numer. Anal.*, 44:699-731, 2006.

[6] Keyes, D.: Domain Decomposition Methods in the Mainstream of Computational Science. In *Domain Decomposition Methods in Science and Engineering*. Fourteenth International Conference on Domain Decomposition Methods, Cocoyoc, Mexico, I. Herrera, D.E. Keyes, O.B. Widlund, and R. Yates, eds. Press of the Universidad Nacional Autónoma de México (UNAM), Mexico, 2003, pp. 79–94.

[7] Kimn, J.-H.: Overlapping Schwarz Algorithms using Discontinuous Iterates for Poisson's Equation. Ph.D. Thesis, Department of Mathematics, New York University, New York, 2001.

[8] Kimn, J.-H.: A convergence theory for an overlapping Schwarz algorithm using discontinuous iterates. *Numer. Math.*, 100:117-139, 2005.

[9] Lions, P.-L.: On the Schwarz alternating method. III: A variant for nonoverlapping subdomains. In T. F. Chan, R. Glowinski, J. Périaux, O. Widlund, eds. *Third International Symposium on Domain Decomposition Methods for Partial Differential Equations*, held in Houston, Texas, March 20-22, 1989. SIAM, 1990, Philadelphia, pp. 202–223.

[10] Loisel, S., Szyld, D.B.: On the convergence of Algebraic Optimizable Schwarz Methods with applications to elliptic problems. Research Report 07-11-16, Department of Mathematics, Temple University, November 2007.

[11] Nabben, R., Szyld, D.B.: Schwarz iterations for symmetric positive semidefinite problems. *SIAM J. Matrix Anal. Appl.*, 29:98–116, 2006.

[12] Quarteroni, A., Valli, A.: *Domain Decomposition Methods for Partial Differential Equations*. Oxford Science Publication, Clarendon, Oxford, 1999.

[13] Quarteroni, A., Valli, A.: *Numerical Approximation of Partial Differential Equations*. Springer, Berlin, 1994. 2nd Corrected Printing 1997.

[14] Smith, B.F., Bjørstad, P.E., Gropp, W.D.: *Domain Decomposition: Parallel Multilevel Methods for Elliptic Partial Differential Equations*. Cambridge University Press, Cambridge, New York–Melbourne, 1996.

[15] St-Cyr, A., Gander, M.J., Thomas, S.J.: Optimized multiplicative, additive and restricted additive Schwarz preconditioning. *SIAM J. Sci. Comput.*, 29:2402-2425, 2007.

[16] Toselli, A., Widlund, O.: *Domain Decomposition Methods—Algorithms and Theory*. Springer Series in Computational Mathematics 34, Springer, Berlin–Heidelberg–New York, 2005.

Editorial Policy

1. Volumes in the following three categories will be published in LNCSE:

i) Research monographs
ii) Lecture and seminar notes
iii) Conference proceedings

Those considering a book which might be suitable for the series are strongly advised to contact the publisher or the series editors at an early stage.

2. Categories i) and ii). These categories will be emphasized by Lecture Notes in Computational Science and Engineering. **Submissions by interdisciplinary teams of authors are encouraged.** The goal is to report new developments – quickly, informally, and in a way that will make them accessible to non-specialists. In the evaluation of submissions timeliness of the work is an important criterion. Texts should be well-rounded, well-written and reasonably self-contained. In most cases the work will contain results of others as well as those of the author(s). In each case the author(s) should provide sufficient motivation, examples, and applications. In this respect, Ph.D. theses will usually be deemed unsuitable for the Lecture Notes series. Proposals for volumes in these categories should be submitted either to one of the series editors or to Springer-Verlag, Heidelberg, and will be refereed. A provisional judgment on the acceptability of a project can be based on partial information about the work: a detailed outline describing the contents of each chapter, the estimated length, a bibliography, and one or two sample chapters – or a first draft. A final decision whether to accept will rest on an evaluation of the completed work which should include

- at least 100 pages of text;
- a table of contents;
- an informative introduction perhaps with some historical remarks which should be accessible to readers unfamiliar with the topic treated;
- a subject index.

3. Category iii). Conference proceedings will be considered for publication provided that they are both of exceptional interest and devoted to a single topic. One (or more) expert participants will act as the scientific editor(s) of the volume. They select the papers which are suitable for inclusion and have them individually refereed as for a journal. Papers not closely related to the central topic are to be excluded. Organizers should contact Lecture Notes in Computational Science and Engineering at the planning stage.

In exceptional cases some other multi-author-volumes may be considered in this category.

4. Format. Only works in English are considered. They should be submitted in camera-ready form according to Springer-Verlag's specifications.
Electronic material can be included if appropriate. Please contact the publisher.
Technical instructions and/or LaTeX macros are available via http://www.springer.com/authors/book+authors?SGWID=0-154102-12-417900-0. The macros can also be sent on request.

General Remarks

Lecture Notes are printed by photo-offset from the master-copy delivered in camera-ready form by the authors. For this purpose Springer-Verlag provides technical instructions for the preparation of manuscripts. See also *Editorial Policy.*

Careful preparation of manuscripts will help keep production time short and ensure a satisfactory appearance of the finished book.

The following terms and conditions hold:

Categories i), ii), and iii):
Authors receive 50 free copies of their book. No royalty is paid. Commitment to publish is made by letter of intent rather than by signing a formal contract. Springer- Verlag secures the copyright for each volume.

For conference proceedings, editors receive a total of 50 free copies of their volume for distribution to the contributing authors.

All categories:
Authors are entitled to purchase further copies of their book and other Springer mathematics books for their personal use, at a discount of 33.3% directly from Springer-Verlag.

Addresses:

Timothy J. Barth
NASA Ames Research Center
NAS Division
Moffett Field, CA 94035, USA
e-mail: barth@nas.nasa.gov

Michael Griebel
Institut für Numerische Simulation
der Universität Bonn
Wegelerstr. 6
53115 Bonn, Germany
e-mail: griebel@ins.uni-bonn.de

David E. Keyes
Department of Applied Physics
and Applied Mathematics
Columbia University
200 S. W. Mudd Building
500 W. 120th Street
New York, NY 10027, USA
e-mail: david.keyes@columbia.edu

Risto M. Nieminen
Laboratory of Physics
Helsinki University of Technology
02150 Espoo, Finland
e-mail: rni@fyslab.hut.fi

Dirk Roose
Department of Computer Science
Katholieke Universiteit Leuven
Celestijnenlaan 200A
3001 Leuven-Heverlee, Belgium
e-mail: dirk.roose@cs.kuleuven.ac.be

Tamar Schlick
Department of Chemistry
Courant Institute of Mathematical
Sciences
New York University
and Howard Hughes Medical Institute
251 Mercer Street
New York, NY 10012, USA
e-mail: schlick@nyu.edu

Mathematics Editor at Springer:
Martin Peters
Springer-Verlag
Mathematics Editorial IV
Tiergartenstrasse 17
D-69121 Heidelberg, Germany
Tel.: *49 (6221) 487-8185
Fax: *49 (6221) 487-8355
e-mail: martin.peters@springer.com

Lecture Notes in Computational Science and Engineering

1. D. Funaro, *Spectral Elements for Transport-Dominated Equations.*
2. H. P. Langtangen, *Computational Partial Differential Equations.* Numerical Methods and Diffpack Programming.
3. W. Hackbusch, G. Wittum (eds.), *Multigrid Methods V.*
4. P. Deuflhard, J. Hermans, B. Leimkuhler, A. E. Mark, S. Reich, R. D. Skeel (eds.), *Computational Molecular Dynamics: Challenges, Methods, Ideas.*
5. D. Kröner, M. Ohlberger, C. Rohde (eds.), *An Introduction to Recent Developments in Theory and Numerics for Conservation Laws.*
6. S. Turek, *Efficient Solvers for Incompressible Flow Problems.* An Algorithmic and Computational Approach.
7. R. von Schwerin, ***M**ulti **B**ody **S**ystem **SIM**ulation.* Numerical Methods, Algorithms, and Software.
8. H.-J. Bungartz, F. Durst, C. Zenger (eds.), *High Performance Scientific and Engineering Computing.*
9. T. J. Barth, H. Deconinck (eds.), *High-Order Methods for Computational Physics.*
10. H. P. Langtangen, A. M. Bruaset, E. Quak (eds.), *Advances in Software Tools for Scientific Computing.*
11. B. Cockburn, G. E. Karniadakis, C.-W. Shu (eds.), *Discontinuous Galerkin Methods.* Theory, Computation and Applications.
12. U. van Rienen, *Numerical Methods in Computational Electrodynamics.* Linear Systems in Practical Applications.
13. B. Engquist, L. Johnsson, M. Hammill, F. Short (eds.), *Simulation and Visualization on the Grid.*
14. E. Dick, K. Riemslagh, J. Vierendeels (eds.), *Multigrid Methods VI.*
15. A. Frommer, T. Lippert, B. Medeke, K. Schilling (eds.), *Numerical Challenges in Lattice Quantum Chromodynamics.*
16. J. Lang, *Adaptive Multilevel Solution of Nonlinear Parabolic PDE Systems.* Theory, Algorithm, and Applications.
17. B. I. Wohlmuth, *Discretization Methods and Iterative Solvers Based on Domain Decomposition.*
18. U. van Rienen, M. Günther, D. Hecht (eds.), *Scientific Computing in Electrical Engineering.*
19. I. Babuška, P. G. Ciarlet, T. Miyoshi (eds.), *Mathematical Modeling and Numerical Simulation in Continuum Mechanics.*
20. T. J. Barth, T. Chan, R. Haimes (eds.), *Multiscale and Multiresolution Methods.* Theory and Applications.
21. M. Breuer, F. Durst, C. Zenger (eds.), *High Performance Scientific and Engineering Computing.*
22. K. Urban, *Wavelets in Numerical Simulation.* Problem Adapted Construction and Applications.
23. L. F. Pavarino, A. Toselli (eds.), *Recent Developments in Domain Decomposition Methods.*

24. T. Schlick, H. H. Gan (eds.), *Computational Methods for Macromolecules: Challenges and Applications.*

25. T. J. Barth, H. Deconinck (eds.), *Error Estimation and Adaptive Discretization Methods in Computational Fluid Dynamics.*

26. M. Griebel, M. A. Schweitzer (eds.), *Meshfree Methods for Partial Differential Equations.*

27. S. Müller, *Adaptive Multiscale Schemes for Conservation Laws.*

28. C. Carstensen, S. Funken, W. Hackbusch, R. H. W. Hoppe, P. Monk (eds.), *Computational Electromagnetics.*

29. M. A. Schweitzer, *A Parallel Multilevel Partition of Unity Method for Elliptic Partial Differential Equations.*

30. T. Biegler, O. Ghattas, M. Heinkenschloss, B. van Bloemen Waanders (eds.), *Large-Scale PDE-Constrained Optimization.*

31. M. Ainsworth, P. Davies, D. Duncan, P. Martin, B. Rynne (eds.), *Topics in Computational Wave Propagation.* Direct and Inverse Problems.

32. H. Emmerich, B. Nestler, M. Schreckenberg (eds.), *Interface and Transport Dynamics.* Computational Modelling.

33. H. P. Langtangen, A. Tveito (eds.), *Advanced Topics in Computational Partial Differential Equations.* Numerical Methods and Diffpack Programming.

34. V. John, *Large Eddy Simulation of Turbulent Incompressible Flows.* Analytical and Numerical Results for a Class of LES Models.

35. E. Bänsch (ed.), *Challenges in Scientific Computing - CISC 2002.*

36. B. N. Khoromskij, G. Wittum, *Numerical Solution of Elliptic Differential Equations by Reduction to the Interface.*

37. A. Iske, *Multiresolution Methods in Scattered Data Modelling.*

38. S.-I. Niculescu, K. Gu (eds.), *Advances in Time-Delay Systems.*

39. S. Attinger, P. Koumoutsakos (eds.), *Multiscale Modelling and Simulation.*

40. R. Kornhuber, R. Hoppe, J. Périaux, O. Pironneau, O. Wildlund, J. Xu (eds.), *Domain Decomposition Methods in Science and Engineering.*

41. T. Plewa, T. Linde, V.G. Weirs (eds.), *Adaptive Mesh Refinement – Theory and Applications.*

42. A. Schmidt, K.G. Siebert, *Design of Adaptive Finite Element Software.* The Finite Element Toolbox ALBERTA.

43. M. Griebel, M.A. Schweitzer (eds.), *Meshfree Methods for Partial Differential Equations II.*

44. B. Engquist, P. Lötstedt, O. Runborg (eds.), *Multiscale Methods in Science and Engineering.*

45. P. Benner, V. Mehrmann, D.C. Sorensen (eds.), *Dimension Reduction of Large-Scale Systems.*

46. D. Kressner, *Numerical Methods for General and Structured Eigenvalue Problems.*

47. A. Boriçi, A. Frommer, B. Joó, A. Kennedy, B. Pendleton (eds.), *QCD and Numerical Analysis III.*

48. F. Graziani (ed.), *Computational Methods in Transport.*

49. B. Leimkuhler, C. Chipot, R. Elber, A. Laaksonen, A. Mark, T. Schlick, C. Schütte, R. Skeel (eds.), *New Algorithms for Macromolecular Simulation.*

50. M. Bücker, G. Corliss, P. Hovland, U. Naumann, B. Norris (eds.), *Automatic Differentiation: Applications, Theory, and Implementations.*

51. A.M. Bruaset, A. Tveito (eds.), *Numerical Solution of Partial Differential Equations on Parallel Computers.*

52. K.H. Hoffmann, A. Meyer (eds.), *Parallel Algorithms and Cluster Computing.*

53. H.-J. Bungartz, M. Schäfer (eds.), *Fluid-Structure Interaction.*

54. J. Behrens, *Adaptive Atmospheric Modeling.*

55. O. Widlund, D. Keyes (eds.), *Domain Decomposition Methods in Science and Engineering XVI.*

56. S. Kassinos, C. Langer, G. Iaccarino, P. Moin (eds.), *Complex Effects in Large Eddy Simulations.*

57. M. Griebel, M.A Schweitzer (eds.), *Meshfree Methods for Partial Differential Equations III.*

58. A.N. Gorban, B. Kégl, D.C. Wunsch, A. Zinovyev (eds.), *Principal Manifolds for Data Visualization and Dimension Reduction.*

59. H. Ammari (ed.), *Modeling and Computations in Electromagnetics: A Volume Dedicated to Jean-Claude Nédélec.*

60. U. Langer, M. Discacciati, D. Keyes, O. Widlund, W. Zulehner (eds.), *Domain Decomposition Methods in Science and Engineering XVII.*

61. T. Mathew, *Domain Decomposition Methods for the Numerical Solution of Partial Differential Equations.*

62. F. Graziani (ed.), *Computational Methods in Transport: Verification and Validation.*

63. M. Bebendorf, *Hierarchical Matrices.* A Means to Efficiently Solve Elliptic Boundary Value Problems.

64. C.H. Bischof, H.M. Bücker, P. Hovland, U. Naumann, J. Utke (eds.), *Advances in Automatic Differentiation.*

65. M. Griebel, M.A. Schweitzer (eds.), *Meshfree Methods for Partial Differential Equations IV.*

66. B. Engquist, P. Lötstedt, O. Runborg (eds.), *Multiscale Modeling and Simulation in Science.*

67. I.H. Tuncer, Ü. Gülcat, D.R. Emerson, K. Matsuno (eds.), *Parallel Computational Fluid Dynamics.*

68. S. Yip, T. Diaz de la Rubia (eds.), *Scientific Modeling and Simulations.*

69. A. Hegarty, N. Kopteva, E. O'Riordan, M. Stynes (eds.), *BAIL* 2008 – *Boundary and Interior Layers.*

70. M. Bercovier, M.J. Gander, R. Kornhuber, O. Widlund (eds.), *Domain Decomposition Methods in Science and Engineering XVIII*

For further information on these books please have a look at our mathematics catalogue at the following URL: `www.springer.com/series/3527`

Monographs in Computational Science and Engineering

1. J. Sundnes, G.T. Lines, X. Cai, B.F. Nielsen, K.-A. Mardal, A. Tveito, *Computing the Electrical Activity in the Heart.*

For further information on this book, please have a look at our mathematics catalogue at the following URL: `www.springer.com/series/7417`

Texts in Computational Science and Engineering

1. H. P. Langtangen, *Computational Partial Differential Equations.* Numerical Methods and Diffpack Programming. 2nd Edition
2. A. Quarteroni, F. Saleri, *Scientific Computing with MATLAB and Octave.* 2nd Edition
3. H. P. Langtangen, *Python Scripting for Computational Science.* 3rd Edition
4. H. Gardner, G. Manduchi, *Design Patterns for e-Science.*
5. M. Griebel, S. Knapek, G. Zumbusch, *Numerical Simulation in Molecular Dynamics.*
6. H. P. Langtangen, *A Primer on Scientific Programming with Python.*

For further information on these books please have a look at our mathematics catalogue at the following URL: www.springer.com/series/5151

GPSR Compliance

The European Union's (EU) General Product Safety Regulation (GPSR) is a set of rules that requires consumer products to be safe and our obligations to ensure this.

If you have any concerns about our products, you can contact us on ProductSafety@springernature.com

In case Publisher is established outside the EU, the EU authorized representative is:

Springer Nature Customer Service Center GmbH
Europaplatz 3
69115 Heidelberg, Germany

Batch number: 10755244

Printed by Printforce, the Netherlands